AF887191

Technology of functional cereal products

Related titles

Bread making: Improving quality
(ISBN 978-1-85573-553-8)
Edited by one of the world's leading authorities in the field, and with a distinguished international team of contributors, *Bread making* reviews key recent research on the ingredients determining bread quality. Part 1 discusses the definition, measurement and improvement of wheat properties, which enhance bread quality. The second part of the book reviews recent research on the properties of flour and dough, which influence bread texture, colour and flavour. The final part of the book looks at the contribution of other ingredients.

Novel food ingredients for weight control
(ISBN 978-1-84569-030-4)
In the Western world, obesity has risen at an epidemic rate in recent years. The scale of the problem is immense, with obesity predicted to become a leading preventable cause of death in many countries in the near future and therefore weight control is a concern of a high percentage of consumers. This important collection, edited by a leader in the field, provides food industry professionals with essential information about particular ingredients that are effective in weight control, their production, use in functional foods and ability to play a role in weight regulation.

Functional foods: Concept to product
(ISBN 978-1-85573-503-3)
Functional foods are widely predicted to become one of the biggest dietary trends of the next twenty-five years. The editors of this book have gathered together leading experts in the field in order to provide the food industry with a single authoritative resource. This book first defines and classifies the field of functional foods, paying particular attention to the legislative aspects in both the USA and EU. It then summarises the key work on functional foods and the prevention of disease. Finally, there is a series of chapters on developing functional products.

Details of these books and a complete list of Woodhead's titles can be obtained by:

- visiting our website at www.woodheadpublishing.com
- contacting Customer Services (e-mail: sales@woodhead-publishing.com; fax: +44 (0)1223 893694; tel.: +44 (0)1223 891358 ext. 30; address: Woodhead Publishing Ltd, Abington Hall, Abington, Cambridge CB21 6AH, England)

Technology of functional cereal products

Edited by
Bruce R. Hamaker

CRC Press
Boca Raton Boston New York Washington, DC

WOODHEAD PUBLISHING LIMITED
Cambridge, England

Published by Woodhead Publishing Limited, Abington Hall, Abington,
Cambridge CB21 6AH, England
www.woodheadpublishing.com

Published in North America by CRC Press LLC, 6000 Broken Sound Parkway, NW,
Suite 300, Boca Raton, FL 33487, USA

© 2008, Woodhead Publishing Limited
The authors have asserted their moral rights.

This book contains information obtained from authentic and highly regarded sources. Reprinted material is quoted with permission, and sources are indicated. Reasonable efforts have been made to publish reliable data and information, but the authors and the publishers cannot assume responsibility for the validity of all materials. Neither the authors nor the publishers, nor anyone else associated with this publication, shall be liable for any loss, damage or liability directly or indirectly caused or alleged to be caused by this book.

Neither this book nor any part may be reproduced or transmitted in any form or by any means, electronic or mechanical, including photocopying, microfilming and recording, or by any information storage or retrieval system, without permission in writing from Woodhead Publishing Limited.

The consent of Woodhead Publishing Limited does not extend to copying for general distribution, for promotion, for creating new works, or for resale. Specific permission must be obtained in writing from Woodhead Publishing Limited for such copying.

Trademark notice: product or corporate names may be trademarks or registered trademarks, and are used only for identification and explanation, without intent to infringe.

British Library Cataloguing in Publication Data
A catalogue record for this book is available from the British Library.

Library of Congress Cataloging in Publication Data
A catalog record for this book is available from the Library of Congress.

Woodhead Publishing ISBN 978-1-84569-177-6 (book)
Woodhead Publishing ISBN 978-1-84569-388-6 (e-book)
CRC Press ISBN 978-1-4200-6673-9
CRC Press order number WP6673

The publishers' policy is to use permanent paper from mills that operate a sustainable forestry policy, and which has been manufactured from pulp which is processed using acid-free and elementary chlorine-free practices. Furthermore, the publishers ensure that the text paper and cover board used have met acceptable environmental accreditation standards.

Typeset in India by Replika Press Pvt Ltd.
Printed by TJ International, Padstow, Cornwall, England

Contents

Contributor contact details

(* = main contact)

Editor
B. R. Hamaker
Whistler Center for Carbohydrate Research
Department of Food Science
Purdue University
West Lafayette
IN 47907-2009
USA

E-mail: hamakerb@purdue.edu

Chapter 1
M. Dean, M. M. Raats and R. Shepherd*
Food, Consumer Behaviour and Health Research Centre
Department of Psychology
University of Surrey
Guildford
Surrey GU2 7XH
UK

E-mail: r.shepherd@surrey.ac.uk

Chapter 2
K. Schmitz* and L. Marquart
Department of Food Science and Nutrition
University of Minnesota
St Paul
MN 55108
USA

E-mail: lmarquar@umn.edu
Kristen.Schmitz@cs-americas.com

J. Willem van der Kamp
TNO Quality of Life
Zeist
NL 3700 AJ
The Netherlands

Chapter 3
J. Miller Jones
The College of St Catherine
4030 Valentine Court
Arden Hills
MN 55112
USA

E-mail: jmjones@stkate.edu

Chapter 4
A. M. Birkett*
National Starch Food Innovation
10 Finderne Avenue
Bridgewater
NJ 08807
USA

E-mail: anne.birkett@nstarch.com

I. L. Brown
Faculty of Health and Behavioural Science
University of Wollongong
Wollongong
NSW 2522
Australia

E-mail: ian.brown@clovercorp.com.au

Chapter 5
A. Kamal-Eldin
Department of Food Science
Swedish University of Agricultural Sciences (SLU)
P. O. Box 7051
750 07 Uppsala
Sweden

E-mail: Afaf.Kamal-Eldin@lmv.slu.se

Chapter 6
M. A. Pereira
Division of Epidemiology and Community Health
University of Minnesota
1300 South Second Street
Minneapolis
MN 55454
USA

E-mail: pereira@epi.umn.edu

Chapter 7
M. Champ
CHU-Hôtel Dieu
HNBI-UMR PhAN & CRNH
Place Alexis Ricordeau
44096 Nantes Cedex 1
France

E-mail: champ@nantes.inra.fr

Chapter 8
P. R. Shewry
Rothamsted Research
Harpenden
Hertfordshire
AL5 2JQ
UK

E-mail: Peter.Shewry@bbsrc.ac.uk

Chapter 9
C. Collar
Cereal Group
Department of Food Science
Instituto de Agroquímica y Technología de Alimentos, IATA-CSIC
Apartado de Correos 73
46100 Burjassot
Spain

E-mail: ccollar@iata.csic.es

Chapter 10
A Kaukovirta-Norja* and P. Lehtinen
Espoo
P.O. Box 1000
FI-02044 VTT
Finland

E-mail: Anu.Kaukovirta-Norja@vtt.fi

Chapter 11

A. Kamal-Eldin* and P. Åman
Department of Food Science
Swedish University of Agricultural Sciences (SLU)
P. O. Box 7051
750 07 Uppsala
Sweden

E-mail: Afaf.Kamal-Eldin@lmv.slu.se

J.-X. Zhang
Nutritional Research
Department of Public Health and Clinical Medicine
University of Umeå
S-9001 87 Umeå
Sweden

K.-E. Bach Knudsen
University of Aarhus
Faculty of Agriculture and Sciences
Department of Animal Health, Welfare and Nutrition
Blichers Allé 20
DK-8830 Tjele
Denmark

K. Poutanen
VTT POB 1000,
FI-02044 VTT
Finland

Chapter 12

A. A. H. Andersson* and P. Åman
Department of Food Science
Swedish University of Agricultural Sciences (SLU)
P. O. Box 7051
750 07 Uppsala
Sweden

E-mail: annica.andersson@lmv.slu.se

Chapter 13

J. R. N. Taylor* and M. N. Emmambux
Department of Food Science
University of Pretoria
Pretoria 0002
South Africa

E-mail: jtaylor@postino.up.ac.za

Chapter 14

C. M. Rosell
Cereal Group. Department of Food Science.
Instituto de Agroquímica y Technología de Alimentos (IATA CSIC)
Apartado de Correos 73
46100 Burjassot
Spain

E-mail: crosell@iata.csic.es

Chapter 15

C. Hall III*
210 Harris Hall
Department of Cereal and Food Sciences
North Dakota State University
Fargo, ND 58105
701-231-6359
USA

E-mail: Clifford.Hall@ndsu.edu

M. C. Tulbek
Northern Crops Institute
North Dakota State University
Fargo, ND 58105
701-231-5493
USA

E-mail: Mehmet.Tulbek@ndsu.edu

Chapter 16
Y. Vodovotz
227 Parker Food Science and Technology Building
2015 Fyffe Court
The Ohio State University
Columbus, OH 43210
USA

E-mail: vodovotz.1@osu.edu

Chapter 17
D. Meyer*
Sensus
Borchwerf 3
4704 RG Roosendaal
The Netherlands

E-mail: Diederick.meyer@sensus.nl

J. de Wolf
Cosun Food Technology Centre
P. O. Box 1308
4700 BH
Roosendaal
The Netherlands

Chapter 18
C. S. Brennan
Institute of Food Nutrition and Human Health
Massey University
Private Bag 11222
Palmerston North
New Zealand

E-mail: C.S.Brennan@massey.ac.nz

Chapter 19
E. K. Arendt and F. Dal Bello*
Department of Food and Nutritional Sciences
University College Cork
Western Road
Cork
Ireland

E-mail: f.dalbello@ucc.ie
e.arendt@ucc.ie

Chapter 20
G. E. Inglett* and D. G. Stevenson
NCAUR/ARS/USDA
1815 N. University Street
Peoria IL 61604
USA

E-mail: george.inglett@ars.usda.gov

S. Lee
Sejong University
98 Gunja-Dong
Gwangjin-Gu
Seoul, 143-747
Republic of Korea

Chapter 21
S. K. Patil
S. K. Patil and Associates
1403 Coventry Lane
Munster
IN 46321
USA

E-mail: sakharam@sbcglobal.net

Chapter 22
G. Zhang, M. Venkatachalam and, B. R. Hamaker*
Whistler Center for Carbohydrate Research
Department of Food Science
Purdue University
West Lafayette
IN 47907-2009
USA

E-mail: gzhang@purdue.edu
hamakerb@purdue.edu

Preface

Cereal grains and the multitude of products made from them have formed a basis of our diet since man began cultivating crops. Today, cereal crops, whether combined together or taken individually as the top three in terms of quantity produced (maize, rice and wheat), far exceed the production in dry weight of any other type of plant or animal food. They provide the principal source of calories and even protein for a large proportion of the world's poor and substantially contribute to the diets of the more fortunate. Yet, in our world of health concerns related to obesity and its associated problems of diabetes and cardiovascular disease, the use of large amounts of cereal-based products has recently been viewed as suspect. Some of this concern has centered on what is thought of as an abundance of products processed from refined flours where the cereal bran and germ have been removed to yield nearly pure starchy endosperm. Partly for this reason, as well as a desire to capitalize on the broader benefits of cereal grains in the diet, a move towards use of whole grains has been emphasized by dietary guideline groups and regulatory agencies in many countries, as well as the European Union, and now has growing consumer demand. Moreover, an array of other benefits of cereal grains and their components, both purported and proven, have been and continue to be researched. These include the important roles of specific dietary fibers in colonic as well as overall physiological health, vitamins and minerals, cereal starches and how they are digested or made resistant, and secondary metabolites such as phenolic compounds that have antioxidant and other functions.

Functional, or health, aspects of cereal grains and flours and technological advances in formulating good and healthy products is the focus of this book. We were very fortunate to attract an excellent group of leading researchers and technologists to contribute their state-of-the-art views of the health benefits of cereal grains and their ingredients and products, regulatory issues surrounding the 'whole grain' initiative, how cereal grains or components can be manipulated either genetically or through processing for better health, and a range of technologies to formulate functional cereal products with improved health benefits for the general public or for specific groups such as the gluten intolerant. This text will be valuable as a useful reference for those developing functional cereal foods, but also to audiences wanting to learn the fascinating fundamentals of cereals and their role in health and wellness. I want to thank this talented group of scientists who gave their valued time to write the chapters, and the excellent editorial staff, particularly Sarah Whitworth, at Woodhead Publishing Limited for their vision and patience. All said, the greatest contribution this book could have would be to improve public health.

B.R. Hamaker
Purdue University

Part I

Introductory issues

1

Consumers and functional cereal products

M. Dean, M. M. Raats and R. Shepherd, University of Surrey, UK

1.1 Why is it important to understand consumer perceptions?

Food is a necessity to sustain life. However, what people choose to eat is not solely based on their biological needs; their choice also addresses many psychological and/or emotional issues (Conner & Armitage, 2002). Eating healthily has increasingly become a means of achieving health with people being advised to look after themselves by eating foods high in nutrients. Nutrition education is considered an important tool in providing knowledge about what constitutes a healthy diet and encouraging the adoption of healthy food choices (Glanz, 1997). The education provided must be informative and persuasive if it is to have any hope of changing behaviour. Recent campaigns to increase fruit and vegetables such as 5-a-day (www.5aday.nhs.uk) and wholegrains such as Nestle's Healthy Heart (www.cerealpartners.co.uk) are some examples of this trend. Although these types of campaigns have been successful in the case of fruit and vegetables (Van Duyn *et al.*, 2001), there are still large differences between the quantities recommended and the amounts consumed in terms of wholegrains (Jones *et al.*, 2002). A study looking at people's perceptions of grain intake (USDA Center for Nutrition Policy and Promotion, 2000) found that, although people perceived that they ate fewer grain servings (2.5–3.2 servings) than they actually ate (4.2–6.2 servings), both were lower than the recommended amount (9 servings). This suggests that the problem is twofold: firstly, people are not aware of their actual intake of wholegrains and secondly, people are eating far less wholegrains than is recommended.

Wholegrain foods are important sources of nutrients and phyto-protective components that have been found to be in short supply in the diets of many consumers (Smith *et al.*, 2003). To increase levels of intake, it is important to consider people's perceptions of grain-based foods, and understand the processes and mechanisms involved in their decision making with regard to these foods.

1.2 Consumption of grain-based foods

The recommended intake of wholegrain for adults in the United Kingdom is 3 servings per day (Richardson, 2003). Data from two nationally representative sample dietary surveys (adults aged 16–64 and adults over 65 years of age) in the United Kingdom (Lang *et al.*, 2003) based on 7 day weighed dietary records found that a third of the adults in both surveys did not consume wholegrain foods on a daily basis and over 95% consumed less than 3 servings a day. There were no significant differences in wholegrain consumption between men and women, and nor was there any seasonal variation. With regard to age, the median consumption was only 2.5 servings per week for those aged 16–64 and 5 servings a week for adults over 65 years of age. Although there were regional differences in the median level of consumption for adults under 65 years of age, this disappeared for those over 65 years. For those under the age of 65, adults in Scotland and Wales had the lowest median number of servings.

In the UK surveys (Lang *et al.*, 2003) the median consumption of wholegrain foods was also found to be significantly greater for non-manual than for manual workers. The number of non-consumers was also higher in the manual worker group independent of all other socio-demographic and lifestyle factors. Furthermore, smokers were found to eat less wholegrain foods than non-smokers. In terms of the types of wholegrain foods consumed, wholemeal bread constituted almost half of the total consumption of wholegrain foods followed by wholegrain breakfast cereals (29%) in adults under 65 years of age; these were also the highest two sources for the over 65 group. Other sources of grains were biscuits, pasta, rice, buns, cakes and pastries. This clearly shows that wholegrain consumption in the United Kingdom is extremely low and comes from a very narrow range of foods. Furthermore, consumer characteristics suggest that wholegrain consumers are more likely to be non-smokers in non-manual jobs.

Data from the National Diet and Nutrition Survey of young British people (Thane *et al.*, 2005) found that the median intake of wholegrain was 7 g per day while the mean intake was 13 g per day. Wholegrain intake was lower for those in a household where the head of household was in manual employment. However, there were no differences between the sexes, youngster's religious affiliation, the region they were from or the seasons. Furthermore, 27% of the participants did not eat any wholegrain. The main sources of wholegrains were breakfast cereals (56%) and bread (25%). This too suggests that the amount of wholegrains consumed by young people was below the recommended level and came from two sources. These UK data further suggest that consumers of wholegrains are more likely to be of higher social status.

In the United States the recommended intake of grain is 6–11 servings a day of which at least 3 servings should consist of a variety of wholegrain foods (USDA, 1997). Disappearance data for the United States show that total grain intake has decreased from 36.2% food energy in 1909 to 23.3% food energy in 1990 (Jacobs *et al.*, 1999). Of the 10 daily grain servings that participants ate, less than one was a wholegrain product. Although wholegrain breads are one of the most accessible wholegrain products in the United States, 20% of adults and 40% of teens and children in a 2001 telephone survey reported that they never ate wholegrain bread (Jones *et al.*, 2002). Similarly, Albertson and Tobelmann (1995) analyzed food consumption data from the Market Research Corporation of America from 9165 adults to find that more than 50% of the young Americans (2–18 years) and 70% of adults (19 and over) reported less than three eating occasions of grain products (whole or refined) per day. Fewer than one eating occasion containing wholegrains were reported by 90% and 73% of the younger and older age categories, respectively. About a fifth of both age groups consumed no wholegrains. US national food survey data (Cleveland *et al.*, 2000) revealed that only 8% of people met the recommended level of at least 3 servings of wholegrains a day and only 36% consumed more than 1 serving a day. Of the 65% consuming less than 1 serving per day, half consumed no wholegrains. The US national food survey data suggest that a consumer of wholegrains meeting the target is more likely to be male, white, of higher socio-economic status, a non-smoker, a frequent exerciser, a user of vitamin–mineral supplements and not overweight.

In Finland more rye bread is consumed compared with countries such as the United Kingdom and the United States, thus increasing the overall intake of wholegrains (Pietinen *et al.*, 1996). Although the consumption of wholegrain is lower in Norway than in Sweden, it is still four times that reported in the United States and thus there is a relatively high level of wholegrain consumption amongst Scandinavians (Jacobs *et al.*, 2001).

A number of surveys found that older people tend to consume more wholegrain than younger people (Johansson *et al.*, 1999; Adams & Engstrom, 2000a; Cleveland *et al.*, 2000; Lang *et al.*, 2001). Furthermore, men consume more wholegrain foods than females (Cleveland *et al.*, 2000; Jacobs *et al.*, 2001; Prattala *et al.*, 2001). This may be due to an increase in overall food consumption rather than specific preference for wholegrain (Lang & Jebb, 2003). Both in the United States and the United Kingdom, income and level of education are also positively associated with wholegrain consumption (Johansson *et al.*, 1999; Adams & Engstrom, 2000a; Cleveland *et al.* 2000; Lang *et al.*, 2001). However, in Finland the highest consumption of rye bread is observed in the lower socio-economic group (Prattala *et al.*, 2001).

In the United States and the United Kingdom, wholegrain consumers have a healthier lifestyle such that they tend to be non-smokers who exercise regularly and consume more fruits and vegetables (Johansson *et al.*, 1999; Adams & Engstrom, 2000a; Cleveland *et al.*, 2000).

1.3 Perceived barriers to eating wholegrain products

Three factors are said to influence a person's food choice: firstly, the food itself; secondly, the consumer; and thirdly, the associated environmental and economic issues (Urala & Lahteenmaki, 2004). In terms of the food, numerous consumer studies have pointed to taste being the primary factor directing food choice (Shepherd, 1990; Richardson *et al.*, 1994; Grunert *et al.*, 2000; Urala & Lahteenmaki, 2003). When speculating as to why consumers eat so few wholegrain products, Adams and Engstrom (2000b) suggested that inferior taste of wholegrain foods, time taken to cook, unfamiliarity with these foods, knowing what constitutes wholegrains and lack of knowledge of health benefits of wholegrains are contributing factors. Smith *et al.* (2001a) reported that a large number of individuals (18–34%) were unaware of the link between wholegrains and reduced risk of chronic diseases. Based on an intervention study, Smith *et al.* (2001b) claim that the barriers to consumption include difficulty when eating outside the home, personal and other household member preferences, and limited variety and availability of wholegrain products. They also report that many females experienced a feeling of bloating after eating wheat-based products. Those who do consume are said to do so by eating mainly bread and breakfast cereals because of convenience, taste and availability.

Seal (2005) reports that few consumers knew the 3 servings a day recommendation for wholegrains and some had difficulty identifying which foods were and were not wholegrain. He suggests that the main barriers to wholegrain consumption are people's knowledge of what is wholegrain, how much to take, what is the portion size, cost, availability, variety in stores, difficulty when eating out, perception of taste and texture, lack of routine at the weekend, preference of family and friends for refined grains and time required for cooking. He proposes that, to increase intake, we need to educate people to increase their awareness of the benefits, to improve their ability to find and identify wholegrains and teach people how to cook wholegrain foods. He also calls for the food industry to produce a greater variety and better quality of wholegrain products. Low levels of wholegrain consumption could thus be a result of negative consumer perceptions. This could further be compounded by consumers having difficulty monitoring their actual intake of wholegrains to ensure they have a realistic perception of their consumption.

1.4 Interventions to increase intake of grain-based foods

The Social Ecological Framework (e.g. Welton *et al.* 1997) can be used to conceptualize the various levels at which interventions to change health behaviours may be targeted. The model suggests that interventions could be aimed at the intrapersonal, interpersonal, organizational, community or at the public policy levels; see Table 1.1 for examples of possible interventions that could be developed to increase wholegrain consumption.

Nutrition labelling in the form of nutritional claims or health claims is provided to consumers in order to help change their eating patterns encouraging people to eat more of certain types of foods and less of others. Nutritional claims highlight particular nutritional properties of foods, i.e. high or low levels of nutrients including fat, salt, fibre and vitamins. Alternatively, a nutritional claim can take the form of a comparison, such as contains 30% more fibre or 5% more calcium. In contrast, health claims are messages that inform consumers about currently recognized links between nutrition and health, including the risks of chronic diseases. A health claim can highlight the fact that the product has a nutritional function (e.g. help develop strong bones), enhances a particular body function (e.g. may improve gut health), reduces the risk of a particular disease (e.g. may help reduce the risk of

Table 1.1 Examples of interventions that could be developed to increase wholegrain consumption based on the Social Ecological Framework, e.g. Welton *et al.* (1997)

Level	Examples of possible influences	Example of a possible intervention
Intrapersonal	An individual's knowledge, skills, attitudes, values, preferences, emotions, values, behaviour	• Individuals are taught to monitor their own intake, set targets and learn how to increase their intake
Interpersonal	An individual's social networks, social supports, family, peers, and neighbours	• Family-based interventions in which parents act as models to their children by increasing their own consumption
Organizational	Businesses, public agencies, churches, service organizations	• Increasing the availability of wholegrain-based foods in school or workplace canteens
Community	Community resources, neighbourhood organizations, social and health services	• Setting up of food co-ops where wholegrain-based foods are made available in communities where few such products are available
Public policy	Legislation, policies, taxes and regulatory agencies, health system, social care system, political/geographic environment	• Legislating for only wholegrain-based cereal products to be used in schools • Making the labelling of wholegrain products clear and consistent

coronary heart disease) or it may be a medical claim (e.g. prevent, treat or cure bone disease).

The efficacy of a nutritional claim relating to wholegrains was investigated in a cross-cultural study carried out by Mialon *et al.* (2002) comparing Australians' and Chinese Malaysians' perceptions of three types of bread and muffins (white low-fibre, white fibre-enriched and wholemeal bread/ multigrain muffin) before and after providing nutritional information. Products were rated for acceptance, sensory, and health and nutrition-related variables. Results showed that there were only minor differences between cultures in their perceptions of the nutritional quality of the different foods. Both groups perceived wholemeal/multigrain products as being healthier. However, the differences in acceptance scores between wholemeal and white products were greater for Australians than for the Chinese Malaysians. Furthermore, Australians rated the multigrain muffin as darkest in colour and strongest in flavour and aftertaste, showing that there are cultural differences in sensory perceptions. Mialon *et al.* (2002) found that the provision of nutritional information strongly and consistently affected the sensory rating for breads. They also found that the mention of an ingredient level in a foodstuff affected the perception of sensory characteristic intensities. Similar changes in sensory property ratings were noted by Westcombe and Wardle (1997) and Shepherd *et al.* (1991/2) in their studies on dairy products with varying levels of fat.

1.5 Functional cereal products

As well as increasing consumption of wholegrains through information provision, increasing the variety and availability of good-tasting, wholegrain foods and the associated marketing and advertising could be another way of achieving the same purpose. Alternatively, new foods could be developed that have all the health-promoting properties of foods like wholegrain foods, but taste and look like the conventional foods that are popular and regularly consumed by the public. For example, people eat a lot of white bread which, due to the processing methods, has had most of the fibre and other important vitamins and minerals that are found in wholegrain products removed. If a loaf of white bread could be produced that has all the goodness of wholemeal bread but which looks and tastes like white bread – as are currently being produced in the United States (e.g. Wonder Whole Grain White and Sara Lee Soft and Smooth Whole Grain White) and in the United Kingdom (Hovis Best of Both) – this might be an easier way of increasing consumers' consumption of foods with the equivalent health-giving properties found in wholegrain foods. These types of new foods are collectively referred to as 'functional foods'.

In simple terms a food can be regarded as functional if it provides health benefits over and above basic nutrition. A more complex definition would be that functional food is similar in appearance to conventional food intended to be consumed as part of a normal diet, but has been modified to aid physiological functions beyond the provision of simple nutrition requirement (Roberfroid, 2000). When the food in question is a cereal – such as wheat, maize, rice, oats, etc. – then these products are known as 'functional cereal products'. The effects of functional foods are usually well defined, such as lowering blood cholesterol levels or strengthening bones. However, although these effects can be measured physiologically, it can be difficult for the consumer to verify them based on their own experience of consuming these products. Although many consumers want to be healthy, they are often reluctant to change their eating habits, even when they know them to be unhealthy; Williamson *et al.* (2000) thus argue that the potential for functional foods is great. Functional foods are seen as a way of enabling the consumer to lead a healthier life without having to change their eating habits (Jonas and Beckmann, 1998). Thus, eating functional cereal foods could possibly be an easy way to increase the intake of nutrients associated with wholegrains without changing eating habits.

1.6 Consumer perceptions of functional grain products

The expected annual growth rate of the functional food market ranged from 15% to 20% at the end of 1990s (Gardette, 2000; Hilliam, 2000; Shah, 2001) to a possible 10% more recently (Weststrate *et al.*, 2002). Despite the decrease in the growth rate, it is impressive when compared with the 2–3% annual growth rate of the whole of the food industry. Functional foods are increasingly available in the shops even though their common feature of being health-beneficial may be more ambiguous (Jonas & Beckmann, 1998; Grunert *et al.*, 2001). Some novel foods, like genetically modified (GM) foods, evoke resistance and doubt whereas others, such as probiotic drinks, are welcomed more readily and incorporated into daily routines. The types of messages associated with the traditional nutritional and functional health claims used with conventional healthy food products could be viewed by consumers as quite different from those associated with functional foods (Saher *et al.*, 2004). The fact that functional foods often require sophisticated food technologies in their production could also impact on their image as compared with conventional healthy foods.

In order to be able to understand and influence the public with regard to their selection of functional grain products, we need to understand consumers' attitudes towards these foods (Gilbert, 1997; Grunert *et al.*, 2000; Weststrate *et al.*, 2002). As there are very few studies conducted in the area of consumer acceptance of functional grain products, it is necessary to look at research conducted on other functional foods in order to understand potential consumer perceptions and acceptance.

Saher *et al.* (2004) argue that most of the research has concentrated on possible health effects while little is known about how consumers respond to them. Early work on attitudes towards functional foods found people to be suspicious about the potential unnaturalness of these foods as a category, although reaction to concrete examples of foods seemed to be more positive (Poulsen, 1999). Blades (2000) claims that the confusion amongst the public about the role of functional foods is due to the relatively new concept. It is suggested that there is confusion as to who should and should not be eating these foods and how they should be used (Consumers' Association, 2000). Studies have found that people lack awareness of functional foods and are sceptical about the producers and their claims concerning these foods (NCC, 1997; Mintel, 2000). Some research (NCC, 1997) suggests that people cannot distinguish between nutritional claims and health claims and believe that the government controls and monitors claims on functional foods. Blades (2000) argues that consumer understanding of the concept of functional foods is poor and the terms 'functional food' and 'nutraceutical' are foreign to many consumers (McCrea, 2001). Childs *et al.* (2000) noted that, in order to understand consumption, it is crucial to regularly assess the level of consumers' knowledge about functional foods.

1.6.1 Associating health with food

The extent of the consumer's search for nutritional information in a given food category has been found to depend on how they perceived that category, such that consumers ignored nutritional information for fun foods such as candy because these foods meet hedonistic (as opposed to health-related) needs (Balasubramanium & Cole, 2002). Cereal bars and other snack products were often seen more as treats and therefore as less serious wholegrain delivery mechanisms. These authors claim that consumers see yoghurt, cereals, bread and juice as credible carriers of functional messages. Poulsen (1999) found that attitude towards enrichment was generally positive when the base product already contained the enriched substance (like calcium in milk). Roe *et al.* (1999) found similar effects for the perception of the healthiness of functional foods where prior beliefs about product healthiness appear to override claim information.

Van Kleef *et al.* (2005) argue that, even though there is increasing scientific evidence that some food components have beneficial physiological and psychological effects over and above the provision of basic nutrients, the development of effective persuasive health claims and successful marketing of functional foods has proven to be rather difficult. They say that health claim studies (such as that of Balasubramanium and Cole, (2002)) have shown that evaluation of health claims is partly determined by the perceptions of healthiness of the base product, which suggests that some health claims combine better with some food products. Van Kleef *et al.* (2005), in their own study, looked at the extent to which consumers perceived specific health claims to be appropriate for particular food products. Ten different base foods were combined with ten different health claims and consumers were asked to rate them on four measures: attractiveness, credibility, uniqueness and intention to buy. Attractiveness was found to be the main driver of intention to buy, followed by uniqueness and credibility. Health claims relating to disease were rated as more attractive than psychological (stress, dementia)- and appearance (youthfulness, skin protection)-related benefits. Health claims, such as giving extra energy and natural defences, were also viewed well both in terms of attractiveness and intention to buy. Margarine and yoghurt featured as attractive carriers for functional claims much more than chewing gum, ice cream and chocolate. Meat replacers were seen as poor functional food carriers. This suggests that people distinguish between foods in terms of whether or not they are suitable as functional food carriers.

From the results, Van Kleef *et al.* (2005) conclude that willingness to try functional food is not only driven by attractiveness but also by how credible the food is and to a lesser extent its uniqueness. Secondly, people evaluate health claims and carriers independently, suggesting flexibility in functional food design and that popular health claims can be applied to several products. Health claims that were popular were addressing relevant disease states, possibly reflecting the products that were available in each country, although prevention of cancer as a health claim showed that the effect was not just based on familiarity. Highly ranked carriers included yoghurt, margarine, brown bread and pills. Given that the first three are often a substantive part of the daily diet, this might reflect the convenience of using these foods as carriers.

In the second study, using yoghurt, Van Kleef *et al.* (2005) looked at how framing, benefit type, focus and relevance of illness affected people's intention to buy functional foods. Results showed that the type of benefit made the largest contribution and also whether or not the health claim related to a personal health problem had an effect. Further, the type of framing had a significant effect on purchase intention, although it also depended on the type of benefit described. Health claims that were related to personally relevant illness were considered more attractive and convincing and were associated with higher purchase intention scores.

Using conjoint analysis, Bech-Larsen and Grunert (2003) looked at the extent to which consumers' perceptions of healthiness depended on the types of claims (none, physiological and prevention), functional enrichments (none, omega3 and oligosaccharides), base products (orange juice, flavoured yoghurt and spread), growing methods (i.e. conventional and organic) and product costs (normal and normal + 20%); comparisons were also made between American, Danish and Finnish consumers. A physiological claim is one that describes how a functional enrichment can affect the body, whereas a prevention claim mentions a disease that can be prevented by enrichment (Poulsen, 1999). Regulation in the European Union and the United States differs with regards to what sort of health claims are allowed (see Chapter 2). Results from Bech-Larsen and Grunert (2003) showed that, as far as health claims were concerned, in all three countries both kinds of health claims had a positive effect on consumers' perceptions of healthiness. Minor differences between Europeans and Americans were attributed to Europeans' unfamiliarity with these products as compared with the Americans who had real experiences of these products. In terms of growing methods, the Europeans were more positive about organic, whereas Americans were indifferent. The cost of the product only had a modest significant effect on perceived healthiness. In addition, interaction effects were found to be larger than the main effects. When spread was fortified with either of the enrichments, it was perceived as healthy, whereas enrichment of yoghurt and orange juice was seen as unhealthy. The authors argued that consumers perceived the yoghurt and juice to be healthy in the first place and the spread to be unhealthy, and hence spread could be made healthier by enrichment. Similarly, for processing, Europeans viewed the enrichment of organic food as being unhealthy, which was not the case for Americans. It is suggested that this is because Europeans see organic as healthy as such and in the United States functional foods have already been successfully marketed. Bech-Larsen and Grunert (2003) suggest that the type of health claims made about these foods, whether it is physiological or preventative, can have an effect on the uptake. Bech-Larsen and Grunert (2003) conclude that consumers' perception of functional foods depended on their perception of the nutritional qualities of the base product rather than on any type of health claim.

1.6.2 The importance of taste and sensory characteristics

Many studies have shown that two of the main conditions for acceptance of functional food are the taste and the trustworthiness of the health claim (Verbeke, 2006). Although increasing the functionality of the food should not necessarily change its sensory quality (Urala & Lahteenmaki, 2004), for grain-based products this could be an area where the functional version of the refined grain product – for example, functional white bread – could be seen as an improvement and a viable alternative to its conventional counterpart, wholemeal bread, which is viewed by many non-consumers as chewy, tough and tasteless (Seal, 2005).

Baskaran and Hardley (2002) also conclude that taste, quality, price/value, convenience and health effects are the key factors in people's intention to buy functional foods. Furthermore, Verbeke (2006) found that consumers are not ready to compromise on taste for health, such that health benefits could not be traded for taste. In a study conducted by Urala and Lahteenmaki (2003) consumers were asked to choose between functional foods and their conventional counterparts in six product categories which were all available in the Finnish marketplace. It was found that people were willing to choose functional foods (51% of choices) but choosing a functional food in one product category did not necessarily mean that consumers chose functional food in another category. Using means-end chain analyses they found that, for all product categories, healthiness, taste and pleasure, security and familiarity, convenience and price described their food choice. Furthermore, they found that the reason was that healthiness was multidimensional and the way it was perceived depended on the product. Sometimes healthiness related to general well-being and at other times to prevention of disease. Healthiness was also linked to improved performance.

In a follow-on study, Urala and Lahteenmaki (in press) asked Finnish consumers to rate eight functional foods and two reference foods on willingness to use, recognition and use, as well as their attitudes towards these functional foods. The results showed that the Finns' willingness to use functional foods could be strongly predicted by perceived reward. This dimension contained items relating to the individual's own personal pleasure as well as positive consequences derived from using functional foods. Urala and Lahteenmaki (in press) also found that consumers differentiated between functional foods and their attitudes, and willingness to use depended on the product in question. They also claim that, unlike previous studies (Jonas & Beckmann, 1998; Backstrom *et al.*, 2003), the meaning of naturalness of functional foods was also product dependent.

1.6.3 The influence of production methods

Perceived healthiness may be influenced by the type of raw material and the processing method used (i.e. origin of the raw material, production date, conservation method, packaging, use of additives, etc. (Bonner & Nelson, 1985)). In terms of functional food, Bech-Larsen and Grunert (2003) argue that perceived healthiness may be related to the enrichment processing methods used, health claims made and the base products used in the process. Cox *et al.* (2004) looked at differences in the acceptance of natural, sweetened and GM-produced functional foods. Results showed that genetic modification was unacceptable even with benefits. Furthermore, there were gender differences as well as differences among products in participants' willingness to consume these functional foods. Sweetened products were unacceptable to people as they were perceived as unnatural. The perceived efficacy of the product was also seen as important. Self-efficacy (i.e. the expectations that people hold about their abilities to accomplish certain tasks) was also found to be a predictor of intention to consume, particularly for a less-preferred product. Susceptibility to certain illnesses was a minor but significant predictor for preferred products, suggesting that perceived susceptibility will not be a driving force for putative remedial products that have disliked ingredients or are produced using unpopular technologies. This suggests that the method of production is an important factor in the acceptability and consumption of functional foods.

Huotilainen *et al.* (2006) asked people to taste three different types of drink after giving them some information about the drinks. Results showed that categorizing the drink as either natural or technological did not affect people's liking for the drinks or their frequency of use. Instead, derived attributes – such as beneficial, artificial, regular and unnecessary – had a significant impact on predicting liking and frequency of use. Furthermore, rating of liking was found to replace the significance of derived attributes in predicting frequency of use, highlighting the importance of food liking. It was noted that, while two of the attributes (artificial and regular) were rooted in the technology or natural category, information on the other two – beneficial and unnecessary – were related to the product or the packaging.

1.6.4 Characterizing consumers

Data from the United States suggest that American consumers who are knowledgeable about functional foods believe that these products give them control over their own health (IFIC, 1999). Knowledge and beliefs are also seen as important determinants for choosing functional foods in the United Kingdom (Hilliam, 1996), although this finding is contested by Pferdekamper (2003) who found health consciousness and preventative health behaviour, rather than knowledge, to be the influencing factors in accepting functional foods. Despite the belief concept being operationalized differently in different studies, most (Hilliam, 1996; Childs, 1997; Verbeke, 2005) found correlations between people's opinions or beliefs about health benefits and acceptance of functional foods. In addition, Verbeke (2005) showed that perceived knowledge and beliefs outweigh the impact of socio-demographic variables on functional food acceptance, confirming that having an ill family member did affect people's choices. However, Verbeke noted that consumer knowledge had an adverse effect on acceptance in the younger age group which disappeared with increasing age. Verbeke claims that, unlike the consumers in the United States, the Belgian (European) consumers believe in the benefits because they may have relatives who are ill and whose critical view of functional food information had lessened with age. He argues that the differences between US and European consumers may be due to the Europeans being more critical towards novel foods and technologies and the lack of trust they have in industry and government. This highlights the importance of conducting market research in different countries and not assuming that because a particular product is acceptable in one country it will also be so in others.

Bech-Larsen and Grunert (2003) claim that, compared with Americans, Europeans are less inclined to accept functional foods as the solution to the dilemma between health and habit. They argue that this may be due to differences in perceptions of what determines healthiness of food. Because certain production methods are considered healthy and natural (e.g. organic), functional enrichment of these foods may have a negative effect on perceived healthiness for foods produced using these methods.

McConnon *et al.* (2004) investigated consumers' perceptions of five different types of healthy foods: foods containing plant sterols – Flora Pro-activ;

foods containing folic acid – breakfast cereal; foods containing probiotics – bioyoghurts; foods with reduced salt content – LoSalt; foods with naturally occurring health benefits – garlic, oily fish. Ratings were given for the following factors: likely benefit to health; worry about potential risk; scientist's knowledge; ease of saying if food contained the substance; whose responsibility to inform the public about health issues (government or industry); how common are the foods; does benefit depend on volume consumed; level of understanding; control over eating the food; frequency of consumption. Results showed that there was a high degree of scepticism with regard to the use of probiotics and plant sterols to optimize health and prevent disease, they were seen as least likely to benefit health and scientists were seen as knowing least about these items, as well as being associated with the most worry and potential risk. Attitude was revealed to be characterized by four dimensions: worry, benefit, control and responsibility. On the whole, functional foods were perceived as positive and the majority expressed the view that such foods benefited their health. Trust expressed in different communicators showed that many consumers viewed science and industry as commercially driven and technology as being imposed on them without consultation. The likelihood of strong public acceptance of functional foods, specifically those that are man-made was found to be lessening. It is argued that consumer perceptions of this new category of foods will be the key element in the success or failure of functional foods.

Backstrom *et al.* (2004) investigated perceptions of ten different foods representing functional foods, nutritionally modified or GM foods, ethnic foods and familiar foods; these included a functional yoghurt, a fat-free yoghurt, a functional ice-cream, calcium-fortified milk, organic bread, organic pork, GM tomato, GM soya, passion fruit, snails, smoked ham and pineapple. Results showed that consumers distinguished between the foods using the dimensions of suspicion, adherence to technology, naturalness, hedonism and necessity. Men and participants with a rural identity or lower education were more suspicious of new foods. Furthermore, urban men were more positive about technology than women. Food was seen as a necessity by people with a lower income, men and participants with a rural identity. Furthermore, individuals on a low income were also found to prefer familiar foods.

Verbeke (2005) pointed out that most of the studies investigating consumer reactions towards functional foods were done during the 1990s in the United States (Wrick, 1992, 1995; Childs, 1997; Childs & Poryzees, 1997; Gilbert, 1997, 2000; IFIC, 1999, 2000) and only a few were focused on European consumer reactions (Hilliam, 1996, 1998; Poulsen, 1999; Bech-Larsen *et al.*, 2001; Makela and Niva, 2002; Pferdekamper, 2003). Verbeke suggested that, despite these studies varying in focus and methodologies, certain socio-demographic characteristics, cognitive and attitudinal factors were found to be potential determinants of consumer acceptance of functional foods. The US data suggest that the functional food consumer is female, well educated, on higher income and over the age of 35 (Childs, 1997; Gilbert, 1997; IFIC, 1999; 2000). European data (Poulsen, 1999; Bech-Larsen *et al.*, 2001) suggest that consumers tend to be female and over the age of 55, but not necessarily highly educated. Hilliam (1996), however, says that acceptance of functional foods indicates a bias towards higher socio-economic groups showing a higher willingness or ability to pay premium prices as well as knowledge and higher awareness. Having young children in the house (Gilbert, 1997) as well as experience with relatives' loss of good health (Childs, 1997; Wrick, 1995; Milner, 2000) have also been shown to act as incentives to adopt disease-preventative food habits.

1.7 Future trends

From wholegrain consumption data in the United States and the United Kingdom it is clear that wholegrain consumers tend on the whole to be male (maybe due to total amount consumed), older in age and from higher socio-economic groups; they are also more likely to be non-smokers who take regular exercise, take vitamin and mineral supplements and are not over weight. This suggests that wholegrain consumers tend to be in general older and wealthier with a healthy lifestyle. One could thus argue that the groups that need to be targeted to encourage the consumption of functional grain products are females, younger in age, from lower socio-economic groups, and those who generally have less healthy diets. Research suggests that the barriers to wholegrain consumption are knowledge of what is wholegrain, knowing how much to eat and what constitutes a portion, cost, availability, perceived taste and texture, and lack of awareness of the benefits. Some of these barriers could potentially be overcome through education and promotion of wholegrain foods. Alternatively the nutritional benefits of wholegrain foods could be gained from eating functional cereal products.

Research on existing functional food products suggests that those who eat functional foods are mainly older women, of higher education, with children and those who have illness in the family. Acceptance of functional food depends on the type of base product that has been altered and the methods used to alter it. The majority of studies suggest that staple foods such as bread can more successfully be transformed into functional foods than hedonistic foods such as cakes and biscuits. However, the perceived barrier of the feeling of bloating after eating wheat products has been identified as a major barrier and could impact on the potential of certain products playing an important role in increasing wholegrain consumption. This suggests that, while certain functional foods such as yoghurt and spreads may have female support, this may not necessarily transfer to all types of grains. As women were found to be lower consumers of wholegrain but higher consumers of functional foods, there is a great potential market to sell functional grain products to women. However, it is important that grains other than wheat, such as rice and oats, are also used as base products to cater for those with wheat intolerances.

Both wholegrain consumers and functional food consumers tend to be older in age. This could be because older people tend to be more health conscious, know others who suffer from health-related problems and also are willing to pay the premium prices. This age group may also have an interest in functional cereal products because they could be providing health benefits regarded as important by this age group. Attracting a younger age group to become consumers of these products may prove to be a challenge, it will be necessary to find appropriate ways of making the potential health benefits of these products be regarded as important by this age group. It may also be important to increase the availability of low-cost products.

The acceptance of functional foods differs between countries, it is thus important to assess and understand consumer perceptions and acceptance in different regions and not assume automatic uniform acceptance across all countries. Studies have shown that consumers are unwilling to compromise on taste even though a product has health benefits (Verbeke, 2006). Thus functional grain products have to be tastier (or at least as tasty) than their conventional counterparts for them to be a success. Furthermore, certain types of health claims and certain production methods have been found to be more acceptable to the general public than others. It is thus important to investigate the appropriateness of claims from particular foods and that the production methods employed are made clear so that the consumers come to trust the products and producers in the longer term.

Some of the main issues identified above are being addressed in a new project called HEALTHGRAIN (www.healthgrain.org) that is funded by the European Commission within the Sixth Framework Programme. The primary objective of the project is to improve and enhance the nutritional value and health benefits of grain products, and to increase the use of wholegrains in modern foods. Consumer expectations and attitudes towards cereal-based products that have been modified to contain more health-promoting components will be investigated in the HEALTHGRAIN project using qualitative and quantitative methodologies. The cross-European project will look into the impact of the modification method on acceptability and how appealing different types of health-promoting aspects of cereal-based product are to consumers in different European countries. It will also explore how these will fit with the existing health image of products. The presented health benefits will be based on the assumed benefits provided by products developed within the HEALTHGRAIN project and are likely to include staple foods such as bread and pasta as well as hedonistic foods such as biscuits, made with grains of wheat and rye. This research will enable industry, government, consumers and other stakeholders to better understand how best to make functional cereal products a success.

Given that eating grains, especially wholegrains, is important to health and that there are many potential barriers to eating these foods, functional grain products may be one way of increasing public consumption of grains. However, to ensure that these products are a success in the marketplace, as well as creating tasty healthy grain products, care needs to be taken in developing suitable ways of communicating the health benefits and information regarding the means of production.

1.8 Acknowledgements

The authors acknowledge the support for the preparation of this manuscript from the European Commission in the Communities Sixth Framework Programme, as part of the project HEALTHGRAIN (FP6-514008). This chapter reflects the authors' views and the Community is not liable for any use that may be made of the information contained in this chapter.

1.9 References

Adams J F and Engstrom A (2000a), 'Dietary intake of whole grain vs. recommendations', *Cereal Foods Worldwide*, **45**, 75–79.

Adams J F and Engstrom A (2000b), 'Helping consumers achieve recommended intakes of whole grain foods', *Journal of the American College of Nutrition*, **19**, 339S–344S.

Albertson A M and Tobelmann R C (1995), 'Consumption of grain and whole-grain foods by an American population during the years 1990–1999', *Journal of American Dietary Association*, **95**, 703–704.

Backstrom A, Prittila-Backman A-M and Tuorila H (2003), 'Dimensions of novelty: a social representation approach to new foods', *Appetite*, **40**, 299–307.

Backstrom A, Prittila-Backman A-M and Tuorila H (2004), 'Willingness to try new foods as predicted by social representations and attitude and trait scales', *Appetite*, **43**, 75–83.

Balasubramanium S K and Cole C (2002), 'Consumers' search and use of nutrition information: The challenge and promise of the nutrition labelling and education act', *Journal of Marketing*, July, 112–127.

Baskaran S and Hardley F (2002), 'Buyers beliefs, attitudes and behaviour: foods with therapeutic claims', *Journal of Consumer Marketing*, **19**, 591–606.

Bech-Larsen T and Grunert K G (2003), 'The perceived healthiness of functional foods: a conjoint study of Danish, Finnish and American consumers' perception of functional foods', *Appetite*, **49**, 9–14.

Bech-Larsen T, Grunert K G and Poulson J B (2001), 'The acceptance of functional food in Denmark, Finland and United States', MAPP working paper No. 73, Aarhus, Denmark: Aarhus School of Business.

Blades M (2000), 'Functional foods and nutraceuticals', *Nutrition and Food Science*, **30**, 273–275.

Bonner G P and Nelson R (1985), 'Product attributes and perceived quality: Foods', in Jacoby J and Olson J C (Eds.), *Perceived Quality – How Consumers View Stores and Merchandise*. Lexington, MA: Lexington Books, pp. 65–79.

Childs N M (1997), 'Functional food and the food industry: consumer, economic and product development issues', *Journal of Nutraceuticals, Functional and Medical Food*, **1** (2), 25–43.

Childs N M and Poryzees G H (1997), 'Foods that help prevent disease: consumer attitudes and public policy implications', *Journal of Consumer Marketing*, **14** (6), 433–447.

Childs N M, Lachance P and Meagher L (2000), 'Advancing nutraceuticals opportunities: priorities for research', *Journal of Nutraceuticals, Functional and Medical Foods*, **2**, 85–103.

Cleveland L E, Moshfegh A J, Albertson A M and Oldman J D (2000), 'Dietary intake of whole grains', *Journal of the American College of Nutrition*, **19**, 331S–338S.

Conner M and Armitage C J (2002), *The Social Psychology of Food*. Buckingham, UK. Open University Press, pp. 1–11.

Consumers' Association (2000), 'Functional food report', http://www.which.net/foodanddrink/contents.html.

Cox D N, Koster A and Russell C G (2004), 'Predicting intentions to consume functional foods and supplements to offset memory loss using an adaptation of protection motivation theory', *Appetite*, **43**, 55–64.

Gardette N (2000), 'Is there a future in Europe for functional food?', MBA thesis, City University Business School, London.

Gilbert L (1997), 'The consumer market for functional food', *Journal of Nutraceuticals, Functional and Medical Foods*, **1** (3), 5–21.

Gilbert L (2000), 'The functional food trend: what's next and what Americans think about eggs', *Journal of the American College of Nutrition*, **19** (5), 507S–512S.

Glanz K (1997), 'Behavioural research contributions and needs in cancer prevention and control: dietary change', *Preventative Medicine*, **26**, S43–S55.

Grunert K G, Bech-Larsen T and Bredahl L (2000), 'Three issues in consumer quality perception and acceptance of dairy products', *International Dairy Journal*, **10**, 575–584.

Grunert K G, Brunso K, Bredahl L and Bech A (2001), 'Food related lifestyle: a segmentation approach to European food consumers', in Frewer L, Risvik E and Schifferstein H (Eds.), *Food People and Society: A European Perspective of Consumers' Food Choices*. Berlin: Springer, pp. 211–230.

HEALTHGRAIN EU Integrated Project 2005–2010. Exploiting bioactivity of European cereal grains for improved nutrition and health benefits, www.healthgrain.org.

Hilliam M (1996), 'Functional foods: the Western consumer viewpoint', *Nutritional Review*, **54** (11), S189–S194.

Hilliam M (1998), 'The market for functional foods', *International Dairy Journal*, **8**, 349–353.

Hilliam M (2000), 'Functional food: how big is the market?', *World of Food Ingredients*, **12**, 50–53.

Huotilainen A, Seppala T, Pirttila-Backman A-M and Tuorila H (2006), 'Derived attributes as mediators between categorization and acceptance of a new functional drink', *Food Quality and Preference*, **17**, 328–336.

IFIC (International Food Information Council (1999), *Functional Foods: Attitudinal Research (1996–1999)*. Washington: IFIC, International Food Information Council Foundation.

IFIC (2000), *Functional Foods: Attitudinal Research August 2000: Quantitative and Qualitative Summary*. Washington, IFIC, International Food Information Council Foundation.

Jacobs D R Jr, Meyer K A, Kushi L H and Folsom A R (1999), 'Is wholegrain intake associated with reduced total and cause-specific death rate in older women?: the Iowa Women's Health Study', *American Journal of Public Health*, **89**, 322–329.

Jacobs D R Jr, Meyer H E and Solvoll K (2001), 'Reduced mortality among whole grain bread eaters in men and women in the Norwegian County Study', *European Journal of Clinical Nutrition*, **55**, 137–143.

Johansson L, Thelle D, Slovoll K, Bjoerneboe G E A and Drevon C H (1999), 'Healthy dietary habits in relation to social determinants and lifestyle factors', *British Journal of Nutrition,* **81**, 211–220.

Jonas M S and Beckmann S C (1998), 'Functional foods: consumer perceptions in Denmark and England', MAPP working paper No. 55, Aarhus, Denmark: Aarhus School of Business.

Jones J M, Reicks M, Adams J, Fulcher G, Weaver G, Kanter M and Marquart L (2002), 'The importance of promoting a whole grain food message', *Journal of the American College of Nutrition*, **21** (4), 293–297.

Lang R and Jebb S A (2003), 'Who consumes whole grains and how much?', *Proceedings of the Nutrition Society*, **62**, 123–127.

Lang R, Thane C W, Bolton-Smith C and Jebb S A (2001), 'Wholegrain food consumption by British adults from two national dietary surveys', *Proceedings of the Nutrition Society*, **60**, 218A.

Lang R, Thane C W, Bolton-Smith C and Jebb S A (2003), 'Consumption of whole-grain foods by British adults: findings from further analysis of two national dietary surveys', *Public Health Nutrition*, **6** (5), 497–484.

Makela J and Niva M (2002), 'Changing views of healthy eating: cultural acceptability of functional foods in Finland', Paper presented at Nordisk Sociologkongress, Pa Island Reyljavic, Iceland.

McConnon A, Fletcher P L, Cade J E, Greenwood D C and Pearman A D (2004), 'Differences in perceptions of functional foods: UK public vs. nutritionists', *Nutrition Bulletin*, **29**, 11–18.

McCrea D (2001), 'Functional foods: in the consumers interest?', *Food Science and Technology Today*, **15**, 35.

Mialon V S, Clark M R, Leppard P I and Cox D N (2002), 'The effect of dietary fibre information on consumer responses to breads and English muffins: a cross-cultural study', *Food Quality and Preference*, **13**, 1–12.

Milner J A (2000), 'Functional Foods: the USA perspective', *American Journal of Clinical Nutrition*, **71**, 1664S–1669S.

Mintel (2000), *Functional Foods*. London: Mintel International Group Limited.

NCC (National Consumer Council) (1997), 'Messages of food; consumers' use and understanding of health claims on food packs'. London: NCC.

Pferdekamper T (2003), 'Determinants of the acceptance of functional food using an example of new probiotic rusk – an empirical analysis', Paper presented at Consumer Perception of Healthiness of Foods and Consumer Acceptance of New Foods, 7–8 April, Middelfart, Denmark.

Pietinen P, Rimm E B, Korhonen P, Hartman A M, Willwett WC, Albanes D and Virtamo J (1996), 'Intake of dietary fibre and risk of coronary heart disease in a cohort of Finnish men – The Alpha Tocopherol, Beta-Carotene Cancer Prevention Study', *Circulation*, **94**, 2720–2727.

Poulsen J B (1999), 'Danish consumers' attitude towards functional foods', MAPP working paper No. 62, Aarhus, Denmark: Aarhus School of Business.

Prattala R, Helasoja V and Mykkanen H (2001), 'The consumption of rye bread and white bread as dimensions of health lifestyles in Finland', *Public Health Nutrition*, **4**, 813–819.

Richardson D P (2003), 'Wholegrain health claims in Europe', *Proceedings of the Nutrition Society*, **62**, 161–169.

Richardson N, MacFie H and Shepherd R (1994), 'Consumer attitude to meat eating', *Meat Science*, **36**, 57–65.

Roberfroid M B (2000), 'Global view on functional foods: European perspectives', *British Journal of Nutrition*, **88** (2), S133–S138.

Roe B, Levy A S and Derby B M (1999), 'The impact of health claims on consumer search and product evaluation outcomes: results from FDA experimental data', *Journal of Public Policy and Marketing*, **18** (1), 89–105.

Saher M, Arvola A, Landeman M and Lahteenmaki L (2004), 'Impressions of functional food consumers', *Appetite*, **42**, 79–89.

Seal C J (2005), 'Whole grain and health', Presented at the Quaker sponsored workshop Whole Grains and Health, Nutrition and Health Conference, 26 November, Olympia Conference Centre, London.

Shah N P (2001), 'Functional foods for probiotics and prebiotics', *Food Technology*, **55**, (11), 43–46.

Shepherd R (1990), 'Attitudes and beliefs as determinants of food choice', in McBride R L and MacFie H J H (Eds.), *Psychological Basis of Sensory Evaluation*. London, Elsevier Applied Science.

Shepherd R, Sparks P, Bellier S and Raats M (1991/2), 'The effects of information on sensory rating and preferences: the impact of attitudes', *Food Quality and Preference*, **3**, 147–155.

Smith A, Kuznesof S, Richardson D P and Seal C J (2001a), 'Effectiveness of dietary intervention strategies aimed at increasing consumption of whole grains or low and reduced fat products in free living volunteers', *Proceedings of the Nutrition Society*, **60**, 178A.

Smith A, Richardson D P, Kuznesof S and Seal C J (2001b), 'Effectiveness and acceptability of a dietary intervention to increase consumption of whole grain products in free living individuals', in Liukkonen K, Kuokka A and Poutanen K (Eds.), *Whole Grain and Human Health*. Espoo, Finland: Technical Research Centre of Finland, pp. 32–35.

Smith, A T, Kuznesof S, Richardson D P and Seal C J (2003), 'Behavioural attitudinal and dietary responses to the consumption of wholegrain foods', *Proceedings of the Nutrition Society*, **62**, 455–467.

Thane C W, Jones A R, Stephen A M, Seal C J and Jebb S A (2005), 'Whole-grain intake of British young people aged 4–18 years', *British Journal of Nutrition*, **94**, 825–831.

Urala N and Lahteenmaki L (2003), 'Reasons behind consumers' functional food choices', *Nutrition and Food Science*, **33** (4), 148–158.

Urala N and Lahteenmaki L (2004), 'Attitudes behind costumers' willingness to use functional foods', *Food Quality and Preference*, **15**, 793–803.

Urala N and Lahteenmaki L (in press), 'Consumers' changing attitudes towards functional foods', *Food Quality and Preference*.

USDA (US Department of Agriculture) (1997), 'Pyramid servings data – results from USDA's 1994–1996 continuing survey of food intake by individuals', Riverdale, California: Food Surveys Research Group.

USDA Center for Nutrition Policy and Promotion (2000), 'Consumption of food group servings: people's perceptions vs. reality', Nutritions, Insights 20.

Van Duyn M A, Kristal A R, Dodd K, Subar A F, Stables G, Nebeling L and Glanz, K (2001), 'Association of awareness, intra-personal and interpersonal factors, and stage of dietary change with fruit and vegetable consumption: a national survey', *American Journal of Health Promotion*, **16**, 69–78.

Van Kleef E, Van Trijp H C M and Luning P (2005), 'Functional foods: health claim–food product compatibility and the impact of health claim framing on consumer evaluation', *Appetite*, **44**, 299–308.

Verbeke W (2005), 'Consumer acceptance of functional foods: socio-demographic, cognitive and attitudinal determinants', *Food Quality and Preference*, **16**, 45–57.

Verbeke W (2006), 'Functional foods: consumer willingness to compromise on taste for health?', *Food Quality and Preference*, **17**, 1–2.

Welton W E, Kantner T A and Katz S M (1997), 'Developing tomorrow's integrated community health systems: a leadership challenge for public health and primary care' *Milbank Quarterly*, **75**, 261–288.

Westcombe A and Wardle J (1997), 'Influence of relative fat content information on responses to three foods', *Appetite*, **28**, 49–62.

Weststrate J A, van Poppel G and Verschuren P M (2002), 'Functional foods, trends and future', *British Journal of Nutrition*, **88**, S233–S235.

Williamson A R, Hunt A E, Pope J F and Tolman N M (2000), 'Recommendations of dieticians for overcoming barriers to dietary adherence in individuals with diabetes', *The Diabetes Educator*, **26**, 272–278.

Wrick K L (1992), 'Consumer viewpoints on 'designer foods', *Food Technology*, March, 100–104.

Wrick K L (1995), 'Consumer issues and expectations for functional foods', *Critical Reviews in Food Science and Nutrition*, **35** (1&2), 167–173.

2

Labelling and regulatory issues related to functional cereal products

K. Schmitz and L. Marquart, University of Minnesota, USA and J. Willem van der Kamp, TNO Quality of Life, The Netherlands

2.1 Introduction

Cereals, especially whole grains, are known to be a rich source of dietary fibre, antioxidants, vitamins, minerals, phenolic acids, lignans and phytosterols. Epidemiological studies indicate that consuming whole grains reduce risk for certain chronic diseases[1] including: weight gain/obesity,[2–6] cardiovascular disease,[7–14] and diabetes.[15–19] Preliminary clinical evidence suggests whole grain intake enhances blood glucose/insulin response,[20–23] lowers blood pressure,[24,25] reduces cholesterol,[26–28] improves biomarkers of bowel health[22] and reduces progression of coronary atherosclerosis.[29]

In response to consumers' interests, regulatory agencies have developed methods to disseminate information about the health aspects of certain foods. In particular, this chapter will explore the regulations for cereal components in both Europe and the United States with an in-depth look at the allowable health claims. First, the discussion focuses on Codex and its definition of health claims, followed by the activities in the European Union related to the recent regulation on nutrition and health claims. Particular attention is paid to the health claims related to cereal components currently in use in the United Kingdom, Sweden and The Netherlands. Next, we will discuss the regulation process in the United States and health claims allowed for cereals. Finally, the chapter discusses issues related to the labelling and definition of whole grains, followed by future trends in regulations.

2.2 Regulation and labelling of functional cereal products from Codex

The Codex Alimentarius Commission, created in 1963 by the Food and Agriculture Organization (FAO) and World Health Organization (WHO), serves to develop food standards and guidelines under the Joint FAO/WHO Food Standards Program. The primary objectives of the Food Standards Program are to protect the health of consumers, ensure fair food trade practices and promote coordination of all food standards work. Within the Codex guidelines there are three different types of health claims: nutrient function claims, other function claims and reduction of disease risk claims.[30] According to Codex a health claim is any representation that states, suggests or implies that a relationship exists between a food or a constituent of that food and health. Nutrient function claims – nutrition claims describing the physiological role of the nutrient in growth, development and normal body functions – are included in health claims. In general, Codex recommends health claims to be consistent with national health policy, including nutrition policy, and to support other policies where applicable. Additionally, health claims should be supported by a sound and sufficient body of scientific evidence, provide truthful and non-misleading information to aid consumers in choosing healthful diets, and be supported by specific consumer education. Aside from defining health claims, Codex does not recommend or take a position on potential health claims.

In 2003, a joint WHO/FAO Expert Consultation group reported on Diet, Nutrition and the Prevention of Chronic Diseases.[31] Here the strength of evidence of factors that might influence chronic diseases is designated into specific categories: convincing, probable, possible or insufficient. This type of listing is also used in assessing the merit of health claims in the United States and in discussions on health claims in the context of the new EU Regulation. The WHO/FAO group has listed whole grain cereals in the probable category for decreasing the risk of developing cardiovascular diseases (CVDs). A high intake of dietary fibre, or non-starch polysaccharide (NSP), is listed as convincing regarding obesity, probable in relation to diabetes type-2 and CVD, and possible in relation to cancer.

In terms of definitions, Codex has not defined whole grains as an official standard and should work to develop an internationally accepted definition. For dietary fibre, the most recent definition was proposed by the Codex Committee on Nutrition and Foods for Special Dietary Uses (CCNFSDU), together with recommendations for labelling. This definition is cited in a range of documents, for instance in a discussion paper on labelling.[32] Key parts of the proposed text are:

> Dietary fibre means carbohydrate polymers with a degree of polymerisation (DP) not lower than 3 which are neither digested nor absorbed in the small intestine. (...) Dietary fibre consists of one or more of the following:

- Edible carbohydrate polymers naturally occurring in the food as consumed.
- Carbohydrate polymers, which have been obtained from food raw material by physical, enzymatic or chemical means.
- Synthetic carbohydrate polymers.

(...), a physiological effect should be scientifically demonstrated by clinical studies and other studies as appropriate, with the exception of non-digestible edible carbohydrate polymers naturally occurring in foods as consumed where a declaration or claim is made with respect to dietary fibre.

Although this definition is the result of many years of discussion, the debate continues as to whether dietary fibre should include – in addition to edible carbohydrate polymers naturally occurring in food – carbohydrate polymers that are synthesized or isolated from food raw materials. A recent overview of dietary fibre and the ongoing debate on definitions is provided by Gray in an ILSI monograph.[33] A significant factor is the observation that intake of dietary fibre is recommended – also in recent guidelines (e.g. Health Council of The Netherlands, 2006[34]) – in particular as an integral part of fruits, vegetables and whole grain products, due to the large number of epidemiology studies showing beneficial effects of such products, whereas less data support health benefits of other types of fibres.

2.3 Regulation and labelling of functional cereal products in the European Union

2.3.1 FUFOSE and PASSCLAIM

Awaiting the EU regulation on health claims, prior health claims in Europe were country specific. The European Commission supported the Functional Foods Science in Europe (FUFOSE) project in 1995 which established a scientific approach to the development of functional food products.[35] After FUFOSE, the European Commission engaged in a 4-year project called 'Process for the Assessment of Scientific Support for Claims on Foods (PASSCLAIM)' due to the importance of scientific substantiation of claims for health claim approval. The project was organized by the International Life Sciences Institute of Europe, to produce a generic tool for assessing the scientific support for health claims on food. At the conclusion of PASSCLAIM criteria for scientific substantiation of claims were determined and laid out in a consensus document.[36] The criteria are as follows.

1 The food or food component to which the claimed effect is attributed should be characterised.
2 Substantiation of a claim should be based on human data, primarily

from intervention studies the design of which should include the following considerations:

(a) Study groups that are representative of the target group.
(b) Appropriate controls.
(c) An adequate duration of exposure and follow up to demonstrate the intended effect.
(d) Characterisation of the study groups' background diet and other relevant aspects of lifestyle.
(e) An amount of the food or food component consistent with its intended pattern of consumption.
(f) The influence of the food matrix and dietary context on the functional effect of the component.
(g) Monitoring of subjects' compliance concerning intake of food or food component under test.
(h) The statistical power to test the hypothesis.

3 When the true endpoint of a claimed benefit cannot be measured directly, studies should use markers.
4 Markers should be:

- Biologically valid in that they have a known relationship to the final outcome and their variability within the target population is known.
- Methodologically valid with respect to their analytical characteristics.

5 Within a study the target variable should change in a statistically significant way and the change should be biologically meaningful for the target group consistent with the claim to be supported.
6 A claim should be scientifically substantiated by taking into account the totality of the available data and by weighing of the evidence.

The criteria describe the standards by which the quality and relevance of the scientific evidence should be judged and the extent to which a claim based on them can be said to be scientifically valid. The substantiation of a claim is heavily weighted towards intervention studies that provide extensive evidence toward a food or food component and function and/or disease relationship. The end result of PASSCLAIM is to bolster consumer confidence in food-related claims and to help consumers make well-informed healthy lifestyle choices.

2.3.2 The European Union regulation on nutrition and health claims made on foods

The European Union first published a draft proposal for regulation of nutrition, function and health claims in July 2002 as an attempt to harmonise health claims and provide scientific guidelines to substantiate claims.[37] The final regulation for nutrition and health claims made on foods was passed on 20 December 2006 and went into effect on 19 January 2007.[38] The regulation covers both food products and food supplements and differentiates between nutrition claims, health claims and reduction of disease claims.

Nutrition claims shall only be permitted if they are listed in the Annex of the Regulation, describing 24 specifications of the conditions for using claims such as: Low Fat, Fat-Free, Source of Protein, High Protein, Source of Vitamin X, High Vitamin X, Light, and Natural. The claim Source of Fibre may only be made where the product contains at least 3 g of fibre per 100 g or at least 1.5 g of fibre per 100 kcal. High Fibre may be used for products with at least 6 g of fibre per 100 g or at least 3 g per 100 kcal. Health claims referring to the reduction of disease risk and to children's development and health under Article 14 may only be made when they have been authorised with the procedure described in Articles 15, 16, 17 and 19.

With regard to health claims, other than those referring to the reduction of disease risk and to children's development and health, the Commission will (Article 13) draw up a positive list of well-established claims, such as 'calcium is good for your bones', if they are based on generally accepted scientific evidence and well understood by the average consumer. These may be used on a label so long as they are proven to apply to the food in question. Member states of the European Union will submit before February 2008 lists of claims along with references to the relevant scientific justification and conditions for use. Before February 2010, the Commission will, after consulting the European Food Safety Authority (EFSA), produce a positive list of health claims. This will enable food producers to introduce products with appropriate composition by using a health claim on this list without submitting a dossier. Any claim submitted for the EU list after this period will have to be examined under Article 5 by the EFSA and approved by the Commission, after consultation of the member states, as described in Article 18. Health claims based on newly developed scientific evidence should undergo an accelerated type of authorisation according to Article 18; the protection of proprietary data will be allowed for a limited, but not yet defined, period of time.

Codex recommends that health claims should be consistent with national health and nutrition policy, and support other policies where applicable. The preamble of the Regulation states that health claims should be scientifically substantiated by taking into account the totality of the available scientific data, and by weighing the evidence. Additionally, health claims should provide truthful and non-misleading information to aid consumers choosing healthful diets and be supported by specific consumer education. An outline on how this statement may be implemented is given by Richardson,[39] taking into account the work of PASSCLAIM,[36] the WHO report[31] and other policy documents.

The aim of the Regulation is to avoid situations where nutrition or health claims mask poor overall nutritional quality of a food product. To this end nutrient profiles will be defined, taking into account the content of different nutrients and substances with a nutritional or physiological effect, such as fat, saturated fat, trans-fatty acids, salt/sodium and sugars, excessive intakes of which in the overall diet are not recommended, as well as poly- and mono-unsaturated fats, available carbohydrates other than sugars, vitamins, minerals, protein and fibre. Two years after the date of regulatory enforcement, the Commission shall, based on scientific advice provided by EFSA, establish specific nutrient profiles, including exemptions, with which food or certain categories of food must comply in order to bear nutrition or health claims. Nutrition claims shall be allowed, where a single nutrient exceeds the nutrient profile, provided that a statement about the specific nutrient appears in close proximity to, on the same side and with the same prominence as the claim.

The Regulation states that claims on foods can be understood by the consumer, taking as a benchmark the 'average consumer, who is reasonably well-informed and reasonably observant and circumspect'. Where appropriate, the Commission shall involve interested parties, in particular food business operators and consumer groups, in order to evaluate the perception and understanding of the claims in question. Due to this Regulation, the issue of consumer understanding is addressed in a growing number of papers and expert groups.[39]

Some products with health claims have already been assessed at the EU level in past years. These are novel food products, containing food ingredients that have not been used to a significant degree for human consumption in the European Union before 15 May 1997. The Novel Foods Regulation (EC) No. 258/97 was adopted in January 1997.[40] In August 2000 the Fazer Company, Finland, made a request for plant sterol-enriched bakery products, grain-based snack products and gum arabic pastilles. The EU Scientific Committee on Food (SCF) expressed concerns on effects of cumulative intakes of such esters from a wide range of foods with added phytosterols. However, at the same time the SCF confirmed with regard to the application of Fazer that the addition of phytosterols to a wide range of bakery products was safe. Fazer agreed to reduce the original application to rye bread. In January 2006 the request was granted for 'rye bread ≥ 50% whole grain rye flour, <30% wheat flour, <4% added sugar to rye bread, and no added fat in the rye bread'.[41] Intake of plant sterols should not exceed 3 g per day. Later that year, the rye product was introduced into the marketplace.

2.3.3 Regulation and labelling of functional cereal products in the United Kingdom

Prior to the EU Regulation, self-regulation of health claims was arranged in a number of EU member states. Sweden was the first country to utilise self-regulation of health claims (1990), followed by The Netherlands (1998) and the United Kingdom (2002). The Joint Health Claims Initiative (JMCI) in the United Kingdom is a joint venture between consumer organisations, enforcement authorities and industry trade associations with a sole purpose to establish a code of practice for health claims on food. The first health claim in the United Kingdom related to cereals was the whole grain health claim in 2002. The official claim reads 'People with a healthy heart tend to eat more whole grain foods as part of a healthy lifestyle' and may be applied to foods deemed appropriate by the committee. The claim is allowed based on the following six conditions as determined in the final report of the Expert Committee to the JHCI Council:[42]

1 The health impact of a diet containing wholegrain foods depends on the rest of the diet as well as other lifestyle factors such as exercise. The claim must be set within this context.
2 The evidence supports an association between a healthy heart and wholegrain consumption but is insufficient to demonstrate cause and effect.
3 The evidence is insufficient to support claims targeted specifically at men.
4 The claim relates to foods containing 51% or more wholegrain ingredients by weight per serving. The term 'whole grain' refers to the major cereal grains including wheat, rice, maize and oats. The structure for all grains is similar and the grain is made up of three components: the endosperm, the germ and the bran.
5 The JHCI strongly recommends that companies seek advice from the Secretariat before using this claim to help ensure that the food product is consistent with good nutrition principles and complies with the JHCI Code of Practice for Health Claims on Food.
6 The wording of the claim has been carefully formulated to reflect the evidence on which the claim has been approved. Wording may be altered, in consultation with the JHCI, as long as the claim does not imply health benefits beyond the scope of the evidence; change the meaning of the claim; or confuse consumers.

Following the whole grains claim, the JHCI released a claim in 2004 for oats and blood cholesterol which reads 'The inclusion of oats as part of a diet low in saturated fat and a healthy lifestyle can help reduce blood cholesterol'.[43] Similar to the first health claim, conditions were developed for use of the claim. The first condition is that the claim relates to whole oats, oat bran, rolled oats and whole oat flour. Beta-glucan soluble fibre may serve as a marker for the oat product because products carrying the claim should contain at least 0.75 g of beta-glucan soluble fibre per serving. In addition, the claim must state what constitutes a serving, the amount of beta-glucan soluble fibre in each serving and the proportion it contributes to a suggested daily intake of 3 g. The claims should also be set in the context of a diet that is low in saturated fat and a healthy lifestyle. The other conditions are for company advice and consumer protection.

2.3.4 Regulation and labelling of functional cereal products in Sweden

The Swedish Code, entitled 'Health Claims in the Labelling and Marketing of Food Products, The Food Industry's Rules (Self-Regulating Programme)', introduced in 1990, was developed by national organisations representing primary production, the food industry and major retail organisations, in close collaboration with relevant authorities.[44] The Swedish Nutrition Foundation (SNF) participated in the development of the Swedish Code and has an advisory and coordinating role. The original Swedish Code allowed generic claims concerning eight different, generally recognised diet–health relationships, closely related to the official nutrition recommendations; no human intervention studies are required for products using such claims. In addition to a labelled health claim, a specific designator – the keyhole symbol – was established to identify lean, fibre-rich alternatives of a food group.[45]

In June of 2003, the whole grain claim was adopted as the ninth generic claim: 'A healthy lifestyle and a balanced diet rich in whole grain products reduces the risk for (coronary) heart disease. Product X is a good source of whole grain (contains Y% whole grain)'. Aside from the product showing substantial effect on the composition of the entire diet, the percentage of whole grain must be at least 50% on a dry weight basis, which is realistic also for soft bread. Furthermore, restrictions regarding fat, sugar and salt content are applied. The latest revision of the Swedish Code[45] is applicable from September 2004.[19]

Constipation and dietary fibre is the second health claim related to cereals in Sweden (www.snf.ideon.se/snf/en/rh/generic_claims.htm). 'A product making a claim regarding the connection between dietary fibre and constipation must meet the criterion for high fibre (3.5 g of dietary fibre per 1000 kJ). As a result, normal daily consumption of the product must provide at least 5 g of dietary fibre. For breakfast cereals, fat content must not exceed 10%. If more than 20% of the fibre content is made up of added fibre concentrates or isolates, documentation on the laxative effect of these fibres must be provided. According to keyhole symbol criteria, the total sugar content (mono- and disaccharides) in breakfast cereals must not exceed 13%. This corresponds to ~10% added sugar in cereals containing only cereal grains. According to the Code, the same criteria for added sugars must also be met for other dry products. For (soft) breads, added sugar content must not exceed 7%. The claim reads: 'A nutritionally balanced diet high in dietary fibre is important for maintaining bowel regularity and reduces the risk of constipation. Product XX is high in dietary fibre'.[45]

The third Swedish health claim concerns cardiovascular disease/atherosclerosis–blood cholesterol levels due to hard fat (primarily saturated) and certain types of dietary fibre. The current recommendation supports claims regarding the connection between particular types of dietary fibre and blood cholesterol levels, and the claims are to be used primarily for oat fibre (beta-glucans). If the food is processed, the Code recommends that the cholesterol-lowering effect must be substantiated after processing. A product making a claim about the connection between oat fibre (beta-glucans) and blood cholesterol levels must contain 0.75 g of beta-glucans per normal serving or provide 3 g per day at a normal amount consumed. The packaging should clearly state how much oat fibre (beta-glucans) the product contains as well as the amount of oat fibre (beta-glucans) that should be eaten to achieve a cholesterol-lowering effect. The claim reads: 'A nutritionally balanced diet high in soluble fibres from oats (beta-glucans) can contribute to lower cholesterol levels in the blood and thereby reduce risk of cardiovascular disease/atherosclerosis/hardening of the arteries. Product Z is high in soluble oat fibres (beta-glucans)'.[45] Recently the scope of the claim has been extended to soluble fibre for barley and for combinations of oats and barley.

In 2001, an approach for the approval of product-specific physiological claims was adopted. Such claims should be based on new/proprietary data, and can be used in the following contexts:

- physiological effects of substances other than nutrients;
- new effects of nutrients;
- glycaemic index (GI) labelling.

The claims must be substantiated by human intervention studies on the final product demonstrating the claimed effect. Animal studies, *in vitro* studies and studies on the active ingredient may be supportive.

GI claims can be formulated as: 'Product X gives a low and slow increase of blood sugar' or: '...has a scientifically documented low GI'. Products should contain at least 15 g of glycaemic carbohydrates per portion and should have a GI below 55, measured by two independent GI determinations. Twelve products, including three breads and two types of muesli have obtained an approved GI claim.

2.3.5 Regulation and labelling of functional cereal products in The Netherlands

The Netherlands Code of Practice for assessing the scientific evidence for health benefits stated in health claims on food and drink products was initiated by The Netherlands Nutrition Centre in 1998 as a voluntary code agreed upon by national organisations representing the food and drink industry, retailers, the advertisers and consumers. Only product-specific claims are allowed. The substantiation of claims needs to be based on relevant scientific data related to human consumption of quantities commonly consumed by the target population. A substantiation of the effectiveness of an ingredient without data related to the product is not sufficient. The health benefit should not be contrary to the recommendations for healthy eating in reports of national authoritative bodies, such as The Netherlands Health Council, and in similar international documents. Applications for health claims are assessed by a committee of at least three independent experts appointed by The Netherlands Nutrition Centre. Three of the eight approved claims are for specific brands of bread: with added Oatwell® oat bran, with Frutafit®-inulin and with n-3 fatty acids, as occurring in fish. The formulation for the first two claims is as follows:

- Product: Pró-FIT bread, containing 2.2 g β-glucan from Oatwell® oat bran per 100 g bread.
  Health benefit: Daily consumption of 140 g of Pró-FIT® bread (4 slices, providing 3 g β-glucan from oat bran per day), has been shown to reduce the serum concentration of low density lipoprotein (LDL) by 3% within three weeks, in persons with elevated cholesterol levels.
- Product: Vitaalbrood® flora.
  Health benefit: Consumption of three slices of Vitaalbrood® flora per day supports a well-balanced gut flora composition and colonic function by selectively stimulating the growth of *Bifidobacterium*. Vitaalbrood® flora contains at least 5 g Frutafit®-inulin per 100 gram.

2.4 Regulation and labelling of functional cereal products in the United States

Nutrition labelling in the United States began in 1990 and since that time numerous health claims and qualified health claims have been approved for labelling on food products. Because fairly strong research has supported the diet–disease relationships associated with eating whole grains, fibre and phytosterols, the United States has several health claims in this area. In addition to the health claims, particular attention lately has been paid to definitions surrounding whole grains.

2.4.1 The Nutrition Labeling and Education Act

The Nutrition Labeling and Education Act (NLEA) of 1990 introduced the use of health claims for food to educate consumers and encourage consumption of healthful foods. In 1993 the Food and Drug Administration (FDA) passed the first seven health claims based on scientific consensus from results of epidemiological, animal and clinical research studies. After passage of the first seven health claims, the FDA established a process to petition for new health claims that addressed substance–disease relationships. The petitioner must provide sufficient scientific support to meet specific regulatory requirements.

2.4.2 The Food and Drug Administration Modernisation Act

While health claims can make a major contribution toward improving health, the standard procedure for approval could take more than a year. This is attributed to the scientific review, publication of a proposed rule and time for public comments, which are all mandated under the FDA's 1993 NLEA food labelling rules for health claims. Therefore, the FDA established the Food and Drug Administration Modernisation Act (FDAMA) of 1997, in part as an attempt to hasten the process for establishing the scientific basis for health claims.[46] This Act allows companies to notify the FDA of their intent to use a new health claim based on existing authoritative statements from one or more federal scientific bodies. The criteria for approving a health claim include that the health claim must:

(a) come from a federal scientific body.
(b) be published by the scientific body and be currently in effect.
(c) state a relationship between a nutrient and a disease or health-related condition.
(d) not be a statement made individually by an employee of a federal scientific body, but rather reflect a consensus of the scientific body.
(e) be based on the scientific body's deliberative review of the scientific evidence.

The first health claim passed after implementation of the FDAMA was the authoritative statement for the connection between whole grains and reduced risk for coronary heart disease and some cancers in July, 1999.[47] The claim is allowed on any product containing 51% whole grain per reference amount customarily consumed (RACC). While 51% whole grain is easily achieved for dry foods such as breakfast cereal, it is more difficult to attain for foods with higher moisture content like bread.

2.4.3 Qualified health claims

In Pearson v. Shalala (164 F.3d 650 (D.C. Cir. 1999)), the United States Court of Appeals ruled that the FDA must consider the possibility of approving health claims that incorporate qualified representations or 'disclaimers'. An example might be 'Preliminary research suggests that X nutrient reduces the risk of Y disease'. Qualified health claims on conventional foods were allowed in December 2002. Currently, there are no qualified health claims related to cereal components or whole grains, but this may be an area for future concentration as related to the phytochemicals found in cereal components.

2.4.4 Current allowable health claims and cereals

Currently the FDA allows for four NLEA health claims related to cereal and grain products and these are found in Title 21 of the United States Code of Federal Regulations.[48] As with all health claims, they are permissible on food products assuming that the food complies with the general requirements for health claims (section 101.14). The FDA has set disqualifying nutrient levels, which are the levels of total fat, saturated fat, cholesterol and sodium in a food above which the food will be disqualified from making a health claim. These levels are 13.0 g for fat, 4.0 g of saturated fat, 60 mg of cholesterol and 480 mg of sodium, per RACC. If the RACC is 30 g or less, then levels of fat, saturated fat, cholesterol and sodium are based on consumption of 50 g.

The four NLEA health claims associated with cereal components, as outlined in Table 2.1, will be discussed in greater depth. The first health claim, fibre-containing grain products, fruits, and vegetables and cancer (section 101.76) is allowed on any products as long as the product contains a good source of dietary fibre, which is equivalent to 10–19% of the recommended daily intake (RDI) or the daily reference value (DRV) (25 g per 2000 calorie diet, section 101.9) per RACC (section 101.54), thus a food company is allowed to make a health claim indicating that grain products may reduce the risk of some cancers. The second health claim is related to fruits, vegetables, and grain products that contain fibre, particularly soluble fibre, and risk of coronary heart disease (section 101.77). The food product must contain a grain product supplying at least 0.6 g of soluble fibre per RACC to use the health claim.

The third NLEA health claim is soluble fibre from certain foods and risk of coronary heart disease (section 101.81). This claim links soluble fibre from certain foods and the risk of coronary heart disease CHD. The original claim in 1997 allowed for oat bran, rolled oats and whole oat flour, and later as part of this claim Oatrim® or *Psyllium* husks used in food products were added. However, the latest legislation approved by the FDA as a component of this health claim allows inclusion of the soluble fibre from barley products. The claim is the same as that for whole oat foods as long as 0.75 g of soluble fibre are delivered in a reference amount of product. After substantial review of the health claim petition submitted by The National Barley Foods Council, the FDA provided an interim final rule extending the health claim on the relationship between oat beta-glucan soluble fibre and reduced risk of coronary heart disease to include products made with dehulled and hulless whole grain barley and other specific dry milled barley products.[49] The FDA concluded that beta-glucan soluble fibre from barley is analytically the same substance as beta-glucan soluble fibre from oat sources. To qualify for the health claim, the barley-containing foods must provide at least 0.75 g of soluble fibre per serving of the food. This allows for barley to elicit health benefits through commonly consumed food products due to its excellent product functionality, and naturally rich nutrient and phytochemical content. The health claim reads: 'Soluble fibre from foods such as [name of food], as part of a diet low in saturated fat and cholesterol, may reduce the risk of heart disease. A serving of [name of food] supplies [X] grams of the soluble fibre necessary per day to have this effect.' Another NLEA health claim is related to plant sterol/stanol esters and risk of coronary heart disease (section 101.83). The health claim is used on any product containing more than either 1.3 g of plant sterol esters or 3.4 g of plant stanol esters. Plant sterols and stanols are

Table 2.1 Health claims permitted for use in the United States related to grains

Code	Title of health claim	Labelling	Eligible sources of grain, fibre or active	Amount of active
101.76	Fibre-containing grain products, fruits, and vegetables and cancer.	Diets low in fat and high in fibre-containing grain products, fruits, and vegetables 'may' or 'might' reduce the risk of some cancers.	The food shall be or shall contain a grain product.	Must contain a 'good source' of dietary fibre (101.54).
101.77	Fruits, vegetables, and grain products that contain fibre, particularly soluble fibre, and risk of coronary heart disease.	Diets low in saturated fat and cholesterol and high in fruits, vegetables, and grain products that contain fibre 'may' or 'might' reduce the risk of heart disease.	The food shall be or shall contain a grain product	At least 0.6 g of soluble fibre per reference amount customarily consumed.
101.81	Soluble fibre from certain foods and risk of coronary heart disease (CHD).	Diets that are low in saturated fat and cholesterol and that include soluble fibre from certain foods 'may' or 'might' reduce the risk of heart disease.	1 Oat bran 2 Rolled oats 3 Whole oat flour 4 Oat trim 5 Psyllium husks	1–3 At least 0.75 g of soluble fibre per reference amount 4 At least 0.75 g of beta-glucan soluble fibre 5 At least 1.7 g of soluble fibre
101.83	Plant sterol/stanol esters and risk of coronary heart disease (CHD).	Diets that include plant sterol/stanol esters 'may' or 'might' reduce the risk of heart disease	Plant sterol or stanol esters	1 1.3 g or more per day of plant sterol esters 2 3.4 g or more per day of plant stanol esters

naturally found in most cereals; however, at much lower levels than is required for health claims.

2.5 Whole grain definitions and health claims – current and emerging issues

2.5.1 Harmonisation of whole grain definition

Over the past few years there has been tremendous interest in whole grains from industry, government and consumers. There are two crucial issues facing the successful introduction of whole grain foods into the food supply, namely: a salient and working definition for both a whole grain ingredient and for whole grain foods. Within the last several years American Association of Cereal Chemists (AACC) International and The Whole Grains Council – a consortium of industry, scientists, chefs and the Oldways Preservation Trust – worked together to formulate a consumer-friendly definition for whole grains and whole grain foods. Another key area missing in the area of whole grains is a harmonised worldwide definition. Countries are slowly recognising this need and have adopted the definition for whole grains as set forth by AACC International due to its thoroughness. The definition was approved by AACC International in 1999 and states: 'Whole grains shall consist of the intact, ground, cracked or flaked caryopsis, whose principal anatomical components – the starchy endosperm, germ and bran – are present in the same relative proportions as they exist in the intact caryopsis.'[50] The entire food industry – including Europe, Asia and so-forth – would benefit from Codex adopting this definition as an official standard followed by implementation to country-specific or EU region food law. Another key issue is to identify the nutrients and other bioactive components in whole grains conferring health benefits, through well-designed human intervention studies. A major new European Commission-supported project is addressing these issues.[51]

General Mills, Inc. submitted a petition on 11 May 2004 to the United States FDA requesting definitions for whole grain content descriptors: 'excellent source', 'good source' and 'made with'. The assumptions made in the petition were that whole grain is a substance, not a broad category of food, and that the descriptors are neither nutrient content nor health claims. For each of the descriptors – 'excellent source', 'good source' and 'made with' – the level of whole grains per serving would be 16 g or more, 8–15 g, and at least 8 g, respectively. In November 2005 the FDA denied the petition on the basis that the FDA cannot classify whole grains due to a lack of regulatory language and definitions (i.e. food category, food ingredient nutrient or other). These particular descriptors are typically used for nutrient content claims for nutrients having established RDIs; however, 'whole grain' does not have a RDI and the FDA's position was that it cannot be classified as a nutrient. Although General Mills, Inc. recommended usage of the fibre content as a marker for whole grains, the FDA felt the use of the descriptors would imply a nutrient content claim for fibre and not necessarily whole grain. The FDA denied the petition on the basis that it needs to carefully consider a definition of whole grains and determine how the term should be classified. However, the FDA did acknowledge a need for action to resolve the myriad of questions, especially regarding the 2005 Dietary Guidelines for Americans recommendation to consume at least three or half of your servings of grain foods as whole grains.

On 15 February 2006 the FDA endorsed the industry-standard definition of whole grain developed by AACC International and by The Whole Grains Council, clearly delineating the specific grains that can be considered whole grains.[52] Cereal grains may include amaranth, barley, buckwheat, bulgur, corn (including popcorn), millet, quinoa, rice, rye, oats, sorghum, teff, triticale, wheat and wild rice. Dehulled barley should be considered a whole grain because only the tough inedible hull or outer covering has been removed, but the bran layer is left intact. In general, the barley that is used for human food in the United States is pearled. Barley that is pearled should not be considered a whole grain because some of the bran layer has been removed. The FDA also emphasized that for corn flour or corn meal to be 'whole grain' it should include the pericarp as well as the other essential fractions. The FDA also delineated some other foods – such as soybeans and seeds – that are not whole grains. The FDA does not consider products derived from legumes (soybeans), oilseeds (sunflower seeds) and roots (arrowroot) as 'whole grains'. Other widely used food products may not meet the 'whole grain' definition. Currently, manufacturers can make factual statements about whole grains on food labels such as '10 g of whole grains' or '1/2 ounce of whole grains'. The FDA subsequently opened a 60-day comment period for further review of consumer education programmes, which seek to help consumers readily identify legitimate whole grain products.[53]

AACC International has developed a task force for defining whole grain foods with the purpose of developing the best possible labelling, research and communications agendas for delivering whole grains to the general public. This effort has encompassed discussions and actions regarding whole grain ingredients and whole grain food definitions involving all major sectors of industry, government and academia while exploring four areas: bioactive components, health claims, the whole grain barley definition and whole grain corn definition. A letter to the FDA dated 17 April 2006 encouraged the FDA to establish rule-making that will: (a) encourage increased consumption of whole grains for better health as indicated by the Dietary Guidelines; (b) give the consumer greater clarity in the marketplace in order to select whole grain foods; (c) eliminate or avoid multiple standards that will paralyse consumers and manufacturers; and (d) support and encourage foods made with a blend of whole and enriched grains, but that contribute significant whole grain content – not just foods that are entirely or almost entirely manufactured with whole grains. AACC International also compiled a more comprehensive list of foods that the FDA might consider using on whole grain labelling statements.[54]

In Europe, the United Kingdom and Sweden are, so far, the only countries with a whole grain health claim. Once the European Union discusses the allowable health claims for all member countries, there may be the potential of extending the whole grain health claim throughout Europe. International harmonisation of health claims and regulatory efforts for greater whole grain consumption would be ideal. Suggested actions for setting international regulatory and policy standards for whole grains include the following.

1 Creating one international definition and symbol of identification for whole grains by committee involvement.
2 Building international whole grain food labelling standards and eliminating confusing terminology.
3 Exploring fortification of whole grains with nutrients low or deficient in the population, e.g. folic acid, phytonutrients.
4 Educating the public about what constitutes a whole grain food and where to find them in the marketplace.
5 Creating meaningful and easy-to-understand portion size descriptors.

2.5.2 Labelling of products for ease of identifying whole grain foods

Various organisations are advocating for labelling of whole grain products for ease of identification. The Whole Grains Council–Oldways Preservation Trust, Boston, Massachusetts has developed a universal stamp (Fig. 2.1) that can be used on whole grain products.[55] These whole grain stamps are now used on hundreds of products in over two dozen countries including the United States and Germany. In order for a company to use the whole grain stamp on retail consumer products, they must be a member of The Whole Grains Council. The Whole Grains Council is a progressive organisation supporting enhanced consumption of whole grain foods. The stamps were first introduced in January 2005 with three different stamps indicating different levels of whole grains (good source, excellent source and 100% excellent source). Since introduction, the stamps have been reduced to two due to the FDA ruling on labelling of whole grains.[49] The first stamp can be used on any product that contains at least 8 g of whole grain per serving. The second stamp is a 100% whole grain stamp for products that contain at least 16 g of whole grain per serving.[56]

2.5.3 Whole grain-related health claims – future developments

The significant epidemiologic support for whole grain intake and reduced risk of major chronic diseases has not yet been complemented by clinical studies demonstrating cause and effect relationships for whole grain and intermediary and/or disease end points. Such relationships will be investigated in Europe, as one of the topics of the major HEALTHGRAIN project, funded

EAT 48g OR MORE OF WHOLE GRAINS DAILY

Whole grain stamp

For products offering a half-serving or more of whole grain.
Contains at least 8 g whole grain per serving.
8 g = 1/2 a MyPyramid serving

EAT 48g OR MORE OF WHOLE GRAINS DAILY

100% whole grain stamp

For products where All of the grain is whole grain.
Contains at least 16 g whole grain per serving.
16 g = 1 full MyPyramid serving

Fig. 2.1 Whole grain stamps provided by The Whole Grains Council for labelling of food products.

by the European Commission.[51] This 5-year project began in mid 2005. The objective is to improve and enhance the nutritional value and health benefits of cereals, and to better use whole grain and its constituents such as vitamins (folate, tocols, choline, etc.), phytochemicals (lignans, sterols, alkylresorcinols, phenolic acids) and indigestible carbohydrates. In addition to *in vitro* and *in vivo* nutrition research, HEALTHGRAIN will study consumer expectations and the sensory quality of bioactive cereal foods and will develop new technologies to enable the production of foods containing health-promoting grain constituents. The project will also generate an overview of the levels of these constituents in a wide range of wheat and rye cultivars grown in different conditions, and will develop a biotechnology toolkit for breeders for these compounds. There is also a comprehensive dissemination programme to communicate the health benefits of whole grains and related products to the European food industry, health professionals and consumers. The work is expected to create a range of opportunities for healthy cereal grain-based products and to strengthen the basis for a set of health claims.

Another collaborative model currently under development is called The Institute for Grains and Health Research (IGHR). This North American whole grain-centred collaborative is dedicated to improving the health and well-being of the population through basic, applied and translational research on grain and grain components. The long-term objectives of the IGHR are to support coordinated research to enhance the knowledge and utilisation of healthful components delivered in grain-based food products. The IGHR will achieve this through collaboration between academic institutions with expertise in various areas of grain research. This collaborative, in partnership with government and industry, will identify and prioritise research projects.

Results will be disseminated through university partners, health organisations and industry trade groups. In addition, this is an opportunity for the North American initiative to learn and work closely with the EU HEALTHGRAIN collaborative to enhance the health attributes of grain-based foods and to promote these novel foods to consumers throughout the world.

Founding members of the IGHR include scientists from four universities with proven expertise in the science of grains and whole grains: (a) The University of Minnesota for nutritional sciences, behavioural health and grain genetics; (b) Kansas State University for baking, milling and feed production; (c) Cornell University for characterisation of grain phytonutrients, grain health benefits, and bioavailability; and (d) The University of Manitoba for experimental systems from grain biochemistry to traditional and novel extraction equipment, to cell, animal and human test facilities. The Wheat Foods Council will be actively involved in the dissemination of research findings to health professionals, commodity groups and consumers.

The success of whole grain-based products in the United States can be attributed to the combination of effective product development and marketing, and use of the FDA-endorsed health claim as a powerful communication tool. A similar approach has been successful in the two EU countries with a whole grain health claim. Other cereal grain-related health claims, on fibres, and soluble fibres/beta-glucans, did not result in similar major shifts in product sales, probably due to the more limited support by marketing and communication efforts. One may conclude that a claim that is well understood by consumers, even if the scientific status is not convincing, combined with a major and effective marketing effort, is crucial for a major shift in consumer behaviour. When, due to sustained marketing efforts, consumer awareness of the health claim and health benefits has grown, product promotion will also be successful when just the nutrient content – '100% whole grain' or 'High in omega-3 and -6 fatty acids' – is mentioned, without the health claim itself.

2.6 Future trends

As the scientific environment continues to evolve, guidelines for establishing food claims will need to be modified within a new regulatory process. A current trend is to modify the composition of raw materials, designed for a healthier nutrient profile combined with consumer friendly sensory characteristics. For cereal grains, new white wheat cultivars are already facilitating the development of white whole grain products. Ongoing research is enhancing the level of amylose in cereal grains; higher levels of resistant starch are created after baking or with other heat treatments The HEALTHGRAIN project is developing tools for breeders to raise the level of phytochemicals and increased anti-oxidative properties of grains.

In the future, evidence may be provided to substantiate new claims based on individual components or groups of components within cereals, such as antioxidants. Such claims would require substantial evidence in support of the health benefit. Previously, the FDA initially denied a health claim for lycopene and cancer, with an eventual decision to reduce the claim to include tomatoes and tomato sauce. This is a strong indicator that petitioners will continue to file subsequent claims on other individual components and their relationship to disease.

Current discussions on the regulation of nutrition and health claims – as in the European Union, on the implementation of the new EU Regulation, and in Canada – indicate a trend towards stricter regulations and a greater emphasis on clinical trials and intervention studies for substantiation of health claims. In the European Union and Canada, the admission of qualified health claims (see also Section 2.4.3) such as in the United States, will be unlikely: it is argued that consumer confidence may decrease if qualified health claims will appear and – in a number of cases – disappear after some years, when the scientific evidence is insufficient. New insights into the health-related functions of specific compounds and combinations of compounds are broadening the scope of health claims beyond the classical areas of cardiovascular health and gastro-intestinal functioning to areas such as natural defence (stimulation of the innate immune system), energy metabolism/weight reduction, insulin sensitivity, skin condition and cognition. New developments in science will enable producers to develop products with levels of specific compounds that are over, for example, ten times higher than found in 'normal' foods – as is the case with the growing range of products enriched with phytosterols/stanols. Such products may be seen as an intermediate class between food and medicine. The present, well-established regulations for admission of new medicines are much stricter than those for functional foods. All these developments will probably enhance the pressure from consumer organisations and others for stricter admission criteria for functional foods.

The European Union and the United States will both need to further enhance dialogue and collaboration among stakeholders from government, industry, academia and consumers alike in order to establish regulatory guidelines that address increasingly complex scientific issues as are outlined above. In due course, a growing international, worldwide consensus on regulations for nutrition and health claims will be highly desirable. A common approach will help to develop a foundation from which to build a more solid scientific and regulatory process.

2.7 Sources of further information and advice

For further information readers are encouraged to review information at The Whole Grains Council,[55] a progressive organisation supporting consumption of whole grain foods. AACC International,[54] a global trade organisation for the cereal industry, also has current information about the progress of whole grain and cereal components. In Europe, the EFSA website will play a growing role in providing information on nutrition and health claims. Also at member state level, the national food safety agencies – such as the Food Standards Agency (FSA) in the United Kingdom and Agence française de sécurité sanitaire des aliments (AFSSA) in France – are providing an increasing amount of information. The HEALTHGRAIN project has established a Regulatory Updates chapter, providing information on cereal grain-related nutrition and health claims as well as general information. ICC, the International Association for Cereal Science and Technology will continue this service after May 2010, when HEALTHGRAIN comes to an end. At present, organisations such as the JHCI in the United Kingdom and SNF in Sweden provide a wealth of information. The future role of such organisations, created to fill a regulations gap at the EU level, may become less prominent. Finally, the FDA code of federal regulations provides industry guidance on labelling and manufacture of food products. Additionally, the Federal Register publishes legislation daily to provide timely dissemination of information to those most interested.

2.8 References

1 Marquart L, Jacobs D, McIntosh G, Reicks M, and Poutanen K (eds) (2007), *Whole Grains and Health*, Blackwell Publishers, Ames, IA.

2 Liu S, Willett W C, Manson J E, Hu F B, Rosner B, and Colditz G (2003), 'Relation between changes in intakes of dietary fiber and grain products and changes in weight and development of obesity among middle-aged women', *American Journal of Clinical Nutrition*, **78**, 920–927.

3 Koh-Banerjee P, Chu N F, Spiegelman D, Rosner B, Colditz G, Willett W C, and Rimm E B (2003), 'Prospective study of the association of changes in dietary intake, physical activity, alcohol consumption, and smoking with 9-y gain in waist circumference among 16 587 US men', *American Journal of Clinical Nutrition*, **78**, 719–727.

4 Koh-Banerjee P, Franz M, Sampson L, Liu S, Jacobs D R Jr, Spiegelman D, Willett W C, and Rimm E B (2004), 'Changes in whole-grain, bran, and cereal fiber consumption in relation to 8-y weight gain among men', *American Journal of Clinical Nutrition*, **80**, 1237–1245.

5 Bazzano L A, Song Y, Bubes V, Good V, Manson J E, and Liu S (2005), 'Dietary intake of whole and refined grain breakfast cereals and weight gain in men', *Obesity Research*, **13**, 1952–1960.

6 McKeown N, Serdula M, and Liu S (2007), 'Whole grains and related dietary patterns in relation to weight gain', in Marquart L, Jacobs D, McIntosh G, Reicks M, and Poutanen K (eds), *Whole Grains and Health*, Blackwell Publishers, Ames, IA.

7 Jacobs D, Meyer K, Kushi L, and Folsom A (1998), 'Whole-grain intake may reduce the risk of ischemic heart disease death in postmenopausal women: The Iowa Women's Health Study', *American Journal of Clinical Nutrition*, **68**, 248–257.

8 Liu S, Stampfer M J, Hu F B, Giovannucci E, Rimm E, Manson J E, Hennekens C H, and Willett W C (1999), 'Whole-grain consumption and risk of coronary heart

disease: results from the Nurses' Health Study', *American Journal of Clinical Nutrition*, **70**, 412–419.
9 Anderson J, Hanna T, Peng X, and Kryscio R (2000), 'Whole grain foods and heart disease risk', *Journal of the American College of Nutrition*, **19**, 291S–299S.
10 Fung T T, Hu F B, Pereira M A, Liu S, Stampfer M J, Colditz F A, and Willett W C (2002), 'Wholegrain intake and the risk of type 2 diabetes: a prospective study in men', *American Journal of Clinical Nutrition*, **76**, 535–540.
11 Truswell A S (2002), 'Cereal grains and coronary heart disease', *European Journal of Clinical Nutrition*, **56**, 1–14.
12 Seal C (2006), 'Whole grains and CVD', *Proceedings of the Nutrition Society*, **65**, 24–34.
13 Flight I, and Clifton P (2006), 'Cereal grains and legumes in the prevention of coronary heart disease and stroke: a review of the literature', *European Journal of Clinical Nutrition*, **60**, 1145–1159.
14 Slavin J (2007), 'Whole grains and cardiovascular disease', in Marquart L, Jacobs D, McIntosh G, Reicks M, and Poutanen K (eds), *Whole Grains and Health*, Blackwell Publishers, Ames, IA.
15 Liu S, Manson J E, Stampfer M J, Hu F B, Giovannucci E, Colditz G A, Hennekens C H, and Willett W C (2000), 'A prospective study of whole grain intake and risk of type 2 diabetes mellitus in U.S. women', *American Journal of Public Health*, **90**, 1409–1415.
16 Montonen J, Knekt P, Jarvinen R, Aromaa A, and Reunanen A (2003), 'Whole-grain and fibre intake and the incidence of type 2 diabetes', *American Journal of Clinical Nutrition*, **77**, 622–629.
17 Venn B J and Mann J I (2004), 'Cereal grains, legumes and diabetes', *European Journal of Clinical Nutrition*, **58**, 1443–1461.
18 Jensen M K, Koh-Banerjee P, Franz M, Sampson L, Grønbæk M, and Rimm E B (2006), 'Whole grains, bran, and germ in relation to homocysteine and markers of glycemic control, lipids, and inflammation', *American Journal of Clinical Nutrition*, **83**, 275–283.
19 Asp N-G and Bryngelsson S (2004), 'The Swedish Code on health claims for foods – Revised version in action September 2004', *Scandinavian Journal of Nutrition*, **48**, 188–189.
20 Leinonen K, Liukkonen K, Poutanen K, Uusitupa M, and Mykkänen H (1999), 'Rye bread decreases postprandial insulin response but does not alter glucose response in healthy Finnish subjects', *European Journal of Clinical Nutrition*, **53**, 262–267.
21 Pereira M A, Jacobs D R J, Pins J J, Raatz S K, Gross M D, Slavin J L, and Seaquist E R (2002), 'Effect of whole grains on insulin sensitivity in overweight hyperinsulinemic adults', *American Journal of Clinical Nutrition*, **75**, 848–855.
22 McIntosh G, Noakes M, Royle P, and Foster P (2003), 'Whole-grain rye and wheat foods and markers of bowel health in overweight middle-aged men', *American Journal of Clinical Nutrition*, **77**, 967–974.
23 Ostman E M, Frid A H, Groop L C, and Bjorck I M (2006), 'A dietary exchange of common bread for tailored bread of low glycaemic index and rich in fibre improved insulin economy in young women with impaired glucose tolerance', *European Journal of Clinical Nutrition*, **60**, 334–341.
24 Pins J, Geleva D, Keenan J, Frazel C, O'Connor P, and Cherney L (2002), 'Do whole-grain oat cereals reduce the need for antihypertensive medications and improve blood pressure control?', *Journal of Family Practice*, **51**, 353–359.
25 Behall K, Scholfield D, and Hallfrisch J (2006), 'Whole grain diets reduce blood pressure in mildly hypercholesterolemic men and women', *Journal of the American Dietetics Association*, **106**, 1445–1449.
26 Davy B, Davy K, Ho R, Beske S, Davrath L, and Melby C (2002), 'High-fiber oat cereal compared with wheat cereal consumption favorably alters LDL-cholesterol

subclass and particle numbers in middle-aged and older men', *American Journal of Clinical Nutrition*, **76**, 351–358.
27 Behall K, Scholfield D, and Hallfrisch J (2004), 'Diets containing barley significantly reduce lipids in mildly hypercholesterolemic men and women', *American Journal of Clinical Nutrition*, **80**, 1185–1193.
28 Behall K, Scholfield D, and Hallfrisch J (2004), 'Lipids significantly reduced by diets containing barley in moderately hypercholesterolemic men', *Journal of the American College of Nutrition*, **23**, 55–62.
29 Erkkila A T, Herrington D M, Mozaffarian D, and Lechtenstein A H (2005), 'Cereal fiber and whole grain intake are associated with reduced progression of coronary artery atherosclerosis in postmenopausal women with coronary artery disease', *American Heart Journal*, **150**, 94–101.
30 Codex Alimentarius Commission (2004), 'Guidelines for use of nutrition and health claims', Guideline 23–1997. Revised 1-2004. FAO/WHO, Rome, Italy.
31 World Health Organization (2004), 'Diet, Nutrition and the Prevention of Chronic Diseases: Report of a Joint FAO/WHO Expert Consultation', WHO Technical Report Series 916, WHO, Geneva.
32 European Communities (2006), Directive 90/496/EEC on nutrition labelling for foodstuffs: Discussion paper on Revision of Technical Issues. https://www.healthgrain.org/download.php?file=ip/regulatory-updates/EC_Nut-Lab_Disc-Paper.pdf.
33 Gray J (2006), *Dietary Fibre, Definition, Analysis, Physiology and Health*, ILSI Europe Concise Monograph Series, pp. 5–12.
34 Health Council of the Netherlands (2006), *Guideline for Dietary Fibre intake*, Publication no. 2006/03E, Health Council of the Netherlands, The Hague.
35 Diplock A T, Aggett P J, Ashwell M, Bornet F, Fern E B, and Roberfroid M B (1999), 'Scientific concepts of functional foods in Europe: Consensus document', *British Journal of Nutrition*, **81**, S(1–27).
36 Aggett P J, Antoine J-M, Asp N-G, Bellisle F, Contor L, Cummings J H, Howlett J, Müller D J G, Persin C, Pijls L T J, Rechkemmer G, Tuijtelaars S, and Verhagen H (2005), 'PASSCLAIM, Process for the Assessment of Scientific Support for Claims on Foods', *European Journal of Nutrition*, **44**, S1–31.
37 Richardson D P (2003), 'Whole grain health claims in Europe', *Proceedings of the Nutrition Society*, **62**, 161–169.
38 Corrigendum to Regulation (EC) No. 1924/2006 of the European Parliament and of the Council of 20 December 2006 on nutrition and health claims made on foods (OJ L 404, 30.12.2006), *Official Journal of the European Union L012*, 18/01/2007, 0003–0018. http://eur-lex.europa.eu/LexUriServ/LexUriServ.do?uri=OJ:L:2007:012:0003:01_REG_2006_1924_18:EN:HTML.
39 Richardson D P (2005), 'The scientific substantiation of health claims with particular reference to the grading of evidence', *European Journal of Nutrition*, **44**, 319–324.
40 Regulation (EC) 258/97/EC of the European Parliament and of the Council (*Official Journal of the European Union L43*, 14/02/1997, 1) of 27 January 1997 concerning novel foods and novel food ingredients,. http://www.fsai.ie/legislation/food/eu_docs/Novel_Foods_and_Ingredients/Reg258.97.pdf.
41 COMMISSION DECISION of 24 January 2006, authorising the placing on the market of rye bread with added phytosterols/phytostanols as novel foods or novel food ingredients under Regulation (EC) No. 258/97 of the European Parliament and of the Council, *Official Journal of the European Union L31*, **49**, 3/02/2006, 18–24.
42 Joint Health Claims Initiative (2002), 'Generic health claim for whole grain foods and heart health'. http://www.jhci.org.uk./approv/wgrainh.htm.
43 Joint Health Claims Initiative (2004), 'Generic health claim for oats and blood cholesterol', http://www.jhci.org.uk/approv/oats.htm.
44 Asp N-G, and Trossing M (2001), 'The Swedish Code on health-related claims in

action – extended to product-specific physiological claims', *Scandinavian Journal of Nutrition*, **45**, 189–192.

45 Swedish Nutrition Foundation (SNF) (2004), 'Health claims in the Labelling and Marketing of Food Products. The Food Sector's Code of Practice'. http://www.hp-info.nu/SweCode_2004_1.pdf.

46 Food and Drug Administration (1997), Food and Drug Administration Modernization Act of 1997, Food and Drug Administration, Washington, DC. http://www.fda.gov/cder/guidance/105-115.htm#sec.%20303.

47 Food and Drug Administration (1999), 'Whole Grain Authoritative Statement Claim Notification', Docket 99P-2209, Food and Drug Administration, Washington, DC. http://www.fda.gov/ohrms/dockets/dailys/070899/070899.htm.

48 United States Code of Federal Regulations (2006), 'Title 21 – Food and Drugs', http://www.accessdata.fda.gov/scripts/cdrh/cfdocs/cfCFR/CFRSearch.cfm.

49 Food and Drug Administration (2006), 'Food labeling: health claims; soluble dietary fibre from certain foods and coronary heart disease', Docket No. 2004P-0512, *Federal Register,* **71**, 98.

50 AACC International (1999), 'Definition of Whole Grain'. http://www.aaccnet.org/definitions/wholegrain.asp.

51 HEALTHGRAIN EU Integrated Project (2005–2010), 'Exploiting bioactivity of European cereal grains for improved nutrition and health benefits'. www.healthgrain.org.

52 Food and Drug Administration (2006), 'Guidance for industry and FDA staff: whole grain label statements', Docket No. 2006D-0066. *Federal Register*, **71**, 85–97. http://www.fda.gov/OHRMS/DOCKETS/98fr/06-1509.pdf.

53 Food and Drug Administration (2006), 'Guidance for Industry and FDA Staff Whole Grain Label Statements: Draft Guidance', Food and Drug Administration, Washington, DC. http://www.cfsan.fda.gov/~dms/flgragui.html#ref1.

54 AACC International (2007), St Paul, MN. http://www.aaccnet.org.

55 The Whole Grains Council (2006), Boston, MA. http://www.wholegrainscouncil.org/.

56 Dahlberg J, Gifford K D and Harriman C W (2007), 'The whole grain stamp program', in Marquart L, Jacobs D, McIntosh G, Reicks M, and Poutanen K (eds), *Whole Grains and Health*, Blackwell Publishers, Ames, IA.

3

Fiber, whole grains, and disease prevention

J. Miller Jones, The College of St Catherine, USA

3.1 Introduction

The dietary fiber hypothesis launched nearly 40 years ago suggested that diets with adequate plant food and dietary fiber were associated with reduced risk of chronic disease (Burkitt and Trowell, 1975; Trowell, 1993). Since this time the role of plant foods (and their key components, such as whole grains and the dietary fiber), in preventing or treating a number of disorders has been researched. Despite much research to try to clarify these roles, there is still controversy regarding dietary fiber and whole grains, and their connection with health and disease prevention. Epidemiological studies that show many positive relationships between dietary fiber and whole grains and disease are also subject to confounding factors. Subjects who eat the most dietary fiber or ingest the highest quantity of whole grains also have other associated positive health habits that could account for the observed health benefits. For example, in the Multiethnic Cohort Study, a representative sample of over 100 000 African-American, Native Hawaiian, Latino, and Japanese-American subjects, the highest fiber eaters also ate less red meat and fat, smoked less, exercised more, were less obese, and had higher fruit and vegetable intakes (Foote *et al.*, 2003). Similar findings and confounding factors are seen in epidemiological studies with very different demographics such as those with the 35 000 Iowa post-menopausal women or the Nurses' Health Study (Jacobs *et al.*, 1999; Liu *et al.*, 2000).

While intervention studies are frequently considered to be superior to epidemiological studies because of their capability to reduce confounding variables, they still are subject to problems that may make interpretation difficult. In fact, large intervention studies in which participants ate additional dietary fiber have also failed to show expected effects. These studies were criticized for aspects of their design such as the use of only a single dietary fiber source for the experimental group and inadequate length of time of the study. Unfortunately the intervention studies that were undertaken to give greater clarity as to the role of fiber in disease and health only resulted in greater confusion.

This lack of clarity leaves many questions regarding the role of dietary fiber and whole grains in health promotion and disease prevention. Some of the questions are: (1) What is the physiological benefit of dietary fiber itself? (2) Does a purified fiber have greater or less benefit than when it is in a natural food matrix and delivers a combination of nutrients and phytochemicals such as in a whole grain food? (3) Does the source of fiber such as that from whole grain offer something unique and different than that offered by fruits and vegetables? (4) Do various fibers have differing effects, and when studies lump these together as a single entity does that add to confusion regarding results of various studies? And (5) Is the inclusion of dietary fiber and whole grain foods in the diet merely a marker for a constellation of other healthful behaviors? This chapter will attempt to answer some of these questions while it reviews recent evidence about the role of dietary fiber and whole grains in weight control, cardiovascular disease, and gut health.

3.2 Fiber, whole grains, and obesity

Obesity and its epidemic rise is a topic concerning most health professionals in developed countries, and is also an issue in urban areas of most developing countries. Dietary fiber and whole grains, like many other dietary constituents, have been named as either part of the problem or part of the solution. Evidence as presented in a recent review indicates that there is an inverse relationship between fiber, whole grains, and obesity (Slavin, 2005). Food consumption data and clinical studies done on different cohorts show this relationship. For example national surveys of over 4000 Dutch subjects indicated that the tertile of the cohort ingesting the most dietary fiber had lower mean energy intakes and lower body mass index values (Lowik *et al.*, 1999). In a crosscultural study with ~13 000 subjects from seven countries, over 90% of the variance in subscapular skinfold (a marker of fatness) could be accounted for by two things: physical activity and dietary fiber intake (Kromhout *et al.*, 2001). These data concur with those from several clinical studies and from a recent epidemiological study of US college students (Rose, 2005), which showed that those ingesting the least fiber tended to be the most obese.

In the Male Health Professionals cohort (Koh-Banerjee *et al.*, 2004), the addition of bran to the diet was associated with less weight gain over the 8-year period of the follow-up. Similar data were seen in a French cohort of over 6000 men and women. Those with the highest total dietary fiber and nonsoluble dietary fiber intakes were less likely to be overweight and to have elevated waist-to-hip ratios (Lairon *et al*., 2005).

Clinical studies also indicated that dietary fiber and whole grains are important for weight maintenance and weight loss. Dietary intakes of 52 normal-weight adults were compared with 52 overweight and obese adults. Those with normal weight consumed an average of 33% more dietary fiber than their overweight counterparts (Davis *et al*., 2006). Fiber intake has been shown to be helpful to those trying to lose weight in some cases (Birketvedt *et al*., 2005), but in other studies appeared to offer no added benefit in weight loss over simple caloric restriction (Gann *et al*., 2003; Howarth *et al*., 2003).

Data for whole grain foods and obesity parallel those for dietary fiber and obesity. Whole grain intake, as deduced from food frequency questionnaires, was inversely related to body mass index and waist-to-hip ratio (McKeown *et al*., 2002) or waist circumference (Newby *et al*., 2003) in the Framingham offspring study and the Baltimore Longitudinal Study of Aging, respectively. Rose (2005) found that college students eating the most whole grain were less likely to be overweight or obese. Data from both the Health Professionals Follow-up Study and the Nurses' Health Study showed those who consumed more whole grain foods at the start of the study consistently weighed less at baseline than those who consumed less whole grain foods. During the 8 to 12-years follow-up in these prospective studies, weight gain was positively related to the intake of refined-grain foods and inversely related to intake of whole grain food products (Koh-Banerjee and Rimm, 2003; Liu *et al*., 2003b; Koh-Banerjee *et al*., 2004).

In terms of whole grain foods as part of an intervention for weight loss, substitution of whole grain bread for white bread in a free-living population of college-age males resulted in weight loss for the whole grain eaters (Mickelsen *et al*., 1979). A recent intervention study, however, found little or no impact on weight loss in the short term of feeding whole grains such as oats. There was, however, some indication of potentially positive impacts on hunger, which might have an effect in longer-term studies (Saltzman *et al*., 2001).

The following mechanisms have been suggested as ways in which dietary fiber and whole grains may help with weight control: (1) Diets with adequate fiber tend to contain more nutrient-rich fruits and vegetables and whole grains, are more nutrient dense and are generally less energy dense. (2) Diets with adequate fiber offer greater food volumes, which may spontaneously curb intake of other foods; large food volume that stays in the stomach longer may enhance feelings of fullness and reduce hunger (Marlett *et al*., 2002). (3) Diets with adequate fiber and whole grains may delay gastric emptying and alter the secretion of gut hormones; this, in turn, may increase overall satiety (Burton-Freeman *et al*., 2002; Benini *et al*., 1995). (4) Whole grain foods provide slowly digested and resistant starch, which may blunt glycemic response and possibly influence satiety (Koh-Banerjee and Rimm, 2003). (5) Intake of whole grain foods, in both clinical trials and observational studies, is associated with reduced plasma biomarkers of obesity, including insulin, C-peptide, and leptin concentrations. Thus, there are a number of plausible ways in which both dietary fiber and whole grains can have an impact on weight. Clinicians and weight loss experts suggest increased intakes of both dietary fiber and whole grains as strategies in the difficult battle to prevent weight gain and to help promote weight loss (Koh-Banerjee and Rimm, 2003; Pereira and Ludwig, 2004; Schaefer *et al.*, 2005).

3.3 Fiber, whole grains, and cardiovascular disease

Dietary fiber intake has been associated with reduced risk for a number of cardiovascular diseases including hypertension, stroke, and heart disease. These diseases will be discussed in this section, with hypertension being addressed first as it is a risk factor for other cardiovascular diseases.

3.3.1 Fiber, whole grains, and hypertension

Blood pressure is inversely related to fiber intake in most epidemiological studies. Higher blood pressures were documented from the over 17 000 participants in the National Health and Nutrition Examination Survey (NHANES III) for those living in the southeastern part of the United States. This is also the area of the country where the least dietary fiber is eaten (Hajjar and Kotchen, 2003). Similarly, in a French cohort of 6000 men and women, those eating the highest total dietary fiber and nonsoluble dietary fiber had the lowest blood pressures (Lairon *et al.*, 2005). In a study of nearly 6000 Spanish university graduates, those with the highest intakes of cereal fiber cut the risk of hypertension by up to 40% compared with those with the lowest intakes of cereal fiber. If vegetable protein was in the diet, the risk was further reduced (Alonso *et al.*, 2006). A meta-analysis of 9 studies including 305 185 subjects found that fiber intake was associated with a 13% decrease in risk for hypertension (Anderson, 2006). For those with elevated blood pressure, a meta-analysis of 25 randomized controlled trials (published in English-language journals before February 2004) showed that dietary fiber added to the diet for at least 8 weeks reduces blood pressure (Whelton *et al.*, 2005).

Whole grains like dietary fiber have been shown to have a beneficial effect on blood pressure. Three-fourths of the participants (n = 88) eating whole grain oats could reduce or stop their blood pressure medication compared with 40% of the control group (Pins *et al.*, 2002). In an Iranian study, higher consumption of whole grains was associated with a reduced risk of hypertension (Esmaillzadeh *et al.*, 2005). Analysis of the DASH (Dietary Approaches to Stop Hypertension) diet and its effectiveness in reducing hypertension showed that whole grains contributed greatly to intakes of fiber as well as magnesium, potassium, zinc, and folate – and that all of these help to lower blood pressure

(Lin *et al.*, 2003). Furthermore, in a subset of 902 women with type 2 diabetes from the Nurses' Health Study, cereal fiber and fruit fiber were significantly associated with an increasing trend of plasma adiponectin concentrations. Adiponectin has anti-inflammatory effects on the cells lining the walls of blood vessels which can influence blood pressure. Adiponectin, released by adipose tissue, influences the body's response to insulin. High blood levels of adiponectin are associated with a reduced risk of heart attack. Low levels of adiponectin are found in people who are obese. Those eating the highest cereal fiber had adiponectin concentrations 24% higher than those ingesting the lowest quintile of cereal fiber (Qi *et al.*, 2005).

Dietary fiber, especially viscous fiber, and whole grains appear to control blood pressure. Magnesium and other minerals and phytochemicals that are part of the whole grain dietary fiber matrix appear to work together in this regard.

3.3.2 Fiber, whole grains, and stroke

Stroke has long been recognized as being related to elevated blood pressure. Thus diets high in dietary fiber and whole grains that reduce the risk of hypertension would most likely be related to reduced the risk of stroke. Several studies bear this out. In a large prospective cohort of apparently healthy women, a diet high in cereal fiber was one of a number of factors associated with a decreased risk of total and ischemic stroke (Kurth *et al.*, 2006). In the 18-year follow-up of the approximately 88 000 female nurses, total and hemorrhagic stroke risk was inversely associated with cereal fiber intake. Total stroke risk decreased by 36% and hemorrhagic stroke risk was cut nearly in half (Oh *et al.*, 2005).

Evidence on whole grains is similar to that with dietary fiber. Whole grain intake was associated with a 36% reduction in ischemic stroke in the approximately 88 000 women in the Nurses' Health Study (Liu *et al.*, 2000). After a review of 121 publications, Ding and Mozaffarian (2006) concluded that whole grains and cereal fiber were likely to be two of several dietary factors that will probably reduce the incidence of stroke.

3.3.3 Fiber, whole grains, and coronary heart disease

Much research points to the role of fiber, especially viscous fiber, in lowering cholesterol in subjects with elevated lipids. The β-glucan of oats and barley has consistently been shown to lower low-density lipoprotein (LDL) and total cholesterol in subjects with elevated cholesterol. The data were sufficient to enable health claims regarding oat and barley β-glucan and cholesterol lowering in the US market (Food and Drug Administration, 2002, 2006).

Experimental evidence on the cardiovascular benefits of dietary fiber, particularly viscous fibers, was in many cases obtained by feeding isolated fibers. While there has been consistency in the data regarding the effects of viscous fibers and the grains containing them, the same consistency is lacking with nonsoluble fibers and grains containing them. For example, feeding of wheat bran in most, but not all studies, has had little impact on serum cholesterol. Feeding of defatted rice bran (that containing only the dietary fiber) is not associated with cholesterol lowering, but fatty acids and oryzanol in the rice bran do lower cholesterol (Most *et al.*, 2005). Corn bran, with its nearly all insoluble fiber, also has been shown to lower cholesterol (Hunninghake *et al.*, 1994; Shane and Walker, 1995).

Myocardial infarction and other measures of coronary disease risk were reduced by an average of 30% in four large prospective studies for the quintile eating the most fiber versus those eating the least fiber (Rimm *et al.*, 1996; Jacobs *et al.*, 1998; Liu *et al.*, 1999; Mozaffarian *et al.*, 2003). Cereal fiber was more effective, in most epidemiological studies, than fruit or vegetable fiber in reducing cardiovascular risk. However, observed reductions in risk could not be entirely accounted for by cereal fiber alone or even by factoring in other grain components known to have cardiovascular impact including folate, and vitamins B_6 and E. This suggests that other factors found in bran and whole grains promote heart health.

Risk reductions for whole grains are similar to those seen with cereal fiber. Large, prospective cohort studies consistently show that ingestion of approximately three servings of whole grain food per day reduces ischemic heart disease risk by 30–36% over those who eat little or no whole grain daily (Jacobs *et al.*, 1999; Liu *et al.*, 1999). In the Health Professionals Follow-up Study, whole grain breakfast cereal, but not refined grain, was associated with reduced risk of cardiovascular disease (Liu *et al.*, 2003b). The strength of these data led to a US health claim for whole grains (Marquart *et al.*, 2003). Studies on patients with existing coronary heart disease also show a positive impact of whole grains. The degree of lumen narrowing in the coronary artery was less for those women (n = 229) who reported eating the most cereal fiber and whole grain (≥6 servings per day) (Erkkila and Lichtenstein, 2005).

Thus the evidence is strong that whole grains and total dietary and cereal fiber are related to reduced risk of cardiovascular disease. Despite the apparent consistency, there are still some issues that remain puzzling. First, whole wheat and wheat bran are the main cereal fibers followed in epidemiological studies, especially those done in the United States and southern Europe. This creates some confusion as most of the documented cholesterol lowering occurs with cereals such as oats that contain viscous fibers such as oat bran (DeGroot *et al.*, 1963; Van Horn *et al.*, 1988; Anderson and Siesel, 1990; Johnston *et al.*, 1998). Second, even with viscous fibers in the diet, there is only an additional 5–8% cholesterol reduction over that observed when subjects add fiber to an American Heart Association low-fat diet. If one applies the formula that there is a reduction in heart disease risk of 2% for every 1% drop in cholesterol, then addition of fiber alone would only account for at most a 16% reduction in cardiovascular risk. However, most epidemiological studies show a risk reduction of 25–40%, reductions much greater than could be attributed to fiber alone. Thus if the effect is due to the ingestion of whole grain, one is forced to deduce that other components in the whole grain are exerting some additive or synergistic effects. This suggests that other grain phytochemicals such as antioxidants and lignans (phytoestrogens) help protect LDL cholesterol from oxidation, affect LDL particle size, improve vascular reactivity, minimize coagulation, alter fibrinolysis, and improve insulin sensitivity. Third, the population that eats the most dietary fiber and whole grain also has stellar, overall positive health habits such as lower intakes of fat and red meat, higher intakes of fruits and vegetables. They are also more likely to participate in regular exercise regimens and less likely to smoke. Thus, the question is whether whole grains themselves are responsible for lowered risk of cardiovascular disease or whether the intake of whole grain is merely a marker of a person who possesses a cluster of positive health behaviors.

3.4 Fiber, whole grains, and the colon and digestive tract

3.4.1 Fiber, whole grains, and digestive health

All the researchers agree that dietary fiber is important for laxation. There are several reasons why fiber can be helpful in increasing movement through the gastrointestinal tract. Dietary fiber increases water holding capacity in the gut and bulking due to the growth of fecal biomass. The net result is greater fecal output, softer, more readily passed stools, and increased fecal frequency (Cummings, 1986; Chen *et al.*, 1998; Sanjoaquin *et al.*, 2004). This, in turn, causes greater regularity, less straining at the stool and reduced incidence of constipation. However, not all people find that fiber proves beneficial in this regard. For those who are inactive and are prone to severe constipation, the addition of dietary fiber may not increase movement through the gut and for these patients the added fiber can create great discomfort (Muller-Lissner *et al.*, 2005).

High-fiber diets appear important for both the prevention and treatment of hemorrhoids and asymptomatic diverticular disease. Data from the Health Professionals Follow-up Study indicate that dietary fiber ingestion is protective. In this cohort of 45 000 males, the quintile eating the most insoluble dietary fiber reduced the relative risk of diverticular disease by 37% (Aldoori and Ryan-Harshman, 2002). Fiber type appears to matter. Diets with the most cellulose reduced the relative risk of diverticular disease by 48%. The best protection against diverticular disease was found when the diet contains fruit and vegetable fiber and is low in total fat and red meat (Aldoori *et al.*, 1998).

3.4.2 Fiber and colon cancer

The role of dietary fiber in preventing colorectal cancer continues to be a topic of heated debate. Animal, case-control, and some epidemiological studies show that dietary fiber reduces the risk of colorectal cancer. Animal studies consistently show risk reduction with dietary fiber. Case-control studies such as the one conducted in a Swiss canton show a 40–50% risk reduction in colorectal cancer associated with total dietary fiber and specific fiber species: cellulose, total soluble fiber, total insoluble fiber, and lignin. In this study there was a slightly greater reduction in risk with vegetable fiber than with grain or fruit fiber (Table 3.1) (Levi *et al.*, 2001). Large, prospective epidemiological studies also show a relationship between dietary fiber intake and reduction in colorectal cancer risk. In the Prostate, Lung, Colorectal, and Ovarian Cancer Screening Trial (n = 33 971), those in the quintile ingesting the most dietary fiber had a 27% lower risk of colorectal adenoma than those in the lowest quintile (Peters *et al.*, 2003). In the European Prospective Investigation into Cancer and Nutrition (EPIC) (n = 519 978 adults), the quintile ingesting the most dietary fiber had a 25% reduced risk of large bowel cancer with the protective effect being greatest for the left side of the colon, and least for the rectum. In this study all fiber types – fruit, cereal and vegetable – were equally protective (Bingham *et al.*, 2003). Dietary fiber remained independently associated with reduced risk when additional analysis considered folate levels in the diet (Bingham, 2006). A Japanese case-control study with 507 patients and 2535 cancer-free outpatients also showed a decreasing risk of colon cancer with increasing intakes of dietary fiber. In this study colon cancer risk was decreased risk by 35% for the quartile eating the most fiber. The reduction was most robust for insoluble dietary fiber (Wakai *et al.*, 2006).

While some studies did show colon cancer risk reduction with high dietary fiber intake, several large studies failed to show any significant relationship. The Nurses' Health Study (n = 88 757 female nurses) (Fuchs *et al.*, 1999) found no significant difference in the incidence of colorectal cancer for those in different quintiles of dietary fiber intake. Results of the Pooling Project of Prospective Studies of Diet and Cancer (13 prospective cohort studies on 725 628 men and women) showed a risk reduction of 16% due to dietary fiber, but the association was attenuated and no longer statistically significant after adjusting for other risk factors (Park *et al.*, 2005). Large intervention studies such as the Wheat Bran Fiber Trial and the Polyp Prevention

Table 3.1 The odds ratios for colorectal cancer with varying fiber sources. After Levi *et al.* (2001)

Fiber source	Odds ratio
Vegetable fiber	0.60
Fruit fiber	0.78
Grain fiber	0.74

Trial with 3209 participants also failed to show that dietary fiber added to the diet reduced polyp recurrence, a putative marker of colorectal cancer risk (Alberts *et al.*, 2000; Schatzkin *et al.*, 2000).

The relationship between dietary fiber and rectal cancer is also unclear. Some studies fail to show any relationship and others do. The EPIC (Bingham *et al.*, 2003) and the Japanese study (Wakai *et al.*, 2006) found little or no association between dietary fiber and cancer of the rectum. However, in the cohort of subjects from western United States, those with the highest intakes of dietary fiber had reduced risk of rectal cancer (Slattery *et al.*, 2004).

The role of dietary fiber in colon cancer may be confusing because the large cohorts used may fail to pick up effects of fiber on various subgroups. A recent re-analysis (Park *et al.*, 2005) from large studies where the data are separated by sexes showed a significant 19% reduction in the risk of polyp reoccurrence for men. No such reduction in risk was observed for women. The authors of the re-analysis suggest that men may benefit more from fiber in the diet than women. This may help to explain some of the discrepant results reported in the literature.

The recent pooled data are not the only results to indicate that different genders (Jacobs *et al.*, 2006) may behave differently. When all subjects were considered, dietary fiber alone was not shown to be significantly associated with risk reduction in the prospective Cancer Prevention Study II Nutrition Cohort (n = 62 609 men and 70 554 women). However, men in the lowest tertile of vegetable and dietary fiber intake had nearly twice the colon cancer risk compared with those with the highest intakes of these two dietary components (McCullough *et al.*, 2003). In a cohort of nearly 80 000 Japanese, Otani and colleagues (2006) found no statistically significant association between dietary fiber intake and colorectal cancer across quintiles. However, further analysis segmenting by gender and intake levels showed that there was increased colon cancer risk for the 7–8% (bottom segment of the lowest quintile) of women eating the lowest amount of fiber.

There are less data on whole grains and colon cancer risk reduction than there are on dietary fiber. Furthermore, as with the data on heart disease, there are confounding factors associated with this relationship. A review by Jacobs and colleagues (1999) showed that whole grains might be associated with reduced cancer risk. More recently in the Swedish Mammography Cohort (n = 61 433), those women reporting the greatest consumption of whole grains (≥4.5 servings per day versus those eating <1.5 servings per day) had lower risk of colon cancer (Larsson *et al.*, 2005). In this cohort, whole grain ingestion was not associated with reduced risk of rectal cancer. In contrast, rectal cancer was inversely associated with whole grain ingestion in a case-control study with subjects from the western part of the United States. In this study there was a 31% decrease in risk for the whole grain consumers; whereas, a high intake of refined-grain products was directly associated with an increased risk of rectal cancer (Slattery *et al.*, 2004).

The role of dietary and whole grains with respect to colorectal cancer remains unclear despite clear research in animal models indicating a benefit and many plausible mechanisms. They include: (1) the delivery to the gut of resistant starch, inulin, and other fermentable fibers to produce butyrate to feed healthy colonic cells; (2) the encouragement of beneficial bacterial species that inhibit growth of pathogens and their attendant production of secondary bile acids; (3) the lowering of colonic pH; (4) the facilitation of faster transit time and dilution of fecal material so carcinogenic moieties have less chance to contact colonic cells; (5) the contribution of important phytochemicals such as antioxidants that can act locally on the cells and contents of the colon; and (6) the contribution of important phytochemicals such as lignans that can act systemically. For example, rye contributes lignans and these lignans increase serum concentrations of substances such as enterodiol, which is associated with a reduction in colorectal adenoma risk (Hallmans *et al.*, 2003; Kuijsten *et al.*, 2006).

Epidemiological data collected on US populations may fail to show positive impacts of dietary fiber and whole grains on colon cancer because dietary intakes in the US fail to meet the recommended levels. In most studies even the quintiles eating the most dietary fiber fail to or barely meet the recommended levels (Jones, 2004). In contrast in the EPIC study, dietary fiber intakes of the highest quintile approached the recommended 38 grams per day in some of the northern European countries, and this study did show a benefit of fiber that remained after considering other variables.

The varying impact on different genders may also be important as well as fiber type. Large studies have looked at cereal fiber versus fruit and vegetable fiber but even this subdivision is simplistic. It implies, for example, that all cereal fiber is alike and yet there is much data to show that the amount and type of fiber varies by cereal grain and that the co-travelers associated with the fibers are quite different (Adom *et al.*, 2003). Further positive effects are often seen in cohorts where the fiber intakes are higher and grains such as oats, barley, and rye make a significant contribution to the dietary fiber and grain intake (Montonen *et al.*, 2003).

3.4.3 Fiber and cancers of the upper gastrointestinal tract

The source of the fiber has been reported to have an impact on the risk reduction for upper digestive tract cancers (Levi *et al.*, 2000). The highest tertile of fiber intake from whole grain offered a 40% decreased likelihood of gastric, esophageal, and other upper digestive tract cancers; those ingesting fiber from the highest tertile of refined grains had nearly a six-fold elevated risk for these upper gastrointestinal tract cancers. In a case-control study of Italians, those in the upper quintile of fiber intake (22 grams per day vs 12 grams per day) had reduced risk of laryngeal cancer. The reduced risk was attributed to fruit and vegetable fiber, and not to cereal fiber, possibly due to the lack of protection from refined grains (Pelucchi *et al.*, 2003). In another Italian study, dietary fiber was inversely related to a number of cancers with odds ratios for the highest category of whole grain cereal consumption and the various cancers as follows: 0.3–0.5 for upper digestive tract and respiratory neoplasms and colon, 0.6 for rectum and liver, 0.4 for gallbladder, 0.8 for pancreas, 0.2 for soft tissue sarcomas, 0.9 for breast and endometrium, 0.7 for ovary, 0.7 for prostate, 0.4 for bladder and kidney, 1.1 for thyroid, and about 0.5 for lymphomas and 0.6 for myelomas (La Vecchia *et al.*, 2003).

The epidemiological studies showing that fiber from whole grains was protective while that from refined grains was not provide support for the idea that the other components of whole grains are needed for fiber to have a positive effect on upper digestive tract cancers. Fibers and the compounds in their matrices may be retained so that they can act along different points in the gut thus having an impact on many types of cancer.

3.5 Fiber, whole grains, and all-cause mortality

Fiber and whole grains have been related to reduction in all-cause mortality in a variety of diverse populations. Ingestion of 25 grams or more of dietary fiber per day in an Israeli cohort reduced the overall mortality risk by 43% compared with those ingesting less dietary fiber (Lubin *et al.*, 2003). This particular study provides important data because it comprised peoples with marked ethnic diversity with wide differences in morbidity and mortality rates and distinctly different lifestyles and eating habits. In the Iowa Women's Health Study, fiber intake *per se* was not important, but the type of fiber had an impact on all-cause mortality. Women who consumed the same amount of total fiber but with over twice as much whole grain fiber as refined-grain fiber had a 17% lower mortality rate than women eating the opposite ratio (Jacobs *et al.*, 2000). This study would indicate that the whole grain fiber has some other important phytochemicals or that having fiber in a matrix is important. In the Health Professionals Follow-up Study, there was a decrease in overall mortality for age for those who chose one or more servings per day of whole grain cereal versus those who ate it not at all or very infrequently (relative risk, RR = 0.80) (Liu *et al.*, 2003a).

3.6 Summary

Fiber of all types has been shown to have an impact on body weight, diabetes, cardiovascular disease, some types of cancer, and overall mortality. In many instances the source of the fiber was important, with cereal fiber or whole grain fiber offering important advantages for cardiovascular disease protection especially. Increased dietary fiber has been shown to be beneficial for blood lipids and cholesterol, and blood glucose hormone profiles. Risk reduction due to fiber alone was often less than that attributed to whole grain foods containing fiber. This reflects whole grains' contribution not just to fiber but to B-vitamins, vitamin E, magnesium, resistant starch, inulin, minerals, antioxidants, lignans, phytoestrogens, unusual fatty acids, phytic acid, sphingomyelins, and other substances.

In maintaining health, many different mechanisms for fiber and whole grains may be at play for a number of different health endpoints. Bulking may have an impact on feelings of fullness and decrease food intake, speed transit time, and dilute carcinogens. Viscous fibers may delay gastric emptying which can have an impact on hunger and may encourage colonic fermentation which produces short-chain fatty acids which in turn inhibits cholesterol synthesis. Fibers may bind bile acids; this impacts serum cholesterol because cholesterol will be used to synthesize bile acids. The binding of bile acids can also decrease the amount of time available in the large gut for the production of troublesome secondary bile acids that could promote cancer growth. Fermentation of fibers in the large bowel produces butyrate, which is the preferred fuel of healthy colonic cells. High-fiber diets can increase in plasma fibrinolytic activity and have a favorable effect on clotting factors by inhibiting platelet aggregation or platelet release of vasoconstrictors. Many of these changes can favorably affect chronic disease risk (Anderson 2006; Bhargava, 2006). Increased dietary fiber, especially viscous fiber, has been shown to be beneficial for blood glucose, lipid, cholesterol, and hormone profiles.

The whole grain package not only provides the fiber but it provides a delivery system so that various substances embedded in the fiber matrix can be delivered throughout the gut. Thus, these antioxidants, phytoestrogens, and other chemicals in the whole grain can act both locally and systemically, helping to give overall benefits to numerous systems in the body.

These data suggest that high intakes of dietary fiber and whole grains do decrease the risk of many chronic diseases. In some cases fiber ingested as isolated fiber has been shown to be effective in risk reduction. However, not all fibers have the same impacts and the impact may be enhanced or impeded when it is part of the fiber matrix of a food such as whole grains. Whole grains appear to offer advantages beyond fiber and different whole grains offer different benefits.

The sad fact is that diets in the United States contain on average 12–14 grams of dietary fiber per day and have less than a serving of whole grain per day. Diets in Europe are slightly better, providing 18–20 grams of dietary fiber per day. Many European countries, particularly those in southern Europe, also have inadequate whole grain intake. Diets that reverse the current trends and increase whole grains and dietary fiber intake are important strategies for reducing chronic disease risk and improving overall health. High-fiber and whole grain diets may be helpful for weight maintenance. Data are strongest linking dietary fiber and whole grain in reducing the risk of diabetes, hypertension, coronary disease, and stroke. Data with respect to dietary fiber and colorectal cancer risk still need further research.

3.7 References

Adom K K, Sorrells M E, Liu R H (2003), 'Phytochemicals and antioxidant activity of wheat varieties', *J Agric Food Chem*, **51** (26), 7825–34.

Alberts D S, Martinez M E, Roe D J, Guillen-Rodriguez J M, Marshall J R, van Leeuwen J B, Reid M E, Ritenbaugh C, Vargas P A, Bhattacharyya A B, Earnest D L, Sampliner R E (2000), 'Lack of effect of a high-fiber cereal supplement on the recurrence of colorectal adenomas. Phoenix Colon Cancer Prevention Physicians' Network', *N Engl J Med*, **342** (16), 1156–62.

Aldoori W, Ryan-Harshman M (2002), 'Preventing diverticular disease. Review of recent evidence on high-fibre diets', *Can Fam Physician*, **48**, 1632–7.

Aldoori W H, Giovannucci E L, Rockett H R, Sampson L, Rimm E B, Willett W C (1998) 'A prospective study of dietary fiber types and symptomatic diverticular disease in men', *J Nutr*, **128** (4), 714–9.

Alonso A, Beunza J J, Bes-Rastrollo M, Pajares R M, Martinez-Gonzalez M A (2006), 'Vegetable protein and fiber from cereal are inversely associated with the risk of hypertension in a Spanish cohort', *Arch Med Res*, **37** (6), 778–86.

Anderson J W (2006), 'Health Benefits of Dietary Fiber', Speech delivered to the Florida Dietetics Association, 10 July 2006.

Anderson J W, Siesel A E (1990), 'Hypocholesterolemic effects of oat products', *Adv Exp Med Biol*, **270**, 17–36.

Benini L, Castellani G, Brighenti F, Heaton K W, Brentegani M T, Casiraghi M C, Sembenini C, Pellegrini N, Fioretta A, Minniti G (1995), 'Gastric emptying of a solid meal is accelerated by the removal of dietary fiber naturally present in food', *Gut*, **36** (6), 825–30.

Bhargava A (2006), 'Fiber intakes and anthropometric measures are predictors of circulating hormone, triglyceride, and cholesterol concentrations in the Women's Health Trial', *J Nutr*, **136** (8), 2249–54.

Bingham S (2006), 'The fibre–folate debate in colo-rectal cancer', *Proc Nutr Soc*, **65** (1), 19–23.

Bingham S A, Day N E, Luben R, Ferrari P, Slimani N, Norat T, Clavel-Chapelon F, Kesse E, Nieters A, Boeing H, Tjonneland A, Overvad K, Martinez C, Dorronsoro M, Gonzalez C A, Key T J, Trichopoulou A, Naska A, Vineis P, Tumino R, Krogh V, Bueno-de-Mesquita H B, Peeters P H, Berglund G, Hallmans G, Lund E, Skeie G, Kaaks R, Riboli E (2003), 'European Prospective Investigation into Cancer and Nutrition. Dietary fibre in food and protection against colorectal cancer in the European Prospective Investigation into Cancer and Nutrition (EPIC), an observational study', *Lancet*, **361** (9368), 1496–501.

Birketvedt G S, Shimshi M, Erling T, Florholmen J (2005), 'Experiences with three different fiber supplements in weight reduction', *Med Sci Monit*, **11** (1), I5–18.

Burkitt D P, Trowell H C (1975), *Refined Carbohydrate and Disease*, New York, Academic Press.

Burton-Freeman B, Davis P A, Schneeman B O (2002), 'Plasma cholecystokinin is associated with subjective measures of satiety in women', *Am J Clin Nutr*, **76** (3), 659-67.

Chen H L, Haack V S, Janecky C W, Vollendorf N W, Marlett J A (1998), 'Mechanism by which wheat bran and oat bran increase stool weight in humans', *Am J Clin Nutr*, **68**, 711–9.

Cummings J H (1986), 'The effect of dietary fiber on fecal weight and composition', in Spiller G (ed.), *Handbook of Dietary Fiber in Human Nutrition*, Boca Raton, FL, CRC Press, pp. 211–81.

Davis J N, Hodges V A, Gillham M B (2006), 'Normal-weight adults consume more fiber and fruit than their age- and height-matched overweight/obese counterparts', *J Am Diet Assoc*, **106** (6), 833–40.

DeGroot A P, Luyken R, Pikaar N A (1963), 'Cholesterol lowering effect of rolled oats', *Lancet*, **2**, 303–4.

Ding E L, Mozaffarian D (2006), 'Optimal dietary habits for the prevention of stroke', *Semin Neurol*, **26** (1), 11–23.

Erkkila A T, Lichtenstein A H (2006), 'Fiber and cardiovascular disease risk, how strong is the evidence?', *J Cardiovasc Nurs*, **21** (1), 3–8.

Esmaillzadeh A, Mirmiran P, Azizi F (2005), 'Whole-grain consumption and the metabolic syndrome, a favorable association in Tehranian adults', *Eur J Clin Nutr*, **59** (3), 353–62.

Food and Drug Administration (2002), Department of Health and Human Services, 'Food labeling, health claims; soluble dietary fiber from certain foods and coronary heart disease. Interim final rule', *Fed Regist*, **67** (191), 61773–83.

Food and Drug Administration (2006), Department of Health and Human Services, 'Food labeling, health claims; soluble dietary fiber from certain foods and coronary heart disease. Final rule', *Fed Regist*, **71** (98), 29248–50.

Foote J A, Murphy S P, Wilkens L R, Hankin J H, Henderson B E, Kolonel L N (2003), 'Factors associated with dietary supplement use among healthy adults of five ethnicities, The Multiethnic Cohort Study', *Am J Epidemiol*, **157** (10), 888–97.

Fuchs C S, Giovannucci E L, Colditz G A, Hunter D J, Stampfer M J, Rosner B, Speizer F E, Willett W C (1999), 'Dietary fiber and the risk of colorectal cancer and adenoma in women', *N Engl J Med*, **340** (3), 169–76.

Gann P H, Chatterton R T, Gapstur S M, Liu K, Garside D, Giovanazzi S, Thedford K, Van Horn L (2003), 'The effects of a low-fat/high-fiber diet on sex hormone levels and menstrual cycling in premenopausal women, a 12-month randomized trial (the diet and hormone study)', *Cancer*, **98** (9), 1870–9.

Hajjar I, Kotchen T (2003), 'Regional variations of blood pressure in the United States are associated with regional variations in dietary intakes, the NHANES-III data', *J Nutr*, **133** (1), 211–4.

Hallmans G, Zhang J X, Lundin E, Stattin P, Johansson A, Johansson I, Hulten K, Winkvist A, Aman P, Lenner P, Adlercreutz H (2003), 'Rye, lignans and human health', *Proc Nutr Soc*, **62** (1), 193–9.

Howarth N C, Saltzman E, McCrory M A, Greenberg A S, Dwyer J, Ausman L, Kramer D G, Roberts S B (2003), 'Fermentable and nonfermentable fiber supplements did not alter hunger, satiety or body weight in a pilot study of men and women consuming self-selected diets', *J Nutr*, **133** (10), 3141–4.

Hunninghake D B, Miller V T, LaRosa J C, Kinosian B, Jacobson T, Brown V, Howard W J, Edelman D A, O'Connor R R (1994), 'Long-term treatment of hypercholesterolemia with dietary fiber', *Am J Med*, **97** (6), 504–8.

Jacobs D R Jr, Meyer K A, Kushi L H, Folsom A R (1998), 'Whole-grain intake may reduce the risk of ischemic heart disease death in postmenopausal women, the Iowa Women's Health Study', *Am J Clin Nutr*, **68** (2), 248–57.

Jacobs D R Jr, Meyer K A, Kushi L H, Folsom A R (1999), 'Is whole grain intake associated with reduced total and cause-specific death rates in older women? The Iowa Women's Health Study', *Am J Public Health*, **89** (3), 322–9.

Jacobs D R, Pereira M A, Meyer K A, Kushi L H (2000), 'Fiber from whole grains, but not refined grains, is inversely associated with all-cause mortality in older Women, the Iowa Women's Health Study', *J Am Coll Nutr*, **19** (3 Suppl), 326S–330S.

Jacobs E T, Lanza E, Alberts D S, Hsu C H, Jiang R, Schatzkin A, Thompson P A, Martinez M E (2006), 'Fiber, sex, and colorectal adenoma, results of a pooled analysis', *Am J Clin Nutr*, **83** (2), 343–9.

Johnston L, Reiss Reynolds, Patz H (1998), 'Cholesterol-lowering benefits of a whole grain ready-to-eat cereal', *Nutr Clin Care*, 1, 6–12.

Jones J M (2004) 'Dietary fibre intake, disease prevention, and health promotion: An overview with emphasis on evidence from epidemiology', in van der Kamp J W, Asp N-G, Miller-Jones J and Schaafsma G (eds), *Dietary Fibre: Bio-active Carbohydrates for Food and Feed*, Wageningen, The Netherlands, Wageningen Academic Publishers.

Koh-Banerjee P, Rimm E B (2003), 'Whole grain consumption and weight gain: a review of the epidemiological evidence, potential mechanisms and opportunities for future research', *Proc Nutr Soc*, **62** (1), 25–9.

Koh-Banerjee P, Franz M, Sampson L, Liu S, Jacobs D R Jr, Spiegelman D, Willett W, Rimm E (2004), 'Changes in whole-grain, bran, and cereal fiber consumption in relation to 8-y weight gain among men', *Am J Clin Nutr*, **80** (5), 1237–45.

Kromhout D, Bloemberg B, Seidell J C, Nissinen A, Menotti A (2001), 'Physical activity and dietary fiber determine population body fat levels, the Seven Countries Study', *Int J Obes Relat Metab Disord*, **25** (3), 301–6.

Kuijsten A, Arts I C, Hollman P C, van't Veer P, Kampman E (2006), 'Plasma enterolignans are associated with lower colorectal adenoma risk', *Cancer Epidemiol Biomarkers Prev*, **15** (6), 1132–6.

Kurth T, Moore S C, Gaziano J M, Kase C S, Stampfer M J, Berger K, Buring J E (2006), 'Healthy lifestyle and the risk of stroke in women', *Arch Intern Med*, **166** (13), 1403–9.

Lairon D, Arnault N, Bertrais S, Planells R, Clero E, Hercberg S, Boutron-Ruault M C (2005), 'Dietary fiber intake and risk factors for cardiovascular disease in French adults', *Am J Clin Nutr*, **82** (6), 1185–94.

Larsson S C, Giovannucci E, Bergkvist L, Wolk A (2005), 'Whole grain consumption and risk of colorectal cancer, a population-based cohort of 60,000 women', *Br J Cancer*, **92** (9), 1803–7.

La Vecchia C, Chatenoud L, Negri E, Franceschi S (2003), 'Session: whole cereal grains, fibre and human cancer: wholegrain cereals and cancer in Italy', *Proc Nutr Soc*, **62** (1), 45–9.

Levi F, Pasche C, Lucchini F, Chatenoud L, Jacobs D R Jr, La Vecchia C (2000), 'Refined and whole grain cereals and the risk of oral, oesophageal and laryngeal cancer.' *Eur J Clin Nutr*, **54** (6), 487–9.

Levi F, Pasche C, Lucchini F, La Vecchia C (2001), 'Dietary fibre and the risk of colorectal cancer', *Eur J Cancer*, **37** (16), 2091–6.

Lin P H, Aickin M, Champagne C, Craddick S, Sacks FM, McCarron P, Most-Windhauser M M, Rukenbrod F, Haworth L: 'Dash-Sodium Collaborative Research Group (2003), Food group sources of nutrients in the dietary patterns of the DASH-Sodium trial', *J Am Diet Assoc*, **103** (4), 488–96.

Liu S, Stampfer M J, Hu F B, Giovannucci E, Rimm E, Manson J E, Hennekens C H, Willett W C (1999), 'Whole-grain consumption and risk of coronary heart disease: results from the Nurses' Health Study', *Am J Clin Nutr*, **70** (3), 412–9.

Liu S, Manson J E, Stampfer M J, Rexrode K M, Hu F B, Rimm E B, Willett W C (2000),. 'Whole grain consumption and risk of ischemic stroke in women: a prospective study', *JAMA*, **284** (12), 1534–40.

Liu S, Sesso H D, Manson J E, Willett W C, Buring J E (2003a), 'Is intake of breakfast cereals related to total and cause-specific mortality in men?', *Am J Clin Nutr*, **77** (3), 594–9.

Liu S, Willett W C, Manson J E, Hu F B, Rosner B, Colditz G (2003b), 'Relation between changes in intakes of dietary fiber and grain products and changes in weight and development of obesity among middle-aged women', *Am J Clin Nutr*, **78** (5), 920–7.

Lowik M R, Hulshof K F, Brussaard J H (1999), 'Patterns of food and nutrient intakes of Dutch adults according to intakes of total fat, saturated fatty acids, dietary fibre, and of fruit and vegetables', *Br J Nutr*, **81** (Suppl 2), S91–S98.

Lubin F, Lusky A, Chetrit A, Dankner R (2003), 'Lifestyle and ethnicity play a role in all-cause mortality', *J Nutr*, **133** (4), 1180–5.

Marlett J A, McBurney M I, Slavin J L (2002), 'American Dietetic Association. Position of the American Dietetic Association: health implications of dietary fiber', *J Am Diet Assoc*, **102** (7), 993–1000.

Marquart L, Wiemer K L, Jones J M, Jacob B (2003), 'Whole grains health claims in the USA and other efforts to increase whole-grain consumption', *Proc Nutr Soc*, **62** (1), 151–60.

McCullough M L, Robertson A S, Chao A, Jacobs E J, Stampfer M J, Jacobs D R, Diver W R, Calle E E, Thun M J (2003), 'A prospective study of whole grains, fruits, vegetables and colon cancer risk', *Cancer Causes Control*, **14** (10), 959–70.

McKeown N M, Meigs J B, Liu S, Wilson P W, Jacques P F (2002), 'Whole-grain intake is favorably associated with metabolic risk factors for type 2 diabetes and cardiovascular disease in the Framlingham Offspring Study', *Am J Clin Nutr*, **76** (2), 390–8.

Mickelsen O, Makdani D D, Cotton R H, Titcomb S T, Colmey J C, Gatty R (1979), Effects of a high fiber bread diet on weight loss in college-age males, *Am J Clin Nutr*, **32** (8), 1703–9.

Montonen J, Knekt P, Jarvinen R, Aromaa A, Reunanen A (2003), 'Whole-grain and fiber intake and the incidence of type 2 diabetes', *Am J Clin Nutr*, **77** (3), 622–9.

Most M M, Tulley R, Morales S, Lefevre M (2005), 'Rice bran oil, not fiber, lowers cholesterol in humans', *Am J Clin Nutr*, **81** (1), 64–8.

Mozaffarian D, Kumanyika S K, Lemaitre R N, Olson J L, Burke G L, Siscovick D S (2003), 'Cereal, fruit, and vegetable fiber intake and the risk of cardiovascular disease in elderly individuals', *JAMA*, **289** (13), 1659–66.

Muller-Lissner S A, Kamm M A, Scarpignato C, Wald A (2005), 'Myths and misconceptions about chronic constipation', *Am J Gastroenterol*, **100** (1), 232–42.

Newby P, Muller D, Hallfrisch J, Qiao N, Andres R, Tucker K L (2003), 'Dietary patterns and changes in body mass index and waist circumference in adults', *Am J Clin Nutr*, **77** (6), 1417–25.

Oh K, Hu F B, Cho E, Rexrode K M, Stampfer M J, Manson J E, Liu S, Willett W C (2005), 'Carbohydrate intake, glycemic index, glycemic load, and dietary fiber in relation to risk of stroke in women', *Am J Epidemiol*, **161** (2), 161–9.

Otani T, Iwasaki M, Ishihara J, Sasazuki S, Inoue M, Tsugane S (2006), 'Dietary fiber intake and subsequent risk of colorectal cancer: The Japan Public Health Center-Based Prospective Study', *Int J Cancer*, **119** (6), 1475–80.

Park Y, Hunter D J, Spiegelman D, Bergkvist L, Berrino F, van den Brandt P A, Buring J E, Colditz G A, Freudenheim J L, Fuchs C S, Giovannucci E, Goldbohm R A, Graham S, Harnack L, Hartman A M, Jacobs D R Jr, Kato I, Krogh V, Leitzmann M F, McCullough M L, Miller A B, Pietinen P, Rohan T E, Schatzkin A, Willett W C, Wolk A, Zeleniuch-Jacquotte A, Zhang S M, Smith-Warner S A (2005), 'Dietary fiber intake and risk of colorectal cancer, a pooled analysis of prospective cohort studies', *JAMA*, **294** (22), 2849–57.

Pelucchi C, Talamini R, Levi F, Bosetti C, La Vecchia C, Negri E, Parpinel M, Franceschi S (2003), 'Fibre intake and laryngeal cancer risk', *Ann Oncol*, **14** (1), 162–7.

Pereira M A, Ludwig D S (2004), 'Dietary fiber and body-weight regulation. Observations and mechanisms', *Pediatr Clin North Am*, **48** (4), 969–80.

Peters U, Sinha R, Chatterjee N, Subar A F, Ziegler R G, Kulldorff M, Bresalier R, Weissfeld J L, Flood A, Schatzkin A, Hayes R B (2003), 'Prostate, Lung, Colorectal, and Ovarian Cancer Screening Trial Project Team. Dietary fibre and colorectal adenoma in a colorectal cancer early detection programme', *Lancet*, **361** (9368), 1491–5.

Pins J J, Geleva D, Keenan J M, Frazel C, O'Connor P J, Cherney L M (2002), 'Do whole-grain oat cereals reduce the need for antihypertensive medications and improve blood pressure control?' *J Fam Pract*, **51** (4), 353–9.

Qi L, Meigs J B, Liu S, Manson J E, Mantzoros C, Hu F B (2005), 'Dietary fibers and glycemic load, obesity, and plasma adiponectin levels in women with type 2 diabetes', *Diabetes Care*, **29** (7), 1501–5.

Rimm E B, Ascherio A, Giovannucci E, Spiegelman D, Stampfer M J, Willett W C (1996), 'Vegetable, fruit, and cereal fiber intake and risk of coronary heart disease among men', *JAMA*, **275** (6), 447–51.

Rose N E (2005), *Whole Grain Intake in College Students and its Association with Body Mass Index*, URN etd-05192005-110138, Masters' Thesis, Virginia Polytechnic Institute.

Saltzman E, Moriguti J C, Das S K, Corrales A, Fuss P, Greenberg A S, Roberts S B (2001), 'Effects of a cereal rich in soluble fiber on body composition and dietary compliance during consumption of a hypocaloric diet', *J Am Coll Nutr*, **20** (1), 50–7.

Sanjoaquin M A, Appleby P N, Spencer E A, Key T (2004), 'Nutrition and lifestyle in relation to bowel movement frequency: a cross-sectional study of 20630 men and women in EPIC-Oxford', *Public Health Nutr*, **7** (1), 77–83.

Schaefer E J, Gleason J A, Dansinger M L (2005), 'The effects of low-fat, high-carbohydrate diets on plasma lipoproteins, weight loss, and heart disease risk reduction', *Curr Atheroscler Rep*, **7** (6), 421–7.

Schatzkin A, Lanza E, Corle D, Lance P, Iber F, Caan B, Shike M, Weissfeld J, Burt R, Cooper M R, Kikendall J W, Cahill J (2000), 'Lack of effect of a low-fat, high-fiber diet on the recurrence of colorectal adenomas. Polyp Prevention Trial Study Group', *N Engl J Med*, **342**, 1149–55.

Shane J M, Walker P M (1995), 'Corn bran supplementation of a low-fat controlled diet lowers serum lipids in men with hypercholesterolemia', *J Am Diet Assoc*, **95** (1), 40–5.

Slattery M L, Curtin K P, Edwards S L, Schaffer D M (2004), 'Plant foods, fiber, and rectal cancer', *Am J Clin Nutr*, **79** (2), 274–81.

Slavin J L (2005), 'Dietary fiber and body weight', *Nutrition*, **21** (3), 411–8.

Trowell H (1993), 'Dietary fibre, ischaemic heart disease and diabetes mellitus', *Proc Nutr Soc*, **32** (3), 151–7.

Van Horn L, Emidy L A, Liu K A, Liao Y L, Ballew C, King J, Stamler J (1988), 'Serum lipid response to a fat-modified, oatmeal-enhanced diet', *Prev Med*, **17** (3), 377–86.

Wakai K, Hirose K, Matsuo K, Ito H, Kuriki K, Suzuki T, Kato T, Hirai T, Kanemitsu Y, Tajima K (2006), 'Dietary risk factors for colon and rectal cancers: a comparative case-control study', *J Epidemiol*, **16** (3), 125–35.

Whelton S P, Hyre A D, Pedersen B, Yi Y, Whelton P K, He J (2005), 'Effect of dietary fiber intake on blood pressure: a meta-analysis of randomized, controlled clinical trials', *J Hypertens*, **23** (3), 475–81.

4

Resistant starch and health

A. M. Birkett, National Starch Food Innovation, USA and
I. L. Brown, University of Wollongong, Australia

4.1 Introduction

Dietary fiber (DF) represents a broad range of carbohydrate compounds varying in structure and chemistry, and nutrition researchers have assessed many of these components individually for their unique role in food formulation and health. From this research some starch sub-types have been identified that escape digestion and absorption as glucose within the upper gastrointestinal tract. Both naturally occurring and synthesized starch sub-types, which are known collectively as resistant starch (RS), have great potential for functional food development because RS ingredients can provide a range of technical benefits in the formulation of foods and assist in the provision of a variety of potential physiological improvements for people consuming those foods.

RS is found naturally in a broad range of starchy foods and can be added as a functional ingredient. RS is an unusual form of DF in that the amount can change during food processing and consumption. Western diets typically contain approximately 5 g of RS (Table 4.1), which originates through the consumption each day of starchy foods such as bread and breakfast cereals. This is much less than in earlier times when foods were less processed and usually contained more legume- and/or grain-based ingredients. Loss of dietary RS has contributed to what has been termed the 'carbohydrate gap', which is the observed deficit in the amount of fermentable nutrients consumed compared with what is required by the intestinal microflora to maintain digestive health. Scientists recommend increasing dietary RS consumption to around 15–20 g/d in order to rectify this 'gap' (Baghurst *et al.*, 1996).

This chapter will focus on RS ingredients that can be added to foods with the emphasis that RS ingredients have two important roles: to restore nutrients

Table 4.1 Estimated resistant starch intakes worldwide

Country	Estimated intake (g/d)	Reference
Australasia		
Australia	5 g female, 5.3 g male (36% cereals, 26% vegetables, 22% fruits)	Baghurst *et al.*, 1996
	3.4–9.4 g range	Roberts *et al.*, 2004
	8 g/10 MJ	Muir *et al.*, 1998
	8.6 g/10 MJ	Walker *et al.*, 1997
New Zealand	5.7 g	Baghurst *et al.*, 1996
Europe		
Europe, mean	4.11 g	Dysseler and Hoffem, 1994
Belgium	3.99 g	Dysseler and Hoffem, 1994
Denmark	3.67 g	Dysseler and Hoffem, 1994
England	3.97 g	Dysseler and Hoffem, 1994
France	3.73 g	Dysseler and Hoffem, 1994
Germany	3.75 g	Dysseler and Hoffem, 1994
Italy	8.5 g (7.2 g north-west, 9.2 g south)	Brighenti *et al.*, 1998
Netherlands	5.29 g	Dysseler and Hoffem, 1994
Norway	3.22 g	Dysseler and Hoffem, 1994
Spain	5.74 g	Dysseler and Hoffem, 1994
Sweden	3.36 g	Dysseler and Hoffem, 1994
	3.2 g (1.3 g bread, 1.2 g potatoes)	Elmstahl, 2002
Switzerland	4.38 g	Dysseler and Hoffem, 1994
United Kingdom	2.76 g	Tomlin and Read, 1990
North America		
United States of America	4.9 g; 2.8–8.0 g range	Murphy and Douglass, 2007
Asia		
China	20 g/10 MJ	Muir *et al.*, 1998
India	10 g	Platel and Shurpalekar, 1994

lost with food processing; and to enhance the physiological benefits that could be gained through the consumption of processed foods. RS ingredients from high-amylose maize or corn (HAM) will be highlighted because globally this is currently the primary source of commercial RS ingredients that is supported by a significant body of published scientific literature.

4.1.1 What is resistant starch?

Starch is the main structure in which plants store glucose. Starch and starch-derived materials can exist in many forms, depending on the plant species and the procedures used in their processing and preparation for consumption. This variation is reflected in the granule morphology and conformation, thermal and textural properties (exhibited during food processing), and nutritional effects. Until the early 1980s it was thought that all cooked starches were treated similarly by the body, namely they were digested within the small intestine and absorbed as glucose. However, it was observed that a portion of dietary starch resisted digestion within the small intestine – hence the name 'resistant starch'. Clinical studies have revealed undigested starch in the ileal effluent and feces, while in animal models starch and starch residues have been observed along the entire length of the large intestine as well as in the feces. These are not isolated observations of individual malabsorption but a physiological feature of selected starches and starchy foods.

An *in vitro* classification scheme has been proposed for starch, based on its digestive fate in the gastrointestinal tract. This scheme has been generally accepted and used widely in describing the nutritional functionality of starch. Fully digested starches are termed rapidly digested (RDS) or slowly digested starch (SDS), based on their rate of digestion, and undigested starches are now termed RS (Englyst *et al.*, 1992).

The initial exploration of RS culminated in the collaborative investigation program organized through EURESTA (1990–1994) titled 'European FLAIR Concerted Action No. 11 on the physiological implications of the consumption of resistant starch in man'. EURESTA involved 36 groups from 10 European countries and its inquiries centered on four major areas, namely: (1) definition and analysis of RS; (2) RS production and technological impact; (3) physiological effects of RS, principally as it relates to the upper and lower gastrointestinal tract; and finally (4) the energy contribution of RS (Brown *et al.*, 2001). Since 1994 a number of RS ingredients have been developed and commercialized and this has provided direction for those groups around the world that have and continue to undertake research with RS. The EURESTA group defined RS as '... the sum of starch and products of starch degradation not absorbed in the small intestine of healthy individuals' (Asp, 1992).

RS is a heterogeneous group that has been divided into four different starch sub-types based on the mechanism by which the starch resists digestion (Brown *et al.*, 1995).

- *Resistant starch type 1 (RS1).* This starch is resistant because the starch is physically trapped within the food matrix. Milling or grinding foods can make the starch more accessible to digestive enzymes.
- *Resistant starch type 2 (RS2).* The RS2 starch granules are inherently resistant due to the composition and conformation of the granule. For most sources of RS2, including potato and banana, normal cooking temperatures and procedures will tend to disrupt the structure of the starch granule, potentially leading to gelatinization, increasing its digestibility. RS2 granules from HAM have a higher gelatinization temperature, making them better suited to commercial food processing conditions.
- *Resistant starch type 3 (RS3).* When starch is cooked and then cooled in water, crystalline regions can develop that are very resistant to digestive enzymes. This usually occurs to a limited extent during normal food

processing of starchy foods. This concept has been used in the manufacture of RS3 ingredients through the utilization of controlled cycles of heating and cooling generally in association with the use of enzymes.

- *Resistant starch type 4 (RS4)*. The digestibility of starch can be altered through the introduction of different chemical structures or through modification of the existing chemical structure of the starch in such a manner as to interfere with the digestive enzymes. Inhibition is dependent upon the type and extent of chemical changes as a result of such processes as dextrinization, etherification, esterification, oxidation, and cross-linking with difunctional reagents. Selected chemistries, alone or in combination, can markedly affect both the functionality and nutritional properties of RS4 when used in foods and other applications.

There may be a further category of RS (potentially RS5) which relies on the ability of starch polymers such as amylose to form inclusion complexes with polar lipids. These V-structures (as determined by X-ray crystallography) have been observed in high-amylose starches (Zobel, 1992) and can also be formed as a result of starch gelatinization in rice (Priestley, 1976) and extruded starch (Mercier *et al.*, 1979). V-structures may be useful for protecting starch-derived material during food processing and transit through the upper gastrointestinal tract in order to deliver physiological benefits.

4.1.2 Recognition of resistant starch

RS has been the subject of research for more than three decades and it has been commercially available as food ingredients for more than 10 years. There have been a variety of *in vitro* analytical methods developed and used by researchers to quantify the content of RS in foods, but currently there is only one method accepted by the Association of Official Analytical Chemists (AOAC), namely AOAC Method 2002.02. However, the gold standard is the *in vivo* quantification of the amount of undigested starch in the ileal effluent. This is a direct measure of starch in the digesta that passes from the small intestine through to the large intestine, and relates to the physiological definition of RS proposed by EURESTA. This approach accounts for potential variations in the amount of RS due to the type of RS, how the food is processed, prepared, and consumed, and the health status of the individual. This is impractical for routine nutrient analysis so *in vitro* methods are typically employed for food analysis prior to ingestion for components such as RS and the broader category of DF. It should be noted that some RS may not be fully quantified using accepted DF methods of analysis and for this reason RS should be measured separately.

Definitions of DF have recently been expanded to include the broader range of indigestible carbohydrates that are found in and/or added to the diet. Definitions proposed by the Institute of Medicine of the National Academies (IOM, 2002) and the American Association of Cereal Chemists (AACC, 2001) in the USA both include RS as a component of DF. The Codex

Alimentarius Commission (FAO/WHO, 2005) has also proposed a definition for DF which if approved would recognize RS.

4.1.3 Sources of resistant starch

Although RS occurs naturally within the diet, the amount consumed is typically low and varies both within and between populations. Table 4.1 lists those countries for which RS intakes have been calculated. Most assessments have been conducted for Western countries, including Australia and various countries in Europe, with limited assessment in Asia. Typically, in Western countries the average RS intake is low relative to some other dietary nutrients, e.g. 5 g in Australia (Baghurst *et al.*, 1996); however, the intake range can vary several fold, e.g. 3.4–9.4 g/d (Roberts *et al.*, 2004). RS intakes are relatively similar between Western countries, irrespective of differences in culture, RS methodologies employed, and dietary interpretation. For example, in Australia, the UK, and the USA the average RS intake is approximately 4–5 g/d. Italy represents an exception, with a higher estimated average intake of 8.5 g/d. Most RS in the Western diet comes from bakery products and vegetables, providing 25% and 26%, respectively (Baghurst *et al.*, 1996). In the two Asian countries where RS intakes were assessed the typical amounts ingested were more than double the typical intakes of Western countries.

Most dietary starch is consumed in cooked foods where the starch is generally fully gelatinized and readily digestible. Typically, around 95% of starch consumed is digested and absorbed as glucose within the upper gastrointestinal tract (Cassidy *et al.*, 1994). For the remaining 5% of starch many extrinsic and intrinsic factors influence the reduced digestibility of the starch in the upper gastrointestinal tract including:

- *food behavior* – the type of starch eaten; the amount, frequency and composition of the food and meal consumed; customs of food preparation and consumption;
- *nature and source of the starch* – amylose content; granularity; conformation; minor components in the starch;
- *food processing* – loss of cellular and native plant structure; processing conditions; extent of starch gelatinization; particle size; density of the food; other food components present including antinutrients (e.g. amylase inhibitors), and the amount and type of viscous fiber(s) and fat(s);
- *individual physiology* – health status; individual physiological differences; extent of chewing; gastric emptying; viscosity; transit time.

There is considerable variation in the RS content of foods (Table 4.2), and the amount of RS is independent of the amount of non-starch polysaccharide (NSP) in the food since it is not fully quantified by the methods used to determine NSP and DF. The RS content of foods can vary from as little as 1% or less of the dry matter in white bread to as much as 18% in boiled haricot beans (Englyst *et al.*, 1992). The amount present in a food can change

Table 4.2 Resistant starch content of selected foods

Food	Dry matter (%)	RS g/100 g dry matter
Common foods		
White bread[1]	54.5	1
White bread [3]		0.9
Wholemeal bread[1]	52.0	1
Cornflakes[1]	95.8	3
Porridge oats[1]	90.7	2
Ryvita crispbread[1]	94.3	3
Peas (frozen, boiled 5 min)[1]	18.3	5
Lentils (boiled 20 min, cold)[1]	28.3	9
Haricot beans (boiled 40 min)[1]	41.4	18
Processing effects		
Boiled potato (hot)[1]	22.8	5
Boiled potato (cold)[1]	23.8	10
Spaghetti (freshly cooked)[1]	28.3	5
Spaghetti (cooked, cooled)[1]	34.7	4
Whole long-grain white rice (cooked)[2]		12
Ground long-grain white rice (cooked)[2]		5
Rolled oats (uncooked)[2]		16
Rolled oats (cooked)[2]		3
Commercial foods containing RS2 ingredients		
White bread[3]		4.3
Breakfast cereal (muesli style)[3]		2.8
Breakfast cereal (expanded style)[3]		1.9
Cracker biscuit (crispbread style)[3]		3.5
Cracker biscuit (savory style)[3]		4.8
Fruit-filled cereal bar[3]		3.3
Fruit-based bar for children[3]		1.3
Pasta (dry)[3]		4.2
Pharmaceutical bulk laxative[3]		16.0

[1] Englyst *et al.* (1992).
[2] Muir *et al.* (1992).
[3] Brown (2004).

during storage, e.g. 5% in hot boiled potato up to 10% in cold boiled potato (Englyst *et al.*, 1992). Processing can have a major impact on the RS content of foods, and typically RS contents are reduced by longer or more severe processing. For example, grinding reduced the RS content of cooked rice from 12% to 5%, and cooking reduced the RS content of oats from 16% to 3% (Muir and O'Dea, 1992).

Starch and starch granule properties (i.e. structure and conformation) are important determinants of RS content. Starches with elevated levels of amylose and lower levels of amylopectin tend to have higher inherent resistance to enzymatic digestion. Amylose and amylopectin differ in their polymeric configuration and granule conformation, contributing to physical properties that include chain length, molecular weight, melting temperature, and three-dimensional structure. Granular morphology and surface properties also differ between starch sources and can affect enzyme interaction and hence digestibility. Many of these differences are lost upon cooking when the starch granules gelatinize, with the exception of HAM starch which tends to retain its granularity at temperatures typically used in food preparation. HAM typically does not fully hydrate and gelatinize until heated to more than 150 °C (Brown *et al.*, 1995).

A number of RS ingredients are available to enrich the diet with RS. The first commercially available RS ingredient was a natural HAM RS2 released in Australia in 1993. It was called Hi-maize™ and contained DF and RS contents of 30% and 45%, respectively. Since that time, the number of RS ingredients available has increased, providing a wide variety of RS and DF contents. Today, commercial and experimental sources of RS1, RS2, RS3, and RS4 are available utilizing technologies that include physical, chemical, and enzymatic modification. These include flour or meal from HAM, barley, and wheat (RS1); extracted and/or hydrothermally prepared starches from HAM (RS2); recrystallized starch from HAM and tapioca (RS3); and chemically modified starch prepared using diphosphate cross-linking from wheat, potato, and corn (RS4) (Gelski, 2006).

Each RS ingredient has its own range of technical and nutritional functionalities.

- RS is not yet recognized by consumers in many countries as being important in its own right. For this reason the DF content of RS ingredients is usually mentioned, although the RS content generally does not equal the amount of DF.
- The relative ratio of digestible to indigestible starch will vary. Using AOAC total DF methods, RS ingredients can differ in their DF contents from 0% (RS3 from tapioca) to more than 60% (RS2 from HAM).
- The RS/DF contribution will depend on whether the RS/DF is preserved in the food as eaten.
- Depending on the country of use, some RS4 ingredients may be considered as novel foods and may require official regulatory approval.
- The physiological outcome can vary depending on the RS ingredient being used. As is the case for other DF components, different types of RS may have different physiological responses.
- The level of supporting evidence will vary between RS ingredients. RS from HAM has a substantial amount of published supporting evidence, whereas some newer sources of RS may have little or no supporting evidence at this time.
- Currently, no commercial RS ingredients are composed only of RS. The non-RS component can contribute significantly to the technical functionality of the RS-rich ingredient within the food. Since the amount and nature of the non-RS component vary in relation to the commercially available RS ingredients, this will result in differences in their impact on the organoleptic, texture, and sensory dimensions of foods.

4.2 Health effects of resistant starch

During the past 30 years there has been a steady increase in our knowledge of RS and its health effects. Research has principally concentrated on the health impact of RS-rich ingredients (extracted from natural sources or processed to increase the RS and/or DF content) with a particular emphasis on the effects of HAM starch and its derivatives. In spite of the differences in clinical methodologies employed and the source of RS used, the substantial body of nutritional and clinical research demonstrates the broad range of health benefits that can be attributed to RS.

Table 4.3 provides a summary of the extensive physiological potential of HAM RS2, based on animal and human research. This table highlights the breadth of evidence supporting a role for dietary RS, including the addition of RS-rich ingredients, in improving digestive health but also indicates that the value of RS in foods extends beyond fermentation to a metabolic role.

4.2.1 Digestive health

Fermentation

Each DF component has a characteristic range of physiological effects determined by its physical and chemical nature. For example, some fibers are viscosifying and slow gastric emptying while others bind bile acids to lower plasma cholesterol and some bind water and effectively function as bulking/laxative agents. Many of the positive physiological changes afforded by RS appear to be connected to its fermentation in the colon (Table 4.3).

Fermentation is the process whereby colonic bacteria utilize a dietary substrate (e.g. RS) to generate energy for their maintenance and growth. Under the anaerobic conditions of the large intestine (colorectal region in humans or the cecal plus colorectal region in animals) the microflora produce a range of metabolic end-products that include short-chain fatty acids (SCFAs), gases, and water that are important for the maintenance of a healthy microflora. SCFAs include acetate, propionate, and butyrate, and are available for absorption by the colon for utilization by the individual or as an energy source for the colonic microflora. SCFAs directly influence the colonic environment, by lowering pH, and improving colonic function.

In clinical studies where diets containing HAM RS2 were consumed, increased SCFAs, particularly butyrate, and lower pH were observed (Van Munster *et al.*, 1994; Phillips *et al.*, 1995; Noakes *et al.*, 1996; Jenkins *et al.*, 1998; Muir *et al.*, 2004). These observations provide a direct indication of microbial RS utilization, and suggest protective outcomes for digestive health. Greater acidity inhibits the growth of potentially pathogenic bacteria, increases the absorption of micronutrients, and inhibits the absorption of compounds with toxic or carcinogenic potential (Bird *et al.*, 2000). Butyrate is the preferred *in vitro* metabolic fuel for colonocytes, enhances growth of normal colon

Table 4.3 Potential health benefits of high-amylose corn RS2 ingredients (Brown *et al.*, 2000; Brown, 2004; Morita *et al.*, 2004a)

Reduced glycemic response	Lower postprandial glucose response; lower insulin response; improved insulin sensitivity; delayed onset of insulin resistance
Reduced energy	Fewer available calories
Improved intestinal health	Fermentable substrate (controlled); cecal and fecal bulking; increased production of short-chain fatty acids; increased butyrate concentration; reduced intestinal pH; reduced levels of secondary bile acids; reduced levels of ammonia; reduced symptoms of diarrhea; increased mucosal barrier strength; improved immune response
Improved colonic tissue health	Increased apoptotic index; decreased DNA damage; decreased levels of cytotoxic compounds
Prebiotic	Selectively utilized by bifidobacteria; promotes growth of beneficial indigenous microbes (lactobacilli and bifidobacteria); promotes probiotic growth and activity (bifidobacteria); reduced pathogenic bacteria levels; increased butyrate production
Culture protagonist	Improves yield of probiotic cultures during growth; improves survival of probiotic cultures during processing, in foods, *in vivo*.
Increased mineral absorption	Calcium; magnesium
Metabolic effects	Increased lipid metabolism
Tolerance	Tolerated at high levels

cells while inhibiting growth of malignant ones, promotes DNA repair and tumor cell differentiation, and is linked with apoptosis of malignant cells (Bird *et al.*, 2000). SCFAs are also thought to enhance colonic function by stimulating colonic blood flow, promoting colonic muscular contraction, which increases tone and nutrient flow, increasing mineral absorption, and promoting colonocyte proliferation to reverse atrophy associated with low-fiber diets (Bird *et al.*, 2000).

RS consumption positively influences other markers of digestive health. In clinical studies HAM RS2 promotes fecal bulking and laxation (Phillips *et al.*, 1995; Hylla *et al.*, 1998; Jenkins *et al.*, 1998; Muir *et al.*, 2004). The laxative effects of RS are less marked than those observed from wheat bran but RS ingredients can be important contributors to digestive health. With an enriched supply of colonic RS, less favorable bacterial metabolic activity is reduced, minimizing the production of potentially cytotoxic compounds. For example, ammonia, phenol, and secondary bile acids become less prevalent metabolites when an RS-enriched diet is consumed (Van Munster *et al.*, 1994; Birkett *et al.*, 1996; Hylla *et al.*, 1998; Grubben *et al.*, 2001; Muir *et al.*, 2004).

The site, rate, quantity, and outcome of fermentation will vary due to the type and source of RS. RS2 from HAM is readily fermented as most RS is utilized prior to excretion in the feces. Animal models have demonstrated that RS is fermented along the entire length of the colon between the cecum and the anus (Morita *et al.*, 1999), a feature that could explain the high tolerance to RS.

Prebiotics and probiotics

The bacterial composition of the colonic microflora reflects the supply and utilization of fermentable substrate within the large intestine. Several bacterial groups have the capacity to ferment starch including *Eubacterium*, *Bacteroides*, *Bifidobacterium*, and *Escherichia* (Bird *et al.*, 2000), and within these groups selected bacteria utilize HAM RS2. *Bifidobacterium* and *Clostridium butyricum* utilized RS *in vitro*, whereas *Escherichia coli*, *Staphylococcus aureus*, and *Lactobacillus* did not (Wang *et al.*, 1999). HAM RS2 is utilized by most *Bifidobacterium in vivo*, shown by increased fecal *Bifidobacterium* numbers after feeding HAM starch to mice (Wang *et al.*, 2002).

With probiotics, a major issue is the delivery of viable organisms to the large intestine. In both laboratory and food processing experiments, HAM RS2 preserved the viability of probiotic strains (Brown *et al.*, 1998). Specific probiotic bacteria attach to the selected starch granules, suggesting a mechanism for enhanced viability because bacteria that adhere to a surface are more resistant to environmentally induced stresses. Probiotic protection was provided during passage through the gastrointestinal tract of mice, pigs, and rats (Brown *et al.*, 1997; Wang *et al.*, 2002; Le Leu *et al.*, 2005) and during food preparation and storage (Brown *et al.*, 1998). Synbiotics containing specific related RS and probiotic(s) provide an alternative delivery format for targeted health effects.

Colonic function

As described previously, many of the colonic effects of RS are mediated via the production of SCFAs during fermentation. One important role of the large intestine is to salvage water and minerals, which is also SCFA mediated. RS-rich ingredients have been found to be useful for inclusion in foods and preparations designed for alleviating the symptoms of diarrhea. In an Indian study HAM RS2 enhanced large bowel salvage of water for people suffering chronic diarrhea from cholera (Ramakrishna *et al.*, 2000). An oral rehydration solution containing RS lowered fecal loss relative to the oral rehydration solution alone within just 12–24 hours. Time to first stool formation was also shortened from 91 hours to 57 hours.

Mineral absorption can occur in both the small and large intestine. In the large intestine absorption is SCFA mediated because minerals are more soluble and more readily absorbed at a lower pH. In animal studies HAM RS2 increased apparent absorption of minerals, particularly calcium and magnesium (Andrieux and Sacquet, 1986; Demigne *et al.*, 1989; Lopez *et al.*, 2001), with Lopez *et al.* (2001) also reporting increased apparent absorption of zinc, iron, and copper. One study has reported an effect of HAM RS2 on mineral absorption in healthy adults. RS2 added to diets alone or in combination with the probiotic *Bifidobacterium lactis* improved absorption of magnesium by 15%, although the levels of absorption of other minerals such as calcium remained stable (Brown *et al.*, 2006). Additional confirmatory studies in humans are required.

An important function of the colonic mucosa is to act as a barrier and protect the body from harmful agents present in the colon. Mucin functions as a protective layer for the mucosa, restricting the adhesion and invasion of pathogenic bacteria. RS2 from HAM was protective in healthy and disease-associated rats by promoting a thicker mucus layer. In healthy rats, those consuming RS had increased colonic mucin density and reduced colonic permeability, affording greater protection (Morita *et al.*, 2004a). Dietary protein has been shown to promote colonic mucus thinning. When rats were fed high-protein diets of differing protein types, the mucus layer was thicker when the diets contained RS (Toden *et al.*, 2005a, 2006). In rats exposed to liver injury via a gut-derived endotoxin, mucin content was higher, with improved mucosal barrier function shown by lower endotoxin translocation when diets contained HAM RS2 (Morita *et al.*, 2004b).

Colonic tissue health

Fiber components have an important physiological role in reducing the impact of a variety of dietary and lifestyle assaults and stresses on our health. Protein is one ingredient that has been linked with higher risk for colorectal cancer. Initial studies that focused on RS2 from HAM suggest that RS may offer some protection. Rats fed RS2-containing diets with high levels of different types of protein – casein, whey, red meat – had less DNA damage, suggesting protection against potential genotoxic agents (Toden *et al.*, 2005a; 2005b, 2006) or an improvement in the biomarkers associated with improved colonic health (Muir *et al.*, 2004).

Apoptosis (controlled cell death) is another biomarker of colonic tissue health that is particularly useful in animal models to simulate the body's response to the initiation of colorectal cancer by a specific carcinogen and enable observations of the role of the diet in protecting against cancer. The SCFA butyrate has been associated with a protective role concerning the prevention of colorectal cancer and the consumption of HAM RS2 has been shown to elevate colonic butyrate concentrations. The ingestion of RS2 from HAM has resulted in significantly increased levels of apoptosis in a rat model that had been exposed to the colon-specific genotoxin azoxymethane (Le Leu *et al.*, 2002). Feeding RS2 to these rats increased the apoptotic index by more than 30% in a dose-dependent manner with no effect on cell proliferation in the genotoxin-exposed colonic cells (Le Leu *et al.*, 2003).

4.2.2 Energy management

Recent research has elucidated several physiological outcomes of RS that are relevant to energy management in at least four ways. The first is that RS has a lower caloric contribution compared with digestible glucose-based carbohydrates (e.g. flour) with lower glucose response and increased energy wastage (increased fecal nutrient excretion). Second, RS lowers insulin response, with increased insulin sensitivity. Third, RS lowers total and regional body fat accumulation, with lower adipocyte volume. Finally, RS is associated with increased lipid oxidation at the expense of carbohydrate oxidation, and decreased lipid production (lipogenesis). All of these physiological properties are important for weight management and the campaign to reduce global obesity.

Caloric contribution

Foods containing RS have a lower dietary caloric contribution because RS does not contribute to available glucose, and increases energy wastage (excretion). However, RS usually does contribute some energy because all the commercial RS ingredients currently available are not 100% resistant and contain some digestible starch. SCFAs produced from RS fermentation can either be utilized by the colonic microflora (providing no energy for the body) or be absorbed and utilized by the body (providing some energy for the body). Typically, the energy contribution of RS ingredients is approximately one-third lower than for digestible starches, but reported values will differ between ingredients, and will depend on the methodologies employed. The reported digestible energy contribution (dietary energy less fecal energy) of HAM RS2 was 67% of that for cornstarch (11.7 kJ/g), when measured in human subjects (Behall and Howe, 1996) and for HAM RS3 was 62% of that for wheat starch when measured in rats (Aust *et al.*, 2001). These studies did not separate energy contribution from glucose relative to that contributed by SCFAs. Others have calculated the energy value of RS alone, representing the energy contribution of SCFAs, but the values differ considerably due to the RS source.

Glucose and insulin response

One of the most consistent effects of RS is the ability to lower the blood glucose response. Many groups have reported a decreased postprandial glucose response when RS ingredients replace digestible glucose-based ingredients like flour (Higgins, 2004). This is simply because RS, being resistant to digestion, does not release glucose in the small intestine. In the large intestine the starch is utilized, with the end-products being gases and SCFAs.

The magnitude of the effect of RS on blood glucose response is particularly dependent on two considerations: firstly, the amount of RS in the RS ingredient, and second, the amount of the RS ingredient relative to other digestible carbohydrates present in the food and diet. Behall and Hallfrisch (2002) fed breads containing starches with varying amylose contents ranging from 30%

to 70%. They noted that breads made with higher amylose starches lowered the postprandial blood glucose response and the bread containing 70% amylose starch provided the greatest reduction. This is due to the higher amylose starches being less digestible and therefore contributing less to blood glucose levels and more to fermentation. Using a parallel approach, Brown *et al.* (1995) demonstrated that the amount of RS ingredient included in bread relative to other carbohydrates present is also an important consideration. White bread made with 5% replacement of flour from HAM RS2 had a relative glycemic impact of 95 vs. 100 for commercial white bread with no flour replacement. Bread with higher flour replacement levels had a marked response, with 10% and 20% flour replacement resulting in breads with relative glycemic responses of 74 and 53, respectively.

Insulin is arguably a more important marker of health than blood glucose concentration. In studies where RS ingredients replaced digestible starch most reported decreased postprandial insulin response (Higgins, 2004). Furthermore, not only lower insulin levels were observed, but also improved utilization of available insulin. Robertson *et al.* (2005) observed increased insulin sensitivity when HAM RS2 was fed over a 4-week period and this was further supported by measurements of higher glucose clearance in both adipose and muscle tissue. The amount of SCFAs may assist to increase insulin sensitivity by lowering free fatty acids (Higgins, 2004); this is supported by Robertson *et al.* (2005), who reported higher circulating SCFAs, higher adipose SCFA uptake, and lower adipose free fatty acid output.

Longer-term benefits of RS on insulin sensitivity have been demonstrated in rat studies. For example, when Byrnes *et al.* (1995) fed rats diets containing either digestible waxy cornstarch or HAM RS2, the rats fed the digestible starch diets developed insulin resistance within 8 weeks. By 12 weeks the insulin response to a glucose challenge for the digestible starch-fed rats was double that of the rats fed the RS diets. Whereas most glycemic/insulin response studies are acute, conducted with only 2-hour postprandial measurements, this study suggests longer-term benefits of RS.

Body composition and lipid storage

Body composition and regional fat deposition are important indicators of chronic disease risk. HAM RS2 lowers body fat levels when replacing digestible carbohydrate (Pawlak *et al.*, 2004), with increased lean body mass (Kiriyama, *et al.*, 1996). More specifically, regional changes in body fat distribution occurred, with reduced epididymal and retroperitoneal fat (Pawlak *et al.*, 2001, 2004). Changes in lipid storage are reflected by lipid metabolism pathways.

Metabolism

Effects of RS on metabolism have been observed both when RS ingredients have replaced digestible carbohydrates, and when the amount of RS/DF is controlled between test diets. Lipogenesis is lower in adipocytes, particularly fatty acid synthase activity, and adipocytes are smaller when diets contain RS (Kabir *et al.*, 1998a, 1998b). A link is now being established between colonic fermentation and metabolism via SCFAs, indicating a potential role for those SCFAs produced during RS fermentation. SCFAs are reported to inhibit adipose tissue lipolysis (Robertson *et al.*, 2003) and in the liver acetate was proposed to inhibit glycogenolysis, with the effect of sparing carbohydrate and increasing lipid oxidation (Higgins, 2004). Propionate appears to have insulin-like effects by stimulating glycolysis, activating glycogen synthase, and reducing gluconeogenesis (Robertson *et al.*, 2003), and has been associated with reduced hepatic cholesterol synthesis (Lopez *et al.*, 2001). Future research will elucidate the nature of the link between colonic fermentation and metabolism.

Changes in metabolic substrate selection can also be demonstrated by measuring the respiratory quotient (RQ), which represents the balance between carbohydrate and lipid oxidation. Studies with RS have demonstrated a shift towards increased lipid oxidation. In particular, Higgins *et al.* (2004) observed a reduction in RQ within just two hours of consuming a meal containing HAM RS2, and the RQ continued to be lower than for the meal with no added RS for up to 6 hours. This was associated with increased total and meal lipid oxidation when meals contained RS (5.4 g) with a 23% increase in meal lipid oxidation during a 24-hour period after the consumption of the test meal.

4.2.3 Resistant starch and nutrient interactions

While the reductionist approach to nutrition research assists in gaining a better understanding of the physiological effects of individual ingredients, people eat foods, and foods are multi-nutrient systems. Therefore, it is particularly important for food developers to understand the health potential of nutrient interactions. Synergies have been observed between RS ingredients and other indigestible carbohydrates (Table 4.4). These synergies include the combined benefits of multiple fibers with differing physiological effects and the ability to regulate the site of fermentation at positions along the large intestine.

Wheat bran

Wheat bran is well known for its fecal bulking and laxation effects; however wheat bran is not well fermented within the colon and therefore, does not represent a good nutrient source to colonic bacteria. Combining RS2 with wheat bran further enhanced the laxation effects of wheat bran and broadened its potential protective effects as measured by improvements in fermentation and cytotoxic biomarkers (Govers *et al.*, 1999; Le Leu *et al.*, 2002; Muir *et al.*, 2004). Digestive health benefits occurred both proximally and distally, which is important because colon cancer is more prevalent in the distal region where fermentable nutrient supply is typically limited.

Table 4.4 Nutrient interactions with high-amylose corn RS2 ingredients

Co-nutrients	Model	Observations									Reference
		F weight, C contents	C weight	F/C pH	F/C SCFA	F/C butyrate	F/C ammonia	F/C phenols	F/C microflora	Colonic apoptosis	
RS2 + wheat bran vs. wheat bran alone	Human	↑		↓	↑	↑	↓	↓			Muir *et al.*, 2004
RS2 + wheat bran vs. wheat bran alone	Rat	↑	↑	↓	↑	↑					Le Leu *et al.*, 2002
RS2 + wheat bran vs. wheat bran alone	Pig	↑	↑	↓	↑	↑					Govers *et al.*, 1999
RS2 + cellulose vs. cellulose alone	Rat	↑	↑	↓	↑	↑					Le Leu *et al.*, 2002
RS2 + psyllium vs. psyllium alone	Rat	↑	↑	↓	↑	↑					Morita *et al.*, 1999
RS2 + FOS vs. FOS alone	Pig								↑ bifido-bacteria		Brown *et al.*, 1998
RS2 + resistant protein vs. resistant protein alone	Rat	↑	↑	↓	↑	↑					Morita *et al.*, 1998
RS2 + probiotic vs. probiotic alone	Rat			↓	↑	↑			↑ Bifido-bacteria Lactobacilli	↑	Le Leu *et al.*, 2005

F, feces; C, cecum; F/C, feces and/or cecum; FOS, Fructo-oligosaccharides.

Cellulose
Similar to wheat bran, cellulose is better known for its fecal bulking than fermentation effects. Combining RS2 with cellulose further enhanced the laxation effects of cellulose and broadened its potential protective effects, with improvements in fermentation biomarkers (Le Leu *et al.*, 2002). Benefits occurred both proximally and distally.

Psyllium
RS2 from HAM and psyllium work together to promote laxation and fermentation (Morita *et al.*, 1999). The benefits occur both proximally and distally and this concept is being used in bulk laxative preparations.

Fructo-oligosaccharide
Fructo-oligosaccharides (FOS) are established prebiotic ingredients. In a pig model, FOS promoted bifidobacteria, but counts were further increased when FOS was combined with HAM RS2 (Brown *et al.*, 1998).

Protein
As with starch, some protein can also reach the large intestine. Known as resistant protein (RP), it can also be fermented; however, cytotoxic by-products such as ammonia and phenols can be produced. Combining RS with RP increased carbohydrate-related fermentation outcomes including significant increases in SCFA production, in particular butyrate (Morita *et al.*, 1998). In another study where both carbohydrate and protein fermentation outcomes were measured, carbohydrate fermentation (as shown by SCFAs) was increased relative to protein fermentation (as shown by urinary phenols) when both RS and RP were fed in diets to rats (Le Leu *et al.*, 2007). This shift in carbohydrate relative to protein fermentation could help to explain the benefits reported by Toden *et al.* (2005b) of including RS in high-protein diets. High-protein diets could result in higher levels of RP reaching the large intestine.

Probiotics
Probiotics are added to foods for their selective benefits on digestive health which are dependent on the probiotic genus and strain. Combining RS with the probiotic *Bifidobacterium lactis* could broaden potential health effects of the probiotic to include fermentation-related events. When RS2 from HAM and *Bifidobacterium lactis* were fed to rats, the synbiotic effectively promoted colonic apoptosis following exposure to a genotoxin at levels not previously attainable by either ingredient alone, and approximately 50% more than observed for the RS ingredient alone (Le Leu *et al.*, 2005).

4.3 Resistant starch for food development

RS-rich ingredients have been available commercially since September 1993, when the HAM RS2 called Hi-maize™ was first launched. Since that time, other RS ingredients have also become available representing different RS types, bases, and chemistries. Each RS ingredient will have its own unique combination of technical and nutritional functionalities, so both should be considered simultaneously when selecting ingredients. Furthermore, some RS ingredients will have more supporting scientific evidence available, which should also be considered when comparing and evaluating ingredients.

The first processed food to contain a commercial RS-rich ingredient was a white bread launched in Australia in April 1994 (Brown, 2004). The RS ingredient called Hi-maize™ increased the DF content of this bread from 2.9% up to 5.6%, enabling the claim 'more dietary fiber than multigrain bread' (Brown *et al.*, 1995). This bread has had continued market success because it has capitalized on the ability of RS ingredients to deliver both technical and nutritional functionality, one aspect of which is to deliver DF. Bread has continued as one of the most common vehicles for the delivery of RS ingredients, with the RS ingredient typically included on a flour replacement basis. Other breads are available that make use of the glycemic response lowering effects of RS ingredients.

RS ingredients have now been incorporated into many other commercial foods and preparations in Asia, Australia, North America, and Europe. RS2 from HAM has found success in a wide range of general purpose foods including baked goods, breakfast cereals, pasta and noodles, and confectionery. Table 4.2 lists the RS content of some commercial foods containing RS2 ingredients, with their RS content measured by AOAC Method 2002.02 (Brown, 2004). The RS content of white bread was almost five-fold higher with the RS2 ingredient, 4.3% vs. 0.9%. Marketing claims relate to fiber, glycemic response, prebiotics, and digestive health. The technical properties that are ideal for these applications are that the product should (Brown, 2004):

- be natural;
- be a source of DF and RS;
- be white;
- have fine particle size (average particle size is <10 μm);
- have high gelatinization temperature;
- have low water absorption compared with some other DFs;
- extrude well;
- have film forming properties e.g. to improve coating crispiness;
- have low bulk density relative to other fibers;
- extend breakfast cereal bowl life.

Tolerance is an important aspect of DF ingredients and RS appears to be well tolerated. Up to 100 g of an RS2 ingredient were consumed each day for one week by subjects consuming the RS2 in cookies and bread (Kendall *et al.*, 2003). No differences in bloating, abdominal pain, and fecal consistency were noted between the control and high-RS2 groups. It is likely that tolerance will differ between RS ingredients. RS2 from HAM is fermented along the entire length of the large intestine, which probably contributes to its high tolerance.

Resistant starch for individuals with special dietary needs
RS ingredients are also useful for special-purpose foods. For texture-modified foods in particular, the combination of technical plus nutritional functionality is highly relevant to product development. HAM RS2 has been included in:

- liquid meal replacements, as a source of fiber in foods where the nutrients are required without the bulk of conventional foods;
- texture-modified foods for people with swallowing difficulty, where RS can contribute insoluble fiber without the fibrous texture or water binding attributable to other insoluble fibers;
- foods for gluten intolerance, as RS ingredients can be made from various starch bases that include corn and tapioca;
- foods for ulcerative colitis, because RS has been shown to improve mucosal healing;
- pharmaceutical applications for laxation, particularly as fiber blends where RS has enhanced the laxative properties of established bulking fibers.

4.4 Conclusion

Dietary starch is being re-assessed for its contribution to health. Not only does digestible starch provide energy via traditional metabolic pathways, but some starch sub-types, collectively called RS, can contribute to health via fermentation-related pathways. A significant body of information has been gathered over the past 30 years to support RS as both a technical and nutritional food ingredient. Consumer research has highlighted the need for functional foods that deliver on taste, quality, convenience, and nutrition, and RS provides this opportunity.

4.4.1 Future research

There are still many unexplored and unanswered questions relating to RS.

- Traditional RS research studies investigate the role of isolated ingredients, from a reductionist perspective, looking largely at associative effects (i.e. cause and effect) via intervention studies. There is a need to understand RS–nutrient interactions and their synergistic effects for health.
- Epidemiological studies are lacking, largely due to methodological and analytical issues. For example, RS is not included in the US Department of Agriculture Nutrient Composition Database, and is not included in the dietary analysis of major prospective studies.
- Long-term clinical intervention studies using RS are lacking. For example, few randomized controlled clinical trials have been conducted for more than 4 weeks.

- Exploratory, mechanistic research is required to understand at the cellular level how RS can affect metabolic processes.
- A 'systems biology'-type approach is required, to understand the role of fermentation on whole body health. Through nutrigenomic and metabolomic research, the technology is now available to elucidate the mechanisms whereby RS can enhance health.

4.4.2 Where to find more information on resistant starch

The number of research articles and reviews on RS continues to grow as more scientific evidence becomes available, and interest in RS as a nutritionally functional ingredient continues to expand. Below is a list of some published reviews that cover both general and specific aspects of RS.

General reviews include the following:

- Sajilata MG, Singhal RS and Kulkarni PR. Resistant starch – a review. *Comprehensive Reviews in Food Science and Food Safety* 2006; **5**: 1–17.
- British Nutrition Foundation – Nugent AP. Health properties of resistant starch. *Nutrition Bulletin* 2005; **30**: 27–54.
- Commonwealth Scientific and Industrial Research Organisation – Baghurst PA, Baghurst KI and Record SJ. Dietary fibre, non-starch polysaccharides and resistant starch. A review. *Food Australia* 1996; **48**: S1–S36.

Specific reviews include the following:

- Brown IL, Yotsuzuka M, Birkett A and Henriksson A. Prebiotics, synbiotics and resistant starch. *Journal of Japanese Association for Dietary Fiber Research* 2006; **10**: 1–9.
- Association of Official Analytical Chemists. *Journal of AOAC International*, Special Guest Editor Section. 2004; **87**: 682–796.
- Bird AR, Brown IL and Topping DL. Starch, resistant starch, the gut microflora and human health. *Current Issues in Intestinal Microbiology* 2000; **1**: 25–37.
- Topping DL and Clifton PM. Short-chain fatty acids and human colonic function: roles of resistant starch and nonstarch polysaccharides. *Physiological Reviews* 2001; **81**: 1031–1064.

4.5 References

AACC (American Association of Cereal Chemists). The definition of dietary fiber. *Cereal Foods World* 2001; **46**: 112–126.

Andrieux C and Sacquet E. Effects of amylomaize starch on mineral metabolism in the adult rat: role of the microflora. *Journal of Nutrition* 1986; **116**: 991–998.

Asp NG. Resistant starch. Proceedings from the second plenary meeting of EURESTA: European FLAIR Concerted Action No. 11 on physiological implications of the consumption of resistant starch in man. *European Journal of Clinical Nutrition* 1992; **46**(Suppl 2): S1.

Aust L, Dongowski G, Frenz U, Taufel A and Noack R. Estimation of available energy of dietary fibers by indirect calorimetry in rats. *European Journal of Nutrition* 2001; **40**: 23–29.

Baghurst PA, Baghurst KI and Record SJ. Dietary fibre, non-starch polysaccharides and resistant starch. A review. *Food Australia* 1996; **48**: S1–S36.

Behall KM and Hallfrisch J. Plasma glucose and insulin reduction after consumption of breads varying in amylose content. *European Journal of Clinical Nutrition* 2002; **56**: 913–920.

Behall KM and Howe JC. Resistant starch as energy. *Journal of the American College of Nutrition* 1996; **15**: 248–254.

Bird AR, Brown IL and Topping DL. Starch, resistant starch, the gut microflora and human health. *Current Issues in Intestinal Microbiology* 2000; **1**: 25–37.

Birkett A, Muir JG, Phillips J, Jones G and O'Dea K. Resistant starch lowers fecal concentrations of ammonia and phenols in humans. *American Journal of Clinical Nutrition* 1996; **63**: 766–772.

Brighenti F, Casiraghi MC and Baggio C. Resistant starch in the Italian diet. *British Journal of Nutrition* 1998; **80**: 333–341.

Brown IL. Applications and uses of resistant starch. *Journal of AOAC International* 2004; **87**: 727–732.

Brown I, Conway P and Topping D. The health potential of resistant starches in foods. An Australian perspective. *Scandinavian Journal of Nutrition* 2000; **44**: 53–58.

Brown IL, McNaught KJ, Andrews D and Morita T. Resistant starch: plant breeding, applications development and commercial use. Chapter 34. In: *Advanced Dietary Fiber Technology*. Eds: BV McCleary and L Prosky. Blackwell Science, Oxford, UK. 2001. pp. 401–412.

Brown IL, McNaught KJ and Moloney E. Hi-maize: new directions in starch technology and nutrition. *Food Australia* 1995; **46**: 272–275.

Brown IL, Wang X, Topping DL, Playne MJ and Conway PL. High amylose maize starch as a versatile prebiotic for use with probiotic bacteria. *Food Australia* 1998; **50**: 603–610.

Brown I, Warhurst M, Arcot J, Playne M, Illman RJ and Topping DL. Fecal numbers of *Bifidobacteria* are higher in pigs fed *Bifidobacterium longum* with a high amylose cornstarch than with a low amylose cornstarch. *Journal of Nutrition* 1997; **127**: 1822–1827.

Brown IL, Yotsuzuka M, Birkett A and Henriksson A. Prebiotics, synbiotics and resistant starch. *Journal of Japanese Association for Dietary Fiber Research* 2006; **10**: 1–9.

Byrnes SE, Brand Miller JC and Denyer GS. Amylopectin starch promotes the development of insulin resistance in rats. *Journal of Nutrition* 1995; **125**: 1430–1437.

Cassidy A, Bingham SA and Cummings JH. Starch intake and colorectal cancer risk: an international comparison. *British Journal of Cancer* 1994; **69**: 937–942.

Demigne C, Levrat M-A and Remesy C. Effects of feeding fermentable carbohydrates on the cecal concentrations of minerals and their fluxes between the cecum and blood plasma in the rat. *Journal of Nutrition* 1989; **119**: 1625–1630.

Dysseler P and Hoffem D. Estimation of resistant starch intake in Europe. Proceedings of the concluding plenary meeting of EURESTA. Including the final reports of the working groups. FLAIR Food-Linked Agro-Industrial Research. Eds: NG Asp, JMM van Amelsvoort and JGAJ Hautvast. 1994. pp. 84–86.

Elmstahl HL. Resistant starch content in a selection of starchy foods on the Swedish market. *European Journal of Clinical Nutrition* 2002; **56**: 500–505.

Englyst HN, Kingman SM and Cummings JH. Classification and measurement of nutritionally important starch fractions. *European Journal of Clinical Nutrition* 1992; **46** Suppl 2:S33–S50.

FAO/WHO (Food and Agriculture Organization/World Health Organization). Codex Alimentarius Commission, 2005. Report of the 26th session of the Codex committee on nutrition and foods for special dietary uses, Bonn, Germany, November 2004.
Gelski J. More revelations on resistant starch. *Food Business News* 2006; 7 March: 50–52.
Govers MJAP, Gannon NJ, Dunshea FR, Gibson PR and Muir JG. Wheat bran affects the site of fermentation of resistant starch and luminal indexes related to colon cancer risk: a study in pigs. *Gut* 1999; **45**: 840–847.
Grubben MJAL, van den Braak CCM, Essenberg M, Olthof M, Tangerman A, Katan MB and Nagengast FM. Effect of resistant starch on potential biomarkers for colonic cancer risk in patients with colonic adenomas. A controlled trial. *Digestive Diseases and Sciences* 2001; **46**: 750–756.
Higgins JA. Resistant starch: metabolic effects and potential health benefits. *Journal of AOAC International* 2004; **87**: 761–768.
Higgins JA, Higbee DR, Donahoo WT, Brown IL, Bell ML and Bessesen DH. Resistant starch consumption promotes lipid oxidation. *Nutrition and Metabolism* 2004; **1**: 8–18.
Hylla S, Gostner A, Dusel G, Anger H, Bartram H-P, Christl SU, Kasper H and Scheppach W. Effects of resistant starch on the colon in healthy volunteers: possible implications for cancer prevention. *American Journal of Clinical Nutrition* 1998; **67**: 136–142.
IOM (Institute of Medicine of the National Academies). Dietary, functional and total fiber. Chapter 7. In: *Dietary Reference Intakes: Energy, Carbohydrates, Fiber, Fat, Fatty Acids, Cholesterol, Protein, and Amino Acids (Macronutrients).* National Academies Press, Washington DC, USA. 2002.
Jenkins DJ, Vuksan V, Kendall CWC, Wursch P, Jeffcoat R, Waring S, Mehling CC, Vidgen, E, Augustin LSA and Wong E. Physiological effects of resistant starches on fecal bulk, short chain fatty acids, blood lipids and glycemic index. *Journal of the American College of Nutrition* 1998; **17**: 609–616.
Kabir M, Rizkalla SW, Champ M, Luo J, Boillot J, Bruzzo F and Slama G. Dietary amylose-amylopectin starch content affects glucose and lipid metabolism in adipocytes of normal and diabetic rats. *Journal of Nutrition* 1998a; **128**: 35–43.
Kabir M, Rizkalla SW, Quignard-Boulange A, Guerre-Millo M, Boillot J, Ardouin B, Luo J and Slama G. A high glycemic index starch diet affects lipid storage-related enzymes in normal and to a lesser extent in diabetic rats. *Journal of Nutrition* 1998b; **128**: 1878–1883.
Kendall CWC, Jenkins DJA and Emam A. Assesment of resistant starch tolerance: a dose response study. Abstract presented at the 9th European Nutrition Conference, October 2003, Rome, Italy.
Kiriyama, S. Physiological function of resistant starch and its use. *Up-to-date Food Processing* 1996; **31**(2): 34–38.
Le Leu RK, Brown IL, Hu Y, Bird AR, Jackson M, Esterman A and Young GP. A synbiotic combination of resistant starch and *Bifidobacterium lactis* facilitates apoptotic deletion of carcinogen-damaged cells in rat colon. *Journal of Nutrition* 2005; **135**: 996–1001.
Le Leu RK, Brown IL, Hu Y, Morita T, Esterman A and Young GP. Effect of dietary resistant starch and protein on colonic fermentation and intestinal tumourigenesis in rats. *Carcinogenesis* 2007; **28**: 240–245.
Le Leu RK, Brown IL, Hu Y and Young GP. Effect of resistant starch on genotoxin-induced apoptosis, colonic epithelium, and lumenal contents in rats. *Carcinogenesis* 2003; **24**: 1347–1352.
Le Leu RK, Hu Y and Young GP. Effects of resistant starch and nonstarch polysaccharide on colonic luminal environment and genotoxin-induced apoptosis in the rat. *Carcinogenesis* 2002; **23**: 713–719.
Lopez HW, Levrat-Verny M-A, Coudray C, Besson C, Krespine V, Messager A, Demigne C and Remesy C. Class 2 resistant starches lower plasma and liver lipids and improve mineral retention in rats. *Journal of Nutrition* 2001; **131**: 1283–1289.

Mercier C, Charbonniere R, Gallant D and Guilbot A. Structural modification of various starches by extrusion cooking with a twin-screw French extruder. Chapter 10. In: *Polysaccharides in Food*. Eds: JMV Blanshard and JR Mitchell. Butterworths, London. 1979. pp. 153–170.

Morita T, Kasaoka S, Hase K and Kiriyama S. *Psyllium* shifts the fermentation site of high-amylose cornstarch toward the distal colon and increases fecal butyrate concentration in rats. *Journal of Nutrition* 1999; **129**: 2081–2087.

Morita T, Kasaoka S, Oh-hashi A, Ikai M, Numasaki Y and Kiriyama S. Resistant proteins alter cecal short-chain fatty acid profiles in rats fed high amylose cornstarch. *Journal of Nutrition* 1998; **128**: 1156–1164.

Morita T, Tanabe H, Sugiyama K, Kasaoka S and Kiriyama S. Dietary resistant starch alters the characteristics of colonic mucosa and exerts a protective effect on trinitrobenzene sulfonic acid-induced colitis in rats. *Bioscience, Biotechnology and Biochemistry* 2004a; **68**: 2155–2164.

Morita T, Tanabe H, Takahashi K and Sugiyama K. Ingestion of resistant starch protects endotoxin influx from the intestinal tract and reduces D-galactosamine-induced liver injury in rats. *Journal of Gastroenterology and Hepatology* 2004b; **19**: 303–313.

Muir JG and O'Dea K. Measurement of resistant starch: factors affecting the amount of starch escaping digestion in vitro. *American Journal of Clinical Nutrition* 1992; **56**: 123–127.

Muir JG, Walker KZ, Kaimakamis MA, Cameron MA, Govers MJAP, Lu ZX, Young GP and O'Dea K. Modulation of fecal markers relevant to colon cancer risk: a high-starch Chinese diet did not generate expected beneficial changes relative to a Western-type diet. *American Journal of Clinical Nutrition* 1998; **68**: 372–379.

Muir JG, Yeow EGW, Keogh J, Pizzey C, Bird AR, Sharpe K, O'Dea K and Macrae FA. Combining wheat bran with resistant starch has more beneficial effects of fecal indexes than does wheat bran alone. *American Journal of Clinical Nutrition* 2004; **79**: 1020–1028.

Murphy MM and Douglass JS. Resistant starch intakes in the United States. Experimental Biology 2007 Meeting Abstracts. 2007. p. A315, No. 533.8.

Noakes M, Clifton PM, Nestel PJ, Le Leu R and McIntosh G. Effect of high-amylose starch and oat bran on metabolic variables and bowel function in subjects with hypertriglyceridemia. *American Journal of Clinical Nutrition* 1996: **64**; 944–951.

Nugent AP. Health properties of resistant starch. *Nutrition Bulletin* 2005; **30**: 27–54.

Pawlak DB, Bryson JM, Denyer GS and Brand-Miller JC. High glycemic index starch promotes hypersecretion of insulin and higher body fat in rats without affecting insulin sensitivity. *Journal of Nutrition* 2001; **131**: 99–104.

Pawlak DB, Kushner JA and Ludwig DS. Effects of dietary glycemic index on adiposity, glucose homoeostasis, and plasma lipids in animals. *Lancet* 2004; **364**: 778–785.

Phillips J, Muir JG, Birkett A, Lu ZX, Jones GP and O'Dea K. Effect of resistant starch on fecal bulk and fermentation-dependent events in humans. *American Journal of Clinical Nutrition* 1995; **62**: 121–130.

Platel K and Shurpalekar KS. Resistant starch content of Indian foods. *Plant Foods for Human Nutrition* 1994; **45**: 91–95.

Priestley RJ. Studies on parboiled rice. I. Comparison of the characteristics of raw and parboiled rice. *Food Chemistry* 1976; **1**: 5–14.

Ramakrishna BS, Venkataraman S, Srinivasan P, Dash P, Young GP and Binder HJ. Amylase-resistant starch plus oral rehydration solution for cholera. *New England Journal of Medicine* 2000; **342**: 308–313.

Roberts J, Jones GP, Rutishauser IHE, Birkett A and Gibbons C. Resistant starch in the Australian diet. *Nutrition and Dietetics* 2004; **61**: 98–104.

Robertson MD, Bickerton AS, Dennis AL, Vidal H and Frayn KN. Insulin-sensitizing effects of dietary resistant starch and effects on skeletal muscle and adipose tissue metabolism. *American Journal of Clinical Nutrition* 2005; **82**: 559–567.

Robertson MD, Currie JM, Morgan LM, Jewell DP and Frayn KN. Prior short-term consumption of resistant starch enhances postprandial insulin sensitivity in healthy subjects. *Diabetologia* 2003; **46**: 659–665.

Sajilata MG, Singhal RS and Kulkarni PR. Resistant starch – a review. *Comprehensive Reviews in Food Science and Food Safety* 2006; **5**: 1–17.

Toden S, Bird AR, Topping DL and Conlon MA. Differential effects of dietary whey and casein on colonic DNA damage in rats. *Australian Journal of Dairy Technology* 2005a; **60**: 44–46.

Toden S, Bird AR, Topping DL and Conlon MA. Resistant starch attenuates colonic DNA damage induced by higher dietary protein in rats. *Nutrition and Cancer* 2005b; **51**: 45–51.

Toden S, Bird AR, Topping DL and Conlon MA. Resistant starch prevents colonic DNA damage induced by high dietary cooked red meat or casein in rats. *Cancer Biology and Therapy* 2006; **5**: e1–e6.

Tomlin J and Read NW. The effect of resistant starch on colon function in humans. *British Journal of Nutrition* 1990; **64**: 589–595.

Topping DL and Clifton PM. Short-chain fatty acids and human colonic function: roles of resistant starch and nonstarch polysaccharides. *Physiological Reviews* 2001; **81**: 1031–1064.

Van Munster IP, Tangerman A and Nagengast FM. Effect of resistant starch on colonic fermentation, bile acid metabolism, and mucosal proliferation. *Digestive Diseases and Sciences* 1994; **39**: 834–842.

Walker KZ, Birkett A, Lu ZX, Jones G, O'Dea K and Muir JG. Development of a simulated Australian diet for adults which may have use as a research tool. *Australian Journal of Nutrition and Dietetics* 1997; **54**: 190–197.

Wang X, Brown IL, Khaled D, Mahoney MC, Evans AJ and Conway PL. Manipulation of colonic bacteria and volatile fatty acid production by dietary high amylose maize (amylomaize) starch granules. *Journal of Applied Microbiology* 2002; **93**: 390–397.

Wang X, Conway PL, Brown IL and Evans AJ. In vitro utilization of amylopectin and high-amylose maize (amylomaize) starch granules by human colonic bacteria. *Applied and Environmental Microbiology* 1999; **65**: 4848–4854.

Zobel HF. Starch granule structure. In: *Developments in Carbohydrate Chemistry*. Eds: RJ Alexander and HF Zobel. AACC, St Paul, Minnesota. 1992. pp. 1–36.

5

Micronutrients in cereal products: their bioactivities and effects on health

A. Kamal-Eldin, Swedish University of Agricultural Sciences, Sweden

5.1 Introduction

Cereals represent an important food category as they contribute a large portion of our daily calorie supply. Besides energy, cereal products are important for nutrition because of their contents of dietary fiber and a wide range of micronutrients and bioactive components including minerals, vitamins, antioxidants, and other bioactive compounds. Epidemiological studies have associated the consumption of whole grain cereal products with reduced incidence of chronic diseases such as diabetes, cardiovascular disease, and certain cancers. Although the compounds responsible for these health benefits and their mode(s) of action have not been identified yet, they might be related, at least partly, to the different minerals, vitamins, and bioactive compounds present in these grains.

Cereals, mostly wheat and rice, are consumed as refined products after milling. Rice milling involves dehulling to produce brown rice that is then milled to remove the bran, which is about 10% of the kernel (Ha *et al.*, 2006). During milling of wheat to produce sifted white flour, about 30% of the wheat kernel is removed (70% extraction rate). It has long been recognized that health-promoting minerals and vitamins are concentrated in the bran and germ of cereal grains (Table 5.1), while the endosperm, the main component of refined products, is virtually devoid of them. Bioactive compounds including antioxidants and phytoestrogens have also been identified in the germ or bran adding an extra nutritional value to these tissues (Craig, 1997). Thus, the refining of whole grain flours to sifted white flours depletes the grain products of several nutritional components (Adrian & Petit, 1970). Although cereal products may contain adequate levels of minerals and vitamins, their

Table 5.1 Approximate proportions (%) of bran and germ fractions in cereal grains

Grain	Bran	Germ
Wheat	14	2.5
Rye	12	3
Barley	30	3
Oats	30	5
Brown rice	7	2.5
Corn	5.5	10.5
Buckwheat	8	4
Millet	8	17

Source: Franz and Sampson (2006).

bioavailability may be low due to the presence of antinutritional factors or matrix effects. Moreover, bioavailability of micronutrients can be influenced by co-occurring antinutrients, technological processing methods, as well as by co-administered inhibitors and enhancers.

This chapter discusses in detail the health effects of whole grain cereal products and associated micronutrients and the levels and bioavailability of nutritional minerals, vitamins, and other phytochemicals, and how milling and processing affect them. Addition of nutrients to cereal products including biofortification are discussed and future trends in relation to micronutrients are briefly reviewed.

5.2 Health effects of naturally occurring micronutrients in cereal products

There are an ample number of studies (Table 5.2) that associate the consumption of whole grain cereals with better metabolic performance and lower incidence of obesity, diabetes, inflammation, cardiovascular diseases, and some types of cancer (Jacobs *et al.*, 1995, 1998, 1999; Slavin *et al.*, 1997, 1999; Pereira *et al.*, 1998, 2001, 2002; Liu *et al.*, 1999, 2000, 2003; Hallfrisch & Behall, 2000; Meyer *et al.*, 2000; Nicodemus *et al.*, 2001; Fung *et al.*, 2002; McKeown *et al.*, 2002; Truswell, 2002; Liese *et al.*, 2003; Montonen *et al.*, 2003; Murtaugh *et al.*, 2003; Sahyoun *et al.*, 2006; Seal, 2006). Epidemiological and experimental animal studies suggest that dietary fiber, a complex mixture of components, exerts protective effects on colon cancer (Jacobs, 1988; Howe *et al.*, 1992; Vargas & Alberts, 1993). The protective effect of dietary fiber may entail a number of mechanisms *inter alia* binding of bile salts, increasing fecal water, diluting carcinogens, and decreasing transit time (Reddy *et al.*, 1989). In addition, fiber enhances bacterial fermentation and production of anticarcinogenic short-chain fatty acids, e.g. butyrate (Jacobs, 1986). Soluble fibers, such as β-glucan of oats and barley, are known for their cholesterol-

Table 5.2 Summary of selected studies on wholegrain intake and health

Study	Subjects (*n*; age (years); % females)	Study design	Results
Jacobs *et al*., 1998	64 492; 55–69; 100%	Prospective cohort	Whole grain intake associated with low body mass index and coronary heart disease risk
Pereira *et al*., 1998	3627; 18–30; 54%	Prospective cohort	Whole grain intake associated with low body mass index and fasting insulin levels
Liu *et al*., 1999	75 521; 36–63; 100%	Prospective cohort	Whole grain intake associated with low coronary heart disease risk
Liu *et al*., 2000	75 521; 36–63; 100%	Prospective cohort	Whole grain intake lowers the risk of type 2 diabetes
Meyer *et al*., 2000	35 988; 55–69; 100%	Prospective cohort	Whole grain intake lowers the risk of type 2 diabetes
Fung *et al*., 2002	42 898; 40–75; 0%	Prospective cohort	Whole grain intake lowers the risk of type 2 diabetes
Van Dam *et al*., 2002	42 504; 45–75; 0%	Prospective cohort	Whole grain intake lowers the risk of type 2 diabetes
McKeown *et al*., 2002	2,941; 54–59; 55%	Cross-sectional	Whole grain intake associated with low body mass index and fasting insulin levels
Pereira *et al*., 2002	11; 25–56; 55%	Intervention study	Whole grain intake lowers fasting insulin
Liu *et al*., 2003	74 091; 38–63; 100%	Prospective cohort	Whole grain intake correlated negatively with weight gain
Montonen *et al*., 2003	4316; 40–69; 47%	Prospective cohort	Whole grain intake lowers the risk of type 2 diabetes
Liese *et al*., 2003	978; 40–69; 55%	Prospective cohort	Whole grain intake increases insulin sensitivity

lowering potential (Brown *et al.*, 1999). Although fiber is an important component of whole grains that contributes to their perceived health effects, other micronutrients including vitamins, phenolic antioxidants, and phytoestrogens seem also to contribute to these health effects (Slavin, 2003). It is most probable that the health effects of whole grain cereals are due to combined effects of dietary fiber and micronutrients.

The obligatory need for minerals and vitamins is beyond question and the contribution of cereal products to vitamin intake is appreciable even if not exactly quantified. The importance of cereals as sources of dietary B- and E-vitamins is often manifested. Notably, cereal products are especially important as sources of tocotrienols for most people not consuming high levels of palm oil. Tocotrienols, especially γ- and δ-tocotrienols, are reported to inhibit hydroxymethylglutaric acid-coenzyme A (HMG-CoA) reductase activity inducing cholesterol-lowering effects (Qureshi *et al.*, 1986, 1989, 1991a,b; Parker *et al.*, 1993; Wang *et al.*, 1993; Raederstorff *et al.*, 2002). They were also reported to reduce endothelial expression of adhesion molecules and adhesion to monocytes (Theriault *et al.*, 2002) and atherosclerotic lesions in ApoE-deficient mice (Qureshi *et al.*, 2001). Tocotrienols may also inhibit growth and proliferation of human breast cancer cells (Nesaretnam *et al.* 1998). Another component of cereals known for cholesterol-lowering properties is γ-oryzanol of rice bran oil (Hegsted *et al.*, 1993; Sugano & Tsuji, 1997).

Oxidative stress is currently recognized as a major factor in inflammation and the above-mentioned chronic diseases. Among minerals, copper and zinc are important constituents of the antioxidant enzyme superoxide dismutase (SOD), an endogenous enzyme that plays an important defensive role against oxidative stress. Vitamin E and other phenolic compounds serve as oxygen radical-scavenging compounds and play major roles in the body as antioxidants protecting cells and cell membranes from oxygen radical attack. A wide range of other phenolic compounds with antioxidant properties and beneficial physiological effects are present in cereal grains. Folates and vitamin B12 are required for the metabolism of homocysteine to methionine, thus preventing its negative contribution to oxidative stress. Cereal grains are rich in a wide range of phenolic antioxidants including hydroxycinnamic acids and derivatives (Graf, 1992; Nardini *et al.*, 1995; Meyer & Frankel, 2001). Phytic acid in cereal grains plays a 'Dr Jekyll, Mr Hyde' role acting as an antinutritional factor by inhibiting mineral bioavailability and as a potential antioxidant (Graf & Eaton, 1990; Zhou & Erdman, 1995; Owen *et al.*, 1998; Slavin *et al.*, 1999; Oatway *et al.*, 2001).

Currently, special interest is being paid to dietary sources of methyl group donors, mainly betaine, in the diet. Betaine – also known as trimethylglycine, glycine betaine, lycine, and oxyneurine – is a zwitterionic quaternary ammonium compound (Craig, 2004). Ingestion of betaine and other methyl donors reduces body levels of homocysteine transforming it to methionine and *N,N*-dimethylglycine. The enzyme betaine homocysteine methyltransferase (E.C. 2.1.1.5) helps in transferring a methyl group from glycine betaine to homocysteine (Malinow, 1994). Whole grain wheat products are rich in betaine, ~600 mg/kg, and enriched in wheat bran at ~5 g/kg, mainly as glycine betaine, i.e. CH_3-$N^+(CH_3)_2$-CH_2COO^- (Slow *et al.*, 2005).

5.3 Minerals and mineral bioavailability

Due to their high consumption, cereal products represent an important source of minerals, especially to individuals of low income. The important minerals for human consumption are shown in Table 5.3. Comparing different whole grain cereal flours, Ragaee *et al.* (2006) showed that barley contains high levels of phosphorus, calcium, potassium, magnesium, sodium, copper, zinc, and iron. Rye was richer than wheat in all minerals and spelt wheat is richer than common wheat in phosphorous and zinc. Unlike millet, sorghum had the poorest mineral profile compared with other cereals (Glew *et al.*, 1997) (Table 5.4). A detailed study by Ruibal-Mendieta *et al.* (2005) comparing whole grain flour and sieved flour ash content and composition of minerals from common wheat (*Triticum aestivum* ssp. *vulgare*) and hard wheat

Table 5.3 Important functions of minerals available from whole grain cereal grains

Mineral	Important for:
P	Metabolism of fats, carbohydrates, and protein, and regulation of growth and production of energy.
K	Regulation of water balance and the distribution of fluids and nutrients on both sides of the cells. Important for growth, proper muscle contraction, normal blood pressure, nerve impulses, and enzyme reactions.
Mg	Regulation of neuromuscular activity and maintenance of normal heart rhythm; proper calcium and vitamin C metabolism; production of DNA, RNA, testosterone and progesterone; and conversion of blood sugar into energy.
Ca	Formation and maintenance of bones, development of teeth and healthy gums, helps blood clotting, and stabilizes different body functions including regular heartbeat, transmission of nerve impulses, muscle growth, protein structuring in DNA and RNA.
Na	Water balance and distribution of fluids and nutrients on both sides of the cells, maintenance of proper blood pH, proper digestion, nerve function, and muscle contractions.
Zn	Absorption and action of B-vitamins, maintenance of acid–base equilibria, protein synthesis, collagen formation, functional immune system, protein and DNA synthesis, muscle contraction, formation of insulin, brain functions, and male testosterone and sperm count.
Fe	Production of hemoglobin, and protein metabolism.
Mn	Co-factor for many enzymes involved in bone formation, energy production, metabolism of protein, carbohydrate, and fat, and promotion of normal growth and development.
Cu	Formation of bones, necessary for the absorption and metabolism of iron and its conversion into hemoglobin, production of RNA, phospholipids, protein and adenosine triphosphate (ATP) metabolism.

Table 5.4 Mineral composition of whole grain cereal flours (mg/kg)

Mineral	Wheat	Rye	Barley	Corn	Oats	Rice	Buck-wheat	Millet	Sorghum
P	1170	3010	2460	990	4110	1030	1550	2400	350
K	1550	4380	3290	1200	4000	1500	2420	2200	240
Mg	250	930	670	470	1170	350	1010	1000	188
Ca	170	330	270	60	530	60	110	100	27
Na	20	50	40	10	40	20	10	—	5
Zn	8	28	13	5	30	17	25	34	3
Fe	12	28	24	11	38	12	15	48	11
Mn	5	22	13	—	58	9	15	7	1
Cu	1	3	1	—	2	2	6	5	0.2

Source: Danish Food Composition Databank (Revision 5.0, 2003), E. Saxholt and A. Møller (Danish Institute for Food and Veterinary Research), http://www.foodcomp.dk/fcdb_default.asp; data for sorghum are from Ragaee *et al.*, (2006).

(*T. aestivum* ssp. *spelta*) showed that minerals are concentrated in the outer parts of cereal grains making products of whole grains much richer in minerals compared with products made of refined flours (Table 5.5), in agreement with the common notion (Hoseney, 1994).

Mineral bioavailability is determined not only by their content in foods but also by the type and content of several other compounds, which might act as inhibitors or enhancers of their bioaccessibility (Sandström *et al.*, 1987). Thus, defined as the amount of ingested compounds that are available for absorption in the gastrointestinal tract, bioaccessibility is a part of bioavailability, referring to availability and luminal changes of minerals prior to absorption. One major antinutrient is phytic acid (myo-inositol 1,2,3,4,5,6-hexakisphosphate), a major form of phosphorus storage in cereal grains that might account for up to 80% of grain total phosphorous (Kelsay, 1987; Larsson *et al.*, 1996). Phytic acid in cereal products is important as it may chelate native as well as fortified metals and lower their intestinal absorption (Wise, 1995; Lopez *et al.*, 2002). Phytic acid binding may be detrimental for the bioavailability of minerals, especially for zinc and iron (Flanagan, 1984; Brune *et al.*, 1992; Hurrell, 1997; Sandström, 1997). Several strategies may be utilized to lower phytate contents in cereal products and improve mineral bioaccessibility, e.g. the use of the phytic acid-degrading enzyme phytase (Egli *et al.*, 2003; Vats & Banerjee, 2004). Some food-processing techniques – e.g. soaking, germination, hydrothermal processing, and fermentation – increase the activity of native phytases of cereals. Bacterial lactic acid fermentation leads to production of organic acids, mainly lactic acid, that lower the pH making it favorable for cereal phytase activity with pH optima in the range of 4.5–5.6 (Sandberg, 2002). Sourdough fermentation, having a similar effect, was found to improve the bioavailability of magnesium, iron, and zinc to rats (Lopez *et al.*, 2003). The inhibitory effects of phytic acid on mineral bioavailability may be counteracted to some extent by

Table 5.5 Ash content and mineral composition (mg/kg) of whole grain and sifted flours of wheat

Mineral	Soft wheat		Hard wheat	
	Whole flour	Sifted flour	Whole flour	Sifted flour
Ash	14 900	4200	18 300	5400
P	2105	729	2923	1022
K	3730	1090	3720	990
Mg	963	180	1274	257
Ca	320	190	284	181
Na	92	88	85	71
Zn	19	6	30	16
Fe	22	6	31	10
Mn	27	5	28	4

Source: Ruibal-Mendieta *et al.* (2005).

absorption enhancers such as organic acids – e.g. acetic, citric, lactic, malic, and ascorbic acids – that form soluble complexes with these minerals in competition with insoluble phytins (Gillooly *et al.*, 1983; Hallberg *et al.*, 1989; Siegenberg *et al.*, 1991; Pabon & Lönnerdal, 1992) as well as proteins (Reddy *et al.*, 1996; Sandström *et al.*, 1989) or carotenoids (Layrisse *et al.*, 2000) that might encapsulate the minerals preventing phytic acid complexation. Utilization of the latter strategy might be useful if the antioxidant properties of phytic acid (Oatway *et al.*, 2001) are found to be of biological relevance.

Other antinutritional factors, besides phytic acid, that may affect the bioavailability of non-heme iron include tannins and some insoluble fibers (Gillooly *et al.*, 1984; Harland, 1989; Lestienne *et al.*, 2005). Certain phenolic compounds, especially those containing catechol and galloyl groups may form insoluble 'iron–tannin' complexes and inhibit iron absorption (Brune *et al.*, 1989, 1991, 1992). A detailed description of the phenolic components in cereal grains will be provided below. Enzymatic degradation of polyphenols in high-tannin cereal grains using polyphenol oxidase during processing has been used as a strategy to improve iron bioavailability (Matuschek *et al.*, 2001). In addition to phytins and polyphenols, the bran fractions of cereals have high soluble and total oxalate contents compared with refined grain cereals flour suggesting that oxalic acid is primarily located in the outer layers of cereal grains (Siener *et al.* 2006). Dietary oxalates are undesirable because of their enhancement of kidney stone formation and because of their inhibitory effects on the absorption of calcium.

5.4 Vitamins and vitamin bioavailability

Cereal products, especially whole grain products rich in bran and germ, are important sources of nutritional B- and E-vitamins. The major B-vitamins in cereal grains (Tables 5.6 and 5.7) include thiamine (B-1), riboflavin (B-2), niacin (B-3), pantothenic acid (B-5), and pyridoxine (B-6) (Adrian & Petit 1970). The bioavailability of thiamine from whole grain wheat bread is relatively high (Ranhotra *et al.*, 1985). Vitamin B-6 represents a group of highly bioavailable 3-hydroxy-2-methyl-pyridine derivatives – i.e. pyridoxine, pyridoxal, and pyridoxamine – which exist in foods in phosphorylated and non-phosphorylated forms. Pyridoxine is the major vitamer in cereal products where it may also exist as a glucosylated adduct, 5′-*O*-(β-D-glucopyranosyl)pyridoxine, of limited bioavailability (Gregory & Kirk, 1981). Cereal products are also a moderate source of folate, which is a generic name for all derivatives of folic acid and is an important B-vitamin. Natural folates are polyglutamates (usually 5–7 glutamyl residues) of pteroic acid and related analogues with similar chemical and physiological properties to those of synthetic folic acid. The predominant folate derivatives in cereal grains are polyglutamyl forms of tetrahydrofolates, 5-methyltetrahydrofolate, and 10-formyltetrahydrofolate. Folic acid is the monoglutamate of pteroic acid and

Table 5.6 Composition of important water-soluble vitamins in whole grain wheat flour and fractions of wheat grain (mg/kg)

Vitamin	Whole grain	Bran	Germ	Sifted flour
Thiamin (B-1)	3.9	8.9	14.5	1.9
Riboflavin (B-2)	0.8	3.6	6.1	0.3
Niacin (B-3)	56	296	58	13
Pantothenic acid (B-5)	6.8	24	12.0	5.0
Pyridoxine (B-6)	3.4	13.8	14.2	0.8
Folate (B-9)	0.5	1.4	1.9	0.3
Biotin (vitamin H)	0.07	0.24	NR	0.02

Sources: Danish Food Composition Databank (Revision S.O, 2003), E. Saxholt and A. Møller (Danish Institute for Food and Veterinary Research), http://www.foodcomp.dk/fcdb_default.asp.

Table 5.7 Composition of important water-soluble vitamins in whole grain cereal flours (mg/kg)

Vitamin	Wheat	Rye	Barley	Corn	Oats	Rice	Buckwheat	Millet
Thiamin (B-1)	3.9	3.3	1.6	3.3	4.2	0.7	3.9	7.3
Riboflavin (B-2)	0.8	1.1	0.7	1.1	1.2	0.3	1.0	3.8
Niacin (B-3)	56	17	55	57	8	14	35	2.3
Pantothenic acid (B-5)	6.8	4.9	5	—	15	5.5	14.5	—
Pyridoxine (B-6)	3.4	2.8	3	3.3	14.4	1.1	4.0	7.5
Folate (B-9)	0.5	0.7	0.3	0.2	0.46	0.3	0.3	0.3
Biotin (vitamin H)	0.07	0.06	—	—	0.19	0.03	—	—

Sources: Danish Food Composition Databank (Revision S.O, 2003), E. Saxholt and A. Møller, (Danish Institute for Food and Veterinary Research), http://www.foodcomp.dk/fcdb_default.asp.

is not a natural form but is a synthetic form of the vitamin that is applied for enrichment of foods including cereal products. Cereal products are also an important source of biotin (vitamin H).

Vitamin E is the generic name of a number of tocol derivatives (Fig. 5.1) having the biological activity of α-tocopherol (Kamal-Eldin & Appelqvist, 1996). Cereal products represent the second most important dietary source of vitamin E after oilseeds and common vegetable oils, and are the most important source of dietary tocotrienols for humans (Lampi *et al.*, 2002). The levels and distribution of tocopherols and tocotrienols in different cereal grains are given in Table 5.8. The vitamin E level in cereal grains is affected by variety, growth location, maturity of seeds, and storage and a double-fold variation is easily encountered (Peterson & Qureshi, 1993). Recently, Falk *et al.* (2004) showed that, of total barley vitamin E, about 13% is present in the

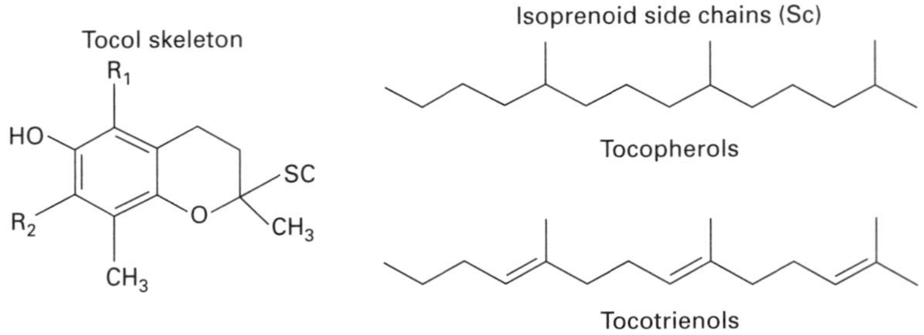

Tocopherol	Tocopherol	R_1	R_2
α-Tocopherol	α-Tocopherol	CH_3	CH_3
β-Tocopherol	β-Tocopherol	CH_3	H
γ-Tocopherol	γ-Tocopherol	H	CH_3
δ-Tocopherol	δ-Tocopherol	H	H

Fig. 5.1 Tocols (tocopherols and tocotrienols) in cereal grains.

Table 5.8 Composition of vitamin E in whole grain cereal flours (mg/kg dry matter)

Vitamin	Wheat[1]	Rye[1]	Barley[2]	Oats[2]	Maize[2]	Rice[3]
α-Tocopherol	10	16	8.6	14.9	3.7	14.6
α-Tocotrienol	4	15	40.3	56.4	5.3	8.7
β-Tocopherol	7	4	0.9	3.0	0.2	1.0
β-Tocotrienol	28	8	8.7	5.4	0	0
γ-Tocopherol	0	0	5.6	0.4	45.0	1.3
γ-Tocotrienol	0	0	10.4	0	11.3	11.9
δ-Tocopherol	0	0	0.7	0	1.0	0.1
δ-Tocotrienol	0	0	0.9	0	0.4	0.5
Total tocols	49	43	76.1	80.1	66.9	38.1

Sources of data: [1]Bramley *et al.* (2000); [2]Panfili *et al.* (2003); [3]Ha *et al.* (2006).

germ (mostly tocopherols), 50% in the pericarp (tocotrienols, and some tocopherols), and 37% in the endosperm (mostly tocotrienols), in agreement with reports on other cereals (Grams *et al.*, 1970; Peterson, 1995; Holasová, 1997). Barley contains less total vitamin E but is higher in tocotrienols, especially the health-protective γ-tocotrienol (Raederstorff *et al.*, 2002).

Wheat germ oil is the vegetable oil richest in vitamin E, up to 2500 mg/kg, which is about 60% α-tocopherol. Milling of brown rice to produce white rice decreases the total tocol level from ~38 to ~12 mg/kg and produces rice bran (Ha *et al.*, 2006). Rice bran, which is about 10% of the dehulled rice grain, contains about 20% oil with ~800 mg/kg of tocols (Qureshi *et al.*, 2000). Yellow corn is unique among cereals for its high content of carotenoids (Egesel *et al.*, 2003).

Vitamin bioavailability from different cereal products is not well understood. This is mainly because of the vast numbers of different structures in cereal products and the significant matrix effects. Another factor that may be problematic in these studies relates to the instability of vitamins and difficulties associated with their analysis, e.g. cereal folates.

5.5 Bioactive phytochemicals other than vitamins

Cereal grains contain a wide range of bioactive phytochemicals including antioxidants of the phenolic type, saponins, sterols, and phytoestrogens. Part of the health effects of cereal fibers might be related to their structural features including linkages to phenolic compounds, lignin, and other bioactive molecules, i.e. to the dietary fiber complex. Phenolic acids are major phenylpropanoid derivatives that exist in cereals in free as well as in bound forms with ester links to fatty acids, sugars, fatty acids, and cell wall polysaccharides (Bily *et al.*, 2004; Bonoli *et al.*, 2004). The most abundant phenolic acids in cereals are ferulic, *p*-coumaric, and sinapic acids (Fig. 5.2), which are mainly concentrated in the bran, aleurone, and pericarp layers. Wheat bran contains up to 7 g/kg, and maize bran up to 30 g/kg, of esterified hydroxycinnamic acids (Clifford, 1999). Some sorghum varieties contain extremely high levels of phenolic acids, especially *p*-coumaric acid, with contents up to 13 g/kg compared with *c.* 65 mg/kg in wheat (Bily *et al.*,

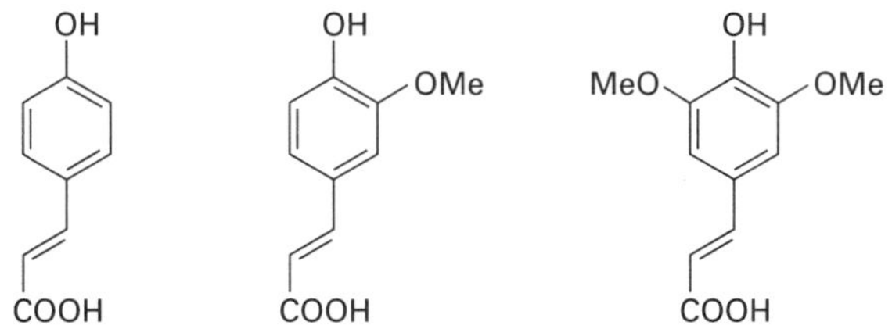

p-Coumaric acid   Ferulic acid   Sinapic acid

Fig. 5.2 Important hydroxycinnamic acids in cereal grains.

2004). Rye grain contains high levels of total phenolic acids mainly in bound forms of ferulic (*c.* 1000 mg/kg), sinapic (*c.* 130 mg/kg) and *p*-coumaric acids (*c.* 60 mg/kg), together with negligible amounts of caffeic, vanillic, *p*-hydroxybenzoic, and protocatechuic acids (Andreasen *et al.*, 1999). Wheat contains lesser amounts of phenolic acids compared with rye. Barley contents of bound ferulic acid (50 mg/kg) and *p*-coumaric acid (3 mg/kg) are much lower than other cereals (Clifford, 1999).

During plant development, ferulic acid undergoes dimerization by oxidative coupling, catalyzed by peroxidases, leading to a number of dehydrodimers (Fig. 5.3) including 8-*O*-4′-dihydro ferulic acid (DiFA), 8,5′-DiFA, 5,5′-DiFA, and 8,8′-DiFA (Ralph *et al.*, 1994; Waldron *et al.*, 1996; García-Conesa *et al.*, 1997; Lempereur *et al.*, 1998). The 8-*O*-4′-DiFA is the most abundant dimer, followed by the 8,5′-DiFA benzofuran form, 8,5′-DiFA, and 5,5′-DiFA (Andreasen *et al.*, 2000). The content of these dimers in the rye bran fraction is 10 to 15-fold higher than that in the flour fraction. Lignin, produced by an enzyme-catalyzed dehydrogenative polymerization of three phenol precursors (*p*-coumaryl, coniferyl, and sinapyl alcohols), may also link to ferulic acid via ether or ester bonds (Bunzel *et al.*, 2000, 2001, 2004). There is also evidence of covalent cross-links between arabinoxylans and proteins through dihydroferulic acid dimers and tyrosine (Piber & Koehler, 2005). In rice, phenylpropanoids are bound to polysaccharides through glucose, arabinose, xylose, galactose, rhamnose, and mannose residues in the cell wall. Two main hydroxycinnamic acid derivatives (β-D-fructofuranosyl-α-D-(6-*O*-(*E*)-feruloylglucopyraniside)-6′-*O*-feruloylsucrose and α-D-

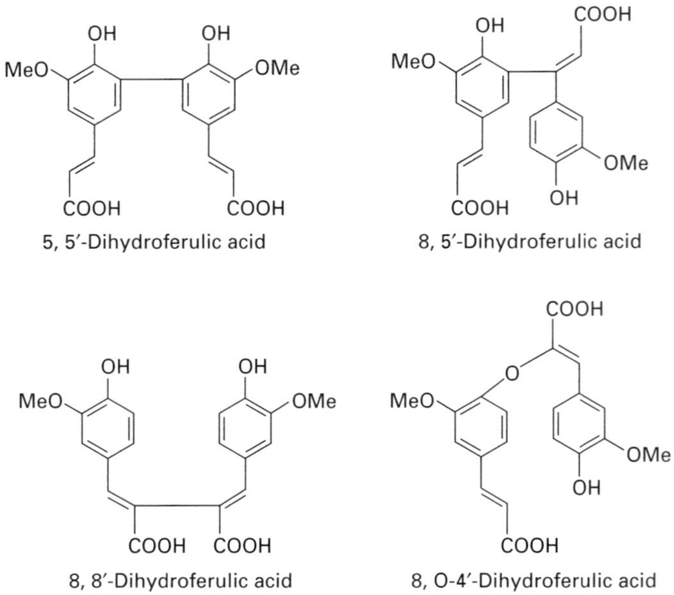

Fig. 5.3 Structures of the major dihydroferulic acid dimers in cereal grains.

fructofuranosyl-α-D-(6-*O*-(*E*)-sinapoyl-glucopyraniside)-6′-*O*-sinapoylsucrose) covalently linked to sucrose through ester bonds, were separated from rice (Tian *et al.* 2005). Polysaccharide cross-linking through the ferulate–arabinoxylan polymers affects fiber solubility in cereal grains (Ishii, 1997; Bunzel *et al.*, 2000, 2001, 2004). Other, more soluble, phenolic acid esters do exist in addition to those just mentioned. For example, rice bran contains a family of oil-soluble ferulic acid esters with sterols and triterpene alcohols – i.e. campesteryl, β-sitosteryl, cycloartanyl, cycloartenyl, and 24-methylene cycloartenyl ferulates – termed γ-oryzanols (Diack & Saska, 1994).

Barley contains a wide range of phenolic antioxidants including benzoic and cinnamic acid derivatives, flavonols, flavones, flavanones, chalcones, proanthocyanidins, quinones, and amino phenolic compounds present in free and bound forms (Hernanz *et al.*, 2001). Oat flour and oat extracts (*Avena sativa*, L.) were among the first antioxidants proposed for the stabilization of lipids and lipid-containing foods (Peters & Musher, 1937; Duve & White, 1991). The antioxidants from oats were identified as esters of caffeic and ferulic acid: monoesters of C_{26} and C_{28} α,ω-diols (Daniels & Martin, 1961); monoesters of hexacosan-1-ol, 26-hydroxyhexacosanoic acid and 28-hydroxyhexacosanoic acid (Daniels *et al.*, 1963; Daniels & Martin, 1967, 1968), and a group of cinnamic acid conjugates, namely avenanthramides (Collins, 1989; Dimberg *et al.*, 1993), avenalumic acids (Collins *et al.*, 1991), and beta-truxinic acid (Dimberg *et al.*, 2001) (Fig. 5.4). Tocopherols/tocotrienols and other phenolic compounds of oats may contribute to the stabilization of lipids during hydrothermal processing such as extrusion cooking (Bryngelsson *et al.*, 2002; Viscidi *et al.*, 2004), and those of barley contribute to beer stability (McMurrough *et al.*, 1996).

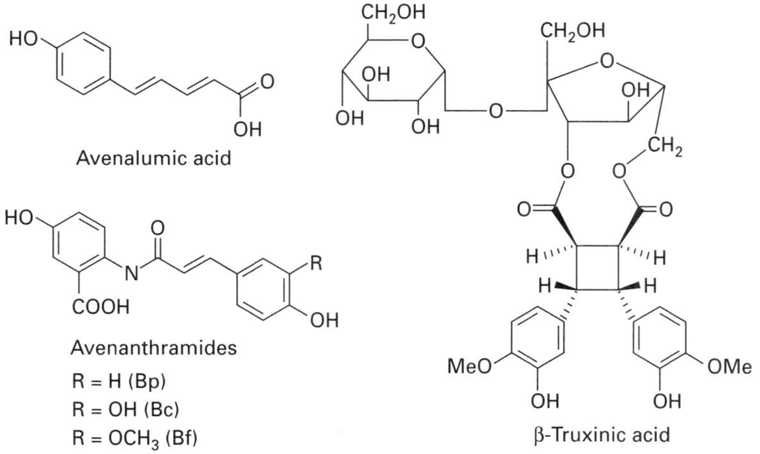

Fig. 5.4 Characteristic phenolic antioxidants in oats. Abbreviations are: Bp, *N*-(4′-hydroxy)-(*E*)-cinnamoyl-5-hydroxyanthranilic acid; Bc, *N*-(4′-hydroxy)-(*E*)-cinnamoyl-5-hydroxyanthranilic acid; Bf, *N*-(4′-hydroxy-3′-methoxy)-(*E*)-cinnamoyl-5-hydroxyanthranilic acid.

Free flavanols (mainly catechin and gallocatechin) as well as flavanol polymers (proanthocyanidins) exist in cereals in free forms. Proanthocyanidin oligomers, including procyanidins B3 and C3 and prodelphinidin B3 and C3, are present in barley (McMurrough and Madigan, 1996). The sorghum grain (*Sorghum bicolor* (L.) Moench), and especially the brans of colored cultivars, contain very high levels of phenolic compounds compared with other cereals. Phenolic acids (ferulic, *p*-coumaric, procatechuic, vanillic, caffeic, *p*-hydroxybenzoic, gallic, and cinnamic) are present in sorghum in free and bound forms (Hahn *et al.*, 1983). Two flavonoids, luteoforol and apiforol, are abundant in the sorghum grain. Sorghum is also unique among cereals for its content of oligomeric polyphenols (proanthocyanidins and condensed tannins) despite a great variability among cultivars (Fig. 5.5). The color of the grain is used to classify sorghum into type I varieties with no tannins or proanthocyanidins, type II varieties with tannins present in pigmented testa, and type III varieties with tannins present in pigmented testa and pericarp (Hahn *et al.*, 1984; Hahn & Rooney, 1986). There is a considerable structural heterogeneity in sorghum condensed tannins with respect to the nature of repeating monomeric units (flavan, flavan-3-ol, and flavanone), type of interflavan bonds (A- and B-types), and patterns of hydroxylation and glucosidation (Reed *et al.*, 2005). Black sorghum varieties contain up to 4500 ppm of anthocyanin pigments, mainly luteolinidin and apigeninidin anthocyanidins, existing in monomeric and polymeric forms (Awika *et al.*, 2005).

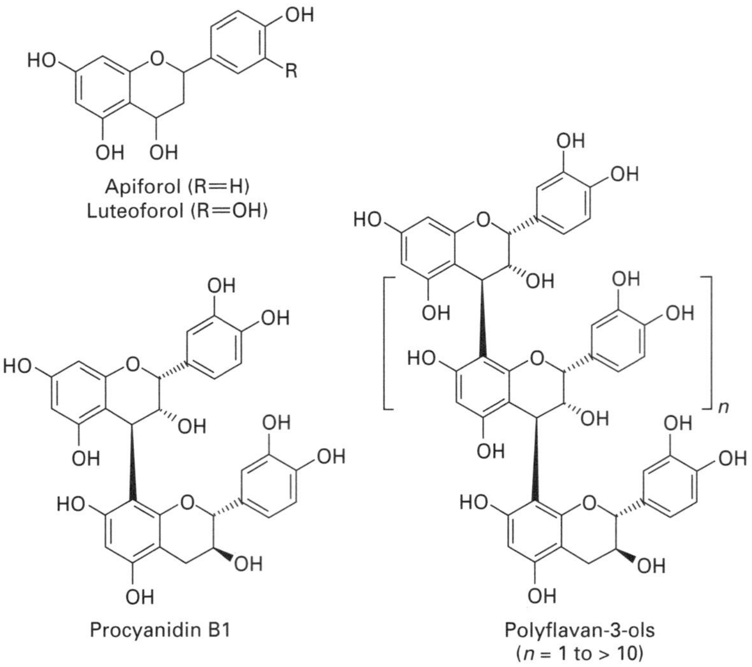

Fig. 5.5 Characteristic phenolic antioxidants in sorghum.

Phytoestrogens, including lignans and isoflavones, present structural similarity to estradiol-17β and exert either estrogenic or anti-estrogen effects through complex mechanisms (Cassidy *et al.*, 2003). Secoisolariciresinol, the major plant lignan in wheat and rye brans (Mazur & Adlercreutz, 1998, 2000; Qu *et al.*, 2005), is converted by the bacteria in the human colon to mammalian lignans enterodiol and enterolactone (Fig. 5.6), with special affinity to the α-estrogen receptor (ER-α) that has been associated with decreased risk of hormone-related cancers (Ingram *et al.*, 1997; Tham *et al.*, 1998; Pietinen *et al.*, 2001; Hultén *et al.*, 2002; Adlercreutz, 2002).

Phytosterols and phytostanols (Fig. 5.7) are concentrated in the outer parts of cereals and are present in free as well as in conjugated forms including steryl esters of fatty or phenolic acids, steryl glycosides, and acylated steryl glycosides (Lampi *et al.*, 2004). Among cereals, free and bound stanols are only found in wheat, rye, triticale, and barley grains (Piironen *et al.*, 2000). The highest phytosterol/phytostanol content was found in rye (*c.* 955 ppm), followed by barley (760 ppm) and wheat (690 ppm), and the lowest was found in oats (447 ppm) (Piironen *et al.*, 2002). Different cereals were shown to contain variable amounts of steryl ferulates. γ-Oryzanol (Fig. 5.8), a commercial product of rice bran oil, is a mixture of cycloartenyl, 24-methylenecycloartanyl, campesteryl, Δ^7-campestenyl, sitosteryl, Δ^7-sitostenyl, stigmasteryl, Δ^7-stigmasteryl, and sitostanyl and campestanyl ferulates (Seitz

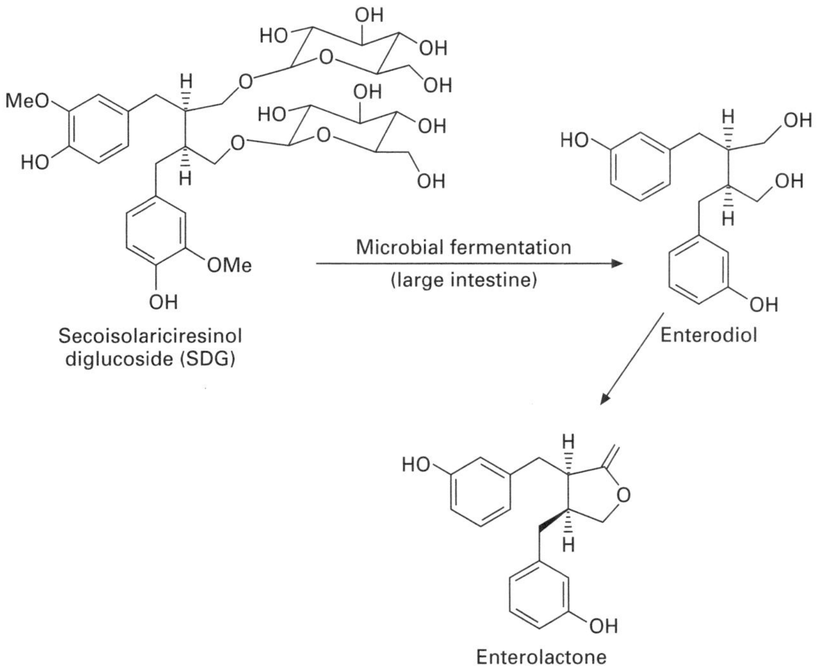

Fig. 5.6 Secoisolariciresinol diglucoside in wheat and rye bran and its transformation to mammalian lignans enterodiol and enterolactone by microbial fermentation (Qu *et al.*, 2005).

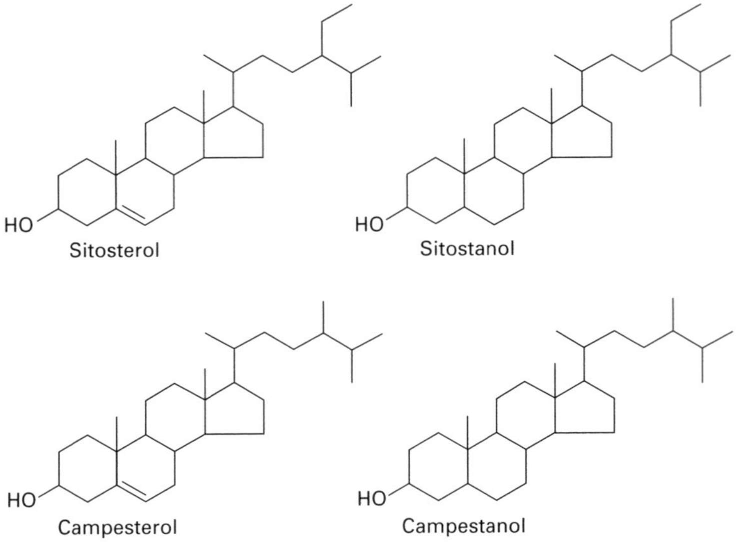

Fig. 5.7 Major phytosterols and phytostanols in cereal grains.

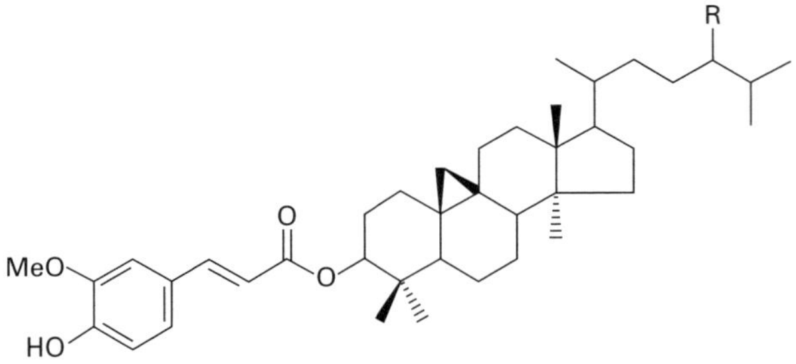

Fig. 5.8 The two major γ-oryzanols in rice bran oil: cycloartanol ferulate (R = H) and 24-methylenecycloartanol ferulate (R = CH_2).

1989; Norton, 1994, 1995; Hakala *et al.*, 2001). The phytosterol content in cereals, albeit being moderate, may contribute to their cholesterol-lowering potential (Piironen *et al.*, 2000), especially when high-sterol cereal products are combined with, for example, sterol-enriched margarines.

Alkylresorcinols (Fig. 5.9) are special alkylphenols that are present in large amounts in the outer layers of wheat and rye and at low levels in barley. Chemically, they represent a group of 5-alkylresorcinol derivatives, with the alkyl chain ranging from 15 to 25 carbon atoms. When ingested, they are metabolized in the same way as tocopherols, i.e. by ω-oxidation followed by β-oxidation, and are reported to have a number of bioactivities (Ross *et al.*, 2004). They have been shown to serve as markers for products containing rye, wheat, and durum wheat (Chen *et al.*, 2004; Landberg *et al.*, 2006) and may be able to serve as biomarkers of intake of whole grain wheat and rye products (Ross *et al.*, 2004).

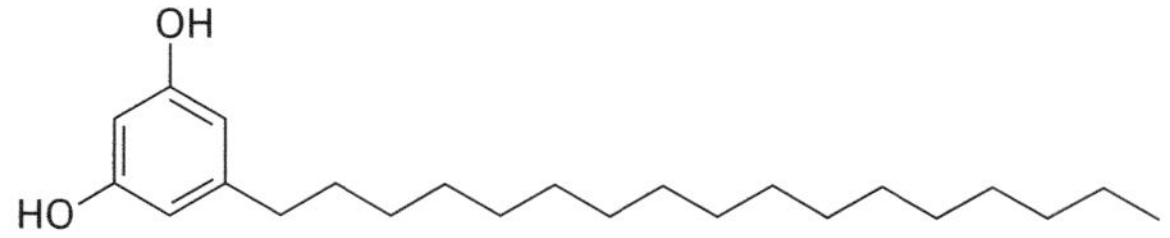

Fig. 5.9 Alkylresorcinols in wheat and rye brans (length of alkyl chain varies between 17 and 25 carbons).

5.6 Micronutrients added to cereal products and their health effects

A major aim of nutritionists is to fill the gap between the individual actual and recommended dietary intake of vitamins and minerals. This is normally achieved by supplementation of staple foods with the micronutrient(s) in question. It has long been known that removal of outer layers of bran for the production of white wheat flour by roller milling led to a range of vitamin B deficiency diseases. Removal of the outer layers of cereals by milling and other refining processes depletes the resulting flour and products of a wide range of vitamins and minerals. The benefits of fortifying white flour to replace those vitamins that have been lost during refining became evident and vitamins B1 and B2 and niacin, calcium, and iron were added to flour and bread. Fortification found impetus during World War II when the UK and US governments recommended voluntary fortification of flour and bread but soon fortified wheat products became mandatory for the armies. Fortification of wheat flour with niacin in the United States proved effective in preventing death from pellagra.

Fortification, to alleviate mineral and vitamin deficiencies, is especially essential in low-income countries where children, women, and the elderly suffer in large numbers from malnutrition and anemia (Johnson *et al.*, 1988). For example, corn and wheat flours are fortified with iron, vitamin A, and B-vitamins in several countries (for more details, see Chapter 14). Fortification of flour with iron (20–30 ppm) contributes about 48% of the recommended dietary allowance (RDA) for iron in Venezuela (WHO, 1999). However, since iron overload may pose a health hazard, the level of iron fortificant should be carefully decided based on consumption, severity of the deficiency, and the chemical form of iron – since this affects its bioavailability. Voluntary fortification of bread, flour, and cereals with folates to prevent *inter alia* neural tube defects has been recommended in ~50 countries despite concerns about the masking of vitamin B12 deficiency by high doses of folates in elderly populations (Kamien, 2006). It has also been argued that fortification with certain nutrients will not provide the variety of nutrients and the complex structure found in whole grain cereals and that whole grain tissues might be important for the health benefits of native cereals (Pereira *et al.*, 2001; Slavin, 2003).

In the last few years, the US Food and Drug Administration (FDA) mandated fortification of cereal flours with folic acid at the level of 140 g/100 g of flour, breakfast cereals, pasta, etc. to protect against neural tube defects in babies (FDA, 1996). Likewise, the Department of Health in the United Kingdom proposes global fortification of cereal flours and products with folic acid at 240 g/100 g (UK Department of Health, 2000). Vitamin E-fortified ready-to-eat breakfast cereals represent a major food source of α-tocopherol in the American diet (Sheppard *et al.*, 1993). A recent study has shown that the bioavailability of vitamin E (400 IU d_9-α-tocopheryl acetate) in a fortified breakfast cereal was 25-fold greater than that from a supplement capsule (Leonard *et al.*, 2004).

Recently, focus has been given to new approaches that depend on genetic and/or technological fortification of foods with the intended micronutrients during production or processing (Blechl, 2004). Breeding programs aiming to enrich cereal grains with trace metals represent one way of enhancing bioactivity of these foods (Gregorio *et al.*, 2000; Grusak, 2002; Raboy, 2002; Welch & Graham, 2004). The term 'biofortification' has thus been coined to describe sustainable biotechnologies leading to increased levels of essential minerals or vitamins (White & Broadley, 2005). For example, the genetic variation in micronutrient composition of cereals can be exploited in conventional breeding programs or through gene technology to produce micronutrient-dense cultivars. *Golden Rice*, rich in β-carotene, was produced by expressing two foreign genes encoding phytoene synthase and carotene desaturase in rice to combat vitamin A deficiency (Ye *et al.*, 2000; Beyer *et al.*, 2002; Zimmermann & Qaim, 2004). Alternative transgenic approaches aiming to improve the levels of other micronutrients have been tried, e.g. overexpression of the iron storage protein ferritin in rice grains to increase its iron concentration (Goto *et al.*, 1999; Lucca *et al.*, 2001). Efforts have also been directed to minimize the levels of compounds that inhibit mineral bioavailability, such as phytates and tannins (Welch & Graham, 2004). Mutational breeding has produced maize and barley mutants with about 95% reduction in phytate levels (Raboy, 2002). cDNAs of homogentisic acid geranylgeranyl transferase (HGGT), the committed enzyme in tocotrienol biosynthesis in monocots, have been isolated for barley, wheat, and rice (GeneBank accession numbers AY222860, AY222861, and AY222862, respectively). In addition, expression of barley HGGT in corn seeds increased their tocotrienol and tocopherol contents by about six-fold (Cahoon *et al.*, 2003).

5.7 Future trends

Although there is ample epidemiological evidence for the beneficial health effects of whole grain cereals (Koh-Banerjee & Rimm, 2003) and a number of studies showing favorable physiological effects of some of their minor components (e.g. cholesterol-lowering and antioxidant effects), these effects have not been proven in confirmatory intervention studies. One confounding factor in the epidemiological studies is the difficulty that the consumer may face in judging what is 'whole grain' due to problems associated with its definition and the illusionary attributes of added colors and flavors. There is thus a strong demand of more objective ways of assessing whole grain intake, e.g. by using biomarkers. Examples of micronutrients that may, after evaluation and validation, serve as specific biomarkers of intake include alkylresorcinols of wheat and rye and avenanthramides of oats. Moreover, more studies directed towards identifying the components responsible for the beneficial health effects, their mechanism of action, and possible synergistic interactions are needed. Appropriate biofortification and processing strategies for the enrichment of cereal products with potential nutritional candidates are under development. Finally, proper definitions for whole grains and developments of the whole grain concept are being sought.

5.8 Sources of further information and advice

- FAO International Network of Food Data Systems (INFOODS), www.fao.org/infoods.
- Danish Food Composition Databank, http://www.foodcomp.dk/fcdb_search.asp.
- US Department of Agriculture(USDA) Nutrient Data Laboratory http://www.nal.usda.gov/fnic/foodcomp/search/.
- International Life Sciences Institute (ILSI) Crop Composition Database (version 2), http://www.cropcomposition.org/.
- Dr Duke's Phytochemical and Ethnobotanical Database, http://www.ars-grin.gov/duke/.

5.9 References

Adlercreutz H (2002), 'Phytoestrogens and breast cancer', *J Steroid Biochem Mol Biol*, **83**, 113–118.

Adrian J and Petit L (1970), 'Vitamins of cereals and their evolution during technologic treatment', *Anna de Nutr Aliment*, **24**, 131–168.

Andreasen M F, Christensen L P, Meyer A S and Hansen A (1999), 'Release of hydroxycinnamic and hydroxybenzoic acids in rye by commercial plant cell wall degrading enzyme preparations', *J Sci Food Agric*, **79**, 411–413.

Andreasen M F, Christensen L P, Meyer A S and Hansen A (2000), 'Content of phenolic acids and ferulic acid dehydrodimers in 17 rye (*Secale cereale* L.) varieties', *J Agric Food Chem*, **48**, 2837–2842.

Awika J M, Rooney L W and Waniska R D (2005), 'Anthocyanins from black sorghum and their antioxidant properties', *Food Chem*, **90**, 293–301.

Beyer P, Al-Babili S, Ye X, Lucca P, Schaub P, Welsch R and Potrykus I (2002), 'Golden Rice: Introducing the beta-carotene biosynthesis pathway into rice endosperm by genetic engineering to defeat vitamin A deficiency', *J Nutr*, **132**, 506S–510S.

Bily A C, Burt A J, Ramputh A L, Livesey J, Regnault-Roger C, Philogene B R and

Arnason J T (2004), 'HPLC-PAD-APCI/MS assay of phenylpropanoids in cereals', *Phytochem Anal*, **15**, 9–15.
Blechl A E (2004), 'Applications of biotechnology for improving the healthfulness and utility of cereals', in Bhalgat M K, Ridely W P, Felsot A S and Seiber J N, *Agricultural Biotechnology: Challenges and Prospects*, ACS Symposium Series No. 866, New York, Oxford University Press, pp. 53–65.
Bonoli M, Marconi E and Caboni M F (2004), 'Free and bound phenolic compounds in barley (*Hordeum vulgare* L.) flours: Evaluation of the extraction capability of different solvent mixtures and pressurized liquid methods by micellar electrokinetic chromatography and spectrophotometry', *J Chromatogr A*, **1057**, 1–12.
Bramley P M, Elmadfa I, Kafatos A, Kelly F J, Manios Y, Roxborough H E, Schuch W, Sheehy P J A and Wagner K-H (2000), 'Vitamin E', *J Sci Food Agric*, **80**, 913–938.
Brune M, Rossander L and Hallberg L (1989), 'Iron absorption and phenolic compounds. Importance of different phenolic structures', *Eur J Clin Nutr*, **43**, 547–558.
Brune M, Hallberg L and Skanberg A B (1991), 'Determination of iron-binding phenolic groups in foods', *J Food Sci*, **56**, 128.
Brune M, Rossander-Hultén L, Hallberg L, Gleerup A and Sandberg AS (1992), 'Iron absorption from bread in humans: inhibiting effects of cereal fiber, phytate and inositol phosphates with different numbers of phosphate groups', *J Nutr*, **122**, 442–449.
Bryngelsson S, Dimberg L H and Kamal-Eldin A (2002), 'Effects of commercial processing on levels of antioxidants in oats (*Avena sativa* L.)', *J Agric Food Chem*, **50**, 1890–1896.
Bunzel M, Ralph J, Marita J and Steinhart H (2000), 'Identification of 4-*O*-5′-coupled diferulic acid from insoluble cereal fiber', *J Agric Food Chem*, **48**, 3166–3169.
Bunzel M, Ralph J, Marita J M, Hatfield R D and Steinhart H (2001), 'Diferulates as structural components in soluble and insoluble cereal dietary fiber', *J Sci Food Agric*, 2001, **81**, 653–660.
Bunzel M, Ralph J, Lu F, Hatfield R D and Steinhart H (2004), 'Lignins and ferulate-coniferyl alcohol cross-coupling products in cereal grains', *J Agric Food Chem*, **52**, 6496–6502.
Brown L, Rosner B, Willett W W and Sacks F M (1999), 'Cholesterol-lowering effects of dietary fiber: a meta-analysis', *Am J Clin Nutr*, **69**, 30–42.
Cahoon E B, Hall S E, Ripp K G, Ganzke T S, Hitz W D and Coughlan S J (2003), 'Metabolic redesign of vitamin E biosynthesis in plants for tocotrienol production and increased antioxidant content', *Nat Biotechnol*, **21**, 1082–1087.
Cassidy A (2003), 'Potential risks and benefits of phytoestrogen-rich diets', *Int J Vitam Nutr Res*, **73**, 120–126.
Chen Y, Ross A B, Åman P and Kamal-Eldin A (2004), 'Alkylresorcinols as markers of whole grain wheat and rye in cereal products', *J Agric Food Chem*, **52**, 8242–8246.
Clifford M N (1999), 'Chlorogenic acids and other cinnamates: Nature, occurrence and dietary burden', *J Sci Food Agric*, **79**, 362–372.
Collins F W (1989), 'Oat phenolics: avenanthramides, novel substituted *N*-cinnamoylanthranilate alkaloids from oat groats and hulls', *J Agric Food Chem*, **37**, 60–66.
Collins F W, McLachlan D C and Blackwell B A (1991), 'Oat phenolics: Avenalumic acids, a new group of bound phenolic acids from oat groats and hulls', *Cereal Chem*, **68**, 184–189.
Craig S A S (2004), 'Betaine in human nutrition', *Am J Clin Nutr*, **80**, 539–549.
Craig W J (1997), 'Phytochemicals: Guardians of our health', *J Am Diet Assoc*, **97**, S199–S204.
Daniels D G H and Martin H F (1961), 'Isolation of a new antioxidant from oats', *Nature*, **191**, 1302.
Daniels D G H and Martin H F (1967), 'Antioxidants in oats: Mono-esters of caffeic and ferulic acids', *J Sci Food Agric*, **18**, 589–595.

Daniels D G H and Martin H F (1968), 'Antioxidants in oats: Glyceryl esters of caffeic and ferulic acids', *J Sci Food Agric*, **19**, 710–712.

Daniels D G H, King H G C and Martin H F (1963), 'Antioxidants in oats: Esters of phenolic acids', *J Sci Food Agric*, **14**, 385–390.

Diack M and Saska M (1994), 'Separation of vitamin E and γ-oryzanols from rice bran by normal-phase chromatography', *J Am Oil Chem Soc*, **71**, 1211–1217.

Dimberg L H, Theander O and Lingnert H (1993), 'Avenanthramides: A group of phenolic antioxidant in oats', *Cereal Chem*, **70**, 637–641.

Dimberg L H, Andersson R E, Gohil S, Bryngelsson S and Lundgren L N (2001), 'Identification of a sucrose diester of a substituted beta-truxinic acid in oats', *Phytochem*, **56**, 843–847.

Duve K J and White P J (1991), 'Extraction and identification of antioxidants in oats', *J Am Oil Chem Soc*, **68**, 365–370.

Egesel C O, Wong J C, Lambert R J and Rocheford T R (2003), 'Combining ability of maize inbreds for carotenoids and tocopherols', *Crop Sci*, **43**, 818–823.

Egli I, Davidsson L, Juillerat M-A, Barclay D and Hurrell R F (2003), 'Phytic acid degradation in complementary foods using phytase naturally occurring in whole grain cereals', *J Food Sci*, **68**, 1855–1859.

Falk J, Krahnstover A, van der Kooij T A W, Schlensog M and Krupinska K (2004), 'Tocopherol and tocotrienol accumulation during development of caryopses from barley (*Hordeum vulgare* L.)', *Phytochem*, **65**, 2977–2985.

FDA (Food and Drug Administration) (1996), 'Food standards: Amendment of standards of identity for enriched grain products to require addition of folic acid', *Fed Register*, **61**, 8781–8797.

Flanagan P R (1984), 'A model to produce pure zinc deficiency in rats and its use to demonstrate that dietary phytate increases the excretion of endogenous zinc', *J Nutr*, **114**, 493–502.

Franz M and Sampson L (2006), 'Challenges in developing a whole grain database: Definitions, methods and quantification', *J Food Comp Anal*, **19**, S38–S44.

Fung T T, Hu F B, Pereira M A, Liu S, Stampfer M J, Colditz G A and Willett W C (2002), 'Wholegrain intake and the risk of type 2 diabetes: A prospective study in men', *Am J Clin Nutr*, **76**, 535–540.

García-Conesa M T, Plumb G W, Waldron K W, Ralph J and Williamson G (1997), 'Ferulic acid dehydrodimers from wheat bran: Isolation, purification and antioxidant properties of 8-*O*-4′-diferulic acid', *Redox Rep*, **3**, 319–323.

Gillooly M, Bothwell T H, Torance J D, MacPhail A P, Derman D P, Bezwoda W R, Mills W, Charlton R W and Mayet F (1983), 'The effect of organic acids, phytates and polyphenols on iron absorption from vegetables', *Br J Nutr*, **49**, 331–342.

Gillooly M, Bothwell T H, Charlton R W, Torrance J D, Bezwoda W R, MacPhail A P, Derman D P, Novelli L, Morral D and Mayet F (1984), 'Factors affecting the absorption of iron from cereals', *Br J Nutr*, **51**, 37–46.

Glew R H, Vanderjagt D J, Lockett C, Grivetti L E, Smith G C, Pastuszyn A and Millson M (1997), 'Amino acid, fatty acid, and mineral composition of 24 indigenous plants of Burkina Faso', *J Food Comp Anal*, **10**, 205-217.

Goto F, Yoshihara T, Shigemoto N, Toki S and Takaiwa F (1999), 'Iron fortification of rice seed by the soybean ferritin gene', *Nat Biotechnol*, **17**, 282–286.

Graf E (1992), 'Antioxidant potential of ferulic acid', *Free Rad Biol Med*, **13**, 435–448.

Graf E and Eaton J W (1990), 'Antioxidant functions of phytic acid', *Free Rad Biol Med*, **8**, 61–69.

Grams G W, Blessin C W and Inglett G E (1970), 'Distribution of tocopherols within the corn kernel', *J Am Oil Chem Soc*, **47**, 337–339.

Gregorio G B, Senadhira D, Htut H and Graham R D (2000), 'Breeding for trace mineral density in rice', *Food Nutr Bull*, **21**, 382–386.

Gregory J F and Kirk J R (1981), 'The bioavailability of vitamin B-6 in foods', *Nutr Rev*, **39**, 1–8.

Grusak M A (2002), 'Enhancing mineral content in plant food products', *J Am Coll Nutr*, **21**, 178S–183S.

Ha, T-Y, Ko S-N, Lee S-M, Kim H-R, Chung S-H, Kim S-R, Yoon H-H and Kim I-H (2006), 'Changes in nutraceutical lipid components of rice at different degrees of milling', *Eur J Lipid Sci Technol*, **108**, 175–181.

Hahn D H and Rooney L W (1986), 'Effect of genotype on tannins and phenols of sorghum', *Cereal Chem*, **63**, 4–8.

Hahn D H, Faubion J M and Rooney L W (1983), 'Sorghum phenolic acids, their high performance liquid chromatography separation and their relation to fungal resistance', *Cereal Chem*, **60**, 255–259.

Hahn D H, Rooney L W and Earp C F (1984), 'Tannins and phenols of sorghum', *Cereal Foods World*, **29**, 776–779.

Hakala P, Lampi A M, Ollilainen V, Werner U, Murkovic M, Wahala K, Karkola S and Piironen V (2001), 'Steryl phenolic acid esters in cereals and their milling fractions', *J Agric Food Chem*, **50**, 5300–5307.

Hallberg L, Brune M and Rossander L (1989), 'Iron absorption in man: ascorbic acid and dose-dependent inhibition by phytate', *Am J Clin Nutr*, **49**, 140–144

Hallfrisch J and Behall K M (2000), 'Mechanisms of the effects of grains on insulin and glucose responses', *J Am Coll Nutr*, **19**, S320–S325.

Harland B F (1989), 'Dietary fiber and mineral bioavailability', *Nutr Res Rev*, **2**, 133–147.

Hegsted M, Windhauser M M, Morris K and Lester S B (1993), 'Stabilized rice bran and oat bran lower cholesterol in humans', *Nutr Res*, **13**, 387–298.

Hernanz D, Nuñez V, Sancho A I, Faulds C B, Williamson G, Bartolomé B and Gómez-Cordovés C (2001), 'Hydroxycinnamic acids and ferulic acid dehydrodimers in barley and processed barley', *J Agric Food Chem*, **49**, 4884–4888.

Holasová M (1997), 'Distribution of tocopherols and tocotrienols in the main products of wheat and rye milling', *Potrav Vedy*, **15**, 343-350.

Hoseney R C (1994), 'Structure of cereals', Chapter 1, in *Principles of Cereal Science and Technology*, St Paul, MN, American Association of Cereal Chemists, pp. 1–28.

Howe G R, Benito E, Castelleto R, Cornee J, Esteve J, Gallagher R P, Iscovich J M, Jiao D A, Kaaks R, Kune G A, Kune S, Labbe K A, Lee H P, Lee M, Miller A B, Peters R K, Potter J D, Riboli E, Slattery M L, Trichopoulos D, Tuyns A, Tzonou A, Whittemore A S, Wuwilliams A H and Zheng S (1992), 'Dietary intake of fiber and decreased risk of cancers of the colon and rectum: evidence from the combined analysis of 13 case-control studies', *J Natl Cancer Inst*, **84**, 1887–1896.

Hultén K, Winkvist A, Lenner P, Johansson R, Adlercreutz H and Hallmans G (2002), 'An incident case-referent study on plasma enterolactone and breast cancer risk', *Eur J Nutr*, **41**, 168–176.

Hurrell R F (1997), 'Bioavailability of iron', *Eur J Clin Nutr*, **51**, S4–S8.

Ingram D, Sanders K, Kolybaba M and Lopez D (1997), 'Case-control study of phyto-oestrogens and breast cancer', *Lancet*, **350**, 990–994.

Ishii T (1997), 'Structure and functions of feruloylated polysaccharides', *Plant Sci*, **127**, 111–127.

Jacobs D R, Slavin J and Marquart L (1995), 'Whole grain intake and cancer: A review of literature', *Nutr Cancer*, **22**, 221–229.

Jacobs D R, Meyer K A, Kushi L H and Folsom A R (1998), 'Whole grain intake may reduce risk of coronary heart disease death in postmenopausal women: The Iowa Women's Health Study', *Am J Clin Nutr*, **68**, 248–257.

Jacobs D R, Meyer K A, Kushi L H and Folsom A R (1999), 'Is whole-grain intake associated with reduced total and cause specific death rates in older women: The Iowa women's health study', *Am J Pub Health*, **89**, 322–329.

Jacobs L R (1986), 'Relationship between dietary fiber and cancer: metabolic, physiologic, and cellular mechanisms', *Proc Soc Exp Biol Med*, **183**, 299–310.

Jacobs L R (1988), 'Fiber and colon cancer', *Gastroenterol Clin North Am*, **17**, 747–760.

Johnson L, Gordon H T and Borenstein B (1988), 'Vitamin and mineral fortification of breakfast cereals', *Cereal Foods World*, **33**, 278.

Kamal-Eldin A and Appelqvist L Å (1996), 'The chemistry and antioxidant properties of tocopherols and tocotrienols', *Lipids*, **31**, 671–701.

Kamien M (2006), 'The repeating history of objections to the fortification of bread and alcohol: from iron filings to folic acid', *Med J Aust*, **184**, 638–640.

Kelsay J L (1987), 'Effects of fiber, phytic acid and oxalic acid in the diet on mineral availability', *Am J Gastroenterol*, **82**, 893–897.

Koh-Banerjee P and Rimm E B (2003), 'Wholegrain consumption and weight gain: A review of the epidemiological evidence, potential mechanisms and opportunities for future research', *Proc Nutr Soc*, **62**, 25–29.

Lampi A, Kamal-Eldin A and Piironen V (2002), 'Tocopherols and tocotrienols from oil and cereal grains', in Shi J, Mazza G and Maguer M-L, *Functional Foods: Biochemical and Processing Aspects,* vol. 2, Boca Raton, FL, CRC Press, pp. 1–21.

Lampi A-M, Moreau R A, Piironen V and Hicks K B (2004), 'Pearling barely and rye to produce phytosterol-rich fractions', *Lipids*, **39**, 783–787.

Landberg R, Kamal-Eldin A, Andersson R and Åman P (2006), 'Alkylresorcinol content and homologue composition in durum wheat (*Triticum durum*) kernels and pasta products', *J Agric Food Chem*, **54**, 3012–3014.

Larsson M, Hulten L R, Sandstrom B and Sandberg A (1996), 'Improved zinc and iron absorption from breakfast meals containing malted oats with reduced phytate content', *Br J Nutr*, **76**, 677–688.

Layrisse M, Garcia-Casal M N, Solano L, Baron M A, Arguello F, Llovera D, Ramirez J, Leets I and Tropper E (2000), 'New property of vitamin A and beta-carotene on human iron absorption: Effect on phytate and polyphenols as inhibitors of iron absorption', *Arch Latinoam Nutr*, **50**, 243–248.

Lempereur I, Surget A and Rouau X (1998), 'Variability in dehydrodiferulic acid composition of durum wheat (*Triticum durum* desf.) and distribution in milling fraction', *J Cereal Sci*, **28**, 251–258.

Leonard S W, Good C K, Gugger E T and Traber M G (2004), 'Vitamin E bioavailability from fortified breakfast cereal is greater than that from encapsulated supplements', *Am J Clin Nutr*, **79**, 86–92.

Lestienne I., Besançon P, Caporiccio B, Lullien-Péllerin V and Trèche S (2005), 'Iron and zinc *in vitro* availability in pearl millet flours (*Pennisetum glaucum*) with varying phytate, tannin, and fiber contents', *J Agric Food Chem*, **53**, 3240–3247.

Liese A D, Roach A K, Sparks K C, Marquart L, D'Agostino R B and Mayer-Davis E J (2003), 'Whole-grain intake and insulin sensitivity: The Insulin Resistance Atherosclerosis Study', *Am J Clin Nutr*, **78**, 965–971.

Liu S, Stampfer M J, Hu F B, Giovannucci E, Rimm E, Manson J E, Hennekens C H and Willett W C (1999), 'Whole-grain consumption and risk of coronary heart disease: results from the Nurses' Health Study', *Am J Clin Nutr*, **70**, 412–419.

Liu S, Manson J E, Stampfer M J, Hu F B, Giovannucci E, Colditz G A, Hennekens C H and Willett W C (2000), 'A prospective study of whole-grain intake and the risk of type 2 diabetes mellitus in US women', *Am J Publ Health*, **90**, 1409–1415.

Liu S, Willett W C, Manson J E, Hu F B, Rosner B and Colditz G A (2003), 'Relation between changes in intakes of dietary fiber and grain products and changes in weight and development of obesity among middle-aged women', *Am J Clin Nutr*, **78**, 920–927.

Lopez H W, Leenhardt F, Coudray C and Remesy C (2002), 'Minerals and phytic acid interactions: is it a real problem for human nutrition?', *Int J Food Sci Technol*, **37**, 727–739.

Lopez H W, Duclos V, Coudray C, Krespine V, Feillet-Coudray C, Messager A, Demigne C and Remesy C (2003), 'Making bread with sourdough improves mineral bioavailability from reconstituted whole wheat flour in rats', *Nutr*, **19**, 524–530.

Lucca P, Hurrell R and Potrykus I (2001), 'Genetic engineering approaches to improve the bioavailability and the level of iron in rice grains', *Theor Appl Genet*, **102**, 392–397.

Malinow M R (1994), 'Homocyst(e)ine and arterial occlusive diseases', *J Int Med*, **236**, 603–617.

Matuschek E, Towo E and Svanberg U (2001), 'Oxidation of polyphenols in phytate-reduced high-tannin cereals: Effect on different phenolic groups and on in vitro accessible iron', *J Agric Food Chem*, **49**, 5630–5638.

Mazur W and Adlercreutz H (1998), 'The scientific basis for risk assessment: naturally occurring oestrogens in food', *Pure Appl Chem*, **70**, 1759–1776.

Mazur W and Adlercreutz H (2000), 'Overview of naturally occurring endocrine-active substances in the human diet in relation to human health', *Nutr*, **16**, 654–658.

McKeown N M, Meigs J B, Liu S, Wilson P W F and Jacques P F (2002), 'Wholegrain intake is favourably associated with metabolic risk factors for type 2 diabetes and cardiovascular disease in the Framingham Offspring Study', *Am J Clin Nutr*, **76**, 390–398.

McMurrough I, and Madigan D (1996b), 'Semipreparative chromatographic procedure for the isolation of dimeric and trimeric proanthocyanidins from barley', *J Agric Food Chem*, **44**, 1731–1735.

McMurrough I, Madigan D and Kelly R J (1996), 'The role of flavonoid polyphenols in beer stability', *J Am Soc Brew Chem*, **54**, 141–148.

Meyer A S and Frankel E N (2001), 'Antioxidant activity of hydroxycinnamic acids on human low-density lipoprotein oxidation', *Methods Enzymol*, **335**, 256–265.

Meyer K A, Kushi L H, Jacob D R, Slavin J, Sellers T A and Folsom A R (2000), 'Carbohydrates, dietary fiber, incident type 2 diabetes mellitus in older women', *Am J Clin Nutr*, **71**, 921–930.

Montonen J, Knekt P, Jarvinen R, Aromaa A and Reuanen A (2003), 'Wholegrain and fiber intake and the incidence of type 2 diabetes', *Am J Clin Nutr*, **77**, 622–629.

Murtaugh M A, Jacobs D R, Jacob B, Steffen L and Marquart L (2003), 'Epidemiological support for the protection of whole grains against diabetes', *Proc Nutr Soc*, **62**, 143–149.

Nardini M, D'Aquino M, Tomassi G, Gentili V, Di-Felice M and Scaccini C (1995), 'Inhibition of human low-density lipoprotein oxidation by caffeic acid and other hydroxycinnamic acid derivatives', *Free Rad Biol Med*, **19**, 541–552.

Nesaretnam K, Stephen R, Dils R and Dabre P (1998), 'Tocotrienols inhibit the growth of human breast cancer cells irrespective of estrogen receptor status', *Lipids*, **33**, 461–469.

Nicodemus K K, Jacobs D R and Folsom A R (2001), 'Whole and refined grain intake and risk of incident postmenopausal breast cancer', *Cancer Causes Control*, **12**, 917–925.

Norton R A (1994), 'Isolation and identification of steryl cinnamic acid derivatives from corn bran', *Cereal Chem*, **71**, 111–117.

Norton R A (1995), 'Quantitation of steryl ferulate and *p*-coumarate esters from corn and rice', *Lipids*, **30**, 269–274.

Oatway L, Vasanthan T and Helm J H (2001), 'Phytic acid', *Food Rev Int*, **17**, 419–431.

Ohta T, Semboku N, Kuchii A, Egashira Y and Sanada H (1997), 'Antioxidant activity of corn bran cell-wall fragments in the LDL oxidation system', *J Agric Food Chem*, **45**, 1644–1648.

Owen R W, Spiegelhalder B and Bartsch H (1998), 'Phytate, reactive oxygen species and colorectal cancer', *Eur J Cancer Preven*, **7**, S41–S54.

Pabon M and Lönnerdal B (1992), 'Effect of citrate on zinc bioavailability from milk, milk fractions and infant formulas', *Nutr Res*, **13**, 103–111.

Panfili G, Fratianni A and Irano M (2003), 'Normal phase high performance liquid chromatography method for the determination of tocopherols and tocotrienols in cereals', *J Agric Food Chem*, **51**, 3940–3944.

Parker R A, Pearce B C, Clark R W, Gordon D A and Wright J J (1993), 'Tocotrienols regulate cholesterol production in mammalian cells by post-transcriptional suppression of 3-hydroxy-3-methylglutaryl-coenzyme A reductase', *J Biol Chem*, **268**, 11230–11238.

Pereira M A, Jacobs D R, Slattery M L, Ruth K J, Van Horn L, Hilner J E and Kushi L H (1998), 'The association of wholegrain intake and fasting insulin in a biracial cohort of young adults: The CARDIA study', *CVD Preven*, **1**, 231–242.

Pereira M A, Pinns J J, Jacobs D R, Marquart L and Keenan J (2001), 'Whole grains, cereal fiber and chronic diseases: Epidemiological evidence', in Spiller G A, *CRC Handbook of Dietary Fiber in Human Nutrition*, Boca Raton, FL, CRC Press, pp. 461–479.

Pereira M A, Jacobs D R, Pins J J, Raatz S K, Gross M D, Slavin J L and Seaquist E R (2002), 'Effect of whole grains on insulin sensitivity in overweight hyperinsulinemic adults', *Am J Clin Nutr*, **75**, 848–855.

Peters F N and Musher S (1937), 'Oat flour as an antioxidant', *Ind Eng Chem*, **29**, 146–151.

Peterson D M (1995), 'Oat tocols: Concentration and stability in oat products and distribution within the kernel', *Cereal Chem*, **72**, 21–24.

Peterson D M and Qureshi A A (1993), 'Genotype and environment effects on tocols of barley and oats', *Cereal Chem*, **70**, 157–162.

Piber M and Koehler P (2005), 'Identification of dehydro-ferulic acid-tyrosine in rye and wheat: Evidence for a covalent cross-link between arabinoxylans and proteins', *J Agric Food Chem*, **53**, 5276–5284.

Pietinen P, Stumpf K, Mannisto S, Kataja V, Uusitupa M and Adlercreutz H (2001), 'Serum enterolactone and risk of breast cancer: A case-control study in eastern Finland', *Cancer Epidemiol Biomarkers Preven*, **10**, 339–344.

Piironen V, Lindsay D G, Miettinen T, Toivo J and Lampi A-M (2000), 'Plant sterols: biosynthesis, biological function and their importance to human nutrition', *J Sci Food Agric*, **80**, 939–966.

Piironen V, Toivo J and Lampi A-M (2002), 'Plant sterols in cereals and cereal products', *Cereal Chem*, **77**, 148–154.

Qu H Y, Madl R L, Takemoto D J, Baybutt R C and Wang W Q (2005), 'Lignans are involved in the antitumor activity of wheat bran in colon cancer SW 480 cells', *J Nutr*, **135**, 598–602.

Qureshi A A, Burger W C, Petterson D M and Elson C E (1986), 'The structure of an inhibitor of cholesterol biosynthesis isolated from barley', *J Biol Chem*, **261**, 10544–10550.

Qureshi A A, Petterson D M, Elson C E, Mangels A R and Din, Z Z (1989), 'Stimulation of avian cholesterol metabolism by α-tocopherol', *Nutr Rep Int*, **40**, 993–1001.

Qureshi A A, Qureshi N, Wright J J K, Shen Z, Kramer G, Gapor A, Chong Y H, Dewitt G, Ong A S H, Petterson D M and Bradlow B A (1991a), 'Lowering serum cholesterol in hypercholesterolemic humans by tocotrienols (palmvitee)', *Am J Clin Nutr*, **53**, 1021S–1026S.

Qureshi A A, Qureshi N, Hasler-Rapacz J O, Weber F E, Chaudhary V, Crenshaw T D, Gapor A, Ong A S H, Chong Y H, Petterson D M and Rapacz J (1991b), 'Dietary tocotrienols reduce concentrations of plasma cholesterol, apolipoprotein B, thromboxane B2, and platelet factor 4 in pigs with inherited hyperlipidemias', *Am J Clin Nutr*, **53**, 1042S–1046S.

Qureshi A A, Mo H, Packer L and Petterson D M (2000), 'Isolation and identification of novel tocotrienols from rice bran with hypocholesterolemic, antioxidant, and antitumor properties', *J Agric Food Chem*, **48**, 3130–3140.

Qureshi A A, Salser W A, Parmar R and Emeson E E (2001), 'Novel tocotrienols of rice bran inhibit atherosclerotic lesions in C57BL/6 ApoE-deficient mice', *J Nutr*, **131**, 2606–2618.

Raboy V (2002), 'Progress in breeding low phytate crops', *J Nutr*, **132**, S503–S505.

Raederstorff D, Elste V, Aebischer C and Weber P (2002), 'Effect of either γ-tocotrienol or a tocotrienol mixture on the plasma lipid profile in hamsters', *Ann Nutr Metab*, **46**, 17–23.

Ragaee S, Abdel-Aal E M and Noaman M (2006), 'Antioxidant activity and nutrient composition of selected cereals for food use', *Food Chem*, **98**, 32–38.

Ralph J, Quideau S, Grabber J H and Hatfield R D (1994), 'Identification and synthesis of new ferulic acid dehydrodimers present in grass cell walls', *J Chem Soc, Perkin Trans 1*, 3485–3498.

Ranhotra G, Gelroth J, Novak F and Bohannon F (1985), 'Bioavailability for rats of thiamin in whole wheat and thiamin-restored white bread', *J Nutr*, **115**, 601–606.

Reddy B, Engle A, Katsifis S, Simi B, Bartram H P, Perrino P and Mahan C (1989), 'Biochemical epidemiology of colon cancer: effect of types of dietary fiber on fecal mutagens, acid, and neutral sterols in healthy subjects', *Cancer Res*, **49**, 4629–4635.

Reddy, M B, Hurrell R F, Juillerat M A and Cook J D (1996), 'The influence of different protein sources on phytate in hibition of nonheme-iron absorption inhumans', *Am J Clin Nutr*, **63**, 203–207.

Reed J D, Krueger C G and Vestling M M (2005), 'MALDI-TOF mass spectrometry of oligomeric food polyphenols', *Phytochemistry*, **66**, 2248–2263.

Ross A B, Kamal-Eldin A and Åman P (2004), 'Dietary alkylresorcinols: Absorption, bioactivities and possible use as biomarkers of whole grain wheat and rye rich foods', *Nutr Rev*, **62**, 81–95.

Ruibal-Mendieta N L, Delacroix D L, Mignolet E, Pycke J M, Marques C, Rozenberg R, Petitjean G, Habib-Jiwan J L, Meurens M, Quetin-Leclercq J, Delzenne N M and Larondelle Y (2005), 'Spelt (*Triticum aestivum* ssp spelta) as a source of breadmaking flours and bran naturally enriched in oleic acid and minerals but not phytic acid', *J Agric Food Chem*, **53**, 2751–2759.

Sahyoun N R, Jacques P F, Zhang X L L, Juan W Y and McKeown N M (2006), 'Whole-grain intake is inversely associated with the metabolic syndrome and mortality in older adults', *Am J Clin Nutr*, **83**, 124–131.

Sandberg A S (2002), '*In vitro* and *in vivo* degradation of phytate', in Reddy N R and Sathe S K, *Food Phytates*, Boca Raton, FL CRC Press, pp. 139–155.

Sandström B (1997), 'Bioavailability of zinc', *Eur J Clin Nutr*, **51**, S17–S19.

Sandström B, Almgren A, Kivisto B and Cederblad A (1987), 'Zinc absorption in humans from meals based on rye, barley, oatmeal, triticale, and whole wheat', *J Nutr*, **117**, 1898–1902.

Sandström B, Almgren A, Kivisto B and Cederblad A (1989), 'Effect of protein and protein source on zinc absorption in humans', *J Nutr*, **119**, 48–53.

Saxholt E and Møller A (2003), Danish Food Composition Databank (Revision 5.0), http://www.foodcomp.dk/fcdb_default.asp.

Seal C J (2006), 'Whole grains and CVD risk', *Proc Nutr Soc*, **65**, 24–34.

Seitz L M (1989), 'Stanol and sterol esters of ferulic and *p*-coumaric acids in wheat, corn, rye and triticale', *J Agric Food Chem*, **37**, 662–667.

Sheppard A J, Pennington J A T and Weihrauch J L (1993), 'Analysis and distribution of vitamin E in vegetable oils and foods', in Packer L and Fuchs J, *Vitamin E in Health and Disease*, New York, Marcel Dekker, pp. 9–31.

Siegenberg D, Baynes R D, Bothwell T H, Macfarlane B J, Lamparelli R D, Car N G, Macphail P, Schmidt U, Tal A and Mayet F (1991), 'Ascorbic acid prevents the dose-dependent inhibitory effects of polyphenols and phytates on nonheme-iron absorption', *Am J Clin Nutr*, **53**, 537–541.

Siener R, Honow R, Voss S, Seidler A and Hesse A (2006), 'Oxalate content of cereals and cereal products', *J Agric Food Chem*, **54**, 3008–3011.

Slavin J L (2003), 'Whole grains are protective: biological mechanisms', *Proc Nutr Soc*, **62**, 129–134.

Slavin, J, Jacobs D and Marquart L (1997), 'Whole-grain consumption and chronic disease: Protective mechanisms', *Nutr Cancer*, **27**, 14–21.

Slavin J L, Martini M C, Jacobs D R and Marquart L (1999), 'Plausible mechanisms for the protectiveness of whole grains', *Am J Clin Nutr*, **70**, 459S–463S.

Slow S, Donaggio M, Cressey P J, Lever M, George P M and Chambers S T (2005), 'The betaine content of New Zealand foods and estimated intake in the New Zealand diet', *J Food Comp Anal*, **18**, 473–485.

Sugano L and Tsuji E (1997), 'Rice bran and cholesterol metabolism', *J Nutr*, **127**, S521–524.

Tham D M, Gardner C D and Haskell W L (1998), 'Potential health benefits of dietary phytoestrogens: A review of the clinical, epidemiological, and mechanistic evidence', *J Clin Endocrinol Metab*, **83**, 2223–2235.

Theriault A, Chao J T and Gapor A (2002), 'Tocotrienol is the most effective vitamin E for reducing endothelial expression of adhesion molecules and adhesion to monocytes', *Atherosclerosis*, **160**, 21–30.

Tian S, Nakamura K, Cui T and Kayahara H (2005), 'High-performance liquid chromatographic determination of phenolic compounds in rice', *J Chromatogr A*, **1063**, 121–128.

Truswell A S (2002), 'Cereal grains and coronary heart disease', *Eur J Clin Nutr*, **56**, 1–14.

UK Department of Health (2000), *Folic Acid and the Prevention of Disease*, London Stationery Office, pp. 1–101.

Van Dam R M, Rimm E B, Willet W C, Stampfer M J, Hu F B (2002), 'Dietary patterns and risk of type 2 diabetes mellitus in U.S. men', *Ann Intern Med*, **136**, 201–209.

Vargas P A and Alberts D S (1993), 'Colon cancer: the quest for prevention', *Oncology*, **7**, 33–40.

Vats P and Banerjee U C (2004), 'Production studies and catalytic properties of phytases (myo-inositolhexakisphosphate phosphohydrolases): An overview', *Enzyme Microb Technol*, **35**, 3–14.

Viscidi K A, Dougherty M P, Briggs J and Camire M E (2004), 'Complex phenolic compounds reduce lipid oxidation in extruded oat cereals', *Food Sci Technol*, **37**, 789–796.

Waldron K W, Parr A J, Ng A and Ralph J (1996), 'Cell wall esterified phenolic dimers: Identification and quantification by reversed phase high performance liquid chromatography and diode array detection', *Phytochem Anal*, **7**, 305–312.

Wang L, Newman R K, Newman C W, Jackson L L and Hofer P J (1993), 'Tocotrienol and fatty acid composition of barley oil and their effects on lipid metabolism', *Plant Foods Hum Nutr*, **43**, 9–17.

Welch R M and Graham R (2004), 'Breeding for micronutrients in staple food crops from a human nutrition perspective', *J Exp Bot*, **55**, 353–364.

White P J and Broadley M R (2005), 'Biofortifying crops with essential mineral elements', *Trends Plant Sci*, **10**, 586–593.

WHO (1999), WHO Regional Office for the Eastern Mediterranean, Alexandria, Egypt, 'Fortification of flour to control iron deficiencies in the Middle East and North Africa', Report of a joint WHO/UNICEF/MI/ILSI workshop, Beirut, Lebanon, 13–16 June 1998, http://www.emro.who.int/nutrition/PDF/Flour_Fortification.pdf.

Wise A (1995), 'Phytate and zinc bioavailability', *Int J Food Sci Nutr*, **46**, 53–63.

Ye X, Al-Babili S, Klöti A, Zhang J, Lucca P, Beyer P and Potrykus I (2000), 'Engineering the provitamin A (beta-carotene) biosynthetic pathway into (carotenoid-free) rice endosperm', *Science*, **287**, 303–305.

Zhou J R and Erdman J W (1995), 'Phytic acid in health and disease', *Crit Rev Food Sci Nutr*, **35**, 495–508.

Zimmermann R and Qaim M (2004), 'Potential health benefits of Golden Rice: A Philippine case study', *Food Policy*, **29**, 147–168.

6

Whole grain consumption and insulin sensitivity

M. A. Pereira, University of Minnesota, USA

6.1 Introduction

The purpose of this chapter is to describe the scientific evidence on the topic of whole grains and insulin sensitivity. Discussion of insulin secretion or frank type 2 diabetes is not the focus of this chapter, and the reader is referred elsewhere for these topics. Although obesity is one of the most important risk markers for type 2 diabetes, there is great potential for physical activity and diet, including whole grain foods, to improve insulin sensitivity independent of energy balance and obesity. For this reason, and because whole grains and obesity are discussed in Chapter 3, this chapter will focus on mechanisms and effects that are independent of obesity. The material in this chapter includes discussion of plausible biological mechanisms linking whole grain intake to insulin sensitivity in a causal manner, observational human studies on this topic, and finally the experimental evidence. Studies of nutrients or components of foods, such as fiber or vitamins and minerals, are discussed in the section on potential mechanisms; the remainder of the focus herein is a discussion of human investigations on the possible independent effect of whole grains, eaten in their whole food form, on insulin sensitivity and therefore, glycemic control and possibly the prevention of type 2 diabetes and cardiovascular disease.

6.2 The global burden of insulin resistance

Diabetes mellitus is a chronic disease that affects almost 16 million Americans – approximately 6% of the population. Diabetes is now recognized as a pandemic, with very high rates around the globe, particularly in regions such as parts of Asia that may have strong genetic predispositions.[1] Of those with diabetes, 90–95% have type 2 diabetes. Of these, roughly one-third are unaware that they have the disease. Diabetes is the seventh leading cause of death in the United States and a major contributor to blindness, kidney disease, cardiovascular disease, nerve disease, and amputations. The total annual economic cost of diabetes appears to exceed $100 billion.[2] The per capita costs of health care for people with diabetes is ~ $10 071 per year, compared with costs for people without diabetes of ~ $2699 per year.

As will be discussed in detail, type 2 diabetes results from insulin resistance, a condition in which the body fails to properly use insulin, combined with relative insulin deficiency. The risk for type 2 diabetes increases with age, usually occurring after age 45. Other important risk factors for type 2 diabetes include genetic factors and obesity. Some have suggested that the average age of diagnosis of type 2 diabetes may be decreasing as a result of the rising prevalence of overweight and obesity in youth.

Nutrition therapy is an important cornerstone in the management of type 2 diabetes[3] but relatively little is understood about the role of diet in the development of this disease. Diet is known to influence body weight and thus is recognized as a modifiable risk factor for type 2 diabetes. High-fat, low-fiber diets have been suggested to increase the risk of insulin resistance and thus lead to the development of type 2 diabetes mellitus. This hypothesis is mainly supported by animal experiments showing a reduction in insulin sensitivity associated with high-fat diets[4] and ecologic studies indicating higher incidences of type 2 diabetes mellitus among populations with high consumption of fats.[5] Until recently, there were few studies of whole foods and dietary patterns in relation to risk of type 2 diabetes.

Dietary recommendations promote the positive health and nutrition benefits of whole grain foods. The *Dietary Guidelines for Americans* recommend that Americans eat at least six servings of grains each day with more servings of whole grain foods because whole grain foods may reduce the risk of coronary heart disease, bowel diseases, and possibly some types of cancer.[6] Evidence is building to support the consumption of whole grain foods as a way to reduce the risk of type 2 diabetes. This chapter reviews current research relating to whole grain consumption and insulin resistance, a precursor and risk factor for type 2 diabetes. 'Whole grain' refers to the presence of all parts of the grain – the bran, endosperm, and germ – in the food. This is important, nutritionally, because nutrients and phytochemicals are not distributed evenly throughout the grain and the traditional milling process results in reduced amounts of key nutrients. In the milling process, the bran and germ are separated from the starchy endosperm and the starchy endosperm is ground into flour. The endosperm makes up about 80% of the whole grain while the germ and bran components vary among different grains. Table 6.1 compares key nutrient levels of whole grain wheat and refined wheat. Whole grains are higher in dietary fiber (insoluble and soluble fibers), vitamins –

Table 6.1 Compositional differences between whole and refined wheat

Component	Whole wheat	Refined wheat
Bran (%)	14	<0.1
Germ (%)	2.5	<0.1
Total dietary fiber (%)	13	3
Insoluble dietary fiber (%)	11.5	1.9
Soluble dietary fiber (%)	1.1	1.0
Protein (%)	14	14
Fat (%)	2.7	1.4
Starch and sugar (%)	70	83
Total minerals (%)	1.8	0.6
Selected minerals		
Zinc (μg/g)	29	8
Iron (μg/g)	35	13
Selenium (μg/g)	0.06	0.02
Magnesium (mg/g)	1.38	0.22
Selected vitamins		
Vitamin B6 (mg/g)	7.5	1.4
Folic acid (mg/g)	0.57	0.11
Phenolic compounds		
Ferulic acid (mg^{-2}/g)	5	0.4
β Tocotrienol (μg/g)	32.8	5.7
Phytate phosphorus (mg/g)	2.9	0.1

Adapted from Thompson (1992).[16]

especially B vitamins – and minerals. About 50–80% of the minerals in grains – including copper, zinc, and magnesium – are found in the bran. Other healthful components found in whole grain include tocotrienols, lignans, phenolic compounds, phytic acid, and phytoestrogens. Consistent with this nutrient-rich profile, diets that are high in whole grains are associated with reduced risk of heart disease, lower rates of some types of cancer, and lower all-cause mortality.[8–11]

Despite health and nutrition recommendations to include more whole grain foods in the diet, Americans fall short of desired levels. A recent diet study reported a mean whole grain intake of only 0.5 times per day for adults. Researchers reported that less than 2% of the study population consumed whole grain more than twice per day and 23% consumed no whole grains over the 14-day collection period.[12] According to a more recent study of dietary intake, using data from the US Department of Agriculture (USDA) 1994–96 Continuing Survey of Food Intakes by Individuals, US adults consumed an average of 6.7 servings of grain products per day, with only 1 serving being whole grain.[13] Thirty-six percent averaged less than 1 whole grain serving per day based on 2 days of intake data and only 8% met the recommendation to eat at least 3 servings per day. Yeast breads and breakfast cereals each provided almost one-third of the whole grain servings, grain-based snacks provided about one-fifth, and less than one-twentieth came from quick breads, pasta, rice, cakes, cookies, pies, pastries, and miscellaneous grains.

6.3 Potential mechanisms

Insulin sensitivity of the peripheral tissues, primarily skeletal muscle and liver, are the primary sites of glucose disposal.[14] Low insulin sensitivity, otherwise know as insulin resistance, of these tissues, is a major determinant of risk for type 2 diabetes as well as cardiovascular disease;[14–16] the other major determinant being decompensation of the insulin producing pancreas beta-cells.[14] How might whole grain foods enhance the action of insulin in stimulating glucose uptake, i.e. improve insulin sensitivity or reduce insulin resistance? At least theoretically, there are myriad mechanisms through which one can hypothesize that whole grain foods may improve insulin sensitivity, including effects of the following components: (1) fiber – especially soluble fiber; (2) glycemic index – especially for those whole grains that are minimally processed and have large particle size or are rich in soluble fiber; (3) magnesium; (4) vitamin E; (5) phytochemicals, including a number of phenolic compounds. Table 6.1 provides an overview of some of the major nutrient and other content differences between typical whole and refined grain products.

Fiber and glycemic index has been found to be important for glycemic control through a number of mechanisms, demonstrated in animals and humans. However, the extent to which the improved glycemic control may be attributed directly to insulin sensitivity versus secretion is unclear. Fiber content and glycemic index are related but are not nearly the same concept. Furthermore, some health benefits attributed to fiber may be confounded by other food attributes or components that are common to high-fiber foods. Nonetheless, it is clear that soluble fiber-rich foods, such as the whole grains oats and barley, have low glycemic index values, and have been shown in experimental studies, some of which are discussed later, to improve insulin sensitivity. One important physiologic factor that appears to contribute to diabetes risk is glucose toxicity, whereby high glucose concentrations may impair the action of insulin at the insulin receptor, causing acute insulin resistance. Therefore, it is possible that those whole grain foods, such as oats and barley, rich in soluble fiber and low in glycemic index, may improve insulin sensitivity through their direct impact on attenuating postprandial glucose spikes and lowering the area under the postprandial glucose curve over several hours following the meal. Certainly, humans in Western culture spend most of their time in the postprandial period, and, in a prediabetic state of impaired glucose tolerance, glucose concentration can be excessively raised on an essentially chronic basis. A defining feature of the Western diet is an abundant intake of refined grain foods and in the context of limited physical activity, a pro-insulin-resistant lifestyle. Figure 6.1 provides a theoretical

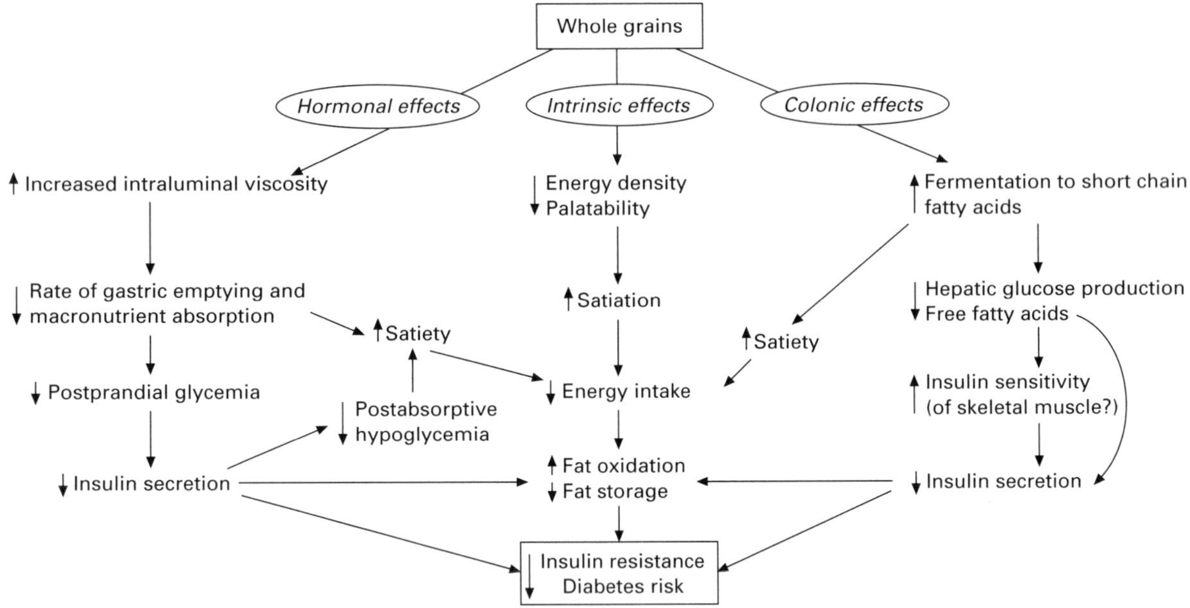

Fig. 6.1 Theoretical pathways through which whole grain intake may improve insulin sensitivity. Adapted from reference 5.

framework for a variety of plausible pathways through which increased whole grain intake, in the context of a prudent carbohydrate- and fiber-rich diet, may improve insulin and glucose metabolism, thereby reducing the risk of type 2 diabetes.

One of the first studies to examine the effect of whole grain on glycemic response was done by David Jenkins and colleagues.[17] Diabetic volunteers consumed bread with varying ratios of milled flour to four including semi-intact grain kernels in a randomized trial. Results indicated that consumption of breads with cracked whole grains resulted in significantly lower postprandial blood glucose and insulin. However, it should be kept in mind that most 'whole grains' as commonly consumed by the general public are not intact whole kernels. These foods as consumed usually have a small particle size, such as whole wheat flour, and therefore a high glycemic index. As suggested in the epidemiologic studies described above, whole grain foods may offer protection from type 2 diabetes through mechanisms other than glycemic index.

Reduced hyperglycemia and insulin demand may explain a protective effect of higher intakes of whole grain foods. The slower intestinal absorption generally reflects the lower glycemic indexes of whole grain foods, with glycemic index being the postprandial blood glucose response following ingestion of a specified food relative to that of pure glucose or white bread. Although whole grain flour may tend to have a similar glycemic index to refined grain flour,[18] research discussed below suggests that habitual consumers of whole grain flour appear to have lower fasting insulin than habitual refined grain flour consumers.

Many characteristics of grains can influence glucose and insulin response. The gel-forming property of soluble fiber sources such as oats and barley has been proposed as a mechanism by which these grains reduce glucose and insulin response. The high viscosity may result in delayed gastric emptying and/or intestinal absorption which in turn reduces the glycemic response.[19] In a review of mechanisms of the effects of grains on insulin and glucose response, Hallfrisch and Behall[20] summarized the influence of particle size, processing, soluble and insoluble fiber content, and amylose and amylopectin content of the grains. The greater the particle size, the lower the glucose and insulin response. The greater the level of processing and refining, the higher the response. Grains with high levels of soluble beta-glucans such as oats, rye and barley are generally more effective in improving insulin sensitivity than wheat, which contains predominantly insoluble dietary fiber. Corn and rice can have either high or low glycemic indices because their amylose and amylopectin contents vary. Higher amylose content results in lower glucose and insulin response.

In addition to the possible postprandial effects of whole grain fiber and lower glycemic index whole grain foods on insulin sensitivity, common mineral and vitamin constituents of many whole grains, such as magnesium and vitamin E are other possible insulin-sensitizing candidates. Some of these possible nutrient effects are shown in Fig. 6.2. Magnesium is an important mineral involved in carbohydrate metabolism, and had been found in some, but not all, studies to be an important modulator of insulin action in the peripheral tissues. The potential role of vitamin E, although less clear in this regard, holds some promise in possibly preventing insulin receptor damage due to oxidative stress secondary to glucose toxicity or other sources of oxidation in the Western diet or lifestyle. As with magnesium, the published literature is inconsistent in its support for vitamin E in this respect.

There has been considerable enthusiasm in nutritional epidemiology over the past decade or so for the topic of so-called 'phytochemicals', what one might refer to as 'candidate nutrients' or the basis for 'functional foods'. Whole grains are a rich source of these substances, particularly some of the polyphenolic compounds, which may serve many roles in the body to prevent chronic disease through a variety of biological systems. Some of these compounds are suspected to be powerful antioxidants, while others are thought to possibly be involved in phosphorylation of the insulin receptor, steps critical to glucose uptake, i.e. insulin sensitivity.

Magnesium depletion has been shown to result in insulin resistance as well as impaired insulin secretion and thereby may worsen control of diabetes, although conclusions have not been drawn about a direct link between magnesium depletion and the etiology of type 2 diabetes mellitus.[21–24] Four prospective cohort studies have found a strong inverse association between magnesium intake and risk of type 2 diabetes.[17,18,20,25] Subjects with higher intakes of magnesium had about a 24–38% lower risk of developing type 2 diabetes, compared with those with lowest intakes. Magnesium is involved

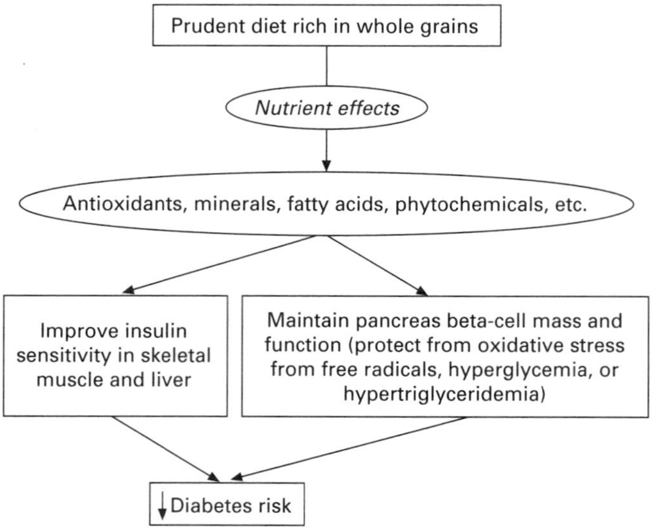

Fig. 6.2 The potential nutrient effects of whole grains on improved insulin sensitivity and beta-cell function.

in glucose homeostasis, specifically by playing a role in the release of insulin by the pancreas.[21–24] It is also thought to be able to increase the number and the affinity of the insulin receptors.[23] Whole grains are a rich source of magnesium. In the experiment described by Pereira *et al.*[26] (see Section 6.5), the whole grain diet had nearly twice the magnesium content compared with the refined grain diet.

The antioxidant theory is another hypothesis related to the mechanisms whereby whole grains may be protective against type 2 diabetes. Whole grains contain many antioxidants, including vitamins, trace minerals, and non-nutrients – such as phenolic acids, lignans, and phytoestrogens – and antinutrients such as phytic acids.[27] Whole grains are also concentrated sources of vitamin E, especially tocotrienols and possibly selenium, depending on the selenium content of the soil where the grains are grown. In fact, recent research indicates that the antioxidant activity of whole grain foods is comparable with fruits and vegetables.[28] Oxidative stress has been associated with reduced insulin-dependent glucose disposal and diabetic complications.[29] Facchini *et al.*[29] propose that decreased dietary intake of liposoluble antioxidant vitamins contributes to a decrease in plasma concentration of carotenoids and tocopherols, which in turn impairs the ability of insulin to stimulate glucose disposal by muscle. Insulin resistance results in an increase in lipid peroxidation, a process accentuated by the associated decrease in plasma antioxidant content. Alternatively, lipid peroxidation, secondary to a decrease in antioxidant vitamins, might impair insulin action. It is also possible that whole grains may protect beta-cell function through enhanced antioxidant content. Beta-cell dysfunction is a critical step in the development of type 2 diabetes. However, much less is known about the possible role of diet in beta-cell function compared with its role in insulin resistance.

It is likely that these and many other possible mechanisms may link whole grain foods intake to insulin sensitivity in a causal manner that may be synergistic and overlapping. No single mechanism in isolation, even if there is such a thing as a mechanism working in isolation in the human body, is thought to play a significant role in glucose modulation and diabetes prevention in most individuals.

6.4 Observational evidence/cross-sectional and prospective epidemiologic studies

Epidemiologic studies have generally supported an inverse association between whole grain intake and type 2 diabetes risk.[30–34] If whole grain intake plays an important role in the etiology of type 2 diabetes, then whole grain intake should be inversely associated with fasting blood concentrations of insulin in adults who are free of type 2 diabetes. In the CARDIA study, a prospective study of cardiovascular disease risk factor evolution in a biracial cohort, fasting insulin was in fact inversely associated with whole grain intake.[35]

Adjusted for other lifestyle factors and body mass index (BMI), fasting insulin tended to be higher in those with higher white bread consumption. In contrast, higher intakes of whole grain bread and whole grain cereals without bran added tended to have inverse associations with fasting insulin despite the relatively high glycemic index of those foods. This study demonstrated a statistically significant inverse association between whole grain intake and fasting insulin and suggests that additional characteristics of whole grains, aside from any beneficial effect on energy balance and body weight, may confer protection from hyperinsulinemia. Further, a dose–response relationship between insulin and whole grain intake is suggested by the lower insulin levels for higher reported consumption of whole grains. These data suggest that replacing two servings per day of white bread with one serving of whole grain bread and one serving of whole grain breakfast cereal would result in a decrease in fasting insulin of approximately 15%.

Recently, McKeown and colleagues[36, 37] examined the association between whole grain intake and factors of the metabolic syndrome in the Framingham Offspring Study. In agreement with the findings of the CARDIA study described above, the authors found that whole grain intake was inversely associated with fasting insulin concentration, after adjustment for other dietary and lifestyle variables. Furthermore, whole grain intake was inversely associated with body mass index, waist–hip ratio, tryglycerides, and blood pressure, while positively associated with high-density lipoprotein (HDL) cholesterol. Although cross-sectional in nature, this study provides good empirical support for the proposal that whole grain intake may improve insulin resistance, as the constellation of metabolic variables included in the metabolic syndrome were associated with whole grain intake.

In a cross-sectional study of dietary habits and urinary glucose excretion, Ekblond and colleagues[38] investigated the association between consumption of certain foods and macronutrients and urinary glucose excretion, which is a predictor of non-insulin-dependent diabetes mellitus. The study population included 14 743 men and 18 064 women aged 50–64 years who were participants in the Danish study 'Diet, Cancer and Health'. After adjustment for age, BMI, activity, and total energy, a negative association with urinary glucose excretion was observed for fruit, cereal, poultry, and olive oil for both men and women. Total fiber intake was inversely associated with glycosuria in persons of both sexes and significantly so in men (odds ratio = 0.68; 95% confidence interval = 0.51–0.91).

Fung *et al.*[39] evaluated the effect of overall dietary patterns (as opposed to individual foods or nutrients) and relationship to biomarkers of obesity and cardiovascular disease risk. The researchers identified two major diet patterns ('Western' and 'prudent') in a subsample of men ($n = 466$) from the Health Professionals Follow Up Study. Foods from the semi-quantitative food frequency questionnaire were classified into 42 food groups on the basis of nutrient content or culinary use. The two major dietary patterns identified were: the 'prudent diet' – characterized by higher intakes of fruits, vegetables, poultry, fish, whole grains, and legumes; and the 'Western diet' – characterized by higher intakes of red meat, processed meat, french fries, eggs, high-fat dairy products, sweets and refined grains. A significant positive correlation was shown between the Western diet and insulin, C-peptide (an indicator of endogenous insulin production), leptin and homocysteine levels. Although this study did not evaluate whole grain consumption alone and the occurrence of diabetes, it provides added evidence that, as part of healthy diet, whole grains may have a positive influence on early biomarkers of insulin resistance.

6.5 Experimental evidence – clinical investigations and controlled trials

In a clinical trial conducted by Fukagawa and colleagues,[40] a euglycemic clamp measured insulin-mediated glucose disposal effects in healthy subjects who were given a high-carbohydrate, high-fiber (HCF) diet with 40% of plant fiber provided by whole grain foods for 21–28 days of diet treatment. Results indicated that following the HCF diet for 21–28 days improved carbohydrate economy by enhanced peripheral sensitivity to insulin. However, as 60% of the fiber was from sources other than whole grains, these results cannot be fully attributed to whole grains.

A recent randomized cross-over clinical trial of 13 subjects with type 2 diabetes showed that subjects who consumed a high intake of dietary fiber, particularly soluble fiber (50 g total fiber with 25 g soluble fiber) had improved glycemic control, decreased hyperinsulinemia, and lower plasma lipid concentrations than those who consumed a moderate amount of fiber (24 g total dietary fiber with 8 g soluble fiber).[41] Both diets had the same macronutrient and energy content, and fiber was provided by grains, fruits, and vegetables. Additional studies are needed to improve our understanding of the role of fiber, and particularly soluble fiber, in the glycemic response.

In a controlled clinical study to test the hypothesis that whole grain consumption would improve insulin sensitivity in overweight adults, Pereira *et al.*[26] compared insulin sensitivity between diets high in whole grain and diets high in refined grain. Following a 6-week feeding period, insulin sensitivity was measured using a euglycemic hyperinsulinemic clamp test. Fasting insulin was 10% lower during the whole grain diet compared with the refined grain diet and was independent of baseline body weight and weight change. In comparison with the refined grain diet, after the whole grain diet the area under the 2-hour insulin curve appeared to be lower, although this finding was not statistically significant. The rate of glucose infusion during the final 30 min of the clamp test was higher following the whole grain diet ($p < 0.05$). These results suggest that insulin sensitivity may be an important mechanism whereby whole grain foods reduce the risk for type 2 diabetes.

6.6 Summary and future trends

Prospective cohort studies have demonstrated inverse associations between whole grain intake and biomarkers and risk factors for type 2 diabetes. Experimental studies have supported the view that these associations may be causal, and therefore that increasing whole grain intake should work to reduce the risk of type 2 diabetes through a variety of pathways, including reduction in obesity and insulin resistance, and improvement in related metabolic pathways and perhaps beta-cell function. Research to date therefore provides good evidence that an increase in whole grain consumption to recommended amounts may reduce the incidence of type 2 diabetes mellitus. Additional epidemiological studies and clinical studies will be important to establish this relationship more firmly and to elucidate possible biological mechanisms. In the meantime, based on the epidemiologic and experimental evidence, it would be prudent for health care providers and public health officials to increase their emphasis on the importance of increasing whole grain in the habitual diets of individuals and in the food supply of the population. Indeed, the *Dietary Guidelines for Americans*[6] recommends at least three servings of whole grain food per day to contribute to a health-optimal dietary pattern.

6.7 Sources of information and advice

In order to supplement the information in this chapter and its references, the reader is referred to other published reviews, textbooks, and scientific organizations for additional information and material on the topic of whole grain intake and insulin sensitivity. Textbooks and chapters include the following:

1 Jacobs D, Pereira MA, Slavin J, Marquart L. Defining the impact of whole grain intake on chronic disease. *Cereal Foods World* 2000; **45**: 51–53.
2 Pereira MA, Pins JJ. Dietary fiber and cardiovascular disease: experimental and epidemiologic advances. *Current Atherosclerosis Reports* 2000; **2**: 494–502.
3 Pereira MA, Jacobs DR Jr, Pins JJ, Marquart L, Keenan J. Whole grains, cereal fiber, and chronic disease: the epidemiologic evidence. In: Spiller GA (ed.). *CRC Handbook of Dietary Fiber in Human Nutrition*, 3rd Edition. New York: CRC Press; 2001, pp. 461–480.
4 Pins JJ, Pereira MA, Jacobs DR Jr, Marquart L, Slaving JE. Whole grains, cereal fiber, and chronic disease: possible biological mechanisms. In: Spiller GA (ed.). *CRC Handbook of Dietary Fiber in Human Nutrition*, 3rd Edition. New York: CRC Press; 2001, pp. 481–498.
5 Pereira MA, Ludwig DS. Dietary fiber and body weight regulation: observations and mechanisms. *Pediatric Clinics of North America* 2001; **48**: 969–980.

6. Pereira MA. Whole grain consumption and body weight regulation. In: Marquart L, Slavin JL, Fulcher RG (ed.). *Whole Grain Foods in Health and Disease*. St Paul: American Association of Cereal Chemists; 2002, pp. 233–242.
7 Pereira MA, Liu S. Types of carbohydrates and risk of cardiovascular disease. *Journal of Women's Health* 2003; **12**: 115–122.
8 Timlin MT, Pereira MA. Breakfast frequency and quality in the etiology of adult obesity and chronic diseases: a review of the literature. *Nutrition Reviews* (in press).
9 Temple NJ, Wilson E, Jacobs DR Jr (eds). N*utritional Health: Strategies for Disease Prevention* (2nd edition). Totowa, New Jersey: Humana Press; 2006.

Scientific organization resources include the following:

- American Diabetes Association, www.diabetes.org.
- American Dietetic Association, www.eatright.org/cps/rde/xchg/ada/hs.xsl/index.html.
- American Heart Association, www.americanheart.org.
- Nutrition Society, www.nutritionsociety.org.
- United States Department of Agriculture, www.usda.gov.
- American Association of Cereal Chemists, www.aaccnet.org.

6.8 References

1 Yoon KH, Lee JH, Kim JW, Cho JH, Choi YH, Ko SH, Zimmet P, Son HY. Epidemic obesity and type 2 diabetes in Asia. *Lancet*, Nov 11 2006; **368**(9548): 1681–1688 (PMID: 17098087).
2 Economic consequences of diabetes mellitus in the U.S. in 1997. American Diabetes Association. *Diabetes Care*, Feb 1998; **21**(2): 296–309 (PMID: 9539999).
3 Sheard NF, Clark NG, Brand-Miller JC, Franz MJ, Pi-Sunyer FX, Mayer-Davis E, Kulkarni K, Geil P. Dietary carbohydrate (amount and type) in the prevention and management of diabetes: a statement by the American Diabetes Association. *Diabetes Care,* Sep 2004; **27**(9):2266–2271 (PMID: 15333500).
4 Storlien LH, James DE, Burleigh KM, Chisholm DJ, Kraegen EW. Fat feeding causes widespread in vivo insulin resistance, decreased energy expenditure, and obesity in rats. *Am J Physiol*, Nov 1986; **251**(5 Pt 1): E576–583 (PMID: 3535532).
5 Rewers M, Hamman R. *Risk Factors for Non-Insulin-Dependent Diabetes*. NIH publication 95–1468. Washington, D.C: National Institutes of Health; 1995.
6 US Departments of Agriculture and Health and Human Services. *Nutrition and Your Health: Dietary Guidelines for Americans*: 2005.
7 Thompson LU. Potential health benefits of whole grains and their components. *Contemp Nutr*, 1992; **17**: 1–2.
8 Jacobs DR, Jr, Marquart L, Slavin J, Kushi LH. Whole-grain intake and cancer: an expanded review and meta-analysis. *Nutr Cancer*, 1998; **30**(2): 85–96 (PMID: 9589426).
9 Jacobs DR, Jr, Meyer KA, Kushi LH, Folsom AR. Whole-grain intake may reduce the risk of ischemic heart disease death in postmenopausal women: the Iowa Women's Health Study. *Am J Clin Nutr*, Aug 1998; **68**(2): 248–257 (PMID: 9701180).

10 Jacobs DR, Jr, Meyer KA, Kushi LH, Folsom AR. Is whole grain intake associated with reduced total and cause-specific death rates in older women? The Iowa Women's Health Study. *Am J Public Health*, Mar 1999; **89**(3): 322–329 (PMID: 10076480).
11 Liu S, Stampfer MJ, Hu FB, Giovannucci E, Rimm E, Manson JE, Hennekens CH, Willett WC. Whole-grain consumption and risk of coronary heart disease: results from the Nurses' Health Study. *Am J Clin Nutr*, Sep 1999; **70**(3): 412–419 (PMID: 10479204).
12 Albertson AM, Tobelmann RC. Consumption of grain and whole-grain foods by an American population during the years 1990 to 1992. *J Am Diet Assoc*, Jun 1995; **95**(6): 703–704 (PMID: 7759751).
13 Cleveland LE, Moshfegh AJ, Albertson AM, Goldman JD. Dietary intake of whole grains. *J Am Coll Nutr*, Jun 2000; **19**(3 Suppl): 331S–338S (PMID: 10875606).
14 DeFronzo RA. Pathogenesis of type 2 diabetes mellitus. *Med Clin North Am*, Jul 2004; **88**(4): 787–835, ix (PMID: 15308380).
15 DeFronzo RA, Ferrannini E. Insulin resistance. A multifaceted syndrome responsible for NIDDM, obesity, hypertension, dyslipidemia, and atherosclerotic cardiovascular disease. *Diabetes Care*, Mar 1991; **14**(3): 173–194 (PMID: 2044434).
16 Reaven GM. Pathophysiology of insulin resistance in human disease. *Physiol Rev*, Jul 1995; **75**(3): 473–486 (PMID: 7624391).
17 Jenkins DJ, Wesson V, Wolever TM, Jenkins AL, Kalmusky J, Guidici S, Csima A, Josse RG, Wong GS. Wholemeal versus wholegrain breads: proportion of whole or cracked grain and the glycaemic response. *BMJ*, Oct 15 1988; **297**(6654): 958–960 (PMID: 3142566).
18 Foster-Powell K, Holt SH, Brand-Miller JC. International table of glycemic index and glycemic load values: 2002. *Am J Clin Nutr*, Jul 2002; **76**(1): 5–56 (PMID: 12081815).
19 Wood PJ, Anderson JW, Braaten JT, Cave NA, Scott FW, Vachon C. Physical effects of beta-D-glucan rich fractions from oats. *Cereal Foods World*, 1989; **34**: 878–882 (PMID:
20 Hallfrisch J, Behall KM. Mechanisms of the effects of grains on insulin and glucose responses. *J Am Coll Nutr*, Jun 2000; **19**(3 Suppl): 320S–325S (PMID: 10875604).
21 White JR, Jr, Campbell RK. Magnesium and diabetes: a review. *Ann Pharmacother*, Jun 1993; **27**(6): 775–780 (PMID: 8329802).
22 Paolisso G, Scheen A, D'Onofrio F, Lefebvre P. Magnesium and glucose homeostasis. *Diabetologia*, Sep 1990; **33**(9): 511–514 (PMID: 2253826).
23 Elamin A, Tuvemo T. Magnesium and insulin-dependent diabetes mellitus. *Diabetes Res Clin Pract*, Nov–Dec 1990; **10**(3): 203–209. (PMID: 2073866).
24 Lopez-Ridaura R, Willett WC, Rimm EB, Liu S, Stampfer MJ, Manson JE, Hu FB. Magnesium intake and risk of type 2 diabetes in men and women. *Diabetes Care*, Jan 2004; **27**(1): 134–140 (PMID: 14693979).
25 Oberley LW. Free radicals and diabetes. *Free Radic Biol Med*, 1988; **5**(2): 113–124 (PMID: 3075947).
26 Pereira MA, Jacobs DR, Pins JJ, Raatz S, Gross M, Slavin J, Seaquist E. The effect of whole grains on body weight and insulin sensitivity in overweight hyperinsulinemic adults. *Diabetes*, 2000; **44**, Suppl: A40.
27 Thompson LU. Antioxidants and hormone-mediated health benefits of whole grains. *Crit Rev Food Sci Nutr,* 1994; **34**(5–6): 473–497 (PMID: 7811379).
28 Miller HE, Rigelhof F, Marquart L, Prakash A, Kanter M. Antioxidant content of whole grain breakfast cereals, fruits and vegetables. *J Am Coll Nutr*, Jun 2000; **19**(3 Suppl): 312S–319S (PMID: 10875603).
29 Facchini FS, Humphreys MH, DoNascimento CA, Abbasi F, Reaven GM. Relation between insulin resistance and plasma concentrations of lipid hydroperoxides, carotenoids, and tocopherols. *Am J Clin Nutr*, Sep 2000; **72**(3): 776–779 (PMID: 10966898).

30 Fung TT, Hu FB, Pereira MA, Liu S, Stampfer MJ, Colditz GA, Willett WC. Whole-grain intake and the risk of type 2 diabetes: a prospective study in men. *Am J Clin Nutr*, Sep 2002; **76**(3): 535–540 (PMID: 12197996).
31 Liu S, Manson JE, Stampfer MJ, Hu FB, Giovannucci E, Colditz GA, Hennekens CH, Willett WC. A prospective study of whole-grain intake and risk of type 2 diabetes mellitus in US women. *Am J Public Health*, Sep 2000; **90**(9): 1409–1415 (PMID: 10983198).
32 Meyer KA, Kushi LH, Jacobs DR, Jr, Slavin J, Sellers TA, Folsom AR. Carbohydrates, dietary fiber, and incident type 2 diabetes in older women. *Am J Clin Nutr*, Apr 2000; **71**(4): 921–930 (PMID: 10731498).
33 Salmeron J, Ascherio A, Rimm EB, Colditz GA, Spiegelman D, Jenkins DJ, Stampfer MJ, Wing AL, Willett WC. Dietary fiber, glycemic load, and risk of NIDDM in men. *Diabetes Care*, Apr 1997; **20**(4): 545–550 (PMID: 9096978).
34 Salmeron J, Manson JE, Stampfer MJ, Colditz GA, Wing AL, Willett WC. Dietary fiber, glycemic load, and risk of non-insulin-dependent diabetes mellitus in women. *JAMA*, Feb 12 1997; **277**(6): 472–477 (PMID: 9020271).
35 Pereira MA, Jacobs DR, Jr, Slattery M, Hilner J, Kushi LH. The association between whole grain intake and fasting insulin in a bi-racial cohort of young adults: The Cardia Study. *CVD Preven*, 1998; **1**: 231–242.
36 McKeown NM, Meigs JB, Liu S, Saltzman E, Wilson PW, Jacques PF. Carbohydrate nutrition, insulin resistance, and the prevalence of the metabolic syndrome in the Framingham Offspring Cohort. *Diabetes Care*, Feb 2004; **27**(2): 538–546 (PMID: 14747241).
37 McKeown NM, Meigs JB, Liu S, Wilson PW, Jacques PF. Whole-grain intake is favorably associated with metabolic risk factors for type 2 diabetes and cardiovascular disease in the Framingham Offspring Study. *Am J Clin Nutr*, Aug 2002; **76**(2): 390–398 (PMID: 12145012).
38 Ekblond A, Mellemkjaer L, Tjonneland A, Suntum M, Stripp C, Overvad K, Johansen C, Olsen JH. A cross-sectional study of dietary habits and urinary glucose excretion – a predictor of non-insulin-dependent diabetes mellitus. *Eur J Clin Nutr*, May 2000; **54**(5): 434–439 (PMID: 10822293).
39 Fung TT, Rimm EB, Spiegelman D. Association between dietary patterns and plasma biomarkers of obesity and cardiovascular disease risk. *Am J Clin Nutr*, 2001; **73**: 61–67 (PMID:
40 Fukagawa NK, Anderson JW, Hageman G, Young VR, Minaker KL. High-carbohydrate, high-fiber diets increase peripheral insulin sensitivity in healthy young and old adults. *Am J Clin Nutr*, Sep 1990; **52**(3): 524–528 (PMID: 2168124).
41 Chandalia M, Garg A, Lutjohann D, von Bergmann K, Grundy SM, Brinkley LJ. Beneficial effects of high dietary fiber intake in patients with type 2 diabetes mellitus. *N Engl J Med*, May 11 2000; **342**(19):1392–1398 (PMID: 10805824).

7

Determining the functional properties of food components in the gastrointestinal tract

M. Champ, INRA, CRNH, France

7.1 Introduction

According to M. B. Roberfroid (1999), a food can be said to be 'functional' if it meets one of the following criteria: '(1) [it] contains a food component (being a nutrient or not) which affects one or a limited number of function(s) in the body in a targeted way so as to have positive effects (Bellisle *et al.*, 1998); (2) [it] has physiological or psychological effect beyond the traditional nutritional effect (Clydesdale, 1997)'. A functional food should then have a relevant effect on well-being and health or result in a reduction in disease risk (Roberfroid, 1999). The component that makes the food 'functional' can be:

- a macronutrient if it has specific beneficial physiological effects [such as resistant starch or (*n*-3) fatty acids];
- an essential micronutrient if its concentration and availability, in the food, allows the consumer to reach the daily recommendations (when it is not the case with a normal and balanced diet); or
- a food component that is not necessarily listed as 'essential' (e.g. some oligosaccharides) or may even be of non-nutritive value (e.g. live micro-organisms or phytosterols).

Functional properties of the food can be intrinsic to the food material or can be linked to a specific treatment and/or supplementation.

Assuming the above definition is consensual, cereals products minimally processed or fractionated can or could be recognized as functional foods. In the United States, the following health claim regarding whole grains: 'Diets rich in whole-grain foods and other plant foods and low in total fat, saturated fat and cholesterol may reduce the risk for heart disease and certain cancers.' was approved by the US Food and Drug Administration in 1999 (FDA Docket 99P-2209). The claim is based on a large number of studies that have demonstrated that consumption of whole grains is inversely associated with morbidity and mortality. Whole grains consumption has consistently been linked with protection from coronary heart disease (CHD) (Anderson *et al.*, 2000; Truswell, 2002). New studies have been performed since this time, and the results have confirmed earlier findings. However, these beneficial effects for health can be attributed to a number of mechanisms that are difficult to prove as each of them contribute to a small part of the overall effect. Nevertheless, in order to reinforce the properties that are thought to be beneficial, measurements have to be performed on the food or food components as eaten, or in a state as close as possible to the state that occurs *in vivo*.

Cereals and more specifically cereal grains are complex in structure and composition. Moreover, they cover a large variety of botanical species. As a consequence, the functional properties of cereal products depend on the botanical origin, the maturation, fractionation and potential technological treatments applied to the grain. Table 7.1 summarizes the major cereal components that are known or are thought to have functional properties in the gastrointestinal tract or to have an effect on the overall metabolism of humans.

Some cereal fractions are already recognized as functional ingredients. Indeed, the Food and Drug Administration has recently (May 2006) allowed a health claim on the relationship between β-glucan soluble fibre from whole oat sources and reduced risk of coronary heart disease (CHD) by adding barley as an additional source of β-glucan soluble fibre eligible for the health claim.

Regarding non-available carbohydrates, in addition to dietary fibre (including β-glucan) and resistant starch, grains contain significant amount of oligosaccharides. These carbohydrates are present as fructans in wheat but also in rye and barley. Van Loo *et al.* (1995) estimated that 78% of the North American intake of oligosaccharides is from wheat. These oligosaccharides, from other sources than cereal grains, have already been proven to be prebiotic. Obviously, such properties of the cereals have to be investigated much more extensively.

In order to claim that a food or a substance has a functional property, a number of trials have to be performed. An evaluation of the strength of the science linking the food/substance to a reduced risk of disease will be made by the authorities assessing the claim (Lupton, 2005). For optimal research efficiency, the petitioner first has to plan the studies that will bring enough scientific evidence for the claim within the shortest period of time and with the minimal cost. These studies usually start with screening steps to evaluate the best candidate among products, formulations and/or vectors. The final studies before obtaining the agreement for the claim have to be intervention

Table 7.1 Major cereal components exhibiting functional properties in the gastrointestinal tract

Fractions	Molecules or sub-fraction	Cereals	Functional properties
Bran/pericarp		Wheat	Regulation of intestinal transit, increase fecal excretion
	β-glucans	Oats, barley, rye	Slowering of gastric emptying, digestion and absorption
	Arabinoxylans	Wheat, rice, rye, barley	May delay digestion because of its viscosity
	Polyphenols	Sorghum, pearl millet	Improved redox state of leucocytes (in mouse) (Alvarez *et al.*, 2006)
	Ferulic acid	Wheat	Decrease viscosity of arabinoxylan
	Minerals	All	Mg may contribute to the protective effects of whole grain on CVD
Albumen			
	Starch	All	From fast to slow glycemic response
	Resistant starch	All	Prebiotic effect Increased fecal excretion
	Cell walls	All	Increased fecal excretion
Germ			
	Lipids – unsaturated fatty acids	All, i.e. wheat germ: ≈10% lipids (5.3% PUFA; 1.2% MUFA)	May decrease oxidative stress in patients with mild hypercholesterolemia (α-linolenic acid) (Alessandri *et al.*, 2006)
	Vitamin B6, folate and vitamin E	All	May contribute to the protective effects of whole grain on CVD
	Dietary fibres	All	Soluble fibre may contribute to a protective effect of whole grain on CHD
	Minerals P, K, Mg, Ca	All	Mg may contribute to the protective effects of whole grain on CVD

studies on the population that is supposed to benefit from the product/food. In between, several investigations are usually necessary: (1) to better specify the optimal properties of the food, ingredient or potential additive; and (2) to explain the mechanisms of action of the product.

This chapter focuses on methodologies that allow the determination or the prediction of the functional properties of foods or food components in the gastrointestinal tract. The chapter considers the different functional properties in cereal products that are either related to the gastrointestinal function or that can be predicted from events occurring in the gastrointestinal tract, including, of course, digestion, fermentation and absorption of nutrients. This chapter will evaluate the strength of the tests and the main criticisms and advantages of most methods will be discussed.

7.2 Functional properties of food components in the gastrointestinal tract

The digestive tract is most frequently the object of functional and health claims and a large market already exists for gut-functional foods worldwide (Table 7.2). As discussed by Cummings *et al.* (2004), it is quite difficult to interpret the benefits to individuals of variation within what is considered to be a normal range. Therefore it is essential to define standards for optimal function for use by the consumer, by industry and by those concerned with public health. Normal function of the gut has been defined by PASSCLAIM (a European Project entitled 'Process for the Assessment of Scientific Support for Claims on Foods')(Cummings *et al.*, 2004).

The main functions of the gut that can be improved by food components are given in Table 7.3. Related biomarkers of these functions, as well as measurement of these markers, are also summarized in Table 7.3.

7.3 *In vitro* and *in vivo* methods for determining the functional properties (in the gastrointestinal tract) of food components

In humans, measurements of the physiological properties of food components are very often indirect as they have to remain non- or poorly invasive. Only blood, faecal and urine analyses remain easily feasible whereas the collection of digestive content can only be performed through intubation techniques or with the participation of patients with digestive stomy. The analysis of a nutrient in the peripheral blood is, of course, the resultant of the absorption of the nutrient from the food but also from its potential uptake by the intestinal mucosa, the liver and the peripheral organs. In some cases (e.g. glucose), it

Table 7.2 Examples of existing claims related to gut health and immunity. Adapted from Cummings *et al.* (2004)

Content claims	Functional claims	Enhanced function claims	Reduction of disease risk or disease risk factor
• Fibre: typical value 2.5 g per 100 g* • 50% of your daily fibre needs in one bowl • Ingredient: dietary fibre* (inulin) • Ingredient: soluble fibre* (inulin) • *Bifidobacteria longum* • *Lactobacillus casei* Shirota • Product X contains special active cultures • Product X contains beneficial active bacteria • Product X supplies 50 million colony forming units of *Lactobacillus plantarum* 299v • 10 billion good bacteria • Product X provides all the good bacteria your body needs	• Helps you stay regular • Aids regular transit • Maintaining the balance of the digestive system • Helps maintain a healthy digestive system • Helps maintain the balance of intestinal flora • To help balance the probiotics in your digestive system • Supports the body's natural defences	• Proven benefits to your digestion • Improves digestion • Promotes natural health digestion • Modulates bowel activity • Promotes natural bowel rhythms • Improves intestinal transit • Active on intestinal comfort • Boosts the body's immune system • Stimulates the immune system • Boosts natural resistance • Actively strengthens the body's natural resistance • Help strengthen your natural defences • Enhances the bodys defences • Helps your body to protect itself • Helps the body to defend itself against external aggressions • Feel fabulous with fibre • Refresh and re-energize your mind and body • Product X stimulates mineral absorption	• Keeps harmful bacteria at bay • Helps reduce the number of harmful micro-organisms • Assists in elimination of harmful bacteria • Produces an antimicrobial substance that fights against pathogens • Product X helps against harmful bacteria, stimulating the immune system, helping reduce the risk of infections and digestive problems • Product X helps calcium absorption and is ideal for those with for osteoporosis or older people in general • Product X promotes recovery from milk allergy, alleviates allergic inflammation and reduces atopic skin symptoms

*There is no currently agreed method for measuring dietary fibre in the European Union.

Table 7.3 Functions of the gut, biomarkers of these functions and potential risk of altered gut functions

Function of the gut	Biomarker of the function	Measurement	Role as risk factor for disease
Digestive functions of the gut			
Gastric function	Gastric emptying rate	Scintigraphy after ingestion of radioactive markers Breath test with stable isotopes	
Digestion, absorption and motility in the small intestine	Glycemic response	Glycemic response after a meal or second meal (120 or 180 min postprandially)	
Colonic fermentation, absorption and motility	Bowel habit frequency, consistency and form, stool weight)	Stool collection	Constipation, bowel cancer, diverticular disease, appendicitis Anal problems (haemorrhoids, fissure) Disordered ano-rectal function
	Gut transit time	Single dose of carmine or chromic oxide (Branch & Cummings, 1978) Single or multiple doses of radio-opaque pellets (Cummings & Wiggins, 1976; Cummings *et al.*, 1976) Single dose of isotope-labelled test meal (Krevsky *et al.*, 1986)	
	Motility		
	Water absorption		
Modulation of gut flora	Composition of the microflora Activity of the microflora	Microflora analysis • Conventional methods (numeration of predominant groups by plating faecal micro-organisms onto selective media)	Possible effect on colonic health

Table 7.3 Continued

Function of the gut	Biomarker of the function	Measurement	Role as risk factor for disease
		• Molecular biological approaches Anaerobic batch culture fermenters Semi-continuous culture Continuous culture (chemostat) • Analysis of metabolites (organic acids (short chain fatty acids, lactic acid, etc.), gases, etc.) • Analysis of enzymes	
Barrier function			
Microflora		Classical methods of bacteriology Advanced techniques in molecular biology	Possible effect on gut infection
Physical mucosal barrier		Measurement of transepithelial and paracellular permeability (using sugars)	Effect on bacteria translocation
Chemical mucosal barrier • Digestive secretions • Antimicrobial peptides		*Ex vivo* methods: Surface hydrophobicity (Lugea *et al.*, 2000) Ussing chamber (Soderholm & Perdue, 2001) *In vivo* methods: Bacterial translocation (Lichtman, 2001; Such *et al.*, 2002) Intestinal permeability (Bjarnason *et al.*, 1995) • Measurement of the ratio of sugars in the urine, one of which is normally impermeable to the intestine Intestinal inflammation • Measurement of various proteins (e.g. cytokines, calprotectin) in faeces	

also reflects endogenous synthesis, which is itself regulated by the composition of the meal and physiological status of the individual. The analysis of urine reflects the absorption of nutrients or other components of the food but also reflects the filtration of nutrients by the kidneys. Finally, faeces are a material containing food residues, unabsorbed metabolites of the nutrients and components of the food, endogenous material such as exfoliated cells, various secretions (including enzymes, mucus or bile acids) bacterial flora and molecules produced by the micro-flora including, for example, metabolites from the fermentation, or vitamins. Only indigestible and specific residues from the food can be quantified in the stools whereas most of the absorption takes place in the proximal part of the small intestine and of the colon.

In order to be able to predict the digestive fate of food components and their physiological effects, *in vitro* methods have been elaborated and, when possible, validated with *in vivo* measurements. This is the case for: (1) *in vitro* digestibility tests, which are able to predict stomachal and small intestinal digestion of carbohydrates, proteins or lipids and absorption of nutrients; (2) *in vitro* fermentation of fibre fractions; (3) Caco-2 cell (or any other cell line) cultures to evaluate nutrient uptake and/or transport. These methods have also been used to describe:

- interactions between components of the food;
- physicochemical properties of the food matrix, before, during and after digestion or fermentation when these properties are known to influence physiology;
- the potential for proteins and peptides to be allergenic.

Between *in vitro* methods – which are often used for screening or to test mechanistic hypotheses – and intervention studies on healthy subjects or patients, experiments on animal models often offer a first *in vivo* evaluation of the physiological responses to a molecule, an extract or a food. These trials can be performed on a large number of animals when rodents are chosen as models, offering the possibility of statistical evaluation of the data obtained with the product to be tested versus a control. At this point, two strategies can be justified consecutively or as alternative choices: (1) to evaluate the physiological response to a 'non-physiological' amount of the product in order to be able to detect a potential effect that would not be measured at a lower dose within a short-term study; (2) to evaluate, in the animal model, the impact of a dose that would be the maximal or average expected dose (often expressed as a percentage of energy intake) to be introduced in the human diet. In any event, in both cases, the animal model has to be carefully chosen depending on the product to be tested. Indeed, transit time, bacterial flora, anatomy of the digestive tract or metabolism characteristics can adversely affect the conclusion of some studies when the animal model is not appropriate for the physiological aspect that has to be investigated (e.g. using a rat model to study the hypocholesterolaemic effect of a substance).

As a consequence, studies on animal models can only be intermediate steps before investigations on humans. However, they can be of great help for screening and especially to investigate for mechanisms. Animal models can be (ethically) killed and digestive contents and/or biopsies can be collected *post mortem* after the ingestion of the substance to be tested (appropriate time of killing has to be selected). Such collection of samples can also be performed on (properly) anaesthetized animals providing that anaesthesia does not interfere with the physiological aspect under study. There are now a large range of animals that can be of interest for specific studies; the main advances in animal models are the knock-out (KO) models (often rodents) in which a specific gene of interest has been invalidated. Large animals such as pigs might also be of interest in a number of nutritional investigations since as their intestinal physiology is much closer to ours than any rodent's. Their size is also much closer to human size, especially when mini-pigs are chosen.

Obviously, humans will remain the best models of humans. As a consequence, any claim will have to be first evaluated on humans and preferably on the precise category of human that will benefit from the additive, ingredient or food under development. In some cases such a study is impossible or unethical; this is often the case when long-term prevention effects are expected and/or when neonates are the target of the product to be tested. In such cases, multiple concordant evidence will be needed.

7.4 *In vitro* gut models

Numerous *in vitro* models have been developed to simulate the mouth, the stomach or more often the small or the large intestine. In the case of *mouth* simulation, the objective is to study specifically chewing through its mechanical action and/or saliva secretion and/or oral digestion of different ingredients or foods. Englyst *et al.* (1992) introduced a meat mincer to simulate chewing in methods able to predict glycaemic response and/or resistant starch content in foods. Secretions are most often simulated by the addition of salivary or pancreatic (less expensive) α-amylase in a buffered medium containing minerals ($CaCl_2$, NaCl, etc.) (Hoebler *et al.*, 2002).

In order to simulate *gastric* digestion, filling and emptying, devices with varying degrees of sophistication have been described and used. Hoebler *et al.* (2002) described an *in vitro* system simulating bucco-gastric digestion to assess physical and chemical changes of foods. *Small intestine* digestion has much been investigated in order to study starch, protein or lipid digestion. Some of the models are very simple: batches in which the substrates to be digested are incubated with pancreatin or several pancreatic and intestinal enzymes in a buffered medium. Of the available methods, the dynamic *in vitro* gastrointestinal model (TIM) is undoubtedly the most sophisticated (Minekus, 1998; Zeijdner & Havenaar, 2000; Minekus *et al.*, 2005). The model has compartments for the stomach, duodenum, jejunum and ileum. Each of the compartments has a flexible inner wall surrounded by water at 37 °C to squeeze the walls and mix the food with the 'secreted' electrolytes and enzymes, and to transport the chyme by peristaltic movements. The pH values, as well as the gastric emptying and small intestinal passage of the food, are controlled according to preset curves. Conditions were simulated according to the *in vivo* situation in adult humans. TIM-2 is the corresponding model for simulating colonic fermentation and absorption of various bioactive molecules or micro-organisms. Another dynamic system has been described by Promchan and Shiowatana (2005) to determine iron bioavailability by simulated gastrointestinal digestion. It is a dynamic continuous-flow dialysis (CFD) method with on-line electro-thermal atomic absorption spectrometric (ETAAS) and pH measurements. The system was described by the authors as a simple, rapid and inexpensive tool for bioavailability studies, especially for minerals at ultra-trace levels.

Finally, *colonic* fermentation has been studied in batches or in semi-continuous systems. The principle of the *in vitro* batch system for fermentation studies is to ferment the substrate or the food in the presence of fresh human faeces (collected in such a way that contact with oxygen is reduced as much as possible) diluted in a complex buffer (see Section 7.6). When foods have to be tested for their colonic fermentability, a pre-digestion has to be performed in order to remove all the materials that would be digested in the upper part of the gut. This pre-digestion involves proteases, α-amylase and lipase (or solvent extraction of lipids). Again, one of the most sophisticated systems is undoubtedly TIM-2 which is able to simulate colonic fermentation and absorption of various bioactive molecules or micro-organisms (Minekus *et al*, 1999). Earlier, in 1995, Molly *et al.* (1993) proposed a five-step multi-chamber reactor as a simulation of the human intestinal microbial ecosystem. The small intestine was simulated by a two-step 'fill and draw' system, the large intestine by a three-step reactor. However, this equipment seems not to have been further used for new investigations.

7.5 *In vitro* evaluation of the bioavailability of nutrients of food components

7.5.1 Carbohydrates

Carbohydrate bioavailability in the digestive tract can be assessed by *in vitro* analysis or *in vivo* evaluation of fractions digested in the small intestine and those that will reach the large intestine. Among carbohydrates that are digested in the upper gut, some fractions are quickly hydrolyzed and absorbed as glucose and/or fructose or galactose. The *Glycaemic index* (GI) has become a very popular index used to evaluate glycaemic and insulin responses to a food and to be able to make dietary recommendations to diabetic patients. It is also thought to be of interest in the overall population and several epidemiological studies show a good correlation between the GI of the diet and glycaemic control (improved by low GI diets), triglycerides and low-density lipoprotein (LDL) cholesterol (positive correlation) (AFSSA, 2004). However, the real impact of a low GI diet or of a low glycaemic load on diabetes, cardiovascular diseases and obesity is not yet clearly proven even if, in both cases, positive studies have been published (AFSSA, 2004). GI is now used for labelling purposes and as a marketing argument in several countries (Australia being the first to have introduced a specific logo regarding GI); however, other countries are not in favour of the use of this index for the general population (e.g. France) (AFSSA, 2004).

Numerous *in vitro* methods have been developed to compare *in vitro* digestibilities of starch in various ingredients and foods. The methods used to predict glycaemic response or GI have varying degrees of sophistication from those involving a single step with pancreatic amylase to those with several steps simulating the various physical, chemical and enzymatic actions of chewing and digestion in the stomach and small intestine. This last type of method also considers the disappearance of the other fractions of the food. As shown in Table 7.4, only some of these methods have been validated using *in vivo* glycaemic responses to the same food. This condition is absolutely essential to guarantee for a large range of foods the appropriate preparation conditions (grinding, mainly) and the amount of enzymes (activity of α-amylase but also amyloglucosidase per gram of sample).

The overall principle of these methods is an *in vitro* α-amylolysis applied to a sample which has to be ground or chewed in order to simulate the action of the teeth in the mouth. This hydrolysis has to be interrupted after an appropriate time of incubation (to fit with values obtained *in vivo*). The hydrolyzed sample is then analyzed to measure the total amount of free glucose and polymers (maltodextrins + maltose + glucose) that have been released during *in vitro* digestion. These carbohydrates are usually extracted in ethanolic solution (80% or 80°GL) and then hydrolyzed into glucose with amyloglucosidase in order to quantify total glucose with a specific method involving a set of enzymes (glucose oxidase/peroxidase or hexokinase/glucose-6P dehydrogenase) and a spectrophotometric determination.

Resistant starch (RS), defined as 'the sum of starch and starch products of starch degradation not absorbed in the small intestine of healthy individuals' (Asp, 1992) and often considered as a dietary fibre, is mostly fermented in the colon and only a very small fraction is usually recovered in faeces. Besides being of lower energy value than available starch (about 2.5 kcal/g instead of 4 kcal/g), most RS ferments in the colon producing short chain fatty acids with a significant proportion of butyrate (the main fuel of colonocytes).

During the 1990s several *in vitro* methods were developed to predict the RS content in foods. These methods have been in some cases validated with values obtained *in vivo* (ileostomy and/or intubation techniques). The main developments have been made by the groups of Englyst (i.e. Englyst *et al.*,

Table 7.4 Evaluation of the bioavailability of starch and other macronutrients – main *in vitro* methods

Reference	*In vitro* measurement	Principle	*In vivo* prediction
Glycemic response			
Englyst *et al.*, 1999	Rapidly available glucose (RAG) Slowly available glucose (SAG)	(1) Pepsin (2) Pancreatin + invertase (3) 20 min hydrolysis ⇒ RAG*; 120 min hydrolysis ⇒ SAG*	Glycaemic response (incremental area under the glucose response curve)
Brighenti *et al.*, 1995	Sugar diffusion indexed to pure maltose (SDI) Reducing sugar released indexed to bread (RSRI) Corrected starch digestion indexed to bread (DIGI)	(1) Salivary α-amylase (2) Pepsin hydrolysis (3) Dialysis (4) Pancreatin + dialysis (0–300 min) ⇒ SDI* 270 min, RSRI* 150 min, DIGI* 120 min	Glycaemic index
Bornet *et al.*, 1989	Starch hydrolysis at 30 min	*In vitro* starch hydrolysis using pancreatic α-amylase	Area under the curves of glycaemic and insulinaemic responses
Granfeldt *et al.*, 1992	Hydrolysis index = area under the hydrolysis curve (% of the correponding area with white wheat bread)	(1) Chewing (2) Pepsin hydrolysis (3) Pancreatin + dialysis	Glycaemic and insulin indices

Table 7.4 Continued

Reference	*In vitro* measurement	Principle	*In vivo* prediction
Resistant starch			
Goñi *et al.*, 1996	%RS	(1) Pepsin hydrolysis (2) α-Amylase hydrolysis (16 h) (3) Quantification of total glucose (/AMG hydrolysis + GOD-POD) in the pellet after centrifugation	Indirect (values from other groups)
Champ, 1996	RS%TS	(1) α-Amylase hydrolysis + AMG (16 h) (2) Ethanol precipitation (3) Quantification of total glucose (/AMG hydrolysis + GOD-POD) in the pellet after centrifugation	Partly indirect (values from own group obtained from healthy subjects using intubation technique and from other groups (mostly ileostomates))
Åkerberg *et al.*, 1998	RS%TS	(1) Chewing (2) Pepsin hydrolysis (3) Pancreatin + AMG (16 h) (4) Ethanol precipitation ⇒ POG and RS	Subjects with ileostomies
Englyst *et al.*, 1992	RS%TS	(1) Grinding in a mincer (2) Sample + guar gum + glass balls (3) Pancreatin + AMG + invertase (4) 20 min hydrolysis ⇒ G20; 120 min hydrolysis ⇒ G120 RS = TS – SDS; (SDS = (G120 – free glucose)* 0.9)	*In vivo* measurement in ileostomates

GOD-POD, Glucose oxidase – peroxidase kit; AMG, amyloglucosidase; POG, potentially available glucose; RS, resistant starch, TS, total starch; SDS, slowly digestible starch, G120, glucose produced from the *in vitro* digestion within 120 min.
*0.9 means that the value obtained from (G120 – free glucose) has to be multiplied by 0.9 in order to calculate the amount of SDS in the sample.

1992; 1996a), Asp and Björck (i.e. Asp *et al.*, 1983; Björck *et al.*, 1986; Åkerberg *et al.*, 1998), Muir (Muir & O'Dea, 1992, 1993), Goñi (Goñi *et al.*, 1996) and within the EURESTA research program (1990–1993) (i.e. Champ, 1992; Englyst *et al.*, 1992) (Table 7.4); later developments were made by McCleary and colleagues who, after a collaborative study (McCleary and Monaghan, 2002), proposed an analytical kit, which is commercially available (Megazyme Ltd, Bray, Republic of Ireland). The Englyst group (Englyst Carbohydrates, Southampton, UK) is also proposing to analyse rapidly available starch (RAS), slowly available starch (SAS) as well as RS in the same analysis. The overall principle involves using *in vitro* α-amylolysis to predict the amount of slowly digestible starch and rapidly digestible starch, since RS is the undigested fraction of starch remaining in the residue once slowly digestible starch has been determined. For GI prediction, methods with varying degrees of sophistication are available reflecting (to a greater or lesser extent) the *in vivo* process of digestion in the mouth, stomach and small intestine.

Proteins and lipid bioavailability has been less extensively studied than carbohydrate bioavailability. Several *in vitro* tests are available even if validations have not necessarily been performed with *in vivo* determinations on healthy humans.

7.5.2 Proteins

In vitro protein digestibility studies are of interest as 'slow' or 'fast' proteins have been shown to have advantages in different physiological situations. Indeed, slowly digested proteins (such as caseins) induce a higher protein gain in young men than those that are rapidly digested (such as whey proteins), whereas 'fast' proteins might be more beneficial than 'slow' proteins in limiting protein losses during aging (Dangin *et al.*, 2003).

The bioavailability of proteins is influenced by its composition (sequence of amino acids), conformation and interactions with other constituent of the food; therefore, several authors designed simple *in vitro* tests to explore these different parameters. Until recently, there were no standardized protocols to assess the digestibility of proteins using simulated gastric fluid. Potential variations in assay parameters include: pH, pepsin purity, pepsin to target protein ratio, target protein purity and method of detection. The International Life Sciences Institute (ILSI) initiated, a few years ago, a multi-laboratory evaluation of a common *in vitro* pepsin digestion assay protocol (Thomas *et al.*, 2004).

Interactions between protein and phytate have recently been investigated, demonstrating the role of gastric pH in the formation of complexes between both components, and the potential role of phytase in preventing the occurrence of such complexes which are responsible for reduced protein digestibilities (Kies *et al.*, 2006). Allergenicity of animal or plant proteins may be influenced by technological treatments applied to the food during its preparation and also by their environment in the food. It is now obvious that, in order to properly predict allergenicity risk, protein and peptides – as digested in the upper part of the digestive tract – have to be tested as they may vary in allergenicity from the whole proteins as present in the food. This is the reason why many of the *in vitro* tests of gastric and duodenal digestion aim to determine the allergenicity of foods or ingredients from animal or plant origins. Certain proteins have been shown to be highly resistant to *in vitro* digestion which helps to explain some sensitization and allergic reactions in a sensitized individual (Murtagh *et al.*, 2003; Moreno *et al.*, 2005abc; Vassilopoulou *et al.*, 2006).

7.5.3 Lipids

As far as lipolysis is concerned, determination is much more complex due to the hydrophobic properties of the molecules and the very wide diversity of their chemical structure. Inhibition of lipolysis by components of the food can be of interest in reducing fat digestion. Molecules with this property are already commercially available from pharmaceutical companies but are also of great interest to the food industry. Several *in vitro* lipolysis models have been described in the literature. One of them, a 'dynamic *in vitro* lipolysis model' was set up to investigate the dissolution of poorly soluble lipophilic drug substances at controlled hydrolysis rates. The model allowed the evaluation of multiple variables such as concentrations of bile salts, Ca^{2+} or lipase activity (Zangenberg *et al.*, 2001ab). Various authors, including the French group of D. Lairon (Juhel *et al.*, 2000), have also studied the effect of various dietary components on the lipolysis of triglycerides using gastric and duodenal media *in vitro*.

7.5.4 Minerals and vitamins

Women are particularly advised to increase their consumption of foods high in calcium and to include foods containing iron such as lean meats and also whole grain products. Indeed, iron losses during menstruation, as well as the increase need during pregnancy, place women at risk of iron deficiency; however, excess of iron can be deleterious for the population. Regarding calcium, an adequate intake from childhood until the late twenties may be the best prevention against osteoporosis. The bioavailability of minerals can be strongly influenced by food matrices and the chemical form of the minerals. Therefore, multiple tests have been published to evaluate this bioavailability.

Mineral bioavailability is very much dependent on the chemical form of the minerals as well as on their interaction with various components of the food. *In vitro* systems, consisting for instance of simulated gastrointestinal digestion and Caco-2 cell culture (Kloots *et al.*, 2004; Remondetto *et al.*, 2004; Perales *et al.*, 2005; Viadel *et al.*, 2006), have been elaborated to estimate the uptake and intestinal absorption of different minerals (calcium, iron and/or zinc) from foods, as eaten, obtained from different preparations.

Simple *in vitro* gastrointestinal methods have also been employed to predict the potential bioavailability of selenium and its species from various plants (Pedrero *et al.*, 2006). These procedures have then to be associated with specific analyses of the mineral-containing compounds. For example, Reyes *et al.* (2006) described a two-dimensional (size exclusion–reversed phase) chromatography approach following off-line identification of low molecular weight compounds – generated during gastrointestinal digestion – using electro-spray tandem mass spectrometry (ESI Q-TOF MS). Recently, Sandberg (2005) reviewed *in vitro* dialyzability methods aiming to evaluate mineral availability in a food or food component. These methods seem preferable, for screening purposes, to the more sophisticated computer-controlled gastrointestinal model. For instance, the calcium bioavailability of various food products has been determined by *in vitro* methods simulating gastrointestinal digestion of the product with pepsin-HCl and pancreatic–biliary salts, and then measuring the fraction of the element that dialyses through a membrane of a certain pore size (Unal *et al.*, 2005).

These *in vitro* dialyzability methods are able to correctly rank iron and zinc bioavailabilities from different meals. The exceptions may be that the effects of milk, certain proteins, tea and also organic acids cannot be predicted. These exceptions do not apparently concern cereal products except, perhaps, those that are fermented with natural yeast (Sandberg, 2005).

Regarding vitamins, the bioavailability of folate has been investigated by TNO (Zeist)(Verwei *et al.*, 2003, 2004; Arkbåge *et al.*, 2003) using the TIM gastrointestinal model.

7.6 *In vitro* evaluation of the fermentability of nutrients of food components

Colonic events have long been underestimated. They obviously contribute to the well-being of the population by influencing colonic transit time and the risk of constipation, flatus or irritable bowel syndrome. The colonic microflora is responsible for the production of short chain fatty acids including the main nutrient of colonocytes. It is also involved in synthesis of amino acids or vitamins but also in the conversion of molecules (such as secondary acids) into substances that have an adverse effect on health. More recently, colonic fermentation was shown to influence the physiology of the upper gut, for example, transit time or nutrient absorption.

This fermentation can be estimated *in vivo* in humans using a breath test to evaluate gas production but also using kinetic studies of stable isotopes to evaluate short chain fatty acid production from carbohydrate fermentation and also colonic nitrogen metabolism (De Preter *et al.*, 2004; Geboes *et al.*, 2005). However, due to the difficulties in carrying out such studies, many *in vitro* tests have been developed to study the extent of fermentation and metabolite production. For instance, a European inter-laboratory study has been performed to test a simple *in vitro* batch system for dietary fibre fermentation studies (Barry *et al.*, 1995). The inoculum was composed of fresh human faeces mixed with a carbonate–phosphate buffer complex supplemented with trace elements and urea. Short chain fatty acids production, residual non-starch polysaccharides (NSP) and pH were determined during fermentation. *In vitro* batch systems can be more or less predictive of *in vivo* fermentation, depending on the batch system but also on the substrates that are tested (Daniel *et al.*, 1997; Wisker *et al.*, 1998). For instance, fermentation of NSPs in mixed diets, but not in barley fibre concentrate, could be predicted by *in vitro* fermentation (Daniel *et al.*, 1997). It was established some time ago that several donors, at least three, should be used to improve the accuracy of *in vitro* estimates of colonic fermentation (McBurney & Thompson, 1989).

TIM-2 is a dynamic, computer-controlled *in vitro* model for simulating conditions in the large intestine – with peristaltic mixing, water absorption and absorption of fermentation products (Minekus *et al.*, 1999).

7.7 *In vivo* evaluation, in animal models, of the bioavailability of nutrients of food components

7.7.1 Carbohydrates

Animal models, mainly rats, have been used to predict the *in vivo* digestibility of starch (Asp and collaborators, i.e. Hagander *et al.*, 1987; Holm *et al.*, 1988; Tovar *et al.*, 1992). However, these studies were able to classify starchy products but could not be used to predict the quantitative glycaemic response to a food nor its RS content. As a consequence, *in vitro* tests or human studies are usually preferred for such determinations.

7.7.2 Proteins

The protein digestibility of an ingredient or a food can be evaluated using animal models. For example, calculation of the protein-digestibility-corrected amino-acid score (PDCAAS) is carried out using young rats (recommended animal model) (Darragh & Hodgkinson, 2000). This model is, however, controversial in some cases, such as foods (or ingredients – especially those containing anti-nutritional factors) specifically intended for the elderly (Gilani *et al.*, 2005). Nevertheless the PDCAAS is now widely used as a routine assay for protein quality evaluation, replacing the more traditional biological methods (e.g. measurement of the protein efficiency ratio (PER) in rats) (Schaafsma, 2005).

The determination of the PDCAAS requires the determination of the digestibility of the protein(s) across the entire digestive tract. This faecal digestibility value is subsequently corrected for endogenous contributions of protein using a metabolic nitrogen value determined by feeding rats a protein-free diet. However, the limitations inherent with this method are well recognized and determining the digestibility of a dietary protein to the end of the small intestine is the preferred alternative. Unlike the faecal digestibility assay, which uses only one basic methodology, ileal digestibility values can be determined in a number of ways (Darragh & Hodgkinson, 2000).

7.8 *In vivo* evaluation, in human volunteers, of the bioavailability of nutrients of food components

7.8.1 Carbohydrates

The determination of GI has been standardized by several groups of scientists (FAO, 1998; Wolever *et al.*, 2003; Brouns *et al.*, 2005). It has to be performed on ten subjects or more (Brouns *et al.*, 2005). The determination requires two different tests (2 hrs per test) on the different subjects for each food that is to be characterized for its GI and for the control – which is either glucose or white bread. The determination in human subjects is thus quite expensive to perform and is only used on foods where a claim is made. In order to compare different formulations and/or technological treatments, *in vitro* tests are faster and less expensive but the method used should have been validated using *in vivo* data obtained on the same foods.

The table published by Brand-Miller and collaborators (Foster-Powell *et al.*, 2002) summarizes most of the data available in the literature (and some unpublished data) at the time of the publication. It indicates that many whole grain cereal products – such as whole grain wheat breads (but not whole meal bread), oatmeal, pasta, high amylose maize and rye bread – have low GI values. As a consequence, these foods are thought to have clear health advantages for people with diabetes, to improve glycaemic control and – according to some of the trials – to reduce total plasma cholesterol and triglycerides (AFSSA, 2004).

Resistant starch can be determined *in vivo* in humans according to two different methods: (1) intubation techniques – these are performed on healthy subjects but require trained investigators and have as a main drawback the fact that the catheters introduced into the digestive tract are responsible for a significant alteration in the transit time; (2) ileostomy studies – these require subjects that have previously been procto-colectomized, usually for ulcerative colitis (Champ *et al.*, 2003b; Champ, 2004). The second technique has the major drawback that it is performed on subjects who cannot be considered as healthy. These subjects usually have a slightly altered transit time but, more importantly, have a microflora in the terminal ileum that is not comparable with the usual ileal bacterial population (Bach Knudsen & Hessov, 1995). As a consequence, it is not possible to have a precise value of RS as defined above. The real value is probably in-between values obtained from intubation techniques and ileostomy studies (Champ *et al.*, 2003b). In any event, it is becoming more and more difficult to find patients with permanent ileostomies and very few groups in the world are still using this method. Intubation techniques are also no longer used due to ethical constraints, which are becoming more and more restrictive worldwide. For these reasons and also due to the costs of these measurements, the *in vitro* methods giving a quantification of RS in food 'as eaten' are now widely used worldwide. Most cereal products have quite low percentages of RS except when high amylose corn starch (native or retrograded) is among the ingredients (Champ *et al.*, 2003b).

7.8.2 Proteins

In vivo protein digestion rates in humans are now mostly investigated through the use of intrinsically labelled protein fractions (often [^{13}C]leucine-labelled proteins) which are ingested as a single meal by the subjects. Post-prandial whole body leucine kinetics are assessed by using a dual-tracer (the second tracer being [$^{2}H_3$]leucine which is i.v. infused) methodology (Boirie *et al.*, 1997; Dangin *et al.*, 2002, 2003). In some cases, one single tracer is used for constant i.v. infusion; in these cases, dietary proteins are not labelled (Dangin *et al.*, 2001; Millward *et al.*, 2002).

7.8.3 Lipids

The digestion and absorption of fat emulsions have been studied in healthy humans through gastric and duodenal aspiration techniques. Smaller droplet size has been shown to facilitate fat digestion by gastric lipase in the stomach and also to increase duodenal lipolysis; in contrast, overall fat assimilation was not affected by this parameter (Armand *et al.*, 1999).

7.8.4 Minerals and vitamins

In healthy volunteers, mineral (e.g. iron, zinc, copper, etc.) and vitamin (e.g. riboflavin, folate, vitamin C, vitamin B_{12}) bioavailability from a complex matrix or a multi-micronutrient dietary supplement can be evaluated by a simple measurement of post-prandial changes in plasma or serum concentrations after consuming the test meal versus the placebo (Navarro & Wood, 2003).

Radioactive isotopes and, more recently, stable isotopes have been used as research tools to determine intestinal mineral absorption in humans under different nutritional and physiological conditions. Feillet-Coudray and Coudray (2005) evaluated the different techniques available to explore the bioavailability and metabolism of magnesium. They concluded that the studies available have clearly demonstrated that stable isotopes provide a useful research tool for determining magnesium absorption. Among the techniques available, the faecal monitoring technique can be used in humans and requires at least 20 mg of magnesium isotope and the collection of faeces over 7 days. This allows a faecal isotopic enrichment of about 10%. In the double-labelling technique, two isotopes – Mg^{26} and Mg^{25} – should be given in high amounts, the first one orally and the other intravenously (80 mg and 40 mg, respectively), to achieve a plasma or urine enrichment of about 10%, 48 hours after isotope administration. Magnesium stable isotopes are now quantified using thermo-ionization (TIMS) or inductively coupled plasma mass spectrometry (ICP/MS) techniques and allow reasonable amounts of magnesium isotopes to be used in human nutrition studies.

The same type of methodologies are available for other minerals, but as discussed in Feillet-Coudray and Coudray (2005), they are not applicable to the real foods for which we would like to evaluate mineral bioavailability.

7.9 *In vivo* evaluation, in human volunteers and animal models, of the fermentability of nutrients, functional properties and/or benefits/risks of food components

7.9.1 *In vivo* evaluation of carbohydrate fermentability and colonic protein metabolism

In vivo evaluation of the fermentability of carbohydrates (the carbohydrate fraction of dietary fibre) can be carried out indirectly by breath tests or using stable isotope techniques. Whereas hydrogen breath tests are still used to evaluate the fermentability of rapidly fermented substrates, more appropriate breath tests have to be used when slowly fermented compounds are being studied. Substrates labelled using the ^{13}C isotope can be employed to measure the pulmonary excretion of labelled carbon dioxide produced by fermentation of the substrate (Christian *et al.*, 2002; Edwards *et al.*, 2002).

Short chain fatty acids production (but mainly acetate) can also be evaluated by stable isotope dilution techniques. Intravenous infusion of [1-^{13}C]acetate at a constant rate while the food or the ingredient under study is eaten allows, by analysis of the tracer in arterialized venous blood, the quantitative estimation of colonic carbohydrate fermentation via evaluation of acetate production (Pouteau *et al.*, 1998).

Colonic nitrogen metabolism can be assessed *in vivo* using lactose-[^{15}N]ureide (Geboes *et al.*, 2005).

7.9.2 Prebiotic properties

The prebiotic properties of a pure fibre or a food can be assessed by quantifying specific bacteria in the stools of humans or the caeco-colon of animals consuming the product for several days or weeks. Classical quantifications on specific media have been used widely whereas molecular biological techniques are now preferred by most microbiologists. As an example, fluorescent *in situ* hybridization using oligonucleotide probes, targeting specific bacteria of interest (e.g. *Bacteroides* spp., *Bifidobacterium* spp., *Clostridium* spp. and *Lactobacillus–Enterococcus* spp.) has been used by Tuohy *et al.* (2001) to evaluate, in human volunteers, the prebiotic effects of foods containing fibre or oligosaccharides. Total bacteria can be simultaneously enumerated using the fluorescent stain 4′,6-diamidino-2-phenylindole (Tuohy *et al.*, 2001).

The simultaneous use of lactose-[^{15}N]ureide and [$^{2}H_4$]tyrosine has been employed to study the effects of pro- and prebiotics on the intestinal flora of human volunteers (De Preter *et al.*, 2004).

7.9.3 Transit time and motility of the digestive tract and faecal excretion

Insoluble fibre present in wheat but also in rice, oat, rye and barley, is associated with improved laxation. The refining processes remove proportionally more of the insoluble fibre than the soluble fibre. Disruption of cell walls can increase the fermentability of dietary fibre. Coarse wheat bran has a greater faecal bulking effect than finely ground bran (Wrick *et al.*, 1983), suggesting that the particle size of the bran but also of the ground whole wheat grain is an important parameter in determining its physiological effect.

Faecal bulking can be simply assessed in standardized conditions by collecting stools and recording time of emission. Several techniques are available to determine orocaecal transit time or total transit time. Orocaecal transit time is usually assessed by breath tests based on the time required for a fermentable substrate to reach the proximal colon (or cecum) and be fermented by the local microflora. Total transit time is often assessed by using multiple coloured particles, which are recovered in the stool. Other authors used radio-opaque pellets (two different shapes) which are counted in stools by radiography, to measure mean transit time (MTT) (El-Oufir *et al.*, 2000). This method was first described by Cummings *et al.* (1976): a constant amount of marker (radio-opaque pellets) was fed to the subjects with each meal over a period of weeks, and its excretion measured in the stools. However, these authors concluded that measurement of MTT is less accurate than MTT measured by giving single doses of similar markers (Cummings & Wiggins, 1976; Cummings *et al.*, 1976).

7.9.4 Effect on colonic carcinogenesis – Animal models

In a meta-analysis of whole grain intake and cancer, whole grains were found to be protective in 43 out of 45 reports of whole grain intake (after exclusion of 6 reports with design/reporting flaws or low intake) (Jacobs *et al.*, 1998). Whole grain and some cereal brans are thought to be protective against colon cancer. The mechanisms thought to be involved in this protection are the increased faecal bulk and decreased transit time, which allow less opportunity for faecal mutagens to interact with the intestinal epithelium. Moreover, fibre can bind or simply dilute bile acids. Secondary bile acids are thought to promote cell proliferation, thereby providing more opportunities for mutations to occur and for abnormal cells to replicate. Fermentation of fibre results in the production of short chain fatty acids, which lowers colonic pH; this in turn inhibits the conversion of primary bile acids into secondary bile acids. Low pH values also reduce the solubility of free bile acids, decreasing their availability for co-carcinogenic activity. Finally, butyrate, one of the three main short chain fatty acids produced by fermentation of carbohydrates in the colon, has been shown to be anti-neoplastic (McIntyre *et al.*, 1993). Wheat bran was the type of fibre most consistently shown to inhibit carcinogenesis (Calvert *et al.*, 1987; Reddy *et al.*, 2000).

Some compounds such as plant lignans present in vegetables and whole grains are converted into enterolactone, which may protect against breast cancer and other hormone-dependent cancers. Indeed, in a study of Danish post-menopausal women, those who were eating the largest amounts of plant lignan had higher concentrations of plasma enterolactone (Johnsen *et al.*, 2004).

Gene expression studies are informative about changes in colon cancer, increase our understanding of the biology of tumourigenesis and aid in developing diagnostic and prognostic markers. Many genes have been found to differ in expression between normal and tumorigenic states, as early as the seemingly normal colonic crypts. Combinations of markers can be used to develop an approach to use molecular screening to follow the progression of this prevalent cancer (Ahmed, 2005).

Aberrant crypt foci (ACF) are thought to be the earliest identifiable neoplastic lesions in the colon carcinogenetic model. Recent data indicate that some ACF bypass the polyp stage in their carcinogenesis thus reinforcing the importance of their early detection and our understanding of their pathogenesis. Various proteomic (Prot) markers may be altered within ACF suggesting possible prospective pathological changes. Genetic mutations as well as epigenetic alterations can also been involved, as discussed in the recent review of Alrawi *et al.* (2006). There is still a lack of validated early and non-invasive markers of cancer risk. Therefore a number of *animal models* have been developed to study the impact of dietary components on colon cancer prevention. These models differ in the animal species, tumour-inducing agents or gene mutation. For instance, to study the impact of ingredients on colon cancer, azoxymethane and dimethylhydrazine induction of ACF in rats has been widely used (Perrin *et al.*, 2001). The model of both familial adenomatous polyposis and sporadic colon cancer, the 'Min mice', which have non-sense mutation in the Apc gene mutation (model of familial polyposis), has also been used by several groups worldwide (Pierre *et al.*, 1997).

7.10 *In vivo* evaluation of satiety and satiation induced by a food component

Cereal products are consumed in large amounts in most countries in the world and therefore it is very difficult to examine the impact of cereal consumption on overweight or obesity. However, changes in whole grain consumption in relation to the prevention of weight gain have recently been investigated within the American Health Professionals Follow-up Study. This specific study was conducted in a prospective cohort of 27 082 men aged 40–75 years at baseline for 8 years (Koh-Banerjee *et al.*, 2004). In a study performed on 74 091 US female nurses, aged 38–63 years in 1984 and followed until 1996, weight gain was inversely associated with intake of high-fibre, whole grain foods but was positively related to the intake of refined grain foods (Liu *et al.*, 2003). The increased consumption of whole grains was inversely related to weight gain, and the associations persisted after changes in added bran or fibre intakes were accounted for. Studies such as those above are, of course, quite difficult to plan and acute studies can be helpful in evaluating specific cereal products. Among these studies, the group of I. Björck in Lund, Sweden has determined satiety scores after an experimental meal or after the second meal (0–180 min after the standardized second meal) (Liljeberg *et al.*, 1999). Subjective questionnaires to assess hunger, satiety, fullness or prospective food consumption are filled in by each subject (Hill *et al.*, 1984; Raben *et al.*, 2003).

7.11 *In vitro* and *in vivo* evaluation of antioxidant bioavailability and properties of food components

Flavonoids, inositols, lignin, phenolics, phytates, phytoestrogens, selenium, tocopherols and zinc are components of whole grains that have been shown to have antioxidant activity. The bioavailability of these molecules can be assessed using an *in vitro* digestion procedure that mimics the physicochemical and biochemical changes that occur in the upper gastrointestinal tract (Vallejo *et al.*, 2004; McDougall *et al.*, 2005ab).

Natural antioxidants present in whole grains may be responsible for a decrease of cancer risk due to whole grain food consumption (Slavin, 2000). A number of *in vitro* tests have been developed to evaluate simultaneously the antioxidant and antiproliferative properties of numerous molecules from plant origin in human (HT-29, HCT116, Caco-2) or murine (CT-26) colon cancer cell lines (Tarozzi *et al.*, 2004; Athukorala *et al.*, 2006; Olsson *et al.*, 2006).

In vivo, the anti-oxidative properties of compounds have been evaluated in various animal models such as dextran sodium sulphate (DSS)-treated rats, which develop intestinal inflammation. Such compounds, when given to the animals prior to DSS-induced inflammation, reduce oxidative DNA damage but also reduce cyclooxygenase-2 levels and restore superoxide dismutase-2 to control levels (Luceri *et al.*, 2004).

7.12 Future trends

Many *in vivo* investigations that have been allowed in Western countries in the recent past might soon be forbidden as they remain invasive. As an example, it is becoming almost impossible to carry out *in vivo* experimentation on newborns and children in advance of the production of new foods that might, however, be expected to be beneficial on the basis of animal experiments and/or *in vitro* tests. Non-invasive tests are being investigated in numerous laboratories in order to obtain more predictive information without any invasive examination of the subject testing the food or the food component.

From the tremendous progress that is being made in various analytical domains – such as 'omics' techniques, analysis of stable isotopes in a large variety of molecules, and nuclear magnetic resonance (NMR) and other imaging techniques, it can be predicted that the determination of the functional properties of food components will soon be drastically changed. Indeed, metabonomics and proteomics will give access to new markers of gastrointestinal functions which will be measured in stools or in urine.

Equipment such as synchrotrons and the associated multiple imaging techniques – such as fluorescent X-ray computed tomography imaging, infrared imaging or diffraction-enhanced imaging – will soon allow new investigations where high resolutions are needed (1–3 μm^2 for infrared analysis or less than 200 nm for fluorescence imaging). Interactions between molecules of the food and intestinal cells, for instance, should be accessible using such techniques.

7.13 Sources of further information and advice

Further information on the functional properties of food components in the gastrointestinal tract can be found in Salminen *et al.* (1998) or Cummings *et al.* (2004). More specific reviews on dietary fibre and resistant starch have been published by Champ *et al.* (2003a,b). Functional foods and the immune system have been widely discussed in Lopez-Varela *et al.* (2002) and in Cummings *et al.* (2004).

In order to achieve worldwide approval for functional foods, there is an urgent need for international consensus on the methods used to assess them; these methods need to be validated on the basis of *in vivo* studies which would be multi-centric (preferably involving multiple continents). In most developed countries, at the level of the National Agencies for Food, strategies to validate new products and new claims are being elaborated; inter-continent and worldwide consensus has then to be obtained on these methodologies and agreement has to be reached in order to guarantee that the strength of the evidence is sufficient to grant an approval of the product. Obviously, we are still far from this level of agreement and some countries are more open to, or tolerant of, new food products.

7.14 References

AFSSA (Agence Française de Sécurité Sanitaire des Aliments) (2004) *Rapport Glucides et Santé : Etat des Lieux, Evaluation et Recommandations*. AFSSA, Maisons-Alfort, France.

Ahmed FE (2005) 'Development of novel diagnostic and prognostic molecular markers for sporadic colon cancer' *Expert Rev Mol Diagn*, **5**, 337–352.

Åkerberg AKE, Liljeberg HGM, Granfeldt YE, Drews AW, Björck IME (1998) 'An *in vitro* method, based on chewing, to predict resistant starch content in foods allows parallel determination of potentially available starch and dietary fiber', *J Nutr*, **128**, 651–660.

Alessandri C, Pignatelli P, Loffredo L, Lenti L, Del Ben M, Carnevale R, Perrone A, Ferro D, Angelico F, Violi F. (2006) 'Alpha-linolenic acid-rich wheat germ oil decreases oxidative stress and CD40 ligand in patients with mild hypercholesterolemia' *Arterioscler Thromb Vasc Biol*, **26**, 2577–2578.

Alrawi SJ, Schiff M, Carroll RE, Dayton M, Gibbs JF, Kulavlat M, Tan D, Berman K, Stoler DL, Anderson GR (2006) 'Aberrant crypt foci' *Anticancer Res*, **26**, 107–119.

Alvarez P, Alvarado C, Mathieu F, Jimenez L, De la Fuente M. (2006) 'Diet supplementation for 5 weeks with polyphenol-rich cereals improves several functions and the redox state of mouse leucocytes' *Eur J Nutr*, **45**, 428–438.

Anderson JW, Hanna TJ, Peng X, Kryscio RJ (2000), 'Whole grain foods and heart disease risk' *J Am Coll Nutr*, **19**, 291S–299S.

Arkbåge K, Verwei M, Havenaar R, Witthöft C (2003) 'Folic acid and (6S)-5-methyltetrahydrofolate bioaccessibility decreases after addition of folate-binding protein to yogurt as studied in a dynamic *in vitro* gastrointestinal model' *J Nutr*, **133**, 3678–3683.

Armand M, Pasquier B, Andre M, Borel P, Senft M, Peyrot J, Salducci J, Portugal H, Jaussan V, Lairon D (1999) 'Digestion and absorption of 2 fat emulsions with different droplet sizes in the human digestive tract' *Am J Clin Nutr*, **70**, 1096–1106.

Asp NG (1992) 'Resistant starch – Proceedings from the second plenary meeting of EURESTA: European FLAIR Concerted Action No. 11 on physiological implications of the consumption of resistant starch in man' *Eur J Clin Nutr*, **46**, Suppl. 2, S1.

Asp NG, Johansson GG, Hallmer H, Siljeström M (1983) 'Rapid enzymatic assay of insoluble and soluble dietary fiber' *J Agric Food Chem*, **31**, 476–482.

Athukorala Y, Kim KN, Jeon YJ (2006) 'Antiproliferative and antioxidant properties of an enzymatic hydrolysate from brown alga, *Ecklonia cava*' *Food Chem Toxicol*, **44**, 1065–1074.

Bach-Knudsen KE, Hessov I (1995) 'Recovery of inulin from Jerusalem artichoke (*Helianthus tuberosus* L.) in the small intestine of man' *Br J Nutr*, **74**, 101–113.

Barry JL, Hoebler C, Macfarlane GT, Macfarlane S, Mathers JC, Reed KA, Mortensen PB, Nordgaard I, Rowland IR, Rumney CJ (1995) 'Estimation of the fermentability of dietary fibre *in vitro*: a European interlaboratory study' *Br J Nutr*, **74**, 303–322.

Bellisle F, Diplock AT, Hornstra G, Koletzkos B, Roberfroid M, Salminen S, Saris WHM (1998) 'Functional food science in Europe' *Br J Nutr*, **80**, S1–S193.

Bjarnason I, MacPherson A, Hollander D (1995) 'Intestinal permeability: an overview' *Gastroenterology*, **108**, 1566–1581.
Björck I, Nyman M, Pedersen B, Siljeström M, Asp NG, Eggum BO (1986) 'On the digestibility of starch in wheat bread – studies *in vitro* and *in vivo*' *J Cereal Sci*, **4**, 1–11.
Boirie Y, Dangin M, Gachon P, Vasson MP, Maubois JL, Beaufrère B (1997) 'Slow and fast dietary proteins differently modulate postprandial protein accretion' *Proc Natl Acad Sci U S A*, **94**, 14930–14935.
Bornet FRJ, Fontvieille AM, Rizkalla S, Colonna P, Blayo A, Mercier C, Slama G (1989) 'Insulin and glycemic responses in healthy humans to native starches processed in different ways: correlation with *in vitro* α-amylase hydrolysis' *Am J Clin Nutr*, **50**, 315–323.
Branch W, Cummings JH (1978) 'Comparison of radio-opaque pellets and chromium sesquioxide as inert markers in studies requiring accurate faecal collections' *Gut*, **19**, 371–376.
Brighenti F, Pellegrini N, Casiraghi MC, Testolin G (1995) '*In vitro* studies to predict physiological effects of dietray fibre' *Eur J Clin Nutr*, **49**, S81–S88.
Brouns F, Bjorck I, Frayn KN, Gibbs AL, Lang V, Slama G, Wolever TMS (2005) 'Glycaemic index methodology' *Nutr Res Rev*, **18**, 145–171.
Calvert RJ, Klurfeld DM, Subramaniam S, Vahouny GV, Kritchevsky D (1987) 'Reduction of colonic carcinogenesis by wheat bran independent of fecal bile acid concentration' *J Natl Cancer Inst*, **79**, 875–880.
Champ M (1992) 'Determination of resistant starch in foods and food products: interlaboratory study' *Eur J Clin Nutr*, **46**, Suppl. 2, S51–S62.
Champ M, Langkilde AM, Brouns F, Kettlitz B, Le Bail-Collet Y (2003a) 'Advances in dietary fibre characterisation. 2. Consumption, chemistry, physiology and measurement of resistant starch; implications for health and food labelling' *Nutr Res Rev*, **16**, 143–161.
Champ M, Langkilde AM, Brouns F, Kettlitz B, Le Bail-Collet Y (2003b) 'Advances in dietary fibre characterisation. 1. Definition of dietary fibre, physiological relevance, health benefits and analytical aspects' *Nutr Res Rev*, **16**, 71–82.
Champ MM (1996) 'The analysis of complex carbohydrates: relevance of values obtained *in vitro*' *Proc Nutr Soc*, **55**, 863–880.
Champ MM (2004) 'Physiological aspects of resistant starch and *in vivo* measurements' *J AOAC Int*, **87**, 749–755.
Christian MT, Amarri S, Franchini F, Preston T, Morrison DJ, Dodson B, Edwards CA, Weaver LT (2002) 'Modeling 13C breath curves to determine site and extent of starch digestion and fermentation in infants' *J Pediatr Gastroenterol Nutr*, **34**, 158–164.
Clydesdale F (1997) 'A proposal for the establishment of scientific criteria for health claims for functional foods' *Nutr Rev*, **55**, 413–422.
Cummings JH, Antoine JM, Azpiroz F, Bourdet-Sicard R, Brandzaeg P, Calder PC, Gibson GR, Guarner F, Isolauri E, Pannemans D, Shortt C, Tuijelaars S, Watzl B (2004) 'PASSCLAIM – Gut health and immunity' *Eur J Clin Nutr* (suppl 2), **43**, II/118–II/173.
Cummings JH, Jenkins DJ, Wiggins HS (1976) 'Measurement of the mean transit time of dietary residue through the human' *Gut*, **17**, 210–218.
Cummings JH, Wiggins HS (1976) 'Transit through the gut measured by analysis of a single stool' *Gut*, **17**, 219–223.
Dangin M, Boirie Y, Garcia-Rodenas C, Gachon P, Fauquant J, Callier P, Ballèvre O, Beaufrère B (2001) 'The digestion rate of protein is an independent regulating factor of postprandial protein retention' *Am J Physiol Endocrinol Metab*, **280**, E340–E348.
Dangin M, Boirie Y, Guillet C, Beaufrère B (2002) 'Influence of the protein digestion rate on protein turnover in young and elderly subjects' *J Nutr*, **132**, 3228S–3233S.
Dangin M, Guillet C, Garcia-Rodenas C, Gachon P, Bouteloup-Demange C, Reiffers-Magnani K, Fauquant J, Ballèvre O, Beaufrère B (2003) 'The rate of protein digestion affects protein gain differently during aging in humans' *J Physiol*, **549**(2), 635–644.

Daniel M, Wisker E, Rave G, Feldheim W (1997) 'Fermentation in human subjects of nonstarch polysaccharides in mixed diets, but not in a barley fiber concentrate, could be predicted by *in vitro* fermentation using human fecal inocula' *J Nutr*, **127**, 1981–1988.

Darragh AJ, Hodgkinson SM (2000) 'Quantifying the digestibility of dietary protein' *J Nutr*, **13**, 1850S–1856S.

De Preter V, Geboes K, Verbrugghe K, De Vuyst L, Vanhoutte T, Huys G, Swings J, Pot B, Verbeke K (2004) 'The *in vivo* use of the stable isotope-labelled biomarkers lactose-[15N]ureide and [2H4]tyrosine to assess the effects of pro- and prebiotics on the intestinal flora of healthy human volunteers' *Br J Nutr*, **92**, 439–446.

Edwards CA, Zavoshy R, Khanna S, Slater C, Morrison DJ, Preston T, Weaver LT (2002) 'Production of 13C labelled pea flour for use in human digestion and fermentation', *Isotopes Environ Health Stud*, **38**, 139–147.

El Oufir L, Barry JL, Flourie B, Cherbut C, Cloarec D, Bornet F, Galmiche JP (2000) 'Relationships between transit time in man and *in vitro* fermentation of dietary fiber by fecal bacteria' *Eur J Clin Nutr*, **54**, 603–609.

Englyst HN, Kingman SM, Cummings JH (1992) 'Classification and measurement of nutritionally important starch fractions' *Eur J Clin Nutr*, **46** (Suppl 2), S33–S50.

Englyst HN, Kingman SM, Hudson GJ, Cummings JH (1996) 'Measurement of resistant starch *in vitro* and *in vivo*' *Br J Nutr*, **75**, 749–755.

Englyst KN, Englyst HN, Hudson GJ, Cole TJ, Cummings JH (1999) 'Rapidly available glucose in foods: an *in vitro* measurement that reflects the glycemic response' *Am J Clin Nutr*, **69**, 448–454.

FAO (Food and Agriculture Organization) (1998) 'Carbohydrates in human nutrition – Report of a Joint FAO/WHO Expert Consultation, Rome, 14–18 April 1997'. FAO Food and Nutrition Paper No. 66. FAO, Rome, Italy.

Feillet-Coudray C, Coudray C (2005) 'The stable isotope use in the exploration of bioavailability and metabolism of magnesium' *Curr Nutr Food Sci*, 1, 63–70.

Foster-Powell K, Holt SHA, Brand-Miller JC (2002) 'International table of glycaemic load values' *Am J Clin Nutr*, **76**, 5–56.

Geboes KP, De Preter V, Luypaerts A, Bammens B, Evenepoel P, Ghoos Y, Rutgeerts P, Verbeke K (2005) 'Validation of lactose[15N,15N]ureide as a tool to study colonic nitrogen metabolism' *Am J Physiol Gastrointest Liver Physiol*, **288**, G994–G999.

Gilani GS, Cockell KA, Sepehr E (2005) 'Effects of antinutritional factors on protein digestibility and amino acid availability in foods' *J AOAC Int*, 88, 967–987.

Goñi I, Mañas E, Garcia-Diz L, Saura-Calixto F (1996) 'Analysis of resistant starch : a method for food and food products' *Food Chem*, **56**, 445–449.

Granfeldt Y, Björck I, Drews A, Tovar J (1992) 'An *in vitro* procedure based on chewing to predict metabolic response to starch in cereal and legume products' *Eur J Clin Nutr*, **46**, 649–660.

Hagander B, Björck I, Asp NG, Efendic S, Holm J, Nilsson-Ehle P, Lundquist I, Schersten B (1987) Rye products in the diabetic diet. Postprandial glucose and hormonal responses in non-insulin-dependent diabetic patients as compared to starch availability *in vitro* and experiments in rats'. *Diabetes Res Clin Pract*, **3**, 85–96.

Hill AJ, Magson LD, Blundell JE (1984) 'Hunger and palatability: tracking ratings of subjective experience before, during and after the consumption of preferred and less preferred food' *Appetite* **5**, 361–371.

Hoebler C, Lecannu G, Belleville C, Devaux MF, Popineau Y, Barry JL (2002) 'Development of an *in vitro* system simulating bucco-gastric digestion to assess the physical and chemical changes of food' *Int J Food Sci Nutr*, **53**, 389–402.

Holm J, Lundquist I, Björck I, Eliasson AC, Asp NG (1988) 'Degree of starch gelatinization, digestion rate of starch *in vitro*, and metabolic response in rats' *Am J Clin Nutr*, **47**, 1010–1006.

Jacobs DR, Meyer KA, Kushi LH, Folsom AR (1998) 'Whole-grain intake may reduce the risk of ischemic heart disease death in postmenopausal women: the Iowa Women's Health Study' *Am J Clin Nutr*, **68**, 248–257.

Johnsen NF, Hausner H, Olsen A, Tetens I, Christensen J, Knudsen KE, Overvad K, Tjonneland A (2004) 'Intake of whole grains and vegetables determines the plasma enterolactone concentration of Danish women' *J Nutr*, **134**, 2691–2697.

Juhel C, Armand M, Pafumi Y, Rosier C, Vandermander J, Lairon D (2000) 'Green tea extract (AR25) inhibits lipolysis of triglycerides in gastric and duodenal medium *in vitro'*, *J Nutr Biochem*, **11**, 45–51.

Kies AK, De Jonge LH, Kemme PA, Jongbloed AW (2006) 'Interaction between protein, phytate, and microbial phytase. In vitro studies' *J Agric Food Chem*, **54**, 1753–1758.

Kloots W, Op den Kamp D, Abrahamse L (2004) '*In vitro* iron availability from iron-fortified whole-grain wheat flour' *J Agric Food Chem*, **52**, 8132–8136.

Koh-Banerjee P, Franz M, Sampson L, Liu S, Jacobs DR, Spiegelman D, Willett W, Rimm E (2004) 'Changes in whole-grain, bran, and cereal fiber consumption in relation to 8-y weight gain among men' *Am J Clin Nutr*, **80**, 1237–1245.

Krevsky B, Malmud LS, D'Ercole F, Maurer AH, Fisher RS (1986) 'Colonic scintigraphy – a physiologic approach to the quantitative measurement of colonic transit in humans' *Gastroenterology*, **91**, 1102–1112.

Lichtman SM (2001) 'Bacterial translocation in humans' *J Pediatr Gastroenterol Nutr*, **33**, 389–395.

Liu S, Sesso HD, Manson JE, Willett WC, Buring JE (2003) 'Is intake of breakfast cereals related to total and cause-specific mortality in men' *Am J Clin Nutr*, **77**, 594–599.

Lopez-Varela S, Gonzales-Gross M, Marcos A (2002) 'Functional foods and the immune system: a review' *Eur J Clin Nutr*, **56**, Suppl 3, S29–S33.

Luceri C, Guglielmi F, Lodovici M, Giannini L, Messerini L, Dolara P (2004) 'Plant phenolic 4-coumaric acid protects against intestinal inflammation in rats' *Scand J Gastroenterol*, **39**, 1128–1133.

Lugea A, Salas A, Casalot J, Guarner F, Malagelada JR (2000) 'Surface hydrophobicity of the rat colonic mucosa is a defensive barrier against macromolecules and toxins' *Gut*, **46**, 515–521.

Lupton JR (2005) 'Determining the strentgh of the relationship between a food, food component, or dietary supplement ingredient and reduced risk of a disease or health-related condition, *J Nutr*, **135**, 340–342.

McBurney MI, Thompson LU (1989) 'Effect of human faecal donor on *in vitro* fermentation variables' *Scand J Gastroenterol*, **24**, 359–367.

McCleary BV, Monaghan DA (2002) 'Measurement of resistant starch' *J AOAC Int*, **85**, 665–675.

McDougall GJ, Dobson P, Smith P, Blake A, Stewart D (2005a) 'Assessing potential bioavailability of raspberry anthocyanins using *in vitro* digestion system' *J Agric Food Chem*, **53**, 5896–5904.

McDougall GJ, Fyffe S, Dobson P, Stewart D (2005b) 'Anthocyanins from red-wine – Their stability under simulated gastrointestinal digestion' *Phytochemistry*, **66**, 2540–2548.

McIntyre A, Gibson PR, Young GP (1993) 'Butyrate production from dietary fibre and protection against large bowel cancer in a rat model' *Gut*, **34**, 386–391.

Millward DJ, Fereday A, Gibson NR, Cox MC, Pacy PJ (2002) 'Efficiency of utilization of wheat and milk protein in healthy adults and apparent lysine requirements determined by a single-meal [1-13C]leucine balance protocol' *Am J Clin Nutr*, **76**, 1326–1334.

Minekus M (1998) 'Development and validation of a dynamic model of the gastrointestinal tract' PhD thesis, University of Utrecht, The Netherlands.

Minekus M, Jelier M, Xiao J, Kondo S, Iwatsuki K, Kokubo S, Bos M, Dunnewind B, Havenaar R (2005) 'Effect of partially hydrolyzed guar gum (PHGG) on the bioaccessibility of fat and cholesterol' *Biosc Biotechnol Biochem*, **69**, 932–938.

Minekus M, Smeets-Peeters M, Bernalier A, Marol-Bonnin S, Havenaar R, Marteau P, Alric M, Fonty G, Huis in't Veld JH (1999) 'A computer-controlled system to simulate conditions of the large intestine with peristaltic mixing, water absorption and absorption of fermentation products' *Appl Microbiol Biotechnol*, **53**, 108–114.

Molly K, Vande Woestyne M, Verstraete W (1993) 'Development of a 5–step multi-chamber reactor as a simulation of the human intestinal microbial ecosystem' *Appl Microbiol Biotechnol*, **39**, 254–258.

Moreno FJ, Mackie AR, Mills EN (2005a) 'Phospholipid interactions protect the milk allergen alpha-lactalbumin from proteolysis during *in vitro* digestion' *J Agric Food Chem*, **53**, 9810–9816.

Moreno FJ, Maldonado BM, Wellner N, Mills EN (2005b) 'Thermostability and *in vitro* digestibility of a purified major allergen 2S albumin (Ses I 1) from white sesame seeds (*Sesamum indicum L.*)' *Biochim Biophys Acta*, **1752**, 142–153.

Moreno FJ, Mellon FA, Wickham MS, Bottrill AR, Mills EN (2005c) 'Stability of the major allergen Brazil nut 2S albumin (Ber e 1) to physiologically relevant *in vitro* gastrointestinal digestion' *FEBS J*, **272**, 341–352.

Muir JG, O'Dea K (1992) 'Measurement of resistant starch: factors affecting the amount of starch escaping digestion *in vitro*' *Am J Clin Nutr*, **56**, 123–127.

Muir JG, O'Dea K. (1993) 'Validation of an *in vitro* assay for predicting the amount of starch that escapes digestion in the small intestine of humans' *Am J Clin Nutr*, **57**, 540–546.

Murtagh GJ, Archer DB, Dumoulin M, Ridout S, Matthews S, Arshas SH, Alcocer MJ (2003) '*In vitro* stability and immunoreactivity of the native and recombinant plant food 2S albumins Ber e 1 and SFA-8' *Clin Exp Allergy*, **33**, 1147–1152.

Navarro M, Wood RJ (2003) 'Plasma changes in micronutrients following a multivitamin and mineral supplement in healthy adults' *J Am Coll Nutr*, **22**, 124–132.

Olsson ME, Andersson CS, Oredsson S, Berglund RH, Gustavsson KE (2006) 'Antioxidant levels and inhibition of cancer cell proliferation *in vitro* by extracts from organically and conventionally cultivated strawberries' *J Agric Food Chem*, **54**, 1248–1255.

Pedrero Z, Madrid Y, Camara C (2006) 'Selenium species bioaccessibility in enriched radish (*Raphanus sativus*): a potential dietary source of selenium' *J Agric Food Chem*, **54**, 2412–2417.

Perales S, Barbera R, Lagarda MJ, Farre R (2005) 'Bioavailability of calcium from milk-based formulas and fruit juices containing milk and cereals estimated by *in vitro* methods (solubility, dialyzability, and uptake and transport by Caco-2 cells)' *J Agric Food Chem*, **53**, 3721–3726.

Pierre F, Perrin P, Champ M, Bornet F, Meflah K, Menanteau J (1997) 'Short-chain fructo-oligosaccharides reduce the occurrence of colon tumors and develop gut-associated lymphoid tissue in Min mice' *Cancer Res*, **57**, 225–228.

Perrin P, Pierre F, Patry Y, Champ M, Berreur M, Pradal G, Bornet F, Meflah K, Menanteau J (2001) 'Only fibres promoting a stable butyrate producing colonic ecosystem decrease the rate of aberrant crypt foci in rats' *Gut*, **48**, 53–61.

Pouteau E, Vahedi K, Messing B, Flourie B, Nguyen P, Darmaun D, Krempf M (1998) 'Production rate of acetate during colonic fermentation of lactulose: a stable-isotope study in humans' *Am J Clin Nutr*, **68**, 1276–1283.

Promchan J, Shiowatana J (2005) 'A dynamic continuous-flow dialysis system with on-line electrothermal atomic absorption spectrometric and pH measurements for *in vitro* determination of iron bioavailability by simulated gastrointestinal digestion' *Anal Bioanal Chem*, **382**, 1360–1367.

Raben A, Agerholm-Larsen L, Flint A, Holst J, Astrup A (2003) 'Meals with similar energy densities but rich in protein, fat, carbohydrate, or alcohol have different effects on energy expenditure and substrate metabolism but not on appetite and energy intake' *Am J Clin Nutr*, **77**, 91–100.

Reddy BS, Hirose Y, Cohen LA, Simi B, Cooma I, Rao CV (2000) 'Preventive potential of wheat bran fractions against experimental colon carcinogenesis: implications for human colon cancer prevention' *Cancer Res*, **60**, 4792–4797.

Reyes LH, Encinar JR, Marchante-Gayon JM, Alonso JI, Sanz-Medel A (2006) 'Selenium bioaccessibility assessment in selenized yeast after '*in vitro*' gastrointestinal digestion

using two-dimensional chromatography and mass spectrometry' *J Chromatogr A*, **1110**, 108–116.
Roberfroid MB (1999) 'Concepts in functional foods: the case of inulin and oligofructose' *J Nutr*, 129, 1398S–1401S.
Salminen S, Bouley C, Boutron-Ruault MC, Cummings JH, Franck A, Gibson GR, Isolauri E, Moreau MC, Roberfroid M, Rowland I (1998) 'Functional food science and gastrointestinal physiology and function' *Br J Nutr*, **80**, Suppl 1, S147–S171.
Sandberg AS (2005) 'Methods and options *in vitro* dialyzability; benefits and limitations' *Int J Vitam Nutr Res*, **75**, 395–404.
Schaafsma G (2005) 'The Protein Digestibility-Corrected Amino Acid Score (PDCAAS) – a concept for describing protein quality in foods and food ingredients: a critical review' *J AOAC Int*, **88**, 988–994.
Slavin JL (2000) 'Mechanisms for the impact of whole grain foods on cancer risk' *J Am Coll Nutr*, **19**, 300S–307S.
Soderholm JD, Perdue MH (2001) 'Stress and gastrointestinal tract. II. Stress and intestinal barrier function' *Am J Physiol Gastrointestinal Liver Physiol*, **280**, G7–G13.
Such J, Frances R, Munoz C, Zapater P, Casellas JA, Cifuentes A, Rodriguez-Valera F, Pascual S, Sola-Vera J, Canicer F, Uceda F, Palazon JM, Perez-Mateo M (2002) 'Detection and identification of bacterial DNA in patients with cirrhosis and culture-negative, nonneutrocytic ascites' *Hepatology*, **36**, 135–141.
Tarozzi A, Marchesi A, Cantelli-Forti G, Hrelia P (2004) 'Cold-storage affects antioxidant properties of apples in Caco-2 cells' *J Nutr*, **134**, 1105–1109.
Thomas K, Aalbers M, Bannon GA, Bartels M, Dearman RJ, Esdaile DJ, Fu TJ, Glatt CM, Hadfield N, Hatzos C, Hefle SL, Heylings JR, Goodman RE, Henry B, Herouet C, Holsapple M, Ladics GS, Landry TD, MacIntosh SC, Rice EA, Privalle LS, Steiner HY, Teshima R, Van Ree R, Woolhiser M, Zawodny J (2004) 'A multi-laboratory evaluation of a common *in vitro* pepsin digestion assay protocol used in assessing the safety of novel proteins' *Regul Toxicol Pharmacol*, **39**, 87–98.
Tovar J., Björck I, Asp NG (1992) 'Digestibility of starch in legumes using the rat' *Eur J Clin Nutr*, **46**(S2), S141–142.
Truswell AS. (2002) 'Cereal grains and coronary heart disease' *Eur J Clin Nutr*, **56**, 1–14.
Tuohy KM, Kolida S, Lustenberger AM, Gibson GR (2001) 'The prebiotic effects of biscuits containing partially hydrolysed guar gum and fructo-oligosaccharides – a human volunteer study' *Br J Nutr*, **86**, 341–348.
Unal G, El SN, Kilic S (2005) '*In vitro* determination of calcium bioavailability of milk, dairy products and infant formulas' *Int J Food Sci Nutr*, **56**, 13–22.
Vallejo F, Gil-Izquierdo A, Perez-Vicente A, Garcia-Viguera C (2004) '*In vitro* gastrointestinal digestion study of broccoli inflorescence phenolic compounds, glucosinolates, and vitamin C' *J Agric Food Chem*, **52**, 135–138.
Van Loo J, Coussement P, de Leenheer L, Hoebregs H, Smits G (1995) 'On the presence of inulin and oligofructose as natural ingredients in the western diet' *Crit Rev Food Sci Nutr*, **35**, 525–552.
Vassilopoulou E, Rigby N, Moreno FJ, Zuidmeer L, Akkerdaas J, Tassios I, Papadopoulos NG, Saxoni-Papageorgiou P, van Ree R, Mills C (2006) 'Effect of *in vitro* gastric and duodenal digestion on the allergenicity of grape lipid transfer protein' *J Allergy Clin Immunol*, **118**, 473–480.
Verwei M, Arkbåge K, Mocking H, Havenaar R, Groten J (2004) 'The binding of folic acid and 5-methyltetrahydrofolate to folate-binding proteins during gastric passage differs in a dynamic *in vitro* gastrointestinal model' *J Nutr*, **134**, 31–37.
Verwei M, Arkbåge K, Havenaar R, van den Berg H, Witthöft C, Schaafsma G (2003) 'Folic acid and 5-methyltetrahydrofolate in fortified milk are accessible as determined in a dynamic *in vitro* gastrointestinal model' *J Nutr*, **133**, 2377–2383.

Viadel B, Barbera R, Farre R (2006) 'Effect of cooking and legume species upon calcium, iron and zinc uptake by Caco-2 cells' *J Trace Elem Med Biol*, **20**, 115–120.

Wisker E, Daniel M, Rave G, Feldheim W (1998) 'Fermentation of non-starch polysaccharides in mixed diets and single fibre sources: comparative studies in human subjects and *in vitro*' *Br J Nutr*, **80**, 253–261.

Wolever TM, Vorster HH, Bjorck I, Brand-Miller J, Brighenti F, Mann JI, Ramdath DD, Granfeldt Y, Holt S, Perry TL, Venter C, Xiaomei W (2003) 'Determination of the glycaemic index of foods: interlaboratory study', *Eur J Clin Nutr*, **57**(3), 475–482.

Wrick K, Robertson JB, Van Soest PJ, Lewis BA, Rivers JM, Roe DA, Hackler LR (1983) 'The influence of dietary fiber source on human intestinal transit and stool output' *J Nutr*, **113**, 1464–1479.

Zangenberg NH, Mullertz A, Kristensen HG, Hovgaard L (2001a) 'A dynamic *in vitro* lipolysis model. I. Controlling the rate of lipolysis by continuous addition of calcium' *Eur J Pharm Sci*, **14**, 115–122.

Zangenberg NH, Mullertz A, Kristensen HG, Hovgaard L (2001b) 'A dynamic *in vitro* lipolysis model. II. Evaluation of the model' *Eur J Pharm Sci*, **14**, 237–244.

Zeijdner EE, Havenaar R (2000) 'The fate of orally administered compounds during passage through the gastrointestinal tract simulated in a dynamic *in vitro* model (TIM)' *Eur. Pharmaceut Contractor*, February, 76–81.

Part II

Technology of functional cereal products

8

Improving the nutritional quality of cereals by conventional and novel approaches

P. R. Shewry, Rothamsted Research, UK

8.1 Introduction

Cereals are major sources of carbohydrate, protein, fibre and other nutrients for humans, being consumed in a wide range of forms in different cultures. In some less developed countries cereals may form the major part of the diet with little supplementation with plant and animal products, while even in more developed countries (e.g. Europe, the Americas, the Pacific rim) cereals may still be major sources of some dietary components such as selenium and some other minerals.

Although three cereal species (wheat, rice and maize) are dominant over much of the world, over ten other 'minor' cereals are also grown worldwide. These species differ in their composition and in their end use properties and utilisation. It is clearly not possible to discuss all cereal species within the confines of this chapter. I will therefore focus on wheat, which is the major cereal consumed by humans in Europe and the Americas, drawing wider comparisons and parallels where appropriate. However, before considering individual targets, I will initially introduce the structure and composition of the cereal grain and the technologies available for grain improvement.

8.1.1 The cereal grain

The mature wheat grain comprises the embryo and endosperm surrounded by a single layer of crushed nucellar cells, the testa (the seed coat) and the pericarp (the fruit coat, a maternal tissue). The endosperm is further differentiated into a single layer of thick-walled aleurone cells surrounding the starchy endosperm cells. The endosperm and embryo together account for over 90% of the grain dry weight: about 80% starchy endosperm, 6–8% aleurone and 2–3% embryo. These three tissues are routinely separated by milling, with the embryo forming the germ, the aleurone about half of the bran fraction and the starchy endosperm the white flour.

The aleurone, starchy endosperm and embryo also differ significantly in their composition and hence nutritional quality. In broad terms the aleurone and embryo have similar compositions, being rich in protein (approximately 20%) and oil (approximately 5–8% in the aleurone and 15–20% in the germ), and containing little or no starch. These two tissues are also the main source of the minerals and vitamins present in the grain, while the bran is also a major source of dietary fibre. These components are consequently lost on milling to give white flour. A similar depletion of nutrients occurs when brown rice is polished to give white rice for human consumption.

In contrast, the starchy endosperm comprises only three main components: starch (70–80% dry weight), protein (approximately 10–12%) and non-starch polysaccharides (approximately 3%). Furthermore, the gluten proteins present in the starchy endosperm cells have lower nutritional quality than those present in the germ and aleurone, which include 7S storage globulins. This is reflected in their total lysine contents which are 8.3 g lysine/100 g protein in germ, 4.8 g% in aleurone and 2.1 g% in starchy endosperm (Pomeranz (1988) taken from Jensen and Martens (1983)).

There is also considerable variation in the composition of cells within the starchy endosperm of cereals, with well-established gradients in starch and protein. Thus, Kent (1966) was able to prepare white flour fractions with protein contents ranging from 8.6 to 54.0%, using air classification and sedimentation in solvent mixtures. These fractions corresponded to the protein-poor central cells of the starchy endosperm and the protein-rich sub-aleurone cells, respectively. Similarly, Millet *et al.* (1991) used pearling to prepare protein-rich sub-aleurone and protein-poor central endosperm fractions from barley, showing that the former were enriched in aggregative B hordeins and the latter in D hordeins.

It is therefore important to consider the location of nutrients within the grain to ensure that they will be present in the required products. Since most wheat products are currently made from white flour this will require focusing on the starchy endosperm tissue rather than on the bran or germ which are the major sites of most nutrients. This also applies to rice, which is mostly consumed after debranning.

8.2 Technologies for grain improvement

8.2.1 Conventional breeding

Cereal improvement has been taking place for millennia with farmers selecting and growing on the most successful genotypes of their crops. However, organised breeding dates only from the nineteenth century, underpinned by the discovery of the principles of genetics and inheritance. Modern plant breeding is a highly scientific process, supported by a range of modern technologies that facilitate the introgression of traits from wild or exotic relatives and the more precise selection of improved progeny using biochemical and molecular markers (see chapters in Bonjean and Angus (2001) for discussions of wheat breeding). However, the application of conventional breeding to crop improvement is limited by the extent to which variation in the trait of interest occurs, either in the crop itself or in related species from which it can be transferred by 'wide crossing'. It is therefore common for crop improvement programmes to include a diversity screen for variation in the trait(s) of interest. This can be a formidable task, for example, over 25 000 varieties of wheat exist (Feldman *et al.*, 1995) and it is usual to focus on 'core collections' known to differ in their region of origin and characteristics.

8.2.2 Mutation breeding

The use of mutagenesis to increase variation in crop plants dates from Stadler (1929) and by 2001 almost 180 durum and bread wheat varieties had been produced and released worldwide from mutation breeding programmes (Maluszynski *et al.*, 2001). A range of mutagens have been used, including chemicals which result in single base changes in DNA (i.e. ethyl methane sulphonate (EMS)) and ionising radiation which can result in major structural changes to chromosomes including deletions and rearrangements. The vast majority of mutations resulting from single base changes are recessive in nature and result in the inactivation of genes or gene products. Hence, they have been most successfully exploited in diploid species and in systems where single genes and proteins regulate important pathways or processes (for example, starch synthesis and the hormonal control of stature).

8.2.3 Transgenesis

Although the first transgenic plants were produced little over 20 years ago, it is now possible to transform all of the major crop plants as well as many minor crops, horticultural species and 'model' plants used as experimental systems for plant research (Curtis, 2004; Peña, 2005). Transgenesis is particularly attractive for crop improvement as it allows the introduction of genes, and hence traits, from any source, including other plants, microbes and animals. It also allows the expression of endogenous genes to be repressed or enhanced in a highly specific fashion, and for their expression to be targeted to specific organs and tissues (and the gene products to specific destinations inside or outside the cell). Furthermore, technologies are available to use transgenesis for gene replacement (Terada *et al.*, 2002) and *in vivo* mutagenesis (Zhu *et al.*, 1999), although these are still not routinely available for most crop species.

Transgenesis clearly has immense potential for exploitation in crop improvement and this has been demonstrated by the massive uptake of varieties of genetically modified (GM) crops around the world. Thus, GM soybean accounted for over 50% of the total world production in 2003, GM cotton for 21% and GM maize and oilseed rape for 11% and 16%, respectively (Halford, 2006). However, the vast majority of these and other GM crops have been modified to improve 'input traits' (notably herbicide resistance and insect resistance) with only small amounts of varieties with modified seed composition being grown. There is also immense public resistance to the use of GM technology, particularly in Europe, and for the manipulation of the composition of food crops. Nevertheless, the demonstration that GM can be used to modify traits of value to consumers (i.e. food and nutritional quality) as well as to crop producers and processors could have a significant impact on future acceptability.

8.2.4 Targeting Induced Local Lesions IN Genomes (TILLING)

McCallum and co-workers (McCallum *et al.*, 2000a,b; Colbert *et al.*, 2001) recently described a new technology that allows the identification of mutations in specific genes of interest. This Targeting Induced Local Lesions IN Genomes (TILLING) has since been adopted widely as a non-GM tool to aid crop improvement. TILLING can be used on mutagenised populations or to identify natural alleles in collections of cultivars or ecotypes, termed ecoTILLING, and requires knowledge of the sequences of the genes of interest in order to design specific primers for PCR amplification. It is a particularly powerful tool for identifying recessive mutations in polyploid species such as wheat, allowing mutations identified in homologous genes on the individual genomes to be identified and combined. This was elegantly demonstrated by Slade *et al.* (2005), who used TILLING to identify novel mutations at the *waxy* (granule-bound starch synthase 1) loci on the A and D genomes of bread wheat. Combination of these mutations with a previously identified *waxy* mutation on the B genome gave a full waxy phenotype. This is discussed further below.

8.3 Improvement of protein quality

8.3.1 Amino acid composition

Although frequently regarded primarily as sources of calories, cereal grains also provide significant amounts of protein to the human diet. For example, Shewry (2000) calculated that the total amount of protein harvested annually in cereals is over twice that harvested in pulses, soybean, groundnut and oilseeds combined.

Cereals are generally consumed as part of a mixed diet and hence deficiencies in their amino acid composition do not usually have serious consequences for human nutrition. However, they do have an economic impact on the utilisation of cereal grain in feed for livestock and hence nutritional quality for animal feed has been the main driver for research on this topic.

The first limiting amino acid in all cereal grains is lysine. Wheat grain contains about 2.6–3.2 g lysine/100 g protein and flour about 1.8–2.7 g lysine/100 g protein (Wrigley and Bietz (1998), adapted from Shoup *et al.* (1966)). The difference between the values for grain and flour results from the removal of the more lysine-rich aleurone and germ on milling; these fractions containing about 4.8 and 8.3 g lysine/100 g protein, respectively (Pomeranz (1988), adapted from Jensen and Martens (1983)). Nevertheless, even the whole grain value falls short of the UN Food and Agriculture Organization (FAO) recommendation of 6.6 g/100 g for infants (the recommended value for adults being lower at 1.6 g/100 g) (FAO/WHO/UNO, 1985). All other cereals are also deficient in lysine, although this deficiency is less severe in rice and oats. The second limiting amino acid in wheat and barley is threonine and in maize tryptophan.

The sulphur-containing amino acids, cysteine and methionine, are often considered together as cysteine can be made from methionine by animals. The combined values for these two amino acids are adequate in most cereals but deficient in legume seeds. However, because cereals are often mixed with legume seeds for livestock feed, there is also interest in increasing the methionine content of cereals.

8.3.2 Exploiting mutant 'high lysine' genes

Interest in increasing grain lysine dates from the early 1960s and particularly from the discovery that several mutant lines of maize had elevated contents of lysine when compared with wild-type grain. The first such line to be identified was *opaque2* (Mertz *et al.*, 1964) followed by *floury2* (Nelson *et al.*, 1965) and a number of other lines previously identified as floury, sugary, opaque, shrunken and brittle based on the visual phenotype of the endosperm (see Bright and Shewry, 1983). Subsequent work led to the identification of one spontaneous mutant and one induced mutant in sorghum (Singh and Axtell, 1973; Axtell *et al.*, 1979) and one spontaneous mutant (Hiproly) (Munck *et al.*, 1970) and a number of induced mutants (Bright and Shewry, 1983; Doll, 1983) in barley.

These high lysine lines have up to two-fold increases in the proportion of lysine in the grain and hence initially proved attractive to breeders. However, in most cases they also have associated effects on the synthesis and accumulation of starch and hence reduced yield. These effects are also responsible for the visual effects on grain phenotype in maize and sorghum.

Consequently, despite much effort over 40 years, only one of these mutant high lysine genes has been successfully exploited and this is the first to be identified, *opaque2*. Lines initially developed from *opaque2* had a soft starchy phenotype, which resulted in more damage during harvest, greater susceptibility to pests and lower acceptance by communities accustomed to growing vitreous types of maize. The breakthrough came with the identification of 'genetic modifier' genes which allowed the development of *opaque2* lines with a hard phenotype (called quality protein maize or QPM) (Geevers and Lake, 1992; Villegas *et al.*, 1992). No equivalent success has been achieved with high lysine mutants in sorghum or barley, although they have proved to be useful tools to study mechanisms in grain development and, as discussed below, to identify nutritionally enriched proteins.

No attempts have been made to identify high lysine genes in wheat, and it is likely that most mutations would be recessive and therefore not result in a phenotype in the tetraploid or hexaploid background of durum and pasta wheats, respectively. However, at least two high lysine mutations in maize are dominant (Coleman and Larkins, 1999) so mutant high lysine wheat is not impossible!

8.3.3 Manipulating free amino acid content

The pools of free amino acids present in cereal grains are usually small (less than 1% of the amounts present in proteins), and are controlled by feedback regulation of key biosynthetic enzymes. In plants, three of the most important essential amino acids – lysine, threonine and methionine – are all synthesised from aspartic acid. Early work on barley showed that it is possible to use mutagenesis to generate lines with feedback-insensitive forms of the first enzyme in the pathway, aspartate kinase (Fig. 8.1) leading to

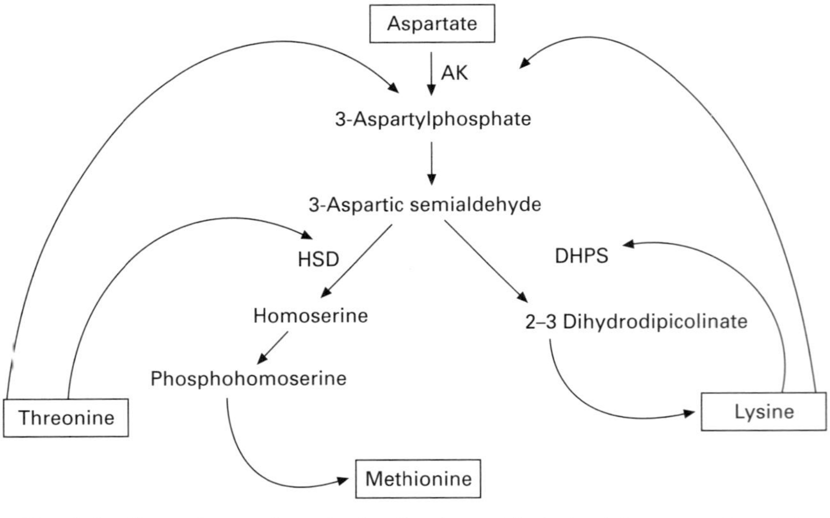

Fig. 8.1 The pathway of synthesis of amino acids derived from aspartic acid. Pathway steps are indicated by lines and feedback steps are shown by loops. AK, aspartate kinase; HSD, homoserine dehydrogenase; DHPS, dihydrodipicolinate synthase.

accumulation of free threonine in the seeds (Bright and Shewry, 1983). It has since been shown that it is necessary to eliminate the feedback sensitivity of the enzyme controlling the entry into the lysine branch of the pathway, dihydrodipicolinate synthase (DHPS) (Fig. 8.1) in order to generate similar increases in free lysine. Furthermore, this can be achieved by genetic engineering rather than by mutagenesis, using feedback-insensitive forms of DHPS from bacteria or maize.

Mazur *et al.* (1999) showed that expression of an enzyme from *Corynebacterium* in maize resulted in significant increases in free lysine when expressed in the embryo or aleurone, and these were sufficient to increase the total lysine content of the grain by up to two-fold. However, no increase was observed when the same enzyme was expressed in the endosperm as lysine breakdown was also increased. Similar breakdown has been observed in some other crops (Falco *et al.*, 1995; Mazur *et al.*, 1999) but Zhu and Galili (2003) have shown that it can be eliminated in Arabidopsis by knocking out enzymes of lysine catabolism. Brinch-Pedersen *et al.* (1996) have also shown that expression of a feedback-insensitive form of DHPS from *Escherichia coli* resulted in a two-fold increase in free lysine in barley grain, while Lee *et al.* (2001) showed that a feedback-insensitive mutant form of the maize enzyme gave 2.5-fold more free lysine when expressed in rice seeds, despite increased catabolism. However, these increases in free lysine are modest when compared with the total amount of lysine in grain proteins.

In September 2006 the first commercial release of GM high lysine corn was announced by the Monsanto company. This line expresses the feedback-insensitive DHPS enzyme from *Corynebacterium* and the increase in free lysine is said to result in an increase in total grain lysine from about 2500–2800 ppm dry weight to about 3500–5300 ppm dry weight (Monsanto petition number 04-CR-114U).

8.3.4 Increasing nutritionally enhanced proteins

Since almost all of the grain amino acids are present in proteins, the most logical approach to increase the proportions of essential amino acids is to increase the amounts of proteins that are rich in these amino acids. The validity of this approach has also been demonstrated by the analysis of Hiproly, a spontaneous high lysine mutant of barley (Munck *et al.*, 1970). In Hiproly the increased lysine content is mainly due to increases in the proportions of four proteins, β-amylase (5.0 g% lys), protein Z (a serpin proteinase inhibitor, 7.1 g% lys) and two inhibitors of chymotrypsin called CI-1 (9.5 g% lys) and CI-2 (11.5 g% lys) (Hejgaard and Boisen, 1980).

In fact, very few lysine-rich proteins have been characterised from plants, with two of the best known being hordothionin and CI-2, both of which are from barley grain. Hordothionin comprises only 45 amino acids including five lysines. Rao *et al.* (1994) used molecular modelling to design mutant forms containing up to 27% lysine and validated the acceptability of these mutations by carrying out studies of synthetic peptides. Zhao *et al.* (2003) also expressed a mutant form containing 12 lysine residues (HT12) in transgenic sorghum, showing increases in total grain lysine of about 50% compared with the wild-type grain. Similar studies of CI-2 were reported by Roesler and Rao (1999, 2000), who designed forms containing up to 25% mol% lysine combined with increases in other essential amino acids. A similar approach was adopted by Forsyth *et al.* (2005) who showed that additional lysines could be inserted into the inhibitory loop region of the protein with only minimal effects on the conformation and stability. A form containing three additional lysine residues (13.1 mol% lys) has also been used to transform grain sorghum in order to improve the nutritional quality (O'Kennedy *et al.*, 2006).

A lysine-rich protein, which is normally expressed in pollen of potato (sb401), has also been expressed in grains of maize (Yu *et al.*, 2004). This protein contains 40 lysines out of 240 amino acids (16.7 %mol lys) and expression in maize resulted in increases in grain protein and grain lysine, the latter by up to about 50%.

RNA interference technology has been used to down-regulate the individual groups of Z19 zeins (Huang *et al.*, 2004) and Z22 zeins (Segel *et al.*, 2003), and these two zein groups simultaneously in the same line (Huang *et al.*, 2006). The down-regulation of only one group of zeins resulted in a modest increase in grain lysine (by 15–20%) but this was approximately doubled (from 2.83% to 5.62%) when both groups were suppressed. In addition, the content of tryptophan was also increased to a similar extent (from 2.83% to 5.62%).

A number of methionine-rich proteins have been characterised in plants, including 2S albumin storage proteins from several dicotyledonous species (Brazil nut, sunflower, amaranthus, cotton) (see reviews by Altenbach and Simpson (1990), and Shewry and Pandya (1999)). The methionine-rich albumin (SFA8) of sunflower contains 16 methionines and 8 cysteines out of only 103 residues and has been shown to increase the methionine content by 27% when expressed in seeds of transgenic rice (Hagan *et al.*, 2003). However, this increase in methionine was accompanied by a decrease in the cysteine content of about 15% and there was little impact on total grain sulphur.

The endogenous prolamins of maize, sorghum and millets also include methionine-rich components; in particular, the δ-zeins of maize contain up to about 27 mol% methionine (Table 8.1). Increases in total grain methionine have been achieved in maize by using transformation to increase the number of expressed genes encoding δ-zein (Anthony *et al.*, 1997) or by modifying δ-zein mRNA stability (Lai and Messing, 2002). Although no data on total grain sulphur were reported, Lai and Messing (2002) did show a small decrease in grain cysteine. Hence, it is likely that the supply of sulphur into grain will need to be increased if substantial increases in total sulphur-containing amino acids (cysteine and methionine) are to be obtained.

Table 8.1 Methionine contents of selected prolamins

Species	Protein	Methionine content (mol%)	Reference
Maize	M_r 18 000 δ-zein	26.9	Swarup *et al.* (1995)
Maize	M_r 10 000 δ-zein	22.6	Swarup *et al.* (1995)
Maize	β-zein	11.4	Pedersen *et al.* (1986)
Sorghum	δ-kafirin	17.0	Izquierdo and Godwin (2005)
Sorghum	β-kafirin	9.3	Chamba *et al.* (2005)
Italian millet	α-setarin	12.7	Naren and Virupaksha (1990)
Italian millet	β-setarin	11.3	Naren and Virupaksha (1990)

M_r, relative molecular mass.

8.4 Developing resistant starch

8.4.1 Starch

Starch is the major component in the cereal endosperm and comprises a mixture of two related polymers. Amylose comprises linear chains of glucose residues with (1→4) α-linkages, while amylopectin also has (1→6) α-linkages, approximately every 20–30 glucose residues resulting in a branched structure. It is probable that the structures and proportions of these polymers and their organisation within the granule are responsible for the functional and nutritional properties and for the differences in these properties that occur within and between species.

Reducing the digestibility of starch in the small intestine may have health benefits in two respects. Firstly, resistant starch contributes to the dietary fibre component of foods, with benefits to colon function. Secondly, reducing the digestion of starch in the small intestine lowers the uptake of glucose, reducing the glycaemic index of the food (and hence the risk of type II diabetes). Amylose and amylopectin show significant differences in their digestibility in cooked foods with amylose forming enzyme-resistant complexes. Hence, high amylose starch is an important target for developing nutritionally enhanced cereals.

The pathways of starch synthesis are now well understood in cereals, with adenosine diphosphate glucose (ADPG) being used for synthesis of both amylose and amylopectin (see reviews by Shewry and Morell (2001), and Tomlinson and Denyer (2003)). ADPG is a substrate for starch synthases which sequentially add single glucose molecules to glucan chains. Two types of starch synthase exist, granule bound-forms (GBSS) which synthesise amylose and non-granule-bound forms (SS) which synthesise the (1→4) α-linkages in amylopectin. The (1→6) α-linkages in amylopectin are then introduced by branching enzymes (SBE). However, these enzymes also usually occur in multiple isoforms, resulting in a higher degree of complexity. Nevertheless studies of mutants and use of transgenesis show that changes in single genes and enzymes can still result in major effects on composition.

Cereal starches generally comprise between about 66% and 75% amylopectin (i.e. 25–34% amylose) but this composition may be manipulated by changing the expression of enzymes catalysing the synthesis of either polymer. The most well known of these 'mutant' starches are *waxy* types, which are high in amylopectin due to reduced activity of GBBS. This phenotype was selected by Slade *et al.* (2005) as proof of concept for the application of TILLING to wheat grain composition as discussed above.

8.4.2 Reducing starch branching enzyme activity (*amylose-extender*)

A natural mutant of barley containing 45–50% amylose (Glacier AC38) was identified some years ago (Walker and Merritt, 1968), while similar *amylose-extender* (*ae*) mutants are known in rice and in maize, where they give amylose contents ranging from about 50% to 90% (Garwood *et al.*, 1976). Analyses of *ae* mutants of maize and rice show that they lack a form of branching enzyme, called SBEIIb (Takeda *et al.*, 1993; Nishi *et al.*, 2001). However, the absence of this enzyme results in the synthesis of a modified form of amylopectin, which is more similar to amylose in structure, having fewer branch points and more long chains.

The *ae* mutations of diploid cereals are recessive but a similar phenotype has recently been obtained in bread wheat using genetic transformation (Regina *et al.*, 2006). This study used RNA interference technology to down-regulate two isoforms of SBE II (SBEIIa and SBEIIb). In contrast to the effects of the *ae* mutations, the suppression of SBEIIb alone was found to have no effect on amylose content. However, suppression of both SBEIIa and SBEIIb resulted in over 70% amylose. Furthermore, incorporation of the high amylose grain into a diet for rats resulted in significant improvements in indices of bowel health including the production of short chain fatty acids.

8.4.3 Reducing starch synthase II activity

Morell *et al.* (2003) identified a novel high amylose line in the hull-less barley variety Himalaya by screening shrunken mutants from plants treated with the mutagen sodium azide (Zwar and Chandler, 1995). This line (Himalaya 292) mapped to the *sex6* (shrunken endosperm) locus and had reduced total starch (17.7% compared with 49% in the control) containing 71% amylose. Two further independent lines with a similar low starch/high amylose phenotype were identified and all were shown to have stop codons preventing translation of starch synthase IIa (SSIIa) transcripts. However, the mutations also had pleiotropic effects on other aspects of starch synthesis including branching enzyme activity.

Himalaya 292 has since been developed into a novel cultivar for the manufacture of 'health foods' with low starch, high amylose and high non-starch polysaccharides (notably β-glucans). The latter are probably increased to compensate for the reduced starch level and, together with the undigested starch, result in increased dietary fibre. Feeding this grain to rats, pigs and humans resulted in improved indices of bowel function, including increased production of short chain fatty acids (Topping *et al.*, 2003; Bird *et al.*, 2004a,b).

Three proteins corresponding to SSII have also been identified in wheat, called starch granule proteins (SGP) A1, B1 and D1. These proteins are encoded by homoeologous genes on the A, B and D genomes and mutations at all three loci have been identified. When mutations at these three loci are combined in a single line, they result in an increase in grain amylose to 30–38% (Yamamori *et al.*, 2000; Shimbata *et al.*, 2005).

8.4.4 Increasing granule-bound starch synthase activity

Itoh *et al.* (2003) have shown that over-expression of a *waxy* (GBSS) gene in transgenic rice can lead to increased synthesis of amylose, with levels of up to 46% being produced.

8.4.5 Conclusions

It is therefore clear that several strategies can be used to generate 'high amylose' cereals, by inhibiting either the SSII enzymes that synthesise the amylopectin chains or the SBE enzymes that introduce branches, or by up-regulating the GBSS enzymes that synthesise amylose. However, it is also clear that these strategies may result in subtle differences in the structure of the starch polymers, which could in turn have different impacts on their nutritional properties.

8.5 Improving dietary fibre composition

8.5.1 Cell wall composition

Although resistant starch also acts as dietary fibre (see Chapter 4), the major fibre components in cereal grains and products are the cell wall polysaccharides. These can account for over 15% of the dry weight in some whole grains such as rye (Vinkx and Delcour, 1996) but for as little as 2–3% in white wheat flour.

The major components in cereal grains are (1→3)(1→4)-β-D-glucans (β-glucans) and arabino-(1→4)-β-D-xylans (arabinoxylans, AX) with smaller amounts of cellulose ((1→4)-β-glucan), callose ((1→3)-β-D-glucan), glucomannans ((1→4)-β-glucomannans) and other components (Stone, 1996). However, the proportions of these cell wall polymers vary between different cereal species and may also vary between aleurone and starchy endosperm cells of the same species (see Table 8.2).

The cell walls of rice starchy endosperm contain substantial amounts of cellulose (approximately 30%) but this is only a minor component in other cereals with the major cell wall polysaccharides being β-glucan and AX.

Table 8.2 Proportions (%) of cellulose, β-glucan and arabinoxylan (AX) in aleurone and starch endosperm cells of various cereals

	Cellulose	β-Glucan	AX
Barley			
Aleurone	2	26	67
Starchy endosperm	2	75	20
Wheat			
Aleurone	2	29	65
Starchy endosperm	4	20	70
Maize			
Starchy endosperm	10	65	15
Rye			
Starchy endosperm	8	12	65
Oats			
Starchy endosperm	?	72	28
Rice			
Starchy endosperm	28	20	27

Other cell wall polysaccharides, proteins and phenolics are not listed. Data are taken from Fincher (1975, 1976); Mares and Stone (1973), Bacic and Stone (1980, 1981), Newton *et al.* (1994), Glitsø and Bach Knudsen (1999), and Pascual and Juliano (1983).

However, whereas the starchy endosperm cells of barley, maize and oats are richer in β-glucans (65–75%), those of wheat and rye are richer in AX (65–70%). The aleurone cell walls of barley also differ from the starchy endosperm cell walls in being richer in AX (67%) than in β-glucan (28%).

It should also be noted that cell walls also contain proteins and phenolics. The latter are particularly important in wheat, rye and maize where two molecules of ferulic acid substituted onto arabinose residues of AX can form cross-links, which render the polymers insoluble (see Vinkx and Delcour, 1996). Finally, the aleurone cell walls are thicker than those of the starchy endosperm resulting in a loss of fibre on milling; for example, 8.8% crude fibre was present in the bran (i.e. the aleurone and other outer layers) abraded from wheat grain containing 3.0% crude fibre (Fellers *et al.*, 1976). Furthermore, the content fell from 3.5% to 2.2%, 1.8%, 1.6% and 1.4% in fractions from five subsequent abrasion steps, indicating the presence also of gradients within the starchy endosperm. Gradients in cell wall composition across the developing starchy endosperm of wheat have also been demonstrated using immunochemical and spectroscopic imaging (Philippe *et al.*, 2006; Toole *et al.*, 2007).

There is a vast literature on the relationship between dietary fibre and health that is outside the scope of this chapter (but see other chapters in this volume). Nevertheless, although it is clear that soluble fibre is particularly important in conferring health benefits, we still know little about structure–function relationships in molecular terms, including the differences between the functional properties of the different fibre components present in cereal cell walls.

8.5.2 Genetic variation in composition

Genetic variation has been reported in the content of β-glucans in barley (Stuart *et al.*, 1988; Kenn *et al.*, 1993) and in the amounts of total and water-soluble AX of wheat (Saulnier *et al.*, 1995; Martinant *et al.*, 1998). In particular, Saulnier *et al.* (1995) showed variation from about 0.4% to 0.8% dry weight in water-soluble AX and from about 5.5% to 7.8% in total AX in ground grains of 22 cultivars grown in France. Martinant *et al.* (1998) subsequently identified a quantitative trait locus (QTL) on the long arm of chromosome 1B that explained 32–37% of the variation in the relative viscosity of extracts that contained water-soluble AX and suggested that the differences related to variation in the ratio of arabinose:xylose (the QTL also explained 35–42% of the variation in AX). Although the variation in AX content in wheat and in β-glucan content in barley is sufficient to be exploited in breeding (and indeed low β-glucan has been a target for malting barleys for many years), there are also strong environmental impacts on the amounts of both components (Stuart *et al.*, 1988; Kenn *et al.*, 1993; Coles *et al.*, 1997; Swanston *et al.*, 1997), which mean that breeding for consistently, high or low amounts may be difficult. An increased content of β-glucans has also been reported in the low amylose barley line Himalaya 292 (see above). Similarly, near infra-red (NIR) analysis of high lysine mutants of barley demonstrated that alleles at the *lys5* locus (Risø mutants 13 and 29) had increased amounts of β-glucan (by 15–20%). However, these increases did not compensate fully for the decrease in starch (reduced by 30%) that also occurred (Munck *et al.*, 2004).

8.5.3 Molecular approaches

Taking a direct molecular approach to manipulating cereal grain cell wall composition is not currently possible due to our incomplete knowledge of the enzymes and genes that control AX and β-glucan synthesis. In particular, the glucosyltransferases that catalyse the synthesis of the xylan chains, the arabinoxyl transferase(s) that add arabinose residues to the xylan backbone and the enzymes that control the esterifcation of ferulic acid to arabinose residues and the formation of diferulate cross-links are all currently not known. Approaches to identifying their activity by a combination of comparative genomics, expression profiling, mutant selection and gene knock-outs are discussed in two recent review articles (Burton *et al.*, 2005; Yong *et al.*, 2005). As with other scientific fields, it is impossible to be certain when breakthroughs will be made, but I would anticipate enough information becoming available to allow the manipulation of cell wall composition in transgenic cereals within the next few years.

8.6 Increasing vitamins and minerals

Cereals are acknowledged to be important sources of a range of vitamins, minerals and phytochemicals including potassium, calcium, magnesium, manganese, iron, molybdenum, selenium, zinc, chromium, thiamine, vitamins B6, E and K, niacin, sterols and phyto-oestrogens (Buttriss, 2003). However, these may be enriched in whole grains rather than white flour products or be poorly available. I will therefore focus on a few selected examples that have been the subject of recent research.

8.6.1 Selenium

Selenium is essential for a number of enzymes in the form of selenocysteine. These enzymes include glutathione peroxidase which prevents oxidative damage by free radicals. However, at high concentrations it is also toxic through competition with sulphur. Bread has traditionally been a major source of selenium accounting for 22% of the total intake in the UK in 1997 (Cassidy, 2003). However, the selenium content of the grain is largely determined by the availability of selenium in the soil. Consequently, the replacement of North American wheat grown on high selenium soils with home-grown wheat grown on low selenium soils has resulted in a decline in the contribution of wheat products to dietary selenium in the UK (Adams *et al.*, 2002; Broadley *et al.*, 2006). Enrichment of wheat by using selenium-containing fertilisers (agronomic biofortification) has been successful in Finland (Yläranta, 1984a,b). However, it may also be possible to increase selenium accumulation in wheat grain by breeding, as both *Aegilops tauschii* (the diploid progenitor of the D genome of bread wheat) and rye are able to accumulate 30–40% more selenium in their grain than bread wheat (Lyons *et al.*, 2005).

8.6.2 Low phytate lines

High proportions of the minerals present in the cereal grain are located in the aleurone cells and are unavailable for human nutrition because they are in the form of phytin (mixed salts of phytic acid, *myo*-inositol-(1,2,3,4,5,6)-hexa*kis* phosphate). These salts account for over 70% of the total phosphate in the grain as well as significant amounts of magnesium, potassium, iron, zinc, calcium and copper. Because phytates cannot be digested, they are excreted, which can result in phosphate contamination of water in areas of intensive livestock production and mineral deficiency in humans. The latter is a particular problem where cereals are consumed as a major part of the diet such as in developing countries, and for women and children.

Mutagenesis has been used to generate lines of barley, maize and rice with lower levels of phytates and inositol and higher levels of inorganic phosphate, although this strategy did not work with hexaploid bread wheat (Raboy *et al.*, 2002). Two mutations (*lpa-1-1*, *lpa-2-1*) were identified in maize and barley and one (*lpa-1-1*) in rice which resulted in decreases of phytate of about 50–

65%. The *lpa-2* gene of maize results from a mutation of an inositol phosphate kinase gene (Shi *et al.*, 2003). In maize and barley the *lpa*-1 mutants have been used to develop 'high available phosphate' (HAP) cultivars with improved nutritional quality (Sugiura *et al.*, 1999; Spencer *et al.*, 2000; Poulsen *et al.*, 2001; Raboy *et al.*, 2002; Veum *et al.*, 2002).

Genetic engineering has also been used to reduce the levels of phytate in cereals by expressing genes encoding phytase. The enzyme from the fungus *Aspergillus niger* has been particularly attractive for this as it has established use as an additive for animal feed. This enzyme has been expressed in wheat under control of the 'constitutive' ubiquitin promoter (Brinch-Pedersen *et al.*, 2000) and the starchy endosperm-specific high molecular weight glutenin subunit 1Dx5 promoter (Brinch-Pedersen *et al.*, 2003) while enzymes from *A. fumigatens* (known to be thermostable), *Selenemonas ruminatium* and *E. coli* have been expressed in rice (Lucca *et al.*, 2001; Hong *et al.*, 2004). However, whereas Lucca *et al.* (2001) used the rice glutelin promoter to confer expression in the developing starchy endosperm, Hong *et al.* (2004) used an α-amylase promoter to confer expression during germination. Drakakaki *et al.* (2005) have also reported the expression of *A. niger* phytase in maize, using the rice glutelin promoter. The expression levels achieved in these studies vary from up to 4-fold increases in phytase activity in wheat (Brinch-Pedersen *et al.*, 2000, 2003) to a 130-fold increase in one rice line (Lucca *et al.*, 2001).

Brinch-Pedersen *et al.* (2006) stated that rat feeding trials with their transgenic wheat are in progress but results from these are not yet available. However, feeding trials with transgenic soybean and canola (oilseed rape) have shown benefits in terms of increased growth and/or reduced excretion of phosphate when fed to chickens and pigs (Denbow *et al.*, 1998; Zhang *et al.*, 2000a,b). Drakakaki *et al.* (2005) have also used a Caco-2 cell model to show that expression of phytase in maize seed resulted in increased uptake of iron by the cells.

8.6.3 Increasing iron

Iron deficiency is the most widespread mineral deficiency in humans, estimated to affect up to 30% of the world population (WHO, 1992). A significant proportion of the iron in cereals is unavailable as it is present in phytates – this should be released by the strategies discussed above. However, this iron is present in the aleurone (i.e. bran) and is therefore absent from products made from white flour. Attempts have therefore been made to increase the content of iron in the starchy endosperm cells by expressing ferritin, an iron-binding protein from soybean or *Phaseolus*. Expression of these proteins in the starchy endosperm of rice using storage protein (glutelin or globulin) promoters resulted in increases in the iron content of up to two- to three-fold, although this level was disappointing in view of the high levels of ferritin expression that have been achieved (Goto *et al.*, 1999; Lucca *et al.*, 2001; Qu *et al.*, 2005).

Qu *et al.* (2005) demonstrated that the expression of soybean ferritin in the grain was associated with decreased iron content in the leaves, indicating that accumulation may have been limited by iron uptake and transport.

Drakakaki *et al.* (2005) also showed that expression of the soybean ferritin, together with fungal phytase (see above), in maize under the control of the rice glutelin promoter resulted in an increase in iron content of 20–70%. This resulted in an increase in the availability of iron to Caco-2 cells which was further increased in the presence of ascorbic acid (200 μM) which is known to promote iron uptake by the human gut.

8.6.4 Vitamin A

Cereals are rich sources of fat-soluble and water-soluble vitamins but these are concentrated in the aleurone and hence depleted by milling (wheat), polishing (rice) or decorticating (millets, sorghum). Hence, high consumption of cereals can lead to vitamin deficiency. Vitamin A deficiency is particularly important in southeast Asia where consumption of white rice has been estimated to contribute to a quarter of a million children losing their sight each year. This led to the well-publicised development of 'Golden Rice', in which two genes from daffodil and one from a bacterium (*Erwinia uredova*) were expressed in the starchy endosperm of rice leading to the accumulation of β-carotene, which can be converted to vitamin A by humans (Ye *et al.*, 2000; Potrykus, 2001; Datta *et al.*, 2003).

More recently, Paine *et al.* (2005) have achieved greatly increased levels of vitamin A accumulation, up to 23-fold greater than in the original Golden Rice. They hypothesised that one of the enzymes, phytoene synthase (*psu*) limited the synthesis of β-carotene and therefore compared *psu* genes from five plant sources (rice, maize, pepper, tomato, daffodil) in transgenic rice. The maize and rice genes proved to be the most effective and the daffodil gene used in previous experiments the least. They calculated that 50% of the recommended daily allowance for a child would be delivered in a 72 g portion of Golden Rice 2, which compares well with a typical child's portion of about 60 g rice/meal.

Golden Rice is a spectacular example of the power of GM technology to increase the amount of an essential nutrient in the part of the plant that is consumed, although this may not be the usual site of accumulation. At present, this cannot be achieved by non-GM means in which the amount but not the location of a nutrient may vary between genotypes and mutant lines.

8.6.5 Folates

An important future target for such work is folic acid, with deficiency leading to defects in neural tube development during pregnancy and to a range of other detrimental effects in adults. There does not appear to be sufficient variation in the folate content of cereals to be exploited by plant breeding but the biosynthetic pathway is now established, allowing genetic engineering approaches to be used (see Basset *et al.* (2005) and Rébeillé *et al.* (2006) for discussion of folate synthesis and metabolism). So far, no work has been carried out in cereals but some success has been achieved in other species.

Hossain *et al.* (2004) and Diaz de la Garza *et al.* (2004) adopted the same strategy based on the hypothesis that the enzyme guanosine triphosphate (GTP) cyclohydrolase 1 catalyses a limiting step in the synthesis of pterins (precursors of folate). They therefore used feedback insensitive forms of the enzyme encoded by an *E. coli* gene or a synthetic gene based on a mammalian gene and expressed them in leaves of *Arabidopsis* and fruits of tomato, respectively. Similar results were observed in the two systems, with pterin precursors accumulating to much higher levels than folates: 1250-fold compared with 2- to 4-fold in Arabiodpsis and 3- to 140-fold compared with 2-fold in tomato. Both authors concluded that further metabolic engineering would be required to achieve more substantial increases in folates. Nevertheless, this is an important 'proof of concept' that manipulation of only one or two enzymes in a complex pathway may be sufficient to result in substantial increases in important nutrients.

8.7 Future trends

Although cereal improvement has been carried out for millennia, interest in nutritional enhancement only dates from the last 50 years. This has been driven by two contrasting needs: to improve the basic nutrition of vast numbers of people in developing countries and, more recently, to combat health concerns related to the consumption, and often over-consumption, of highly refined foods in industrialised societies. Much of the research on the latter has only been initiated over the past decade.

Bearing in mind the complexity of most cereal genomes, the progress so far has been satisfying but not spectacular. However, several substantial research programmes have recently been initiated that should lead to rapid progress in many of the objectives discussed here as well as topics that have not been discussed.

The HARVESTPLUS programme led by the CGIAR institutes is focusing on micronutrients – particularly iron, zinc and vitamin A – using large-scale screening of germplasm to identify novel variation in a range of crops including cereals, to facilitate exploitation in plant breeding (www.harvestplus.org/micronut/html). Complementary to this is the 'HEALTHGRAIN' programme supported by the European Union under FP6 (www.healthgrain.org/pub/). This is focusing on resistant starch, dietary fibre and a range of phytochemicals (folate, phenolics including lignans, tocols, sterols, alkylresorcinols) in wheat, using a combination of germplasm screening, mutagenesis and transformation to identify and generate new variation for exploitation in breeding. The 'Food Futures' programme is a flagship programme for the CSIRO in Australia and includes the nutritional enhancement of cereals in a wide portfolio of projects on crop and animal products (www.csiro.au). Finally, in 2005 the Bill and Melinda Gates Foundation (www.gatesfoundation.org) announced a massive investment to improve the delivery of essential amino acids, vitamins A and E, iron and zinc in sorghum varieties for growth in arid and semi-arid regions of Africa.

It should be noted that all of these programmes will need to deliver their benefits via conventional plant breeding, to ensure that the improved traits are incorporated into lines that can compete with the best current cultivars in terms of yield, resistance to pests, pathogens and abiotic stresses. Hence, we are probably looking at a 5–10 year time scale for delivering many improvements. Nevertheless, the future looks bright for those who currently rely on cereals to maintain their nutritional status.

8.8 Sources of further information and advice

Books

- *Wheat: Chemistry and Technology*, Volumes I and II, Khan K and Shewry PR (eds.) (ed.), St Paul, MN, American Association of Cereal Scientists, 4th edition, In Press.
- *Seed Proteins*, Shewry PR and Casey R (eds.), Dordrecht, Kluwer Academic, 1999.
- *Cereal Grain Quality*, Henry RJ and Kettlewell PS (eds.), London, Chapman and Hall, 1996.

Journals

Special issue of the *Journal of Cereal Science* 'The Role of Cereals in a Healthy Diet' to be published in 2007.

Websites

The following websites provide information on major research programmes:

- HEALTHGRAIN, www.healthgrain.org/pub
- HARVESTPLUS, www.harvestplus.org/micronut.html.
- Food Futures, www.csiro.au.
- Gates Foundation, www.gatesfoundation.org

Additional websites include:

- Wheat The Big Picture, www.wheatbp.net/, provides information on wheat grain development and structure.
- Grain Genes, www.wheat.pw.usda.gov/GG2/index.shtml, for access to genomic and genetic resources for the Triticeae and Aveneae.

8.9 References

Adams M L, Lombi E, Zhao F-J and McGrath S P (2002), 'Evidence of low selenium concentrations in UK bread-making wheat grain', *Journal of the Science of Food and Agriculture*, **82**, 1160–1165.

Altenbach S B and Simpson R B (1990), 'Manipulation of methionine-rich protein genes in plant seeds', *Trends in Biotechnology*, **8**, 156–160.

Anthony A, Brown W, Buhr D, Ronhovde G, Genovesi D, Lane T, Yingling R, Rosato M and Anderson P (1997), 'Transgenic maize with elevated 10 kDa zein and methionine', in Cram W G (ed.), *Sulphur Metabolism in Higher Plants,* Leiden, Backhuys Publishers, p. 295.

Axtell J D, Van Scoyoc S W, Christensen P J and Ejeta G (1979), 'Current status of protein quality improvement in grain sorghum', in *Seed Improvement in Cereals and Grain Legumes*, Vol. 1, Vienna, IAEA, pp. 357–365.

Bacic A and Stone B A (1980), 'A (1→3)- and (1→4)-linked β-D-glucan in the endosperm cell walls of wheat', *Carbohydrate Research,* **82**, 372–377.

Bacic A and Stone B A (1981), 'Chemistry and organisation of aleurone cell wall components from wheat and barley', *Australian Journal of Plant Physiology*, **8**, 475–453.

Basset G J C, Quinlivan E P, Gregory III J F, Hanson A D (2005), 'Folate synthesis and metabolism in plants and prospects for biofortification', *Crop Science*, **45**, 449–453.

Bird A R, Flory C, Davies D A, Usher S and Topping D L (2004a), 'A novel barley cultivar (Himalaya 292) with a specific gene mutation in starch synthesis IIa raises large bowel starch and short-chain fatty acids in rats', *Journal of Nutrition,* **134**, 831–835.

Bird A R, Jackson M, King R A, Davies D A, Usher S and Topping D L (2004b), 'A novel high-amylose barley cultivar (*Hordeum vulgare* var Himalaya 292) lowers plasma cholesterol and alters indices of large-bowel fermentation in pigs', *British Journal of Nutrition*, **92**, 607–615.

Bonjean A P and Angus W J (2001), *The World Wheat Book. A History of Wheat Breeding*, Paris, Lavoisier Publishing.

Bright S W J and Shewry P R (1983), 'Improvement of protein quality in cereals', *Critical Reviews in Plant Science*, **1**, 49–93.

Brinch-Pedersen H, Galili G, Knudsen S and Holm P B (1996), 'Engineering of the aspartate family biosynthetic pathway in barley (*Hordeum vulgare* L) by transformation with heterologous genes encoding feedback-insensitive aspartate kinase and dihydrodipicolinate synthase', *Plant Molecular Biology*, **32**, 611–620.

Brinch-Pedersen H, Olesen A, Rasmussen S K and Holm P B (2000), 'Generation of transgenic wheat (*Triticum aestivum* L) for constitutive accumulation of an *Aspergillus* phytase', *Molecular Breeding*, **6**, 195–206.

Brinch-Pedersen H, Hatzack F, Sorensen L D and Holm P B (2003), 'Concerted action of endogenous and hererologous phytase on phytic acid degradation in seed of transgenic wheat (*Triticum aestivum* L)', *Transgenic Research*, **12**, 649–659.

Brinch-Pedersen H, Dionisio G and Holm P B (2006), 'Seeds for improved phosphate and mineral bio-availability', in Jaiwal P K (ed.), *Plant Genetic Engineering*, Vol. 7: *Metabolic Engineering and Molecular Farming 1*, Houston, TX, Studium Press, pp. 321–337.

Broadley M R, White P J, Bryson R J, Meacham M C, Bowen H C, Johnson S E, Hawkesford M J, McGrath S P, Zhao F–J, Breward N, Harriman M and Tucker M (2006), 'Biofortification of UK food crops with selenium', *Proceedings of the Nutrition Society*, **65**, 1–14.

Burton R A, Farrokhi N, Bacic A and Fincher G B (2005), 'Plant cell wall polysaccharide biosynthesis: real progress in the identification of participating genes', *Planta*, **221**, 309–312.

Buttriss J (2003), 'Plant foods and health', in Goldberg G (ed.), *Plants: Diet and Health.*

The Report of a British Nutrition Foundation Task Force, Oxford, Blackwell Science, pp. 1–26.

Cassidy A (2003), 'Cereals, nuts and pulses', in Goldberg G (ed.), *Plants: Diet and Health. The Report of a British Nutrition Foundation Task Force*, Oxford, Blackwell Science, pp. 134–146.

Chamba E B, Halford N G, Wilkinson M and Shewry P R (2005), 'Molecular cloning of β-kafirin, a methionine-rich protein of sorghum grain', *Journal of Cereal Science*, **41**, 381–383.

Colbert T, Bradley J T, Tompa R, Reynolds S, Steine M N, Yeung A T, McCallum C M, Comai L and Henikoff S (2001), 'High-throughput screening for induced point mutations', *Plant Physiology*, **126**, 480–484.

Coleman C E and Larkins B A (1999), 'The prolamins of maize', in Shewry P R and Case R (eds.), *Seed Proteins*, Dordrecht, Kluwer Academic Publishers, pp. 109–139.

Coles G D, Hartunian-Sowa S M, Jamieson P D, Hay A J, Atwell W A and Fulcher R G (1997), 'Environmentally-induced variation in starch and non-starch polysaccharide content in wheat', *Journal of Cereal Science*, **26**, 47–54.

Curtis I S (2004), *Transgenic Crops of the World: Essential Protocols*, Dordrecht, Kluwer Academic Publishers.

Datta K, Baisakh N, Oliva N, Torrizo L, Abrigo E, Tan J, Rai M, Rehana S, Al-Babili S, Beyer P, Potrykus I and Datta S K (2003), 'Bioengineered '*golden*' indica rice cultivars with β-carotene metabolism in the endosperm with hygromycin and mannose selection systems', *Plant Biotechnology Journal*, **1**, 81–90.

Denbow D M, Grabau E A, Lacy G H, Kornegay E T, Russel D R and Umbeck P F (1998), 'Soybeans transformed with a fungal phytase gene improve phosphorus availability for broilers', *Poultry Science*, **77**, 878–881.

Diaz de la Garza R, Quinlivan E P, Klaus S M J, Basset G J C, Gregory III J F and Hanson A D (2004), 'Folate biofortification in tomatoes by engineering the pteridine branch of folate synthesis', *Proceedings of the National Academy of Sciences of the USA*, **101**, 13720–13725.

Doll H (1983), 'Barley seed proteins and possibilities for their improvement', in Gottschalk W and Muller H P (eds.), *Seed Proteins: Biochemistry, Genetics, Nutritive Value*, The Hague, Martinus Nijhof, pp. 205–223.

Drakakaki G, Marcel S, Glahn R P, Lund E K, Pariagh S, Fischer R, Christou P and Stoger E (2005), 'Endosperm-specific co-expression of recombinant soybean ferritin and *Aspergillus* phytase in maize results in significant increases in the levels of bioavailable iron', *Plant Molecular Biology*, **59**, 869–880.

Falco S C, Guida T, Locke M, Mauvais J, Sandres C, Ward R T and Webber P (1995), 'Transgenic canola and soybean seeds with increased lysine', *Biotechnology*, **13**, 577–582.

FAO/WHO/UNO (1985), *Necessidades de Energia y Proteinas*, Geneva, World Health Organization.

Feldman M, Lupton F G H and Miller T E (1995), 'Wheats', in J Smartt and N W Simmonds (Eds.), *Evolution of Crop Plants*, 2nd edition, Harlow, UK, Longmans, pp. 184–192.

Fellers D A, Mossman A P, Johnston P H and Wheeler E L (1976), 'Study of relationships between wheat protein contents of two UK varieties and derived flour protein fractions at varying extraction rates. II Studies by hand-dissection of individual grains', *Cereal Chemistry*, **51**, 66–74.

Fincher G B (1975), 'Morphology and chemical composition of barley endosperm cell walls', *Journal of the Institute of Brewing*, **81**, 116–122.

Fincher G B (1976), 'Ferulic acid in barley cell walls: a fluorescence study', *Journal of the Institute of Brewing*, **82**, 347–349.

Forsyth J L, Beaudoin F, Halford N G, Sessions R, Clark A R and Shewry P R (2005), 'Design, expression and characterisation of lysine-rich forms of the barley seed protein CI-2', *Biochimica et Biophysica Acta*, **1747**, 221–227.

Garwood D L, Shannon J C and Creech R G (1976), 'Starches of endosperms possessing different alleles at amylose-extender locus in *Zea mays* L', *Cereal Chemistry*, **53**, 355–364.

Geevers H O and Lake J K (1992), 'Development of modified opaque2 maize in South Africa', in Mertz E T (ed.), *Quality Protein Maize*, St Paul, MN, American Association of Cereal Chemists, pp. 49–78.

Glitsø L V and Bach Knudsen K E (1999), 'Milling of whole grain rye to obtain fractions with different dietary fibre characteristics', *Journal of Cereal Science*, **29**, 89–97.

Goto F, Yoshihara T, Shigemoto N, Toki S and Takaiwa F (1999), 'Iron fortification of rice seed by the soybean ferritin gene', *Nature Biotechnology*, **17**, 282–286.

Hagan N D, Upadhyaya N, Tabe L M and Higgins T J V (2003), 'The redistribution of protein sulphur in transgenic rice expressing a gene for a foreign, sulphur-rich protein', *Plant Journal*, **34**, 1–11.

Halford N G (2006), 'From primitive selection to genetic modification, ten thousand years of plant breeding', in Halford N G (ed.), *Plant Biotechnology*, New York, John Wiley and Sons Ltd, pp. 3–27.

Hejgaard J and Boisen S (1980), 'High-lysine proteins in Hiproly barley breeding: identification, nutritional significance and new screening methods', *Hereditas,* **93**, 311–320.

Hong C Y, Cheng K J, Tsent T H, Wang C S, Liu L F and Yu S M (2004), 'Production of two highly active bacterial phytases with broad pH optima in generating transgenic rice seeds', *Transgenic Research*, **13**, 29–39.

Hossain T, Rosenberg I, Selhub J, Kishore G, Beachy R and Schubert K (2004), 'Enhancement of folates in plants through metabolic engineering', *Proceedings of the National Academy of Sciences of the USA*, **101**, 5158–5163.

Huang S, Adams W R, Zhou Q, Malloy K P, Voyles D A, Anthony J, Kriz A L and Luethy M H (2004), 'Improving nutritional quality of maize proteins by expressing sense and antisense zein genes', *Journal of Agricultural and Food Chemistry*, **52**, 1958–1964.

Huang S, Frizzi A, Florida C A and Kruger D E (2006), 'High lysine and high tryptophan transgenic maize resulting from the reduction of both 19- and 22kDa α-zeins', *Plant Molecular Biology*, **61**, 525–535.

Itoh K, Ozaki H, Okada K, Hori H, Takeda Y and Mitsui T (2003), 'Introduction of *Wx* transgene into rice *wx* mutants leads to both high and low-amylose rice', *Plant Cell Physiology*, **44**, 473–480.

Izquierdo L and Godwin I D (2005), 'Molecular characterization of a novel methionine-rich δ-kafirin seed storage protein gene in sorghum (*Sorghum bicolour* L)', *Cereal Chemistry*, **82**, 706–707.

Jensen S A and Martens H (1983), 'The botanical constituents of wheat and wheat milling fractions. II Quantification by amino acids', *Cereal Chemistry*, **60**, 172–177.

Kenn D A, Dagg A H S and Stuart I M (1993), 'Effect of environment and genotype on the fermentability of malt produced from four Australian barley varieties', *American Society of Brewing Chemists Journal*, **51**, 119–122.

Kent N L (1966), 'Subaleurone endosperm cells of high protein content', *Cereal Chemistry*, **43**, 585–601.

Lai J and Messing J (2002), 'Increasing maize seed methionine by mRNA stability', *Plant Journal*, **30**, 395–402.

Lee S I, Kim H U, Lee Y H, Suh S C, Lim Y P, Lee H Y and Kim H I (2001), 'Constitutive and seed-specific expressions of a maize lysine feedback-insensitive dihydrodipicolinate synthase gene leads to increased free lysine levels in rice seeds', *Molecular Breeding*, **8**, 75–84.

Lucca P, Hurrell R and Potrykus I (2001), 'Genetic engineering approaches to improve the bioavailability and the level of iron in rice grains', *Theoretical and Applied Genetics,* **102**, 392–397.

Lyons G, Ortiz-Monasterio I, Stangoulis J and Graham R (2005), 'Selenium concentration in wheat grain: is there sufficient genotypic variation to use in breeding?', *Plant and Soil*, **269**, 369–380.

Maluszynski M, Szarejko I and Maluszynska J (2001), 'Induced mutations in wheat', in Bonjean A P and Angus W J (eds.), *The World Wheat Book. A History of Wheat Breeding*, Paris, Lavoisier Publishing, pp. 959–977.

Mares D and Stone B A (1973), 'Studies on wheat endosperm. I. Chemical composition and ultrastructure of the cell walls', *Australian Journal of Biological Sciences*, **26**, 793–812.

Martinant J P, Cadalen T, Billot A, Chartier S, Leroy P, Bernard M, Saulnier L and Branlard G (1998), 'Genetic analysis of water-extractable arabinoxylans in bread wheat endosperm', *Theoretical and Applied Genetics*, **97**, 1069–1075.

Mazur B, Krebbers E and Tingey S (1999), 'Gene discovery and product development for grain quality traits', *Science*, **285**, 372–375.

McCallum C M, Comai L, Greene E A and Henikoff S (2000a), 'Targeted screening for induced mutations', *Nature Biotechnology*, **18**, 455–457.

McCallum C M, Comai L, Greene E A and Henikoff S (2000b), 'Targeted Induced Local Lesions IN Genomes (TILLING) for plant functional genomics', *Plant Physiology*, **123**, 439–442.

Mertz E T, Bates L S and Nelson O E (1964), 'Mutant gene that changes protein composition and increases lysine content of maize endosperm', *Science*, **145**, 279–280.

Millet M O, Montembault A and Autran J C (1991), 'Hordein composition differences in various anatomical regions of the kernel between two different barley types', *Sciences des Aliments*, **11**, 155–161.

Morell M K, Kosar-Hashemi B, Cmeil M, Samuel M S, Chandler P, Rahman S, Buleon A, Batey I L and Li Z (2003), 'Barley *sex6* mutants lack starch synthase IIa activity and contain a starch with novel properties', *Plant Journal*, **34**, 173–185.

Munck L, Karlsson K E, Hagberg A and Eggum B O (1970), 'Gene for improved nutritional value in barley seed protein', *Science*, **168**, 985–987.

Munck L, Møller B, Jacobsen S and Søndergaard I (2004), 'Near infrared spectra indicate specific mutant endosperm genes and reveal a new mechanism for substituting starch with (1→3,1→4)-β-glucan in barley', *Journal of Cereal Science*, **40**, 213–222.

Naren P A and Virupaksha T K (1990), 'α- and β-setarins: methionine-rich proteins of Italian millet (*Setaria italica* (L) Beauv)', *Cereal Chemistry*, **67**, 32–34.

Nelson O E, Mertz E T and Bates L S (1965), 'Second mutant gene affecting the amino acid pattern of maize endosperm proteins', *Science*, **150**, 1469–1470.

Newton J, Stark J R and Riffkin H L (1994), 'The use of maize and wheat in the production of alcohol', in *Proceedings of the 4th Conference on Malting, Brewing and Distilling*, London, Institute of Brewing.

Nishi A, Nakamura Y, Tanaka N and Satoh H (2001), 'Biochemical and genetic analysis of the effects of amylose-extender mutation in rice endosperm', *Plant Physiology*, **127**, 459–472.

O'Kennedy M M, Grootboom A and Shewry P R (2006), 'Harnessing sorghum and millet biotechnology for food and health – review', *Journal of Cereal Scienc*, **44**, 224–235.

Paine J A, Shipton C A, Chaggar S, Howells R M, Kennedy M J, Vernon G, Wright S Y, Hinchcliffe E, Adams J L, Silverstone A L and Drake R (2005), 'Improving the nutritional value of Golden Rice through increased pro-vitamin A content', *Nature Biotechnology*, **23**, 482–487.

Pascual C G and Juliano B O (1983), 'Properties of cell wall preparations of milled rice', *Phytochemistry*, **22**, 151–159.

Pedersen K, Argos P, Naravana S V L and Larkins B A (1986), 'Sequence analysis and characterisation of a maize gene encoding a high-sulfur zein protein of M_r 15,000', *Journal of Biological Chemistry*, **261**, 6279–6284.

Peña L (2005), 'Transgenic plants – methods and protocols', in Peña L (ed.), *Methods in Molecular Biology*, Vol. 286, Totowa, N J, Humana Press.

Philippe S, Saulnier L and Guillon F (2006), 'Arabinoxylan and (1→3)(1→4)-β-glucan deposition in cell walls during wheat endosperm development', *Planta*, **224**, 449–461.

Pomeranz Y (1988), 'Chemical composition of kernel structures', in Pomeranz Y (ed.), *Wheat: Chemistry and Technology*, St Paul, MN, American Association of Cereal Chemists, pp. 97–158.

Potrykus I (2001), 'Golden Rice and beyond', *Plant Physiology*, **125**, 1157–1161.

Poulsen H D, Johansen K S, Hatzack F, Boisen S and Rasmussen S K (2001), 'Nutritional value of low-phytate barley evaluated in rats', *Acta Agriculturae Scandinavica Section A – Animal Science*, **51**, 53–58.

Qu L Q, Yoshihara T, Ooyama A, Goto F and Takaiwa F (2005), 'Iron accumulation does not parallel the high expression level of ferritin in transgenic rice seeds', *Planta,* **222**, 225–233.

Raboy V, Young K A, Larson S R and Hayes J E (2002), 'Genetics of phytic acid synthesis and accumulation', in Reddy A R and Sathe S K (eds.), *Food Phytates,* Boca Raton, FL, CRC Press, pp. 68–84.

Rao A G, Hassan M and Hempel J C (1994), 'Structure–function validation of high lysine analogs of α-hordothionin designed by protein modelling', *Protein Engineering*, **7**, 1485–1493.

Rébeillé F, Ravanel S, Jabrin S, Douce R, Storozhenko S and Van Der Straeten D (2006), 'Folates in plants: biosynthesis, distribution and enhancement', *Physiologia Plantarum*, **126**, 330–342.

Regina A, Bird A, Topping D, Bowden S, Freeman J, Barsby T, Kosar-Hashemi B, Li Z, Rahman S and Morell M (2006), 'High-amylose wheat generated by RNA interference improves indices of large-bowel health in rats', *Proceedings of the National Academy of Sciences of the USA*, **103**, 3546–3551.

Roesler K R and Rao A G (1999), 'Conformation and stability of barley chymotrypsin inhibitor-2 (CI-2) mutants containing multiple lysine substitutions', *Protein Engineering*, **12**, 967–973.

Roesler K R and Rao A G (2000), 'A single disulfide bond restores thermodynamic and proteolytic stability to an extensively mutated protein', *Protein Science*, **9**, 1642–1650.

Saulnier L, Peneau N and Thibault J-F (1995), 'Variability in grain extract viscosity and water-soluble arabinoxylans content in wheat', *Journal of Cereal Science*, **22**, 259–264.

Segal G, Song R and Messing J (2003), 'A new opaque variant of maize by a single dominant RNA-interference-inducing transgene', *Genetics*, **165**, 387–397.

Shewry P R (2000), 'Seed proteins', in Black M and Bewley J D (eds.), *Seed Technology and its Biological Basis*, Sheffield, UK, Sheffield Academic Press, pp. 42–84.

Shewry P R and Morell M (2001), 'Manipulating cereal endosperm structure, development and composition to improve end-use properties', in *Advances in Botanical Research Incorporating Advances in Plant Pathology*, Vol. **34**, pp. 165–236.

Shewry P R and Pandya M J (1999), 'The 2S albumin storage proteins', in Shewry P R and Casey R (eds.), *Seed Proteins*, Dordrecht, Kluwer Academic Publishers, pp. 587–615.

Shi J R, Wang H Y, Wu Y S, Hazebroek J, Meeley R B and Ertl D S (2003), 'The maize low-phytic acid mutant *lpa2* is caused by mutation in an inositol phosphate kinase gene', *Plant Physiology,* **131**, 507–515.

Shimbata T, Nakamura T, Vrinten P, Saito M, Yonemaru J, Seto Y and Yasuda H (2005), 'Mutations in wheat *starch synthase II* genes and PCR-based selection of a SGP-1 null line', *Theoretical and Applied Genetics*, **111**, 1072–1079.

Shoup F K, Pomeranz Y and Deyoe C W (1966), 'Amino acid composition of wheat varieties and flours varying widely in bread-making potentialities', *Journal of Food Science*, **31**, 94–101.

Singh R and Axtell J D (1973), 'High-lysine mutant gene (*hl*) that improves protein quality and biological value of grain sorghum', *Crop Science*, **3**, 535–539.

Slade A J, Fuerstenberg S I, Loeffler D, Steine M N and Facciotti D (2005), 'A reverse genetic, non-transgenic approach to wheat crop improvement by TILLING', *Nature Biotechnology*, **23**, 75–81.

Spencer J D, Alle G L and Sauber T E (2000), 'Phosphorus bioavailability and digestibility of normal and genetically modified low-phytate corn for pigs', *Journal of Animal Science*, **78**, 675–681.

Stadler L J (1929), 'Chromosome number and the mutation rate in *Avena* and *Triticum*', *Proceedings of the National Academy of Sciences of the USA*, **15**, 876–881.

Stone B A (1996), 'Cereal grain carbohydrates', in Henry R J and Kettlewell P S (eds.), *Cereal Grain Quality*, London, Chapman and Hall, pp. 251–288.

Stuart I M, Loi L and Fincher G B (1988), 'Varietal and environmental variations in (1→3)(1→4)-β-glucanase potential in barley: relationships to malting quality', *Journal of Cereal Science*, **7**, 61–71.

Sugiura S H, Raboy V, Young K A, Dong F M and Hardy R W (1999), 'Availability of phosphorus and trace elements in low-phytate varieties of barley and corn for rainbow trout (*Oncorhynchus mykiss*), *Aquaculture*, **170**, 285–296.

Swanston J S, Ellis R P, Pérez-Vendrell A, Voltas J and Molina-Cano J L (1997), 'Patterns of barley grain development in Spain and Scotland and their implications for malting quality', *Cereal Chemistry*, **74**, 456–461.

Swarup S, Timmermans M C P, Chaudhuri S and Messing J (1995), 'Determinants of the high-methionine trait in wild and exotic germplasm may have escaped selection during early cultivation of maize', *Plant Journal*, **8**, 359–368.

Takeda C, Takeda Y and Hizukuri S (1993), 'Structure of the amylopectin fraction of amylomaize', *Carbohydrate Research*, **246**, 273–281.

Terada R, Urawa H, Inagaki Y, Tsugane K and Lida S (2002), 'Efficient gene targeting by homologous recombination in rice', *Nature Biotechnology*, **20**, 1030–1034.

Tomlinson K and Denyer K (2003), 'Starch synthesis in cereal grains', in *Advances in Botanical Research Incorporating Advances in Plant Pathology*, Vol. **40**, pp. 1–61.

Toole G A, Wilson R, Parker M L, Wheeler T R, Shewry P R and Mills E N C (2007), 'The effect of growing environment on endosperm cell wall development and composition studied by infrared spectroscopic imaging techniques', *Planta*, **225**, 1393–1403.

Topping D L, Morell M K, King R A, Li Z Y, Bird A R and Noakes M (2003), 'Resistant starch and health – Himalaya 292, a novel barley cultivar to deliver benefits to consumers', *Starch-Starke*, **55**, 539–545.

Veum T L, Ledoux D R, Bollinger D W, Raboy V and Cook A (2002), 'Low-phytic acid barley improves calcium and phosphorus utilisation and growth performance in growing pigs', *Journal of Animal Science*, **80**, 2663–2670.

Villegas E, Vasal S K and Bjarnarson M (1992), 'Quality Protein Maize – what is it and how was it developed?', in Mertz E T (ed.), *Quality Protein Maize*, St Paul, MN, American Association of Cereal Chemists, pp. 27–48.

Vinkz C J A and Delcour J A (1996), 'Rye (*Secale cereale* L) arabinoxylans: a critical review', *Journal of Cereal Science*, **24**, 1–14.

Walker T J and Merritt N R (1968), 'Genetic control of abnormal starch granules and high amylose content of a mutant of 'Glacier' barley', *Nature*, **221**, 482–484.

WHO (1992), 'National strategies for overcoming micro-nutrient malnutrition', Document A45/3, Geneva, WHO.

Wrigley C W and Bietz J A (1988), 'Proteins and Amino Acids', in Pomeranz Y (ed.), *Wheat: Chemistry and Technology,* St Paul, MN, American Association of Cereal Chemists, pp. 159–275.

Yamamori M, Fujita S, Hayakawa K and Matsuki J (2000), 'Genetic elimination of a starch granule protein, SGP–1, of wheat generates an altered starch with apparent high amylose', *Theoretical and Applied Genetics*, **101**, 21–29.

Ye X, Al-Babili S, Klöti A, Zhang J, Lucca P, Beyer P and Potrykus I (2000), 'Engineering pro-vitamin A (β-carotene) biosynthetic pathway into (carotenoid-free) rice endosperm', *Science,* **287**, 303–305.

Yläranta T (1984a), 'Raising the selenium content of spring wheat and barley using selenite and selenate', *Annales Agriculturae Fenniae*, **23**, 75–84.

Yläranta T (1984b), 'Effect of selenium fertilization and foliar spraying at different growth stages on the selenium content of spring wheat and barley', *Annales Agriculturae Fenniae*, **23**, 85–95.

Yong W, Link B, O'Malley R, Tewari J, Hunter C T, Lu C-A, Li X, Bleecker A B, Koch K E, Mccann M C, McCarty D R, Patterson S E, Reiter W-D, Staiger C, Thomas S R, Vermerris W and Carpita N C (2005), 'Genomics of plant cell wall biogenesis', *Planta*, **221**, 747–751.

Yu J, Peng P, Zhang X, Zhao Q, Zhy D, Sun X, Liu J and Ao G (2004), 'Seed-specific expression of a lysine rich protein *sb401* gene significantly increases both lysine and total protein content in maize seeds', *Molecular Breeding*, **14**, 1–7.

Zhang Z B, Kornegay E T, Radcliffe J S, Denbow D M, Veit H P and Larsen C T (2000a), 'Comparison of genetically engineered microbial and plant phytase for young broilers', *Poultry Science*, **79**, 709–717.

Zhang Z B, Kornegay E T, Radcliffe J S, Wilson J H and Veit H P (2000b), 'Comparison of phytase from genetically engineered *Aspergillus* and canola in weanling pig diets', *Journal of Animal Science*, **78**, 2868–2878.

Zhao Z-Y, Glassman K, Sewalt V, Wang N, Miller M, Chang S, Thompson T, Catron S, Wu E, Bidney D, Kedebe Y and Jung R (2003), 'Nutritionally improved transgenic sorghum', in Vasil I K (ed.), *Plant Biotechnology 2002 and Beyond*, Dordrecht, Kluwer Academic Publishers, pp. 413–416.

Zhu X and Galili G (2003), 'Increased lysine synthesis coupled with a knockout of its catabolism synergistically boosts lysine content and also transregulates the metabolism of other amino acids in *Arabidopsis* seeds', *Plant Cell*, **15**, 845–853.

Zhu T, Peterson D J, Tagliani L, St Clair G, Baszczynski C L and Bowen B (1999), 'Targeted manipulation of maize genes *in vivo* using chimeric RNA/DNA oligonucleotides', *Proceedings of the National Academy of Sciences of the USA*, **96**, 8768–8773.

Zwar J A and Chandler P M (1995), 'α-amylase production and leaf protein synthesis in a gibberellin-responsive dwarf mutant of 'Himalaya' barley (*Hordium vulgare* L)', *Planta*, **197**, 39–48.

9

Novel high-fibre and whole grain breads

C. Collar, Instituto de Agroquímica y Tecnología de Alimentos (IATA-CSIC), Spain

9.1 Introduction

High-fibre and whole grain breads play a prominent role in creating a healthy diet. The rapid and general increase in the number of overweight people in Western society and heightened consumer concerns about good nutrition have refocused technical research and the food industry on the merits of including some form of dietary fibre in baked goods. Dietary fibre remains a nutritional concept that can be accommodated within the ambit of the terms 'functional foods' or 'nutraceuticals'. With encouragement from health groups and government agencies, the public increasingly is choosing healthy foods naturally rich in fibre (whole grains), as well as the fast-growing array of fibre-enriched foods. The development and use of a wide variety of innovative dietary fibres from conventional and non-conventional sources as multi-functional ingredients are making the production of many of these enriched baked products possible.

In this chapter, an updated overview is given of novel high-fibre breads from both whole grains and fibre-enriched refined grains. The nutritional benefits, the technological drawbacks to be overcome by cereal science and technology, and consumer attitudes towards high-fibre breads and whole grain breads are also highlighted and discussed.

9.2 Whole grains, refined grains, and fibre-fortified refined grains in breadmaking

The US Food and Drug Administration (FDA) considers 'whole grain' the cereal grains that consist of the intact, ground, cracked, milled or flaked fruit of the grains whose principal components – the starchy endosperm, germ, and bran – are present in the same relative proportions as they exist in the intact grain (Fig. 9.1). The following definition was developed and ratified by the Whole Grains Council:

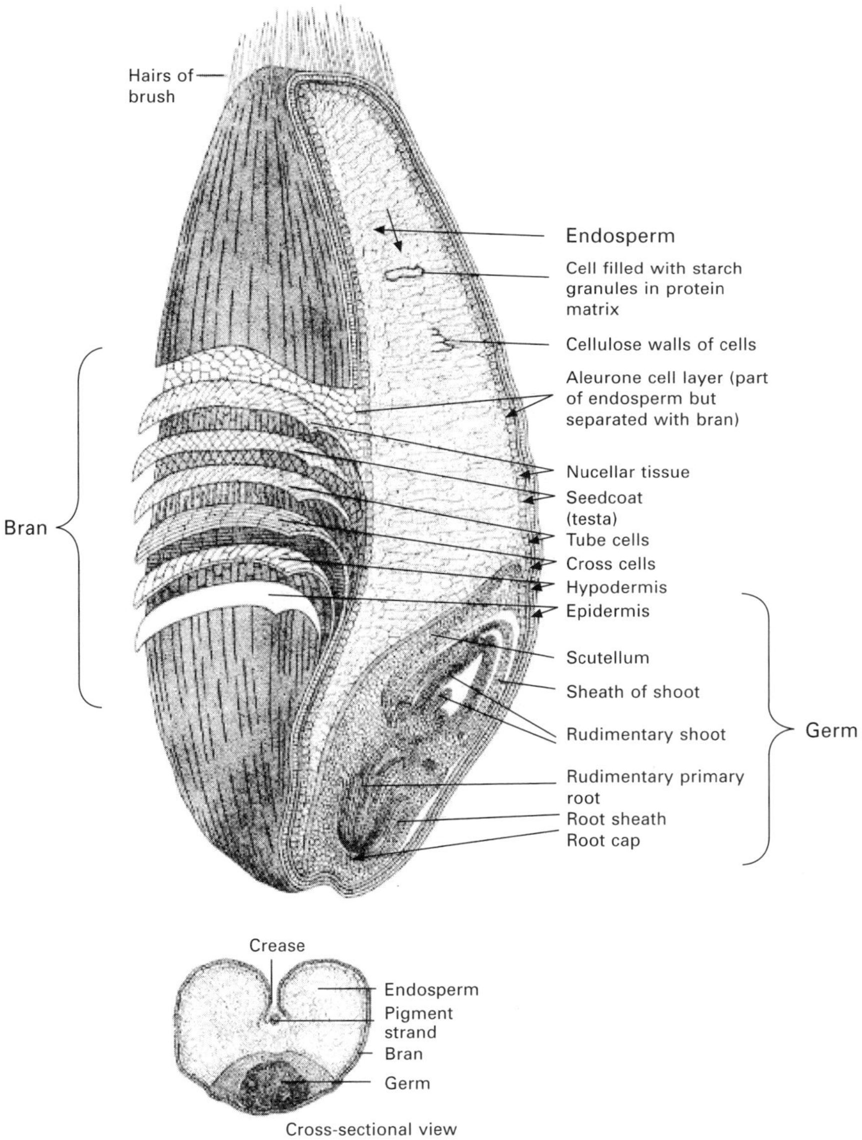

Fig. 9.1 The structure of the wheat grain – longitudinal and cross-sectional views. From Pomeranz (1987).

> Whole grains or foods made from them contain all the essential parts and naturally occurring nutrients of the entire grain seed. If the grain has been processed (e.g. cracked, crushed, rolled, extruded, lightly pearled and/or cooked), the food product should deliver approximately the same rich balance of nutrients that are found in the original grain seed.

Examples of generally accepted whole grain foods and flours are: amaranth, barley (lightly pearled), brown and coloured rice, buckwheat, bulgur, corn and whole cornmeal, emmer, farro, grano (lightly pearled wheat), Kamut® grain, millet, oatmeal and whole oats, popcorn, quinoa, sorghum, spelt, teff, triticale, whole rye, whole or cracked wheat, wheat berries, and wild rice. The FDA states that rolled and 'quick oats' can be called whole grains, because they contain all of their bran, germ, and endosperm. Other widely used food products may not meet the whole grain definition such as those derived from legumes (soybeans), oilseeds (sunflower seeds) and roots (arrowroot). For a whole grain food to meet the whole grain health-claim standards, the food must include 51% whole grain flour by weight of final product and must contain 1.7 g dietary fibre, all components of the intact grain, and a minimum of 16 g whole grain per serving. Studies show health benefits at about three servings per day. The new *Dietary Guidelines for Americans* (USDA, 2005) recommend the following: everyone age 9 and up should eat at least three servings of whole grains per day and men aged 14–50 and active women should eat an additional serving or two of whole grains, depending on their age and activity level. Components in whole grains associated with improved health status include lignans, tocotrienols, phenolic compounds, and antinutrients including phytic acid, tannins, and enzyme inhibitors. In the grain-refining process outer layers are removed, resulting in the loss of dietary fibre, vitamins, minerals, lignans, phyto-oestrogens, phenolic compounds, and phytic acid (Fig. 9.2); the resulting refined grains being more concentrated in starch. Wholemeal foods contain whole grains that have been milled to a finer texture rather than leaving the grain intact. Wholemeal contains all the components of the grain, therefore wholemeal foods are also whole grain. Wholemeal bread and rye bread are typical examples of products made with wholemeal. A food may either be naturally high in fibre such as wholemeal bread, or have fibre added such as high-fibre white bread. A point of difference is that high-fibre foods do not always contain all the outer layers of the grain. Australian, European, and Japanese food regulations require that a food must contain at least 1.5 g of fibre per serving before it can state that it contains fibre or that it is a source of fibre; and 3 g per serving before it can claim to be high fibre. In addition, a food can state that it is a fibre-enriched product if it contains +25% more fibre than the reference product, the difference being at least 3 g/100 g. According to a recent report from the Department of Food Science and Nutrition, University of Minnessota, worldwide whole grain and high-fibre foods intake is lower than recommended across populations and cultures. While traditional diets in some parts of the world may utilize whole grains,

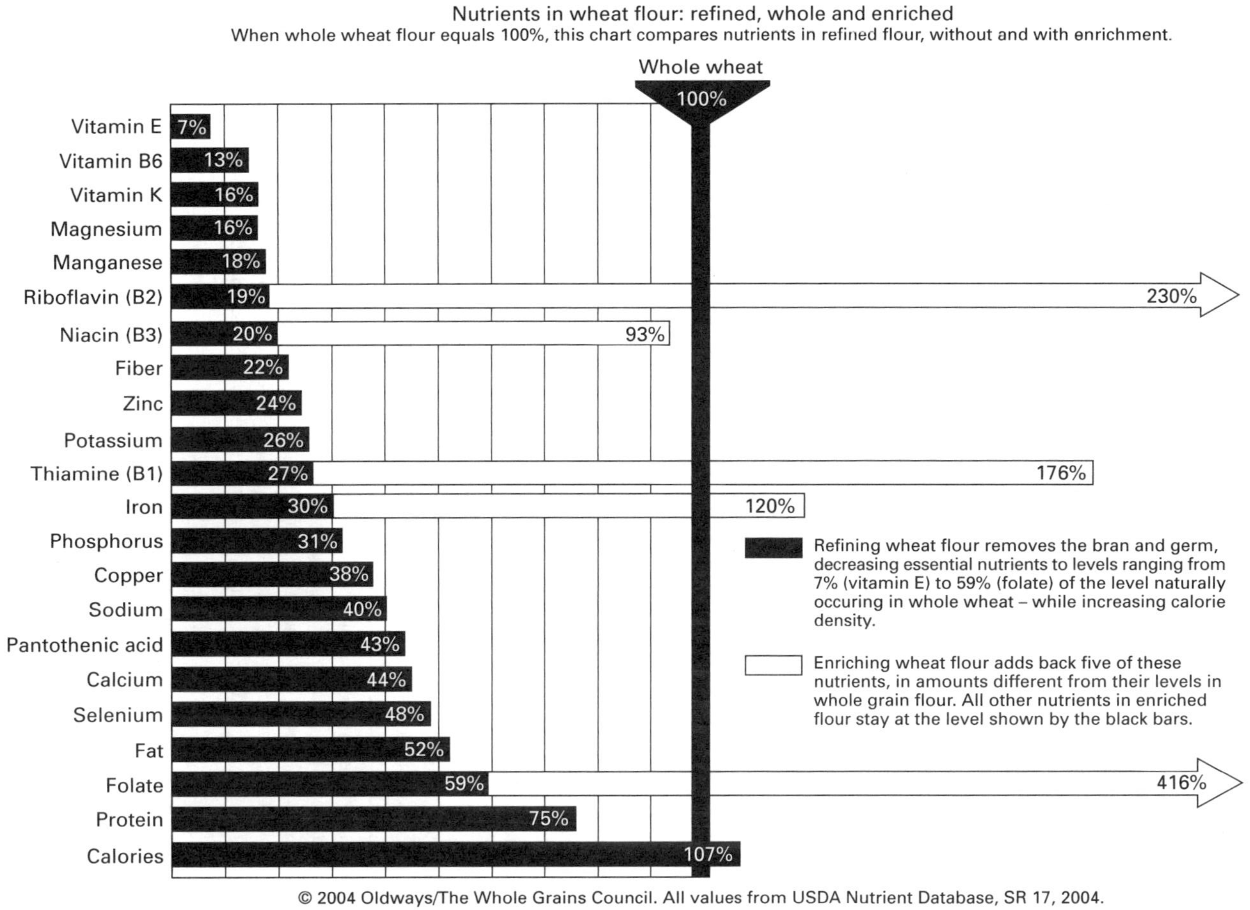

Fig. 9.2 Nutrients for whole wheat flour, refined wheat flour (wheat with almost all bran and germ removed), and enriched wheat flour (refined flour with some nutrients added back in). From The Whole Grains Council (2004) (www.wholegrainscouncil.org).

when more westernized habits are adopted, diets often lose their whole grain component. In Western cultures, those who are more highly educated and wealthy are more prone to eat whole grains, but consumption is still low, regardless of demographics (Table 9.1).

9.3 Health benefits of high-fibre and whole grain white breads: protective nature in the diet

9.3.1 Whole grain health studies

Recent research studies add strong evidence that whole grain intake is protective against many diseases. Based on epidemiological studies and biologically plausible mechanisms, scientific evidence shows that regular consumption of whole grain foods provides health benefits in terms of reduced rates of cardiovascular disease (Truswell, 2002) and several forms of cancer (Chatenoud *et al.*, 1998). It may also help regulate blood glucose levels (Bruce *et al.*, 2000) and obesity (Slavin, 2000, 2004). Whole grains intake was also linked with a lower overall risk of mortality from cardiovascular disease in older adults (Sahyoun *et al.*, 2006), with slower build-up of artery-narrowing plaque in post-menopausal women already diagnosed with a heart condition (Erkkila *et al.*, 2005) (more than 34 g of fibre/day), and with a reduction in rectal cancer by 66%, in a study of over 2000 people eating a high-fibre diet (Slattery *et al.*, 2004). Whole grains are recommended by the American Diabetes Association for diabetes mellitus prevention (Franz *et al.*, 2002), the American Heart Association, the *Dietary Guidelines for Americans*, and Healthy People 2010 (www.wholegrain.umn.edu). Whole grains have high

Table 9.1 Fibre intake recommendations versus current fibre intake across the globe. From University of Minnesota (www.wholegrain.umn.edu)

Fibre intake recommendations:
- World Health Organization recommends more than 25 g/day.
- American Hearth Association recommends 25–30 g/day.
- US National Academy of Sciences Institute of Medicine recommendations: women <50 years, 25 g/day; >50 years, 21 g/day; men <50 years, 38 g/day; >50 years, 30 g/day.

Current fibre intake across the globe:
- US (2001) = 15 g/day (less African-Americans).
- Ireland (2001) = men, 23.2 g/day; women, 17.4 g/day.
- UK (2001) = 77% of population below the UK recommendations of 18 g/day.
- Denmark (1999) = men, 22 g/day; women, 18 g/day.
- Germany (2001) = men, 21.9 g/day; women: 19.5 g/day.
- Spain (2001) = men, 20.1 g/day; women: 15.7 g/day.
- Sweden (1999) = mean intake, 18.3 g/day.
- Brazil (2000) = 24 g/day mean.
- Costa Rica (2001) = 23.8 g/day, rural $\gg$ urban.

concentrations of dietary fibre, resistant starch, and oligosaccharides but also they are rich in nutrients and phytochemicals with known health benefits. The wide range of protective components in whole grains and potential mechanisms for protection have been recently described (Slavin, 2003). The bran and germ fractions derived from conventional milling provide a majority of the biologically active compounds found in a grain. Specific nutrients include high concentrations of B vitamins (thiamin, niacin, riboflavin, and pantothenic acid) and minerals (Ca, Mg, K, P, Na, and Fe), elevated levels of basic amino acids (for example, arginine and lysine), and elevated tocol levels in the lipids, which have been linked to disease prevention. Numerous phytochemicals, some common in many plant foods (phytates and phenolic compounds) and some unique to grain products (avenanthramides, avenalumic acid), are responsible for the high antioxidant activity of whole grain foods (Miller *et al.*, 2002).

Published whole grain feeding studies report improvements in biomarkers with whole grain consumption – for example, weight loss (Liu *et al.*, 2003), blood-lipid improvement (Gerhardt and Gallo, 1998), and antioxidant protection (Adom and Liu, 2002). Although it is difficult to separate the protective properties of whole grains from dietary fibre and other components, the disease protection seen from whole grains in prospective epidemiological studies far exceeds the protection from isolated nutrients and phytochemicals in whole grains.

The need for more research on the mechanisms for this protection is strongly emphasized. In addition, some components in whole grains may be most important in this protection and should be retained in food processing. Dietary intakes of whole grains fall short of current recommendations to eat at least three servings daily. The whole grain health claim should increase the consumption of whole grain foods in the American population (Marquart *et al.*, 2003). This is in keeping with the Food Pyramid educational materials, which recommend that a minimum of six servings of grain foods be eaten daily, with at least one half or three of those servings as whole grains. Efforts to develop health claims for whole grains in Europe are still underway. The UK Joint Health Claims Initiative published an endorsement that whole grain foods are associated with a healthy heart (Richardson, 2003). However, the current EU food labelling legislation prohibits claims that attribute to a foodstuff the property of preventing, treating, or curing a human disease. Thus, reduction of risk of disease claims are not permitted at present (Redgwell and Fischer, 2005).

9.3.2 Health benefits of dietary fibre

The importance of dietary fibres in the human diet has been recognized since epidemiological and human studies showed the connection between dietary fibre intake and occurrence of diseases in modern civilization (Table 9.2). The intake of fibre and fibre-containing foods is well below the recommended levels in all Western countries (Table 9.1). While more research is needed to

Table 9.2 Diseases resulting from poor dietary fibre intakes. Hypotheses based on epidemiological studies. From Trowell *et al.* (1985)

- Constipation.
- Diverticular disease of the colon.
- Cancer of the large bowel.
- Appendicitis.
- Crohn's disease and ulcerative colitis.
- Functional gastrointestinal disorders: irritable bowel and other syndromes.
- Duodenal ulcer.
- Hiatal hernia and gastro-oesophageal reflux.
- Obesity: interaction of environment and genetic predisposition.
- Diabetes mellitus (insulin-independent diabetes).
- Gallstones.
- Lipid metabolism and coronary heart disease.
- Varicose veins, haemorrhoids, deep-vein thrombosis, and pelvic phleboliths
- Renal stones.

clarify details of the nutritional benefits of dietary fibre, several studies have shown that its consumption has been inversely associated with the risk of a number of chronic diseases such as obesity (Howarth *et al.*, 2001), diabetes, metabolic syndrome, cardiovascular diseases, cancer, and intestinal disorders (Malkki, 2004). Generally accepted health claims for dietary fibre include: (a) improvement of colonic function (transit time reduction, decrease of constipation, increased faecal bulking); (b) lowering of blood low-density lipoprotein (LDL) cholesterol (restricted to soluble fibres with water-holding capacity); (c) reduction of the blood glucose level (glycaemic index restricted to soluble fibres with water-holding capacity) (Schaafsma, 2004). The insoluble fraction of the fibre seems to be related to the intestinal regulation, whereas the soluble fibre is associated with the decrease in cholesterol levels and the absorption of intestinal glucose (Rodríguez *et al.*, 2006). Also, several fibres have demonstrated, *in vitro* and *in vivo*, their capacity for adsorbing carcinogenic agents, consequently it is recommended to consume plant foods with lignified or suberized cell walls that are the most effective in linking hydrophobic carcinogenic agents (Slavin, 2001). Until now, the only generic health claim that is supported by scientific evidence is the improvement of colonic function. The development of new validated biomarkers will help to identify beneficial biological effects of the heterogeneous components of dietary fibre as a whole. Currently, there is no clear picture on the cause and effect relationship between a component of dietary fibre or compound associated with dietary fibre, and perceived health benefits. Some, or even all, of the proposed benefits may result from the small amount of associated plant substances, which are present in many dietary fibre preparations currently in use. According to Redgwell and Fischer (2005), this may be the case for the more chemically complex whole cereal grain fibres and, for this reason, the authors stress it would seem a wise option for the food industry to continue to search for and develop technologies that allow the incorporation of chemically unmodified whole grain-based fibre into an increasing range of popular product categories.

9.4 Novel high-fibre and whole grain breads

9.4.1 Fibres and fibre blends to make high-fibre breads

The consumer is already aware of the advantages of a diet rich in cereals, fruit, and vegetables, the major daily sources of dietary fibre differing from food to food. Several non-starch foods provide up 20–35 g of fibre/100 g dry weight and others containing starch provide about 10 g/100 g dry weight. The fibre content of fruits and vegetables is 1.5–2.5 g/100 g dry weight. Among the different fibre-rich foods, cereals are one of the main sources of dietary fibre, contributing about 50% of the fibre intake in Western countries; 30–40% dietary fibre may come from vegetables, about 16% from fruits, and the remaining 3% from other minor sources (Rodríguez *et al.*, 2006). The continued emphasis on the importance of dietary fibre to the Western diet and the need for products with a lower caloric content and density are forcing food companies to allocate more resources to the development of fibre-enriched products, which provide a broad spectrum of bioactive and texture modulating properties. The challenge for the industry is to accomplish this goal without sacrificing the organoleptic appeal, i.e. taste, flavour, and texture – of the final food product.

Industrial perspective and product opportunities

The food industry can only benefit from including dietary fibre in products with authenticated claims on the short- and long-term gains to be derived from their consumption. As future research details the specific nutritional benefits of individual components of dietary fibre, food companies will need flexible alternatives in order to validate new 'functional' food claims and to respond rapidly to emerging trends in fibre-enriched products. Water binding, network formation, reduction of calories due to a reduced consumption, and actions as a low caloric filler and as a substitute for high-caloric ingredients like fat, sugar, or starch are advantages in different applications (Endress and Fischer, 2000). The authors state that two types of products can be produced: (a) fibre-enriched products, which do not differ significantly in sensory and optic properties from traditional ones; and (b) new products, in which additional properties of fibres, such as colour or texture, are used to design innovative products that differ from traditional products. Both ways are appropriate to reduce the deficit in fibre of about 10 g per capita per day, as indicated by nutrition scientists (Table 9.1), and could help to reduce the risk of certain diseases. Additionally, the dietary fibre content can be used as a marketing element. Bread is still the main source of fibre intake, but the fibre content in starchy endosperm is comparably low related to the seed coat. The dietary fibre content of the flour normally used is about 4 g/100 g, which leads to a dietary fibre content of below 3 g/100 g bread. The range in fibre contents of white bread and wholemeal bread is shown in Fig. 9.3 (Endress and Fischer, 2000). For fortification of white bread with dietary fibre, the use of bran and a dietary fibre isolate are compared. It is necessary to add about 5% of fibre

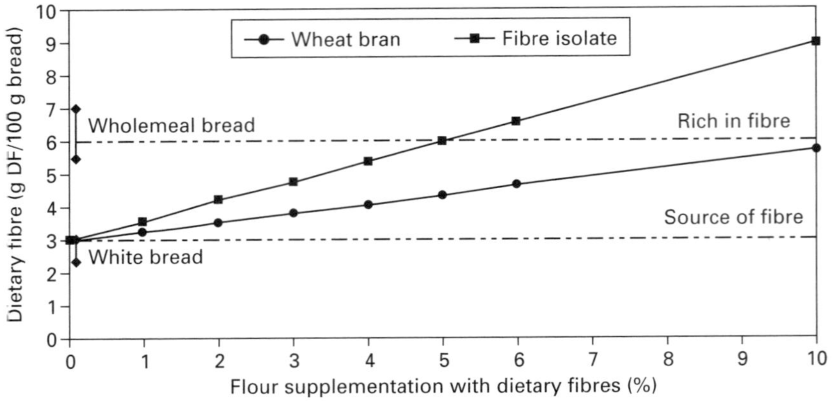

Fig. 9.3 Dietary fibre (DF) fortification of bread. From Endress and Fischer (2000).

isolate, with 88% dietary fibre, to the flour to reach the fibre content of wholemeal bread. Using bran with 45% dietary fibre, more than 10% is needed. Additionally, with 10% cereal bran the sensory properties differ significantly from the white bread. This is not the case if dietary fibre isolates are used. The application of dietary fibres in bread and bakery products has several aims. Besides the traditional fortification of dietary fibre, targets such as calorie reduction and water binding are of interest for keeping products fresh for longer and higher yields. Depending on the bakery product, different effects on texture might be of interest. As can be seen from rheological measurements, there are big differences in the interaction of dietary fibres with the dough matrix (Table 9.3). A long dough stability and a low degree of dough softening under high shear forces can be achieved with wheat grain fibre, but not with wheat stem fibre (Endress and Fischer, 2000). An increased water uptake indicates prolonged freshness and shelf-life of a bakery product (Armero and Collar, 1998). With special fibre blends, physiological as well as functional requirements can be met. To develop new dietary fibres or to improve traditional ones, different raw materials can be used. Classification of the raw material is often made according to its traditional use for eating which also has the strongest influence on the image and rating of the derived dietary fibre. The first group in this classification represents parts of plants that are separated during food processing, whereas traditionally the whole fruit or seed was consumed. Examples of this group are dietary fibres from fruits produced as by-products of juice production – like apple or citrus pectin; fruit fibres from apple, blackcurrant, chokeberry, and raspberry; but also fibres from beet root, carrot, or tomato, which are processed to vegetable juice. Finally by-products from starch, sugar, oil, or protein isolation are used, resulting in wheat grain fibre, potato fibre, and cotyledonous fibres of grain legumes like lupin or pea that are not identical to the husk fibres of the same plants. A second group of dietary fibres are derived from plants or parts of plants that are usually not eaten by humans, like cereal stems or straw,

Table 9.3 Influence of dietary fibre on dough properties. From Endress and Fischer (2000)

	Standard	1.5% wheat grain fibre	1.5% wheat stem fibre	6% fibre blend*
Water uptake	55.6%	58.0%	57.2%	58.1%
Dough development time	2.0 min	1.5 min	2.0 min	2.0 min
Dough stability	5.5 min	8.0 min	3.0 min	9.0 min
Degree of softening	70 FE	55 FE	80 FE	50 FE
Calculated water binding of fibres in dough		2.8 g	2.3 g	1.1 g

Rheological data according to ICC Standard 115 (Farinogram).
* Inulin, Herbacel AQ Plus Citrus, pea fibre, and Herbapekt apple pectin extract.

bamboo, trees, cotton plants, and nut shells. The image of these products is poor and hence the marketing potential of this second group is limited (Endress and Fischer, 2000).

One decade ago, various commercial dietary fibre products were added into bakery product formulations to test their suitability as processing aids and flavour enhancers. Fibres included: Fibrex (Fibrex AB, Sweden), produced from desugared, dried sugar beet and having high moisture absorption and retention characteristics, enabling it to be incorporated into bread; Novelose (National Starch & Chemical GmbH, Neustadt, Germany), a concentrated resistant starch product with a neutral taste and virtually fat-free, designed for long-life foods such as snacks and expanded and extruded cereals; Herbacel (Herbafood Nahrungsmittel GmbH, Werder, Germany), a family of products from fruit and vegetable fibres for use as dietary fibre enhancers; Primafibre (E. Timm & Son, UK), a concentrated wheat bran product with a high moisture absorption capacity and a mild cereal aroma, for use in bakery products; Exafine, a pea fibre product consisting of 70% cellulose, for use in frozen dough; and Swelite, a pea fibre product made from the internal cell walls, and comprising two-thirds insoluble cellulose and one-third soluble pectin (Anon. 1995).

The availability of a wide range of dietary fibres – produced and promoted as innovative, healthy, and multi-functional – is increasing. Fibres produced under the Vitacel® brand (J. Rettenmaier & Söhne, Germany) include wheat, oat, tomato, apple, and orange fibres. The manufacturers report that the fibres have a good water-binding capacity and special sensory properties due to the technology used and to the very fine milling (particle size 130 μm). The fibres do not negatively affect the taste of the enriched product when compared with dietary fibre from cereal-derived brans or beets, which can confer a particular flavour and colour and can lead to consumer rejection. A fibre-rich mixed rye/wheat bread (70:30) was developed with added Vitacel® apple fibre (5.5%) with more than 6% of dietary fibre in the final product. A low-calorie sandwich bread including 12% Vitacel® wheat fibre and 88%

white wheat flour as raw materials – and milk powder, gluten, and baking improver among other ingredients – has been manufactured (Table 9.4). It has a sensory quality comparable with regular sandwich bread in all parameters, slightly finer pores, softer and brighter crumb, and an extended shelf-life of up to 5 days without a preservative. Of particular interest is a high-fibre white bread made with Litesse, a polydextrose from Danisco Sweeteners Ltd (Thomson, 2005). Litesse behaves like a resistant polysaccharide, being fermented in the lower gastrointestinal tract, and has shown prebiotic effects in human subjects given doses of 4 g/day (Zhong *et al.*, 2000). The formulation has a fibre content of 6.1 g/100 g serving, the same fibre content as most wholemeal breads, and a white bread crumb, keeping the same appearance as a standard white loaf. Litesse can be dry blended with a bread improver or other dry ingredients of a standard white bread dough requiring approximately 2–3% extra water at 8% Litesse addition. According to the manufacturers, breads can be made to the standard Chorleywood bread process, and to the sponge and dough processing method, producing very similar results to a standard white bread (Table 9.5). The suitability of Novelose, a resistant starch produced by National Starch, Manchester, UK has been discussed when used in combination with other fibre sources in multi-fibre strategies to produce high-fibre foods (Anon., 2002). Successful production of white bread has been achieved using Novelose as part of a multi-fibre approach; the bread had good sensory properties, a total dietary fibre content of 18%, and improved processing characteristics.

Experimental high-fibre breads

Tangkanakul *et al.* (1995) investigated breads made from wheat flour substituted at up to 30% levels with fibre from red kidney bean, rice bran, soybean residue, or white sesame seed. Wheat flour substitution decreased loaf volume

Table 9.4 Nutritional information for a standard and low-calorie sandwich bread including Vitacel® wheat fibre WF 101 from J. Rettenmaier & Söhne GmbH

Nutrition information	Standard (per 100 g)	Low-calorie sandwich bread (per 100 g)
Calorific value	1129 kJ 270 kcal	830 kJ 198 kcal
Protein	7.5 g	8.1 g
Carbohydrates, total	50 g	38.6 g
Of which:		
Starch		36.7 g
Sugar		1.9 g
Fat, total	4.10 g	0.94 g
Saturated fatty acid	2.10 g	0.12 g
Dietary fibre	2.80 g	8.50 g
Sodium	0.50 g	0.47 g

Table 9.5 Nutritional information (average values) for a control white pan bread and a high fibre white bread including Litesse at 4.51% total weight (8% on flour weight) from Danisco Sweeteners Ltd

	High fibre white bread		Control white pan bread	
	Per 100 g	Per 40 g	Per 100 g	Per 40 g
Energy (kcal)	232	93	241	96
Energy (kJ)	972	389	1012	405
Protein (g)	7.1	2.8	7.6	3.0
Carbohydrates (g)	45.1	18.0	47.9	19.0
Of which sugars (g)	1.0	0.4	1.0	0.4
Fat (g)	2.0	0.8	2.0	0.8
Of which saturates (g)	0.6	0.2	0.6	0.2
Fibre, g	6.1	2.4	2.0	0.8
Of which polydextrose (g)	4.3	1.7	0	0
Sodium (g)	0.6	0.2	0.6	0.2

and increased the density of the breads. Bread enriched with soybean residue had the lowest bread volume and the highest density. The water absorption index of most fibre-enriched breads was lower than that of pure wheat flour bread. Fibre enrichment increased hardness and decreased springiness of breads. Breads made with 100% wheat flour (no substitution) had a fibre content of 3.06 g/100 g. The fibre content of fibre-enriched breads that was considered acceptable was in the range 7.4–13.4 g/100 g.

Barley fibres have been extensively tested for flour replacement and/or flour supplementation in baking applications. Waxy hulless barley (12% dietary fibre, 7% β-glucan) was used in yeast breads made with 50% (by wt) barley flour with cracked barley (Hawrysh *et al.*, 1996). High-β-glucan barley breads were reported to show good shelf-life, showing only slight deterioration in attributes such as overall aroma and flavour intensity, sweet flavour, grain-like/musty odour, grain-like flavour and aftertaste. Waxy hulless barley bread thus showed good potential as a high-fibre product. Barley flour has been added to wheat flour at a replacement level of 10% without major adverse effects on bread quality, and with an increase in dietary fibre content of the bread from 3% to 5%. In bread made with 10% barley flour, soluble dietary fibre content did not change but insoluble dietary fibre content increased during the baking process (Cho and Lee, 1996a). High fibre bread, in which the dry milled/sieved barley flour fraction replaced 20% of the standard flour, contained 4.2 times the total dietary fibre, 7.6 times the total β-glucans, and 0.8 times the kilocalories/serving compared with the control (Knuckles *et al.*, 1997). This bread was judged acceptable in laboratory acceptance tests, although loaf volume was reduced and colour was slightly darker than the control (Tables 9.6 and 9.7). When the water-extracted barley fraction was substituted in breads, colour scores and acceptability improved over those containing the dry milled/sieved barley fraction. Wheat bread enriched

Table 9.6 Bread evaluation of baking and sensory quality characteristics and colour measurements. From Knuckles *et al.* (1997)

Evaluation parameters	Control, plain bread	5% Barley WS fraction	20% Barley fraction D	40% Barley fraction D
Baking quality[a]				
Absorption (%)[b]	59.1 (59)	66.2 (66)	81.0 (81)	98.1 (90.1)
Mix time (min)	2.5 (~3)	4.0 (~7°)	4.0 (~5)	10–12 (~12°)
Loaf volume (cm)[3]	870	792	607	302
Sensory quality characteristics[d]				
Crust colour	8.1	8.3	7.8	5.1
Grain	6.8	6.4	6.7	5.0
Crumb colour	7.3	7.3	7.1	4.8
Texture	6.5	6.6	6.9	4.9
Flavour/aroma	6.8	7.0	6.4	4.8
Overall acceptability[e]	6.8	6.6	6.4	4.6
Colour measurements[f]				
L score	68.34	68.08	60.17	nd[g]
a score	–1.30	–1.02	1.02	nd
b score	9.96	10.79	11.85	nd

WS, water soluble.
[a]Averages of four to six batches of 200 g flour basis, with dough divided into two loaves per batch.
[b]Farinograph results are given in parentheses.
[c]Second of two peaks.
[d]Means, n = 30; maximum score = 10.
[e]Nine-point verbal hedonic scale ratings: 9 = like extremely; 5 = neither like nor dislike; 1 = dislike extremely.
[f]Measured at five places/slice for three to five slices from different loaves of each type.
[g] nd = Not determined.

with whole barley flakes was found to be suitable in terms of sensory properties for inclusion at up to 10–15% in a high fibre diet, providing 2.5 times more dietary fibre than control wheat bread (Gorecka *et al.*, 1998). The effect on bread quality of the addition of high-fibre barley to wheat flour has been investigated by Kawka and Konieczna (2003). The authors reported that addition of 30% high-fibre barley to a white wheat flour reduced loaf volume, but sensory scores were comparable with those given for control samples.

Dietary fibre contents of breads containing 10% soy curd residue and breads containing 10% makkolli (rice wine) residue were three-fold and two-fold higher, respectively, than bread made with 100% wheat flour but the breads were found unacceptable by sensory panels (Cho and Lee, 1996b).

The addition of soluble dietary fibre from soymilk residue extract or polydextrose effectively retarded bread staling compared with control bread with no added fibre (Kim and Moon, 1999). The authors suggested that soluble dietary fibre may be used as a functional ingredient in bread, providing soluble dietary fibre with desirable physiological effects.

An optimal formulation – involving white flour with equal proportions of coarse and fine bran at 20% replacement, germ at 7.5% – and stearoyl-2-

Table 9.7 Composition of breads containing β-glucan-enriched barley fractions. From Knuckles *et al.* (1997)

Component (%)[a]	Control, plain bread	5% Barley WS fraction	20% Barley fraction D	40% Barley fraction D
Dry matter[b]	69.49	64.39	59.68	51.38
Protein (N × 6.25)	15.22	14.63	14.93	14.46
Fat	2.50	1.14	2.29	2.56
Ash	1.61	1.79	2.13	2.61
Carbohydrate[c]	80.66	82.45	80.65	80.36
Dietary fibre[d]				
TDF	3.49	4.66	14.67	23.40
SDF	1.77	3.27	5.56	9.63
IDF	2.42	1.65	7.90	12.43
β-Glucans				
TBG	0.47	1.85	3.58	7.47
SBG	0.31	0.85	1.14	3.39
IBG	0.16	1.17	2.44	4.08
Calories per serving (kcal)[d]	77.6	76.8	61.2	46.9

[a]Mean of duplicate analyses, dry weight basis.
[b] Fresh weight basis.
[c] Calculated by difference.
[d]Fat = 8 kcal/g, protein = 4 kcal/g, carbohydrate = 4 kcal/g.
Abbreviations are: T, total; S, soluble; I, insoluble; DF, dietary fibre; BG, β-glucans.

lactylate at 0.5% – has been developed for toast bread containing supplementary dietary fibre, but retaining a light crumb colour and superior sensory properties (Sidhu *et al.*, 1999). The effects of bran, wheat germ, and bakery additives on texture and quality of white bread, used for toasting, were determined by Al-Saqer *et al.* (2000). Bread was prepared using whole wheat (control) or white flour supplemented with 10–30% bran or with 5–10% wheat germ. Results showed a decrease in specific loaf volume with increasing bran levels, but at levels up to 20% added bran, values were still superior to those of the control bread. Addition of up to 7.5% wheat germ also produced breads with a higher specific loaf volume than the control bread. Breads with up to 20% added bran and 7.5% added wheat germ had a softer texture than the control bread. Breads containing up to 20% red coarse bran, 30% red fine bran, or 7.5% wheat germ had higher sensory scores than the control bread. Addition of 50 ppm ascorbic acid and 0.5% sodium stearoyl-2-lactylate further improved the quality of the test breads. The authors concluded that an acceptable high-fibre toast bread with a softer texture and of a higher sensory quality than whole wheat bread can be produced using white flour with added bran, germ, and bakery additives.

The utilization of sugar beet fibre of coarse, medium, or fine particle size in the production of high-fibre bread has been investigated at 2.5–7.5% supplementation rates in terms of its effects on bread volume, Hunter colour measurements, and Instron texture values (Ozboy and Koksel, 1999). Breads supplemented with laboratory-prepared coarse and medium-sized sugar beet fibre gave better bread volume; however, Hunter L lightness values and the Instron firmness values were not affected by sugar beet fibre particle size. The addition of corn grits resulted in considerable increases in Hunter yellowness values, which resulted in an improvement in the crumb colour of the sugar beet fibre-supplemented bread. Baking studies with a composite sugar beet fibre preparation (medium and coarse particles, added at 5–15%) indicated that the production of sugar beet fibre-supplemented breads does not generally require vital wheat gluten addition up to the 10% sugar beet fibre level. The addition of an improver and vital wheat gluten resulted in softer bread crumb texture.

The use of Jerusalem artichoke powder as a source of soluble dietary fibre in breadmaking has been assessed at 10 or 12% addition to wheat/rye sourdough bread formulations and compared with control bread and bread containing 10 or 12% commercial inulin, in terms of sensory properties and changes in soluble carbohydrates during baking (Praznik *et al.*, 2002). Bread made with Jerusalem artichoke powder was judged to have high sensory quality, comparable with that of control bread and bread containing Raftiline or Raftilose (Orafti Group, Belgium).

The development of high-fibre white bread containing added inulin (approximately 3–5%), as well as bread containing inulin plus other functional ingredients (Ca (as calcium lactate) or linseeds and Ca), has been described by Draganov *et al.* (2004). Results showed that it was possible to produce fibre-enriched white bread with good sensory, nutritional, and physicochemical properties. Bread containing inulin, Ca, and linseed had a Ca content of 250 μg/100 g and dietary fibre and essential fatty acid contents of 2.5 and 3.9 g/ 100 g, respectively.

A low-calorie bread (677 kJ/100 g) with a high fibre content (11.12%) has been developed by Yanev and Terzieva (2005), based on the formulation of a whole grain mix containing predominantly micronized wheat bran and blending the mix with white flour. Five whole grain mixes were used (standard or with added onion, pumpkin seeds, sunflower seeds and linseeds, or comminuted soybeans) and blended with white flour at a mix:flour ratio of 1:5. According to the authors, all bread types obtained had good sensory properties, with the bread containing sunflower seeds and linseeds generally receiving the highest sensory scores.

A multi-fibre strategy has been recently applied by Rosell and Collar (2006) to develop a low-calorie, high-fibre bread (Fig. 9.4) with high sensory acceptability and good keeping properties for up to 10 days of storage (Fig. 9.5); the work was based on the formulation of a fibre mix containing soluble and insoluble dietary fibres from roots and peas and blending the mix with white flour. The resulting KTG bread contains five times the dietary fibre and provides half of the energy of the regular white pan bread (Table 9.8).

Fig. 9.4 KTG low-calorie, high-fibre bread. From Rosell and Collar (2006).

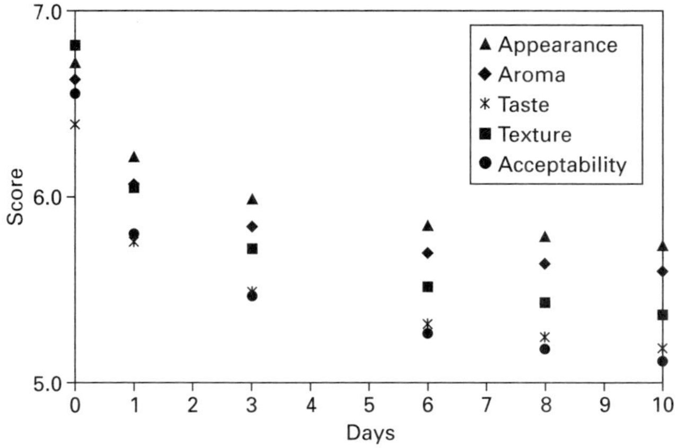

Fig. 9.5 Evolution of sensory scores of KTG low-calorie, high-fibre bread during storage. Nine-point verbal hedonic scale ratings: 5 = neither like nor dislike; 6 = like slightly; 7 = like moderately.

Table 9.8 Nutritional information for KTG low-calorie, high-fibre bread and control pan bread. From Rosell and Collar (2006)

Nutrition information	KTG low-calorie, high-fibre bread (per 100 g)	Control pan bread (per 100 g)
Moisture (g)	43.3	36.1
Fat (g)	1.21	1.45
Protein (g)	13.51	13.90
Ash (g)	2.08	1.69
Digestible carbohydrates (g)	16.66	42.33
Total dietary fibre (g)	23.24	4.53
Soluble dietary fibre (g)	2.79	1.28
Insoluble dietary fibre (g)	19.37	2.84
Energy (kcal)	132	238

9.4.2 Commercial whole grain and high-fibre breads

Commercial high-fibre white breads

During the last decade, several baking companies have launched whole grain and high-fibre breads into the market, under different brands, to meet the demands of consumers who prefer the texture and taste of white bread, but are attracted to the health benefits offered by wholemeal bread. While white bread remains by far the largest segment of the bread category, eight out of ten white bread consumers are interested in bread that has real whole grain nutrition, but retains the taste of white bread, according to research conducted by Harman Atchison Research. The following examples illustrate the array of commercial products available.

Wonder White® (Buttercup®) is a pre-sliced, packaged white bread that was developed by Quality Bakers Australia Ltd under the Buttercup® brand in 1994. The discovery of Hi-maize™, a natural ingredient milled from a specially developed maize grown in Australia, provided the opportunity to produce an innovative high-fibre bread that looks and tastes like regular white bread, but contains twice the fibre of regular white bread and significantly more dietary fibre than multi-grain bread. The bread was made using regular commercial breadmaking techniques and contains resistant starch, a type of dietary fibre that differs from wheat fibre and other insoluble dietary fibre found in regular bread, including wholemeal. It is classified as 'a good source of dietary fibre' with a dietary fibre content of 5.7 g/100 g. As a result of on-going research, the dietary fibre content of Wonder White® has been increased in order to substantiate the claims 'Now with twice the fibre' and 'Double the fibre of regular white bread'.

Nature's Gold™ is a natural flour product, rich in nutrients and dietary fibre, derived from a novel milling process developed by Goodman Fielder Milling in Australia. In 1995, Quality Bakers used Nature's Gold™ to produce VitaGold®, a pre-sliced, packaged bread under the Uncle Tobys brand. It produces a soft, fine-textured golden bread with no visible fibrous bits.

Products made from Nature's Gold™ have a milder flavour with no aftertaste. Nature's Gold™ is richer in dietary fibre, protein, vitamins, and minerals compared with wholemeal flour. Bread made from Nature's Gold™ (e.g. VitaGold®) contains 7.4 g dietary fibre/100 g – higher than typical white bread (2.7 g/100 g), mixed grain bread (5.1 g/100 g), or wholemeal bread (6.5 g/100 g).

Hyfibe White® bread is a soft and smooth, tasty, sliced white bread with increased fibre (4.4 g per serving). The loaf weight is 700 g, with 19 slices (17 + 2 crusts) per package, and contains soy fibre, wheat fibre, wheat bran, and inulin.

Commercial whole grain breads
In 2005, the New Product Pacesetters 2004 study by Interstate Bakeries Corporation noted the consumer trend toward healthier choices, stating that it sees 'significant growth potential among foods and beverages offering whole grains, natural and organic ingredients, and fortification with select vitamins, minerals, antioxidants, and soy as consumers seek improved health and vitality'. The company has introduced White Bread Fans 100% Whole Grain from Wonder®, a 100% whole grain bread created specifically for people who love the mild flavour and soft feel of white bread. It provides a good source of fibre and calcium, and nine vitamins and minerals.

The *Dietary Guidelines for Americans* (USDA, 2005), released in January 2005, advises consumers to choose foods with higher fibre, less sugar, and fewer trans-fats. Anticipating consumers' renewed attention to the guidelines, Boulder, Colorado-based Rudi's Organic Bakery introduced whole grain higher-fibre breads in late 2004 under its new Whole Grains & Fiber label – e.g. 7 Grain with Flax, and Wheat and Oat. At present, new organic whole grain breads – Honey Sweet Whole Wheat, 14 Grain, and Apple 'n Spice – have been launched into the market (Fig. 9.6). One serving size (40–48 g) of the Rudi's Organic Whole Grains & Fiber breads – free of hydrogenated oils, trans fats, and artificial colours and preservatives – provides 3–5 g dietary fibre and 90–120 cal (Table 9.9).

9.4.3 Patented high-fibre breads

Single and associated mixtures of dietary fibres from different sources have been used in the formulations of high-fibre breads under patent. Examples comprise wheat bran (Thompson, 1978), fibres from soy bean and rice hulls (Tsantir and Gorman, 1973), oat and pea hulls (Hodgson and Watkins, 1989), blends of cellulose and cellulose derivatives (Thompson, 1978), fruit paste (Fahlen, 1990), and yellow field pea hulls (Morton, 1979). The deleterious effects of dietary fibre addition on the technological and sensory performance of fibre-enriched breads have been partially or totally counteracted by the presence of surfactants with a hydrophylic–lipophylic balance (HLB) ≥ 9 (Yoko, 1990), thickening polymers (Silva, 1991), blends of oxidizers,

Honey Sweet Whole Wheat	
Nutrition Facts	
Serving Size: 1 slice (43g/1.5oz)	
Servings Per Container: 15	
Amount per serving	
Calories 110 Calories from Fat 10	
	% Daily value*
Total fat 1g	2%
Saturated Fat 0g	0%
Trans Fat 0g	
Polyunsaturated Fat 0.5g	
Monounsaturated Fat 0g	
Cholesterol 0mg	0%
Sodium 170mg	7%
Total Carbohydrate 19g	6%
Dietary Fiber 3g	11%
Sugars 2g	
Protein 5g	
Vitamin A 0%	• Vitamin C 0%
Calcium 0%	• Iron 6%
Thiamin 6%	• Riboflavin 2%
Niacin 6	• Folate 2%
Vitamin B6 4%	
*Percent Daily Values are based on a 2000 calorie diet. Your daily values may be higher or lower depending on your calorie needs.	

14 Grain	
Nutrition Facts	
Serving Size: 1 slice (40g/1.4oz)	
Servings Per Container: 14	
Amount per serving	
Calories 90 Calories from Fat 10	
	% Daily value*
Total fat 1g	2%
Saturated Fat 0g	0%
Trans Fat 0g	
Polyunsaturated Fat 0.5g	
Monounsaturated Fat 0g	
Cholesterol 0mg	0%
Sodium 150mg	6%
Total Carbohydrate 19g	6%
Dietary Fiber 4g	15%
Sugars 2g	
Protein 4g	
Vitamin A 0%	• Vitamin C 0%
Calcium 0%	• Iron 6%
Thiamin 15%	• Riboflavin 10%
Niacin 6%	• Folate 0%
Vitamin B6 0%	
*Percent Daily Values are based on a 2000 calorie diet. Your daily values may be higher or lower depending on your calorie needs.	

Apple 'n Spice	
Nutrition Facts	
Serving Size: 1 slice (48g/1.7oz)	
Servings Per Container: 14	
Amount per serving	
Calories 120 Calories from Fat 10	
	% Daily value*
Total fat 1g	2%
Saturated Fat 0g	0%
Trans Fat 0g	
Polyunsaturated Fat 0.5g	
Monounsaturated Fat 0g	
Cholesterol 0mg	0%
Sodium 200mg	8%
Total Carbohydrate 25g	8%
Dietary Fiber 5g	15%
Sugars 6g	
Protein 4g	
Vitamin A 0%	• Vitamin C 0%
Calcium 2%	• Iron 8%
Thiamin 8%	• Riboflavin 2%
Niacin 6%	• Folate 2%
Vitamin B6 2%	
*Percent Daily Values are based on a 2000 calorie diet. Your daily values may be higher or lower depending on your calorie needs.	

Wheat and Oat	
Nutrition Facts	
Serving Size: 1 slice (40g/1.40oz)	
Servings Per Container: 14	
Amount per serving	
Calories 90 Calories from Fat 10	
	% Daily value*
Total fat 1g	2%
Saturated Fat 0g	0%
Trans Fat 0g	
Polyunsaturated Fat 0.5g	
Monounsaturated Fat 0g	
Cholesterol 0mg	0%
Sodium 190mg	8%
Total Carbohydrate 18g	6%
Dietary Fiber 4g	16%
Sugars 2g	
Protein 3g	
Vitamin A 0%	• Vitamin C 0%
Calcium 0%	• Iron 4%
Thiamin 6%	• Riboflavin 2%
Niacin 6%	• Folate 2%
Vitamin B6 2%	
*Percent Daily Values are based on a 2000 calorie diet. Your daily values may be higher or lower depending on your calorie needs.	

7 Grain with Flax	
Nutrition Facts	
Serving Size: 1 slice (40g/1.4oz)	
Servings Per Container: 14	
Amount per serving	
Calories 100 Calories from Fat 15	
	% Daily value*
Total fat 2g	3%
Saturated Fat 0g	0%
Trans Fat 0g	
Polyunsaturated Fat 0.5g	
Monounsaturated Fat 0g	
Cholesterol 0mg	0%
Sodium 150mg	6%
Total Carbohydrate 18g	6%
Dietary Fiber 4g	15%
Sugars 2g	
Protein 4g	
Vitamin A 0%	• Vitamin C 0%
Calcium 2%	• Iron 8%
Thiamin 30%	• Riboflavin 35%
Niacin 4%	• Folate 2%
Vitamin B6 2%	
*Percent Daily Values are based on a 2000 calorie diet. Your daily values may be higher or lower depending on your calorie needs.	

Fig. 9.6 Nutritional facts for organic whole grain, high-fibre breads under the Whole Grains & Fiber label from Rudi's Organic Bakery (www.rudisorganicbakery.com).

Table 9.9 Ingredients of organic whole grain, high-fibre breads under the Whole Grains & Fibre label from Rudi's Organic Bakery

Honey Sweet Whole Wheat
Organic whole wheat flour, purified water, organic evaporated cane juice, organic vital wheat gluten, organic high-oleic sunflower oil, cultured whole wheat flour, sea salt, unfiltered honey, yeast, organic molasses, soy lecithin, organic oat flour, organic barley malt syrup, vinegar, natural enzymes, ascorbic acid.

Whole Grains & Fiber – 14 Grain
Organic whole wheat flour, water, organic evaporated cane juice, organic oat fibre, organic vital wheat gluten, organic cracked barley, organic rolled oats, organic cracked wheat, organic high-oleic sunflower oil, organic cornmeal, organic quinoa flour, sea salt, cultured wheat flour, yeast, organic millet, organic ground flaxseeds, organic oat flour, organic rye flour, soy lecithin, organic barley malt, organic whole spelt flour, organic brown rice flour, organic Kamut® flour, organic amaranth flour, organic triticale flour, organic barley flour, organic buckwheat flour, vinegar, natural enzymes. CONTAINS WHEAT AND FLAX SEEDS

Whole Grains & Fiber – Apple 'n Spice
Organic whole wheat flour, water, organic dried apples, organic dark brown sugar, organic oat fibre, organic vital wheat gluten, organic rolled oats, yeast, organic yellow cornmeal, organic cinnamon, organic high-oleic sunflower oil, sea salt, cultured wheat flour, soy lecithin, organic oat flour, vinegar, natural enzymes. CONTAINS WHEAT

Whole Grains & Fiber – Wheat and Oat
Organic whole wheat flour, purified water, organic oat fibre, organic evaporated cane juice, organic vital wheat gluten, organic yellow cornmeal, organic high-oleic sunflower oil, sea salt, cultured wheat flour, organic oat bran, organic whole rye flour, soy lecithin, yeast, organic barley malt, natural enzymes, vinegar, organic rolled oats, organic wheat bran. CONTAINS WHEAT

Whole Grains & Fiber – 7 Grain with Flax
Organic whole wheat flour, purified water, organic evaporated cane juice, organic oat fibre, organic quinoa, organic Kamut® flour, organic whole spelt, organic natural sesame seeds, organic vital wheat gluten, organic sunflower seeds, organic potato flour, organic high-oleic sunflower oil, organic brown flax seed, cultured wheat flour, organic blackstrap molasses, yeast, soy lecithin, organic millet, sea salt, natural enzymes, vinegar, organic oat bran, organic cracked wheat. CONTAINS WHEAT, SESAME, SUNFLOWER, AND FLAX SEEDS.

surfactants, gluten and α-amylase (Kiyoto, 1994), fibres added in a liquid state (Kazuyoshi, 1995), and the addition of dextrins and cellulose (Kimihito, 2001) to bread formulations.

9.5 Consumer attitudes towards high-fibre breads and whole grain white breads

As stated by Redgwell and Fischer (2005), food companies are responding to the need for foods with enhanced satiety and lower energy density by allocating more resources to the development of dietary fibre-enriched products.

The industry challenge is to ensure that such foods are both palatable and attractive to the consumer, while maintaining nutritional benefits. The authors claim that such food company requirements can be met through basic research on the physicochemical basis for the functionality of dietary fibre, and by developing a broad spectrum of novel products. In parallel, the health benefits associated with the regular consumption of whole grain foods are encouraging the industry to develop a wide variety of palatable and healthy whole grain breads to offer to consumers. Although benefits are observed at relatively low levels of intake (between two and three servings per day), the consumption of whole grain foods in some Western countries is less than one serving per day. The main sources of whole grain are wholemeal and rye breads and whole grain breakfast cereals. Typical consumers of whole grain foods tend to be older, from a high socio-economic group, are less likely to smoke and are more likely to exercise than non-consumers (Lang and Jebb, 2003). Some of these attributes may contribute to the observed health benefits. However, whole grain foods are an important source of a range of nutrients as part of a healthy eating plan. There is considerable scope for strategies to promote increased consumption of whole grain foods to reduce the risk of a variety of chronic diseases. As Camire (2004) recently reported, whole grains gained new stature in 1999 when General Mills successfully petitioned the US FDA to authorize a health claim: 'Diets rich in whole grain foods and low in total fat, saturated fat and cholesterol, may help reduce the risk of heart disease and certain cancers'. Foods that bear this health claim for whole grains must contain 51% or more whole grain ingredients by weight per reference amount customarily consumed (RACC) as specified by the FDA. A RACC could be a slice of bread, a cup of breakfast cereal, or another realistic serving size. Attempting to incorporate at least 51% whole grain ingredients in a formulation has led to creative product development. The petition specified that the total dietary fibre (TDF) of whole wheat (11 g of fibre/100 g of wheat) should be used as reference to assess compliance, despite the fact that some whole grains have lower TDF levels than wheat. The FDA used the following formula to specify the qualifying amount of TDF required for a food to carry the claim: 11 g × 51% × RACC/100.

For a healthy diet, the recommended daily dose of dietary fibre is between 25 and 30 g/day (Miller Jones, 2004). As reported by Redgwell and Fischer (2005), at the beginning of the 1990s 10–25% of Americans ate no vegetables, and 47% not a single serving of fruit, on a given day. For the authors, there was no doubt that palatability, convenience, and price were tending to overwhelm healthy eating decisions in spite of the fact that consumers placed nutrition second in importance to taste in factors for food selection. More recent surveys indicate that public awareness of the benefits of consuming dietary fibre is growing (Sloan, 2001). According to a 2001 Healthfocus International Trends report, more than 40% of American and Australian shoppers and 33% of Western European shoppers said that 'high fibre' was an extremely important label claim. The relevant news for food manufacturers was that consumers put dietary fibre content third (behind how to limit fat and calories) as the most sought after topic for health information in supermarkets.

9.6 Particular difficulties in the development of high-fibre and whole grain white breads: effects of formulation and processing technology

9.6.1 The industrial perspective

Currently, there are sources of dietary fibre that possess a much broader spectrum of technological functionality than those coming from cereal brans with limited functionality in terms of water control, giving rise to negative sensory attributes in the final product. From an industrial perspective the sources can be separated into three broad classes: (a) hydrocolloids, mostly soluble polysaccharides; (b) bioactive oligosaccharides; and (c) whole plant cell wall materials derived from cereal grains, fruits, and vegetables (Redgwell and Fischer, 2005). All possess both physiological and technological functionalities to varying degrees, which determine their respective applications. In particular, hydrocolloids swell and produce a viscous solution or dispersion when exposed to water. The largest range of components is polysaccharide-based, including such polymers as galactomannans, glucomannans, pectinaceous materials, arabinogalactans, seaweed-based extracts, the microbials (gellan and xanthan), and other macromolecular entities such as cellulosics, glucans, and starches (Phillips and Williams, 2000). The consumer does not see these in the same positive light as fibres from more natural sources, such as whole grains or fruits and vegetables. When used in food products, it is mandatory that hydrocolloids – such as gellan gum, xanthan gum, galactomannans, alginates, the cellulose-derived gums, etc. – are labelled by an E-number, denoting a chemical additive, and this can be perceived by the public as a negative factor. Their presence in fibre-fortified food is therefore not altogether consistent with the healthy, natural image that food companies wish to promote by publicizing the presence of dietary fibre in their products. In part this is due to the fact that the consumer may see soluble hydrocolloids as industrially expedient texturizers, rather than as dietary fibres with health benefits. This is unfortunate, because alginates, pectic polymers, galactomannans, and carrageenans are obtained from natural sources that are themselves consumed as whole plants in many parts of the world. These hydrocolloids are dietary fibres that possess physiological functionalities that in some cases are equivalent to those from whole grain-, fruit-, and vegetable-derived plant cell walls. The manufacturer is unable to highlight this feature in products that contain hydrocolloids, because most are used below the threshold concentration at which a health claim is permitted. Nevertheless, with the rise in the use of soluble hydrocolloids in prepared foods and the increasing consumption of the latter, the cumulative content of the hydrocolloids in several products could make a significant contribution to the daily dietary fibre intake. To take advantage of this fact, the physiological functionality and potential health benefits of hydrocolloids as dietary fibres needs more emphatic validation. This will enable the food industry to convey to the consumer a much stronger message on the positive attributes of these industrially important food components.

An integral part of a healthy lifestyle is the consumption of fresh, natural food and this is often associated with images of fruit and vegetables and whole grain cereals. Although the health benefits of these foods can be attributed to many different nutrients, the dietary fibre content has become a key component in promoting the belief that diets based on a significant intake of fresh fruit and vegetables have long-term health benefits. The dietary fibres in these foods are derived almost entirely from the plant cell walls, the major structural components of fruit and vegetables (Waldron *et al*., 2003). The plant cell walls consist of a mixture of several different types of polysaccharides, proteins, lignin, and associated substances. The complex chemistry of this structure of 'soluble' and 'insoluble' polysaccharides and other macromolecules has only been partially characterized. Their technological and physiological functionality, although generally accepted, awaits further clarification in terms of the individual contribution of each type of macromolecule present in the cell wall. From an industrial standpoint, this is not a major setback and may in fact be an advantage, as the consumer already accepts that dietary fibres in their entirety are beneficial for a healthy diet. Despite their positive image, the use of dietary fibres purified from whole plant cell walls in manufactured food products is low compared with the soluble hydrocolloids, due to some limitations in terms of their supply and technological functionality.

The major non-starchy polysaccharide components of the primary cell wall of fruit and vegetables differ markedly from those in cereals. Both contain cellulose but the latter are dominated by mixed linkage glucans and arabinoxylans, while the parenchymatous tissues of fruit consist predominantly of pectin and xyloglucan. This may be an important difference in terms of physiological functionality, but, until more is known about the polysaccharide-specific effects of dietary fibre, its significance cannot be assessed. The cell wall material of fruit basically remains a 'pure' polysaccharide matrix with a neutral flavour impact. In whole grain-based fibres, a diverse range of chemical components are held in a highly insoluble matrix, which contains the highly cross-linked compound lignin as well as range of phytochemicals – including lignans, tocotrienols, phenolic compounds, phytosterols, vitamins, and antinutrients such as phytic acids, tannins, enzyme inhibitors, proteins, and lipids (Slavin, 2003). These are not easily removed by chemical extraction and have a marked impact on the sensory profile of any product to which whole grain fibre is added. Overall, the physicochemical properties of whole grain-based fibres are more hydrophobic than the fruit cell wall material and the dietary fibre is more difficult to disperse in an aqueous environment, which raises problems of a textural nature. On the basis of its dietary fibre content, cell wall material of fruits and vegetables can contain between 60 and 90% dietary fibre, whereas the bran fraction of some whole grain cereals – such as oats, wheat, and rice – contain 16–32%, 35–45%, and 20–33% total dietary fibre, respectively (Dreher, 1987).This is offset by the fact that the yield of cell wall material from fruit and vegetables on a fresh weight basis is very low (1–4%).

9.6.2 The technological perspective: drawbacks and solutions

The demand from consumers for healthy and low-calorie foods has stimulated food technologists to develop innovative ways to bring dietary fibre into new appealing high-fibre food products that contribute to the fibre recommendation intake and to fulfil consumer's expectations. In breadmaking, the incorporation of a high amount of fibre to ensure beneficial physiological impact often causes disruption of the continuous matrix causing adverse technological effects in bread manufacture, and thus negatively affecting dough viscoelasticity and the functional quality of the resulting fortified breads. Decreased functional quality of the final bread has been ascribed to a dilution of functional gluten proteins (Pomeranz *et al.*, 1977) and to a reduced starch content that impairs viscometric dough behaviour during pasting and gelling (Collar *et al.*, 2006). In addition, fibre replacement of flour disrupts the starch–gluten matrix and restricts the force required for gas cells to expand in a particular dimension (Gan *et al.*, 1992) affecting dough viscoelastic behaviour and constraining dough machinability and gassing power. The deleterious effects of fibre on bread are mainly related to decreased loaf volume from lowered gas retention, objectionable gritty texture, and unsuitable taste and mouth feel (Lai *et al.*, 1989). The fibre source and the type and degree of processing are the main factors influencing the functional properties of high-fibre ingredients – solubility, viscosity, gelation, water-binding – and oil-binding capacities, mineral and organic molecule binding – and thus affecting the functional quality of the intermediate manufacturing and end products (Wang *et al.*, 2002). Changes in the swelling and water-absorption properties of the dough often result in different processing behaviour and related production problems with a slack, sticky, or excessively stiff dough, leading to bread of inferior quality (Haseborg and Himmelstein, 1988).

The adverse effects described with high-fibre contents in breads can be counteracted to a variable extent with the addition of vital gluten, surfactants, and surfactant/shortening blends (Shogren *et al.*, 1981), with the use of high-grade flours (Sosulski and Wu, 1988), and with the incorporation of hemicellulases (Haseborg and Himmelstein, 1988; Laurikainen *et al.*, 1998). The combination of bran sourdough and an enzyme mixture (α-amylase, xylanase, and lipase) has recently been shown to significantly improve the volume, texture, and shelf-life of wheat bread supplemented with wheat bran (20 g bran/100 g of flour) (Katina *et al.*, 2006). It is reported that the use of fermented bran improves the structure of the gluten network and may alter water migration between starch, protein, and bran particles during storage of high-fibre wheat bread, being the effects of enzymes mainly due to reduced starch crystallization during storage. The antistaling effect of the combined use of bran sourdough and the enzyme mixture was related to reduced starch retrogradation rate and to a slowing of the increase in the rigidity of the polymer structure, and was connected to degradation of cell wall components leading to altered water distribution between the starch–protein matrix (Fig. 9.7). The effects of a combination of different fibres have recently been reported as a suitable strategy for overcoming the deficiencies of individual fibres and counteracting their deleterious effect, and probably improving dough handling properties (Collar *et al.*, 2005, 2007; Rosell *et al.*, 2006).

It has been established that dough should meet certain rheological requirements in order to produce high-quality bread with a prolonged shelf-life. These requirements concern the correct combination of uniaxial compression, empirical extension, and surface rheological properties, especially during the first stages of breadmaking.

High flour replacement at different levels (from 6% up to 34%) by fibres from different sources significantly changes the mixing properties, machinability, and extensional behaviour of the resulting hydrated flour–fibre blends (Fig. 9.8). Soluble, insoluble, and partially soluble fibres included inulin (Fibruline), sugar beet fibre (Fibrex), pea cell wall fibre (Swelite), and

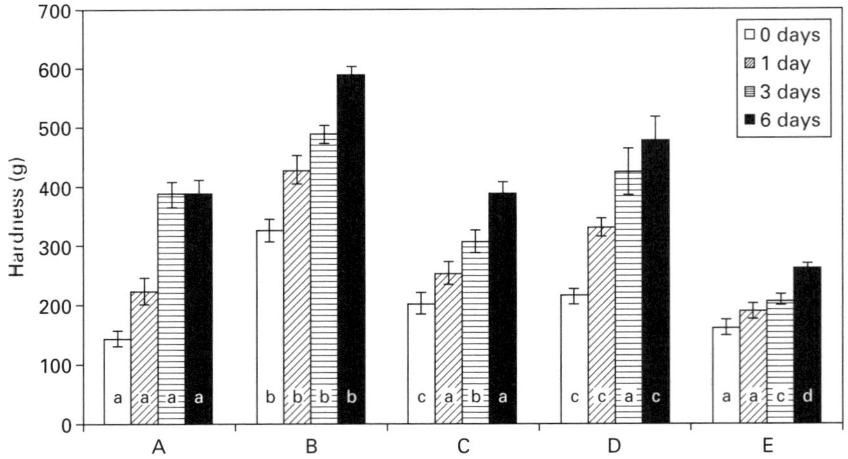

Fig. 9.7 Staling of bran breads measured as crumb hardness during 6 days of storage. Columns marked by the same letter are not statistically different at $P < 0.05\%$ (comparisons made at the same measuring points; e.g. after 1 day of storage). A, white wheat bread; B, reference bran bread; C, bran bread with enzyme mixture of α-amylase, xylanase, and lipase; D, bran sourdough bread; E, bran sourdough bread with enzyme mixture of α-amylase, xylanase, and lipase. From Katina *et al.* (2006).

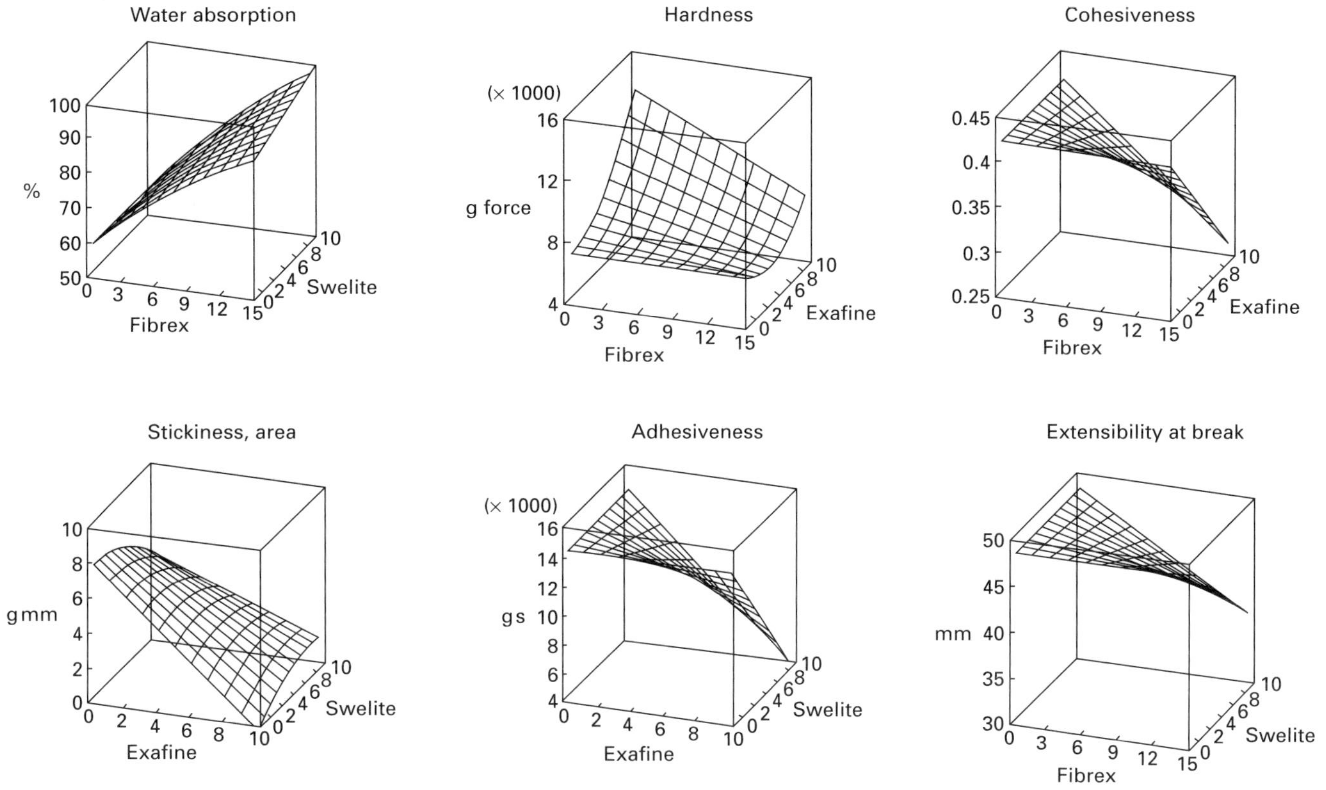

Fig. 9.8 Response surface plots of mixing, mechanical, surface-related, and extensional wheat dough parameters dependent on fibre formulation. From Collar *et al.* (2005).

pea hull fibre (Exafine). The trend and the extent of the effects on dough viscoelastic parameters were variable (beneficial/ adverse) and were closely dependent on the nature of the fibres in the blend and on the extent of the flour substitution. Fibrex provided the most beneficial effects on dough mixing and textural profiles, when added singly, in particular the prominent increase in water absorption (Rosell *et al.*, 2005), as well as a suitable decrease in gumminess with no significant adverse effects on the main mechanical, surface, and extensional properties. Caution should be paid to both the Fibrex–Swelite pair and the single Exafine because of the adverse shortening and hardening effects they respectively induce. The Addition of Exafine to Fibrex-flour blends is not recommended, because of the deleterious effect of the pair on mechanical properties and extensional behaviour, particularly in reducing springiness and cohesiveness. The association of Exafine and Swelite, is strongly encouraged to minimize adhesiveness and stickiness (Collar *et al.*, 2005).

9.7 Future trends

The successful implementation of the recommendations to increase whole grains and fibre intake will require the cooperative efforts of industry, government, academia, non-profit health organizations, and the media. Future revelations based on basic research and clinical human studies are needed to confirm the health benefits of whole grains, to develop processing techniques that will improve the palatability of whole grain products, and to educate consumers about the benefits of high fibre and whole grain consumption. With the significant progress in technology, the industry can provide a diverse range of prepared foods, with increasing consumer appeal. Dietary fibre could be included in many of these recipes providing that it is compatible with these technologies and does not compromise the sensory quality of the product. With this gradual approach, industry can help consumers move along the path to healthier choices and shift dietary patterns. A coalition of government agencies, academics, cereal scientists and technologists, health professionals, and industry professionals should work together to: (a) expand the number of high-fibre and whole grain products on the market that appeal to the consumer, by ensuring that products are affordable, convenient, and tasty; (b) identify products that can start the gradual transition by partial replacement of flour with whole grain flour in bread combined with a strong education and marketing message; (c) reacquaint consumers with the taste of whole grains and high-fibre breads, and (d) introduce new and lesser used whole grains to provide more choices for consumers.

The extent to which whole grain and enriched dietary fibre products will contribute to sustained growth in the sector of functional foods and functional ingredients, depends on the results of research in validating the perceptions of the health benefits of these products and on the efforts of the food industry in formulating and promoting products that carry these benefits to health-conscious consumers.

9.8 Sources of further information and advice

Websites of interest:

- American Association of Cereal Chemists International, AACCI, www.aaccnet.org.
- Food Science and Nutrition Department at the University of Minnesota, www.wholegrain.umn.edu/.
- The Wheat Foods Council, www.wheatfoods.org/.
- US Department of Agriculture, www.ars.usda.gov/.
- US Food and Drug Administration, www.fda.gov.
- Whole Grains Council, www.wholegrainscouncil.org.
- World Health Organization, www.who.int.

9.9 References

Adom KK, Liu RH (2002), 'Antioxidant activity of grains', *J Agric Food Chem*, **50**, 6182–6187.

Al-Saqer JM, Sidhu JS, Al-Hooti SN (2000), 'Instrumental texture and baking quality of high-fibre toast bread as affected by added wheat mill fractions', *J Food Process Preserv* **24** (1), 1–16.

Anon. (1995), 'In flake and fibre form', *Lebensmitteltechnik*, **27** (11), 22–23.

Anon. (2002), 'Health and convenience – can they co-exist?', *Innov Food Technol*, **14**, 74–75.

Armero E, Collar C (1998), 'Crumb firming kinetics of wheat breads with anti-staling additives', *J Cereal Sci*, **28** (2), 165–174.

Bruce B, Spiller GA, Klevay LM, Gallagher SK (2000), 'A diet high in whole and unrefined foods favorably alters lipids, antioxidant defenses, and colon function', *J Am Coll Nutr*, **19**, 61–67.

Camire ME (2004), 'Technological challenges of whole grains', *Cereal Foods World*, **49** (1), 20–22.

Chatenoud L, Tavani A, La Vecchia C, Jacobs DR, Negri E, Levi F, Franceschi S (1998), 'Whole-grain food intake and cancer risk', *Int J Cancer*, **77**, 24–28.

Cho MK, Lee WJ (1996a), 'Preparation of high-fibre bread with barley flour', *Korean J Food Sci Technol*, **28** (4), 702–706.

Cho MK, Lee WJ (1996b), 'Preparation of high-fibre bread with soybean curd residue and makkolli (rice wine) residue', *J Korean Soc Food Sci Nutr*, **25** (4), 632–636.

Collar C, Santos E, Rosell CM (2005), 'Providing healthier baked goods through high fibre optimised formulations: functional requirements', in P Fito and F Toldrá (Eds), *Proceedings of EFFOST Intradfood 2005, Innovations in Traditional Foods*, Vol. II, Elsevier, London, pp. 879–882.

Collar C, Santos E, Rosell CM (2006), 'Significance of dietary fibre on the viscometric pattern of pasted and gelled flour-fibre blends', *Cereal Chem*, **83** (4), 370–376.

Collar C, Santos E, Rosell CM (2007), 'Assessment of the rheological profile of fibre-

enriched bread doughs by response surface methodology', *J Food Eng*, **78**, 820–826.
Draganov L, Atanasova E, Gadzheva M (2004), 'Bread production adding inulin', *Khranitelno-vkusova Promishlenost*, **6**, 24–26.
Dreher ML (1987), *Handbook of Dietary Fibre: An Applied Approach*. Marcel Dekker, New York.
Endress H-U, Fischer J (2000), 'Fibre and fibre blends for individual needs – a physiological and technological approach – Herbstreith & Fox Corporate Group and Herbafood Ingredients GmbH'. Presentation at *Dietary Fibre 2000*, May 13–18 2000, Dublin.
Erkkila AT, Herrington DM, Mozaffarian D, Lichtenstein AH (2005) 'Cereal fibre and whole-grain intake are associated with reduced progression of coronary-artery atherosclerosis in postmenopausal women with coronary artery disease', *Am Heart J*, **150** (1), 94–101.
Fahlen AE (1990), 'High fibre natural bread' US Patent 4,971,823.
Franz MJ, Bantle JP, Beebe CA, Brunzell JD, Chiasson JL, Garg A, Holzmeister LA, Hoogwerf B, Mayer-Davis E (2002), 'Evidence-based nutrition principles and recommendations for the treatment and prevention of diabetes and related complications', *Diabetes Care*, **25**, 148–198.
Gan Z, Galliard T, Ellis PR, Angold RE, Vaughan JG (1992), 'Effect of the outer bran layers on the loaf volume of wheat bread', *J Cereal Sci*, **15**, 151–163.
Gerhardt AL, Gallo NB (1998), 'Full-fat rice bran and oat bran similarly reduce hypercholesterolemia in humans', *J Nutr*, **128**, 865–869.
Gorecka D, Kawka A, Gasiorowski H, Sroczynska B, Weglerska-Smolarkiewicz E (1998), 'Dietary fibre in bread containing barley flakes', *Przeglad Piekarski i Cukierniczy*, **46** (2), 2–3.
Haseborg E, Himmelstein A, (1988), 'Quality problems with high-fibre breads solved by use of hemicellulase enzymes', *Cereal Foods World*, **33** (5), 419–422.
Hawrysh ZJ, Pick ME, Gee MI, Brudler J (1996), 'Sensory and instrumental evaluation of stored waxy hulless barley bread', in *IFT Annual Meeting*, *Book of Abstracts*, New Orleans, LA, June 1996 p. 185.
Hodgson WW, Watkins TR (1989), 'Low calorie high fibre bread', US Patent 4,824,683.
Howarth NC, Saltzman E, Roberts SB (2001), 'Dietary fibre and weight regulation', *Nutr Rev*, **59**, 129–139.
Katina K, Salmenkallio-Marttila M, Partanen R, Forssell P, Autio K (2006), 'Effects of sourdough and enzymes on staling of high-fibre wheat bread', *Lebensm Wiss Technol*, **39**, 479–491.
Kawka A, Konieczna E (2003), 'The effects of high dietary fibre barley product on quality and chemical composition of bread', *Zywnosc*, **9** (4), 71–81.
Kazuyoshi O (1995), 'Dietary fibre-containing bread and cake and their production', JP Patent 7,327,583.
Kim H, Moon TW (1999), 'Quality attributes of bread with soymilk residue dietary fibre', *Food Sci Biotechnol*, **8** (4), 245–250.
Kimihito W (2001), 'Bread enriched with dietary fibre and its production', JP Patent 2001045960.
Kiyoto S (1994), 'Production of dietary fibre-containing breads, composition for bread production and improver for bread production', JP Patent 6,070,670.
Knuckles BE, Hudson CA, Chiu, MM, Sayre RN (1997), 'Effect of beta-glucan barley fractions in high-fibre bread and pasta', *Cereal Foods World*, **42** (2), 94–96, 98–99.
Lai CS, Hoseney RC, Davis AB (1989), 'Effects of wheat bran in breadmaking', *Cereal Chem*, **66** (3), 217–219.
Lang R, Jebb SA (2003), 'Who consumes whole grains, and how much?' *Proc Nutr Soc*, **62** (1): 123–127.
Laurikainen T, Härkönen H, Autio K, Poutanen K (1998), 'Effects of enzymes in fibre-enriched-baking', *J Sci Food Agric*, **76**, 239–249.

Liu S, Willett WC, Manson JE, Hu FB, Rosner B, Colditz G (2003), 'Relation between changes in intakes of dietary fibre and grain products and changes in weight and development of obesity among middle-aged women', *Am J Clin Nutr*, **78**, 920–927.
Malkki Y (2004), 'Trends in dietary fibre research and development: a review', *Acta Aliment*, **33**, 39–62.
Marquart L, Wiemer KL, Jones JM, Jacob B (2003), 'Whole grain health claims in the USA and other efforts to increase whole-grain consumption', *Proc Nutr Soc*, **62**, 151–160.
Miller G, Prakash A, Decker E (2002), 'Whole-grain micronutrients', in, L Marquart, JL Slavin and RG Fulcher (Eds), *Whole-Grain Foods in Health and Disease*, Eagan Press, St Paul, MN, 243–258.
Miller Jones J (2004), 'Dietary fibre intake, disease prevention, and health promotion: an overview with emphasis on evidence from epidemiology', in JW Van der Kamp, NG Asp, J Miller Jones and G Schaafsma (Eds), *Dietary Fibre: Bioactive Carbohydrates for Food and Feed*, Wageningen Academic Publishers, The Netherlands, pp. 143–164.
Morton S (1979), 'High fibre white bread', Patent CA1,048,848.
Ozboy O, Koksel H (1999), 'Utilization of sugar beet fibre in the production of high fibre bread', *Zuckerindustrie*, **124**, 712–715.
Phillips GO, Williams PA (2000), *Handbook of Hydrocolloids*, Woodhead Publishing Limited, Cambridge, UK.
Pomeranz Y (1987), *Modern Cereal Science and Technology*, pp. 25–39. VCH Publlishers, Weinheim, Germany, pp. 25–39.
Pomeranz Y, Shogren MD, Finney KF, Bechtel DB (1977), 'Fibre in breadmaking. Effects on functional properties', *Cereal Chem*, **54**, 25–41.
Praznik W, Cieslik E, Filipiak-Florkiewicz A (2002), 'Soluble dietary fibres in Jerusalem artichoke powders: composition and application in bread', *Nahrung*, **46**, 151–157.
Redgwell R J, Fischer M (2005), 'Dietary fibre as a versatile food component: An industrial perspective', *Mol Nutr Food Res*, **49**, 421–535.
Richardson DP (2003), 'Whole grain health claims in Europe', *Proc Nutr Soc*, **62**, 161–169.
Rodríguez R, Jiménez A, Fernández-Bolaños J, Guillén R, Heredia A (2006), 'Dietary fibre from vegetable products as source of functional ingredients', *Trends Food Sci Technol*, **17**, 3–15.
Rosell CM, Collar C (2006), Patent pending. Composición de fibras para la obtención de productos bajos en calorias y ricos en fibras. No. de registro 200601668. País de prioridad: España. Fecha de prioridad: 2006.
Rosell CM, Santos E, Collar C (2006), 'Mixing properties of fibre enriched wheat bread doughs: a response surface methodology study', *Eur Food Res Technol*, **223** (3), 333–340.
Sahyoun, NR, Jacques PF, Zhang XL, Juan W, McKeown NM (2006), 'Whole-grain intake is inversely associated with the metabolic syndrome and mortality in older adults', *Am J Clin Nutr*, **83** (1): 124–133.
Schaafsma G (2004),'Health claims, options for dietary fibre', in JW Van der Kamp, NG Asp, J, Miller Jones and G Schaafsma (Eds), *Dietary Fibre: Bioactive Carbohydrates for Food and Feed*, Wageningen Academic Publishers, The Netherlands, pp. 27–38.
Shogren MD, Pomeranz Y, Finney KF (1981), 'Counteracting the deleterious effects of fibre in breadmaking', *Cereal Chem,* **58** (2), 142–144.
Sidhu JS, Al-Hooti SN, Al-Saqer, JM (1999), 'Effect of adding wheat bran and germ fractions on the chemical composition of high-fibre toast bread', *Food Chem*, **67** (4), 365–371.
Silva RF (1991), 'High fibre bakery products', US Patent 5,035,903.
Slattery ML, Curtin KP, Edwards SL, Schaffer DM, (2004), 'Plant foods, fibre, and rectal cancer', *Am J Clin Nutr*, **79** (2), 274–281.
Slavin JL (2000), 'Whole grains, refined grains and fortified refined grains: what's the difference?' *Asia Pac J Clin Nutr*, **9**, S23–S27.

Slavin J L (2001), 'Dietary fibre and colon cancer', in S Cho and M. Dreher (Eds), *Handbook of Dietary Fibre*, Marcel Dekker, New York, pp. 31–45.
Slavin JL (2003), 'Why whole grains are protective: biological mechanisms', *Proc Nutr Soc*, **62**, 129–134.
Slavin JL (2004), 'Whole grains and human health', *Nutr Res Rev*, **17**, 99–110.
Sloan A E (2001), 'Growing demand for dietary fibre', *Funct Food Nutraceut*, Sept/Oct, 12–13.
Sosulski FW, Wu KK (1988), 'High-fibre breads containing field pea hulls, wheat, corn, and wild oat brans', *Cereal Chem*, **65**(3), 186–191.
Tangkanakul P, Tungtrakul P, Vatanasuchart N, Auttaviboonkul P, Niyomvit B (1995), 'Physical and chemical properties of high fibre breads and cookies', *Food*, **25** (2), 95–107.
The Whole Grains Council (2004), *Overview of the Science of Whole Grains*, The Whole Grains Council/Oldways Preservation Trust, Boston, MA.
Thomson P (2005), 'White breads can also be high fibre', *Wellness Foods Europe*, November 30–31.
Thompson JB (1978), 'Low calorie diet bread', US Patent 4,109,018.
Trowell HJ, Burkitt D, Heaton K (1985), *Dietary Fibre, Fibre-Depleted Foods and Diseases*, Academic Press, New York.
Truswell AS (2002), 'Cereal grains and coronary heart disease', *Eur J Clin Nutr*, **56**, 1–14.
Tsantir K, Gorman RA (1973), 'Low calorie bread', US Patent 3,767,423.
USDA (United States Department of Agriculture) (2005), *Dietary Guidelines for Americans*, Dietary Guidelines Advisory Committee 2005, www.ars.usda.gov/dgac/.
Waldron KW, Parker ML, Smith A C (2003), 'Plant cell walls and food quality', *Compr Rev Food Sci Food Safety*, **2**, 101–119.
Wang J, Rosell CM, Benedito de Barber C (2002), 'Effect of the addition of different fibres on wheat dough performance and bread quality', *Food Chem*, **79**, 221–226.
Yanev A, Terzieva V (2005), 'Possibilities for production of new low-energy breads', *Khranitelno-vkusova Promishlenost*, **4**, 14–15.
Yoko M (1990), 'Preparation of dietary fibre-containing bread', Patent JP2227020.
Zhong S, Luo B, Xiang M, Liu H, Shai Z, Wang T, Craig S (2000), 'Studies on the effects of polydextrose on physiological function in Chinese people', *Am J Clin Nutr*, **72**, 1503–1509.

10

Traditional and modern oat-based foods

A. Kaukovirta-Norja and P. Lehtinen, VTT, Finland

10.1 Introduction: oats as a food raw material

The beneficial health effects of oats are well known and documented. Oats have traditionally been used mainly as oatmeal, bran or flakes, which are used to produce porridge, bread and breakfast cereals. Apart from porridge, which can be made from 100% oats, in most consumer products oats are added as an ingredient, to provide increased customer value. For example, oat ingredients can be added into a variety of consumer products in order to provide health-promoting properties, to adjust the flavour and visual appearance or to achieve technological goals, such as water binding.

Most of the novel oat ingredients are bran or β-glucan-enriched oat fractions. Furthermore, in addition to the health-promoting and technologically versatile fibre fraction, oats contains high amount of fats with nutritionally beneficial fatty acid and lipid class composition, proteins rich in valuable amino acids, and unique phenolic compounds, such as avenanthramides, among other minor nutrients. However, only a small minority of oat crops are processed to food and other valuable products, and most of the harvested oats are used for animal feeding purposes.

The kinetics of β-glucan solubility and the tendency to form very viscous shear thinning gels varies in different oat products. Thus modern oat ingredients are suitable for various types of food applications without the adverse effects that were evident in conventional oat products. These novel ingredients are available in various forms with different composition, appearance and taste, and different technological functionality.

This chapter introduces oats as a food raw material, its composition and its technological properties. Furthermore, different types of oat products and some examples of production technologies are presented. In addition, this chapter summarizes current knowledge on the health-promoting effects of oats and finally the future trends for oat products are discussed.

10.2 Oat production and composition

The world production of oats was 24.5 million tons in the crop season 2005/2006 (International Grain Council (IGC), GMR352). The main producers of oats are Argentina, Australia, Canada, the EU and the USA. Even though oats have a reputation as a health-promoting cereal, their main use is still as feed for livestock. Worldwide, about 50% of oats are fed to cattle and it is especially used as feed for horses, dairy cattle, breeding animals, poultry and young livestock (National Grain and Feed Association, htpp://www.ngfa.org). It is estimated that 25% of oats are used for food, seed and industrial products. However, in Finland for example, which is one of the main producers of oats in the EU, only 6% of oats are used for food purposes with the majority of the crop being used directly on farms as feed.

The main technological parts of the oat kernel are the hull, cell wall (i.e. bran) and endosperm fractions (Fig. 10.1). The oat kernel has a hard outer hull, exterior to the endosperm and bran, which can comprise up to 30–40% of its weight. Oats are normally de-hulled before use for human consumption and therefore the oat hull is a large by-product of the milling process. Oat hull can be cost-effectively burned to produce, for example, heating energy. However, the main use of oat hull is to incorporate it with animal feed mixtures where it acts like straw. In addition, oat hull can be used as a raw material for furfural or pentoses which can be further used in many industrial applications, such as nylon, synthetic rubber solvents, insecticides, resins, pharmaceuticals, preservatives and special sugars.

The main cell wall component, 2–7% of the total kernel dry weight, in the oat endosperm is a mixed linked (1→3),(1→4)-β-D-glucan. Oat β-glucan

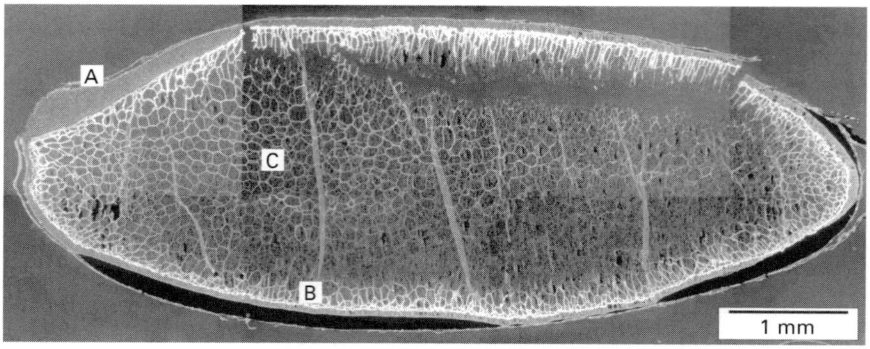

Fig. 10.1 The structure of the oat kernel: a longitudinal cut, stained with calcofluor. A, hull; B, bran; C, endosperm.

has a higher molecular weight than barley or rye β-glucan (Beer *et al.* 1997). In addition to β-glucan, oat cell walls contain pentosans (around 2% of total kernel weight). Nutritionally, β-glucan and pentosans are dietary fibre components that are not degraded by digestive enzymes. Both hemicellulose polymers are easily extractable and form viscous shear thinning solutions (Welch 1995, Colleoni-Sirghie *et al.* 2004).

Starch is the main storage compound in oat grains constituting 60% of the dry weight (Wood *et al.* 1991). The starch molecules amylopectin and amylose are arranged in compound granules within the endosperm cells (Bechtel and Pomeranz 1981). Oat starch granules are relatively small, around 3–5 μm in diameter. Around one-third of the oat starch is amylose (Zhou *et al.* 1998).

Oats contain 15–20% protein (Robbins *et al.* 1971, McMullen 1991). The nutritional value of oat proteins is higher than that of other cereal proteins due to the lower content of prolamine protein (Peterson 1976, Hamad and Fields 1979). Compared with other cereals oats are known to contain high amounts of lipids. Many technological properties of oats – such as their pasting properties and the behaviour during milling, extrusion and baking – can be attributed to the high lipid content. The lipid content of oats is normally 5–9% (Sahasrabudhe 1979, Youngs 1986, Saastamoinen *et al.* 1989). In oats, lipids are located throughout the kernel, while in other cereals they are concentrated mainly in the embryo (Lockhart and Hurt 1986). Most of the oat fatty acids are unsaturated. The oleic acid portion of oat lipids is exceptionally high among cereal lipids (Youngs 1986, Saastamoinen *et al.* 1989). The three most abundant fatty acids – palmitic (C16:0), oleic (C18:1) and linoleic (C18:2) acids – account for 90–95% of the total fatty acids in oats. Furthermore, oats are exceptional among cereals in that native quiescent grains already possess a significant lipid hydrolysing (lipase) activity (Urquhart *et al.* 1983, 1984, Ekstrand *et al.* 1993, Liukkonen *et al.* 1993). Even so, lipids of dry, intact oat grains can stay stable for years when stored under proper conditions (Welch 1977). The general composition of whole grain oat flour and the oat bran fraction is presented in Table 10.1.

10.3 The range of oat food products

In the following section, oat products are divided into traditional and novel oat products. Both of these categories include ingredients (Table 10.2) and consumer products (Table 10.3).

10.3.1 Oat ingredients

The majority of whole grain oats that are intended for human consumption are processed into oat flakes by roller mills or into cut grains by cutting the kernel into three to four pieces using steel cutters. Compared with oat flour,

Table 10.1 The general composition of whole grain oat flour and oat bran

	Whole grain oat flour (%)	Oat bran (%)
Protein	15–17%	15–18%
Starch and sugars	59–70%	10–50%
Fat	4–9%	5–10%
Total dietary fibre	5–13%	10–40%
β-Glucan	2–6%	5–20%

Table 10.2 Oat ingredients

Ingredient category	Ingredient description
Whole grains	
Oat flakes	Flaked (kilned/steamed) oat grains
Malted grains	Malted oat
Oat flour	Milled oat grains
Oat fractions	
Oat bran	Outer layers of the kernel including the aleurone layer
Endosperm flour	Ground starchy endosperm
Hull fractions	Cellulose fibre from hulls
Oat bran concentrate	Enriched oat fibre • 10–20% β-glucan in milled products • up to 90% β-glucan in concentrates from wet processes
Oat extracts	Oat malt extract, syrup or dried

the handling and further processing of these flakes or cut grains is easier, as oat flour tends to form lumps. Oat flakes can be used as such for baking processes as they will disintegrate readily once mixed with water. For applications in which the disintegration is unwanted, the flakes are available with higher thickness. Cut grains will retain part of their structure throughout the baking process and can thus provide a grainy appearance in the final product.

The health effects associated with oats relate mainly to the total dietary fibre and β-glucan contents of oat products. However, the antioxidants are also concentrated mainly in the outer layers of the oat kernel, and thus the total dietary fibre, soluble β-glucan and antioxidants can be enriched in the bran products (Peterson 2001). Bran is prepared from the whole grain oat flour by removing the starchy endosperm using sieving or air classification. The amount of starchy endosperm remaining in the bran varies from product to product. Regular oat bran contains typically 6–8% β-glucan, whereas novel oat bran concentrates from dry milling processes can contain up to 22% β-glucan. By wet extraction processes very pure β-glucan extracts (β-glucan content even over 90%) can be produced. However, in many cases these fractions are too expensive for food uses. In addition to β-glucan, oat bran also contains other dietary fibre components, pentosans and insoluble

Table 10.3 Consumer products containing oats

Consumer product type	Typical oat material used for the product	Remarks
Bread	All oat products appropriate	Novel bread making technology is needed to produce oat bread without the addition of wheat gluten
Porridge	Oat flakes, oat bran	May be combined with other non-gluten flours
Pasta	Oat flour, oat bran	Most oat pasta products contain large proportions of wheat
Snacks	Extruded products	Oat provides flavour and enhances nutritional status
Drinks	Oat flakes	(a) Milk-type oat extracts that are free of particulate material; (b) yoghurt-type colloidal products containing particulate material
Muesli	Oat flakes, extruded oat products	
Breakfast cereals	Extruded oat products	Rice or maize starch is usually added to extruded oat products to improve the structure
Confectionery	Oat extract	
Biscuits	Oat flakes	
Processed food	Oat flakes, oat bran concentrates	Oat products can be used to replace wheat-based water binding agents, or fat

fibre, so that the total dietary fibre content in oat bran products is 17–35%; this is considerably higher than in oat flakes which typically contain 10–12% dietary fibre. Oat bran products are available in different forms including native bran flour and extruded products. The extrusion process is applied to modify the flavour by introducing a slightly roasted flavour and to pre-gelatinize the oat starch remaining in the bran.

Oat bran concentrates are suitable for a wide range of food products, because as a neutral and non-ionic polymer its viscosity is not affected by pH (Dawkins and Nnanna 1995). NaCl at concentrations used in food (1–3%) decreased the viscosity of oat (Dawkins and Nnanna 1995) and barley β-glucan (Gómez *et al.* 1997). β-Glucan preparations with lower molecular weights and of lower viscosity are easier to add into a food product in amounts sufficient for physiological functionality than high molecular weight preparations that give higher viscosities.

The ability of β-glucans to form viscous solutions/high viscosity in solutions makes them potential alternatives as thickening agents in different food applications, e.g. ice creams, sauces and salad dressings (Wood 1986). Compared with other thickening agents, oat gum was less viscous than xanthan and guar at 0.5%, but slightly more viscous than locust bean gum at the same concentration (Dawkins and Nnanna 1995). Oat flakes and steel cut oats are also available as commercial premixes with other baking ingredients. These premixes are aimed at simplifying logistics and making the launch of new oat products easier.

In addition to β-glucan-enriched bran concentrates, other functional fractions can also be produced from oats. Wu and Stringfellow (1995) reported a single process to produce a β-glucan fraction and a protein-rich fraction from hexane-extracted, low-fat oat. Forssell *et al.* 1992 separated polar lipids from oat oil. The polar lipid fraction of oats was found to be a more powerful antioxidant than similar fractions from soybean or turnip rape.

10.3.2 Consumer products containing oats

Porridge is a very traditional way to enjoy the pleasant, mild taste of oats. A new type of porridge mixture, oat flake blends with berries and fruit, has been introduced e.g. in Scandinavia over the last 5–7 years and this has been successful in the marketplace. By adjusting the flake thickness and by pregelatination of flakes, the cooking time of the product can be adjusted. In addition, instant types of the porridge are available.

In bread making processes, oats behave in a distinctly different way to wheat. Oat proteins do not form similar gluten networks to wheat and thus the wheat bread making process cannot be used as such for oats. However, around 10% of wheat flour can easily be replaced with oat flakes. As the oat content is increased, the gas retention of the dough deteriorates due to the weakening of the gluten network. This results in reduced loaf volume (Zhang *et al.* 1998). Despite these difficulties, oat flakes are often added to bread (up to 20% of total flavour content) to adjust the flavour and to increase the dietary fibre content. However, also 100% oat bread is possible to produce with modified baking processes.

The fate of oat β-glucan has been intensively studied and it has been shown that many baking ingredients have endogenous β-glucanase activity that decreases the molecular weight of β-glucan and thereby reduces its visocity (Flander *et al.* 2002, Åman *et al.* 2004). This may potentially decrease the cholesterol-lowering properties of β-glucan, as the high viscosity of intact β-glucan molecules is suggested to boost the cholesterol-lowering effect. Anttila *et al.* 2004 have developed an *in vitro* method to predict the gastric viscosity of oat-containing foods.

In addition to porridge and bread, oats are also used in various forms as an ingredient in pasta, biscuits, snack bars and beverages. In these products oats are often added to introduce the cholesterol-lowering effect, to affect the absorption of glucose, to introduce technological functionality such as water binding or to provide oat flavour. Many new products still remain to be developed to increase the food uses of oats. Even so, active research and development has already created new liquid or high-moisture, oat-based foods: oat milk (Lindahl *et al.* 1997, Önning *et al.* 1998, Chronakis *et al.*

2004), oat ice cream, oat pancake mix and meal replacement drinks (Mikola 2004). The studies of Lyly *et al.* (2004) showed that it is technologically feasible to use oat β-glucans as thickening agents in soups. One promising group of oat products is fermented foods such as oat 'yoghurt' (fermented non-dairy oat pudding) (Salovaara and Kurka 1993, Salovaara 1997). In addition to cholesterol-lowering effects, fermented oat-based products have been reported to stimulate bifidobacteria flora in humans (Mårtensson *et al.* 2005).

Oat fractions have also been introduced into milk-based products. Oat β-glucan fractions have especially been used in low-fat milk products such as low-fat ice creams, yoghurts and cheeses (Brennan *et al.* 2002). The β-glucan fractions seem to have beneficial effects on the structure and mouthfeel of these products. Bekers *et al.* 2001 reported a new-type fermented oat product where fat-free milk was added to oat mash to increase the protein content and β-glucan stability in the mixture. The final product was rich in β-glucan, contained living *Lactobacillus* and *Bifidobacteria* strains and was low in calories. In addition, oat bran concentrate has been used to replace fat in meatballs (Yilmaz and Dağlioğlu 2003). The sensory properties of bran-enriched meatballs did not differ from control samples and they had high acceptability.

10.4 Bioactive compounds and health benefits of oat and oat products

10.4.1 Health benefits related to β-glucans

Most of the health benefits of oats are related to the soluble fibre fraction, the β-glucans. Health claims connected to β-glucan-containing foods are allowed in the USA (FDA 2004), Sweden (Anonymous 2004), Finland (Anonymous 2000) and the UK (Joint Health Claims Initiative 2004). The health effects of oat β-glucan, including the cholesterol-lowering capacity and the ability to balance post-prandial blood glucose and insulin response, have been intensively studied since the 1970s and summaries such as the meta-analysis of Ripsin *et al.* (1992) and Brown *et al.* (1999) are available. The cholesterol-lowering effect of β-glucan is generally accepted and is documented in several separate studies (Anderson *et al.* 1990, 1991, Kestin *et al.* 1990, Davidson *et al.* 1991, Keenan *et al.* 1991, Whyte *et al.* 1992, Braaten *et al.* 1994a, Önning *et al.* 1999, Amundsen *et al.* 2003, Kerckhoffs *et al.* 2003). Furthermore, the positive effects of β-glucan on post-prandial blood glucose has been reported in several studies (Wood *et al.* 1990, Braaten *et al.* 1991, 1994b, Tappy *et al.* 1996, Tapola *et al.* 2005). However, the positive health effects of β-glucans seem to strongly depend on several features – molecular weight, solubility, concentration and viscosity (Wood 2002) – as well as the food matrix carrying the β-glucan. This is supported by several studies showing no significant cholesterol-lowering effects of oat β-glucan (Leadbetter *et al.* 1991, Törrönen *et al.* 1992, Uusitupa *et al.* 1992, Beer *et al.* 1995, Lovegrove *et al.* 2000, Robitaille *et al.* 2005) or no effects on blood glucose (Wood *et al.* 1994a). These have created confusion and inspired the still on-going debate on the factors behind the effectiveness of oat β-glucan.

Maier *et al.* (2000) compared five different ways of providing extra fibre in the diet and found that an oat bran muffin *increased* low-density lipoprotein (LDL) cholesterol levels while the other four intervention foods (amaranth muffin, oat bran, oat bran flakes and a variety of oat bran products in the diet) *reduced* LDL cholesterol as expected. Similarly, a study comparing bread, cookies and a drink as carrier products showed that orange juice with β-glucan lowered LDL cholesterol while bread and cookies did not (Kerckhoffs *et al.* 2003). The baking process of bread decreased the molecular weight of β-glucan but this was not a clear reason for the lack of effect and, in particular, did not explain the ineffectiveness of cookies that contained a relatively high molecular weight β-glucan. Furthermore, it has recently been reported by Naumann *et al.* (2006) that low molecular weight β-glucans (80 kDa) can have a significant lowering effect on both total and LDL serum cholesterol when incorporated into a fruit drink.

Previously, the health claims and regulations concerning β-glucan-containing foods have been based only on the amount of β-glucan in the product. The properties of the β-glucan molecule (such as molecular weight) or the effect of processing on its physiological efficacy have not been taken into account in the labelling rules for foods containing β-glucan. However, currently, in the Swedish health claim Code of Practice for the food sector (Anonymous 2004) this is taken into consideration by requiring a substantiation of the cholesterol-lowering effect after processing of raw materials rich in β-glucan.

In conclusion, the clinical data on the health benefits of whole grain cereals and cereal fibres are indisputable. In addition to cholesterol-lowering effects, several other health effects have been related to cereal fibres and dietary fibres in general, such as reduction of bowel transit time (Feldheim and Wisker 2000), reduction in risk of colorectal cancer (Hill 1997), production of short chain fatty acids (Karppinen *et al.* 2000, Velasquez *et al.* 2000, Wisker *et al.* 2000) and promotion of the growth of beneficial gut microflora (prebiotic effect) (Crittenden *et al.* 2002, Tungland *et al.* 2003). However, the clinical data on the physiological effects of β-glucans are still somewhat contradictory and some open questions related to the health effects of β-glucans are still waiting for answers. More information is needed to understand the correlation between the amount of β-glucans, the molecular weight of β-glucans, the extractability of β-glucans, the ability of β-glucan to form viscous gels, the influence of food matrix on β-glucans and finally the physiological effects of β-glucans in humans. Precise information on these technological and physiological correlations will help the food industry to develop oat products with tailored health benefits.

10.4.2 Other bioactive compounds and their possible health benefits

Even though β-glucan is the most well-known health-promoting compound in oats, the nutritionally attractive fatty acids and sterols in lipids, and other minor bioactive compounds such as phenolics can enhance the healthy image of oats. Phenolic compounds and tocols are mainly regarded as being responsible for oat antioxidant activity (Hammond 1983, Forssell *et al.* 1990). Oat sterols have also been reported to have antioxidative properties (White and Armstrong 1986, Kahlon 1989). Forssell *et al.* (1992) reported that oat lecithin is a better antioxidant than soy or turnip rape lecithin. Emmons *et al.* (1999) found that the total phenolic content of oats was significantly correlated with antioxidant activity.

Oats contain mainly lipid-soluble tocols (vitamin E) and water-soluble vitamin B. The main tocols in oats are α-tocopherols and α-tocotrienols (Hammond 1983). Oats and oat products contain high amounts of thiamin and pantothenic acid (Lockhart and Hurt 1986) and are especially rich in biotin compared with other cereals (Lásztity 1998).

Phenolic compounds act as a part of the defence mechanisms of plants, protecting plants against pathogens, pests and other stress conditions (drought, ultraviolet light, etc.). In general, they have antimicrobial, antioxidative and anticarcinogenic properties (Pratt 1992). Oats contain quinones of benzoic acid, cinnamic acids, flavones, flavonols, flavanones, anthocyanidins, aminophenols and precursors of these compounds (Collins 1986). In total, native oat flour contains 87 mg/kg phenolic acids, the main component being ferulic acid (Sosulski *et al.* 1982).

Avenanthramides are special oat phenolics, which are regarded as central compounds of the defence mechanism (Collins 1989, Dimberg *et al.* 1993). Avenanthramides have been shown to have strong antioxidant activities (Peterson *et al.* 2002) and avenanthramides occur as constitutive components of oat grains (Mattila *et al.* 2005). In oat leaves, avenanthramides are not detected until the leaves are inoculated with an incompatible race of the oat crown rust pathogen, *Puccinia coronata* f. sp. *avenae* (Miyagawa *et al.* 1995) or other fungal elicitor, such as victorin. Germination increased the amount of avenanthramides significantly (Kaukovirta-Norja *et al.* 2001).

10.4.3 The suitability of oats for people with coeliac disease

Coeliac disease is an intolerance to certain protein fractions of cereals. Individuals with coeliac disease have a sensitivity to prolamin fractions of wheat (gliadin), rye (secalin) and barley (hordein). Previously, oats were also included in the list of cereals causing symptoms in coeliac patients. However, over the last 10 years several independent studies have shown that oats do not cause any harmful effects in coeliac patients (Picarelli *et al.* 2001, Janatuinen *et al.* 2002, Størsrud *et al.* 2003a,b, Högberg *et al.* 2004). These studies have shown that oats are suitable for both adult and child patients. Only in one study did oats seem to be risky for coeliac patients

(Lundin *et al.* 2003); in this study 1 of the 19 patients had dramatic symptoms from the oat-containing diet. The debate on the safety of oats for coeliac patients is still ongoing worldwide.

10.5 Challenges in the development of functional oat products

Even though oat ingredients can be used in a variety of applications without any technological problems, there are certain factors that should be taken into account when choosing the right oat product and processing conditions for different applications. Firstly, the solubility and rheological behaviour of oat β-glucans varies from one ingredient and one use to another. Whereas fast solubilization of β-glucan may be considered beneficial for the physiological responses, in certain applications it can be considered as an unwanted property leading to a structure that is too viscous. Secondly, overall stability and sufficient shelf-life are important success criteria for all oat products and should therefore always be addressed. Thirdly, in the applications in which the health effect is related to the high molecular weight of β-glucan, special attention must be paid to avoid any processing steps that would lead to depolymerization and the loss of health-promoting properties. High molecular weight β-glucan is particularly sensitive to processing. Freezing was not found to affect the molecular weight of β-glucan (Beer *et al.* 1997, Suortti *et al.* 2000, Kerckhoffs *et al.* 2003), but it decreased its solubility (Beer *et al.* 1997). On the other hand, heating made β-glucan more soluble (Bhatty 1992, Jaskari *et al.* 1995) and enhanced its physiological efficacy.

Furthermore, in order to avoid unwanted depolymerization of β-glucan during processing, other ingredients should be checked carefully for the presence of β-glucanase activity. Otherwise, molecular weight- and viscosity-dependent health benefits may be lost. For example, a molecular weight degradation has been observed during juice and fresh pasta processing, and during bread making (Åman *et al.* 2004, Andersson *et al.* 2004).

The kinetics of water binding, solubilization of β-glucan and the formation of viscosity are important properties of oat ingredients. To a certain extent, the producer of the oat ingredients can control these properties without jeopardizing the health-promoting properties by adjusting the particle size and the degree of starch damage and by introducing tailored crystal and amorphous structures with the desired properties.

The high lipid content and high lipolytic enzyme activity of oats can cause serious deterioration that, during processing and storage, can affect the sensory properties of oat products, and even lead to a loss of nutritive value. Therefore, special attention should be paid to ensure sufficient shelf-life. Oat lipids deteriorate by two distinctly different mechanisms: enzymatic lipid hydrolysis and non-enzymatic lipid oxidation. As the majority of oat products that are aimed for human consumption are heat denaturated to inactivate enzymes, the latter mechanism is likely to predominate in commercial oat products. During storage of oat products the concentration of volatile oxidation products can increase by a factor of 2–100, resulting in paint- or cardboard-like odours associated with the rancidity products (Sjövall *et al.* 1997, Heiniö *et al.* 2002, Lehtinen *et al.* 2003). The oxidation can be minimized by storing the product at low humidity and by avoiding contact with oxygen and light. The endogenous antioxidants in oats are not only important due to the possible health effects but also for technological reasons as they can improve the storage stability of oat products. According to Dimberg *et al.* (1996), and Molteberg *et al.* (1996) avenanthramides and caffeic acid were significantly correlated with low rancidity, a lack of bitterness, and freshness, whereas the opposite was found for most of the simple phenolic compounds.

10.6 Future trends

Oats are well known among consumers and are perceived as a tasty cereal. The approved health claim status of oats is important and is associated with one of the cutting edge properties of oats over other cereals, i.e. its soluble fibre, β-glucan. Barley is the only other cereal to contain as much or even more β-glucans than oats. Clinical data suggest that the physiological effects of barley β-glucans are very similar to that of oat β-glucans. However, some differences can be found in their β-glucan structure (Wood *et al.* 1994b, Cui *et al.* 2000, Tosh *et al.* 2004) and oats are known to contain more soluble β-glucan than barley. Furthermore, oat β-glucans generally have a higher molecular weight than barley β-glucans (Wood 2001); this can have an influence on β-glucan extractability and availability, and therefore also on physiological effects.

Lyly *et al.* (2007) studied the factors affecting consumers' willingness to use beverages and ready-to-eat frozen soups containing oat β-glucan. The consumer studies were carried out in Finland, France and Sweden. The presence of a health claim (either cholesterol-lowering or glucose-retarding) gave a significant but small added value to beverages and soups. However, the desirable taste of the samples was the most predominant factor affecting the willingness to use them. In general, taste and pleasantness have been shown to be the most important factors affecting food choice of new foods (Arvola *et al.* 1999, Urala and Lähteenmäki 2003, Urala 2005). Oat flavour is highly accepted by customers; thus oat ingredients can be considered to be ideal for delivering the health-promoting properties in a variety of consumer products.

Oats have great potential as a health-promoting raw material in several types of foods. Convenience is an important consumer trend and the snacks market has been an increasing food sector for some years. Results of trials indicate that snack beverages are a beneficial form in which to deliver the health-promoting properties of oats to consumers (Salovaara, 1997, Mikola 2004, Naumann *et al.* 2006). A modified malting process can be used to produce oat ingredients containing a pleasant, nut-like aroma for dry oat snacks (Heiniö *et al.* 2001, 2002, Wilhelmson *et al.* 2001).

When estimating the future potential for oats and oat products, one very promising features is that oats are suitable for people suffering from coeliac disease. This makes oats unique among the common cereals used, for example, for baking. Oats have already proven to be useful in the diets of coeliac patients in several countries.

10.7 References

Åman P, Rimsten L and Andersson R (2004), 'Molecular weight distribution of β-glucan in oat-based foods', *Cereal Chem*, **81**, 356–360.

Amundsen Å L, Haugum B and Andersson H (2003), 'Changes in serum cholesterol and sterol metabolites after intake of products enriched with an oat bran concentrate within a controlled diet', *Scand J Nutr*, **47**, 68–74.

Anderson J W, Spencer D B, Hamilton C C, Smith S F, Tietyen J, Bryant C A and Oeltgen P (1990), 'Oat-bran cereal lowers serum total and LDL cholesterol in hypercholesterolemic men', *Am J Clin Nutr*, **52**, 495–499.

Anderson J W, Gilinsky N H, Deakins D A, Smith S F, O'Neal D S, Dillon D W and Oeltgen P R (1991), 'Lipid responses of hypercholesterolemic men to oat bran and wheat bran intake', *Am J Clin Nutr*, **54**, 678–683.

Andersson A, Armoe E, Fredriksson H, Andersson R and Aman P (2004), 'Molecular weight and structure units of (1→3,1→4)-β-glucans in dough and bread made from hull-less barley milling fractions', *J Cereal Sci*, **40**(3), 195–204.

Anonymous (2000), *Valvonta 8/2000.* A handbook to control health claims related to blood cholesterol in food package markings and in other marketing. (Kolesteroliväitteiden valvontaopas. Elintarvikkeiden pakkausmerkinnöissä ja muussa markkinoinnissa käytettäviä veren kolesteroliin viittaavia väitteitä koskeva opas.) Helsinki, National Food Agency Finland [In Finnish].

Anonymous (2004), *Health Claims in the Labelling and Marketing of Food Products. The Food Sector's Code of Practice, Appendix 2*, Lund, SNF Swedish Nutrition Foundation (http://www.hp-info.nu/SweCode_2004_1.pdf; accessed 12.5.2005).

Anttila H, Wood P J, Brummer Y, Sontag-Strohm T, Tosh S and Wang Q (2004), 'The effect of processing on the molecular weight and extractability of oat β-glucan using a method to simulate physiological digestion', in AACC/TIA Joint Annual Meeting 2004, Abstract published online at www.aaccnet.org/meetings/2004/abstracts, St Paul, Minnesota, American Association of Cereal Chemists.

Arvola A, Lähteenmäki L and Tuorila H (1999), 'Predicting the intent to purchase unfamiliar and familiar cheeses: the effects of attitudes, expected liking and food neophobia', *Appetite*, **32**, 113–126.

Bechtel D B and Pomeranz Y (1981), 'Ultrastructure and cytochemistry of mature oat (*Avena sativa* L.) endosperm. The aleurone layer and starchy endosperm', *Cereal Chem*, **58**, 61–69.

Beer M U, Arrigoni E and Amando R (1995), 'Effects of oat gum on blood cholesterol levels in healthy young men', *Eur J Clin Nutr*, **49**, 517–522.

Beer M U, Wood P J, Weisz J and Fillion N (1997), 'Effect of cooking and storage on the amount and molecular weight of (1-3)(1-4)-β-D-glucan extracted from oat products by an in vitro digestion system', *Cereal Chem*, **74**, 705–709.

Bekers M, Marauska M, Laukevics J, Grube M, Vigants A, Karklina D, Skudra L and Viesturs U (2001), 'Oat and fat-free milk based functional product', *Food Biotechnol*, **15**, 1–12.

Bhatty R S, (1992), 'Total and extractable β-glucan contents of oats and their relationship to viscosity', *J Cereal Sci*, **15**, 185–192.

Braaten J T, Wood P J, Scott F W, Riedel K D, Poste L M and Collins M W (1991), 'Oat gum lowers glucose and insulin after an oral glucose load', *Am J Clin Nutr*, **53**, 1425–1430.

Braaten J T, Wood P J, Scott F W, Wolynetz M S, Lowe M K, Bradley-White P and Collins M W (1994a), 'Oat beta-glucan reduces blood cholesterol concentration in hypercholesterolemic subjects', *Eur J Clin Nutr*, **48**, 465–474.

Braaten J T, Scott F W, Wood P J, Riedel K D, Wolynetz M S, Brulé D and Collins M W (1994b), 'High β-glucan oat bran and oat gum reduce postprandial blood glucose and insulin in subjects with and without type 2 diabetes', *Diab Med*, **11**, 312–318.

Brennan C S, Tudorica C M and Kuri V (2002), 'Soluble and insoluble dietary fibres (non-starch polysaccharides) and their effects on food structure and nutrition', *Food Ind J*, **5**, 261–272.

Brown L, Rosner B, Willett W W and Sacks F M (1999), 'Cholesterol-lowering effects of dietary fiber: a meta-analysis', *Am J Clin Nutr*, **69**, 30–42.

Chronakis I S, Öste Triantafyllou A and Öste R (2004), 'Solid-state characteristics and redispersible properties of powders formed by spray-drying and freeze-drying cereal dispersions of varying (1→3,1→4)-β-glucan content', *J Cereal Sci*, **40**, 183–193.

Colleoni-Sirghie M, Jannink J-L, Kovalenko I V, Briggs J L and White P J (2004), 'Prediction of β-glucan concentration based on viscosity evaluations of raw oat flours from high β-glucan and traditional oat lines', *Cereal Chem*, **81**(4), 434–443.

Collins F W (1986), 'Oat phenolics: structure, occurrence and function', in Webster F H (Ed.), *Oats: Chemistry and Technology*, St Paul, Minnesota, American Association of Cereal Chemists, pp. 227–295.

Collins F W (1989), 'Oat phenolics: avenanthramides, novel substituted *N*-cinnamoyl-anthranilate alkaloids from oat groats and hulls', *J Agric Food Chem*, **37**, 60–66.

Crittenden R, Karppinen S, Ojanen S, Tenkanen M, Fagerström R, Mättö J, Saarela M, Mattila-Sandholm T and Poutanen K (2002), 'In vitro fermentation of cereal dietary carbohydrates by probiotic and intestinal bacteria', *J Sci Food Agric*, **82**, 781–789.

Cui W, Wood P J, Blackwell B and Nikiforuk J (2000), 'Physicochemical properties and structural characterization by two dimensional NMR spectroscopy of wheat β-D-glucan: comparison with other cereal β-D-glucans', *Carbohydr Polym*, **41**, 249–258.

Davidson M H, Dugan L D, Burns J H, Bova J, Story K and Drennan K B (1991), 'The hypocholesterolemic effects of b-glucan in oatmeal and oat bran. A dose-controlled study', *JAMA*, **266**, 1833–1839.

Dawkins N L and Nnanna I A (1995), 'Studies on oat gum [(1→3,1→4)-β-D-glucan]: composition, molecular weight estimation and rheological properties', *Food Hydrocolloids*, **9**, 1–7.

Dimberg L, Theander O and Lingnert H (1993), 'Avenanthramides – A group of phenolic antioxidants in oats', *Cereal Chem*, **70**, 637–641.

Dimberg L H, Molteberg E L, Solheim R and Frølich W (1996), 'Variation in oat groats due to variety, storage and heat treatment. I: Phenolic compounds', *J Cereal Sci*, **24**, 262–272.

Ekstrand B, Gangby I, Akesson G, Stollman U, Lingnert H and Dahl S (1993), 'Lipase activity and development of rancidity in oats and oat products related to heat-treatment during processing', *J Cereal Sci*, **17**, 247–254.

Emmons C L, Peterson D M and Paul G L (1999), 'Antioxidants capacity of oat (*Avena sativa* L.) extracts. 2. *In vitro* antioxidant activity and content of phenolic and tocol antioxidants', *J Agric Food Chem*, **47**, 4894–4898.

FDA (Food and Drug Administration) (2004), '*Code of Federal Regulations,* Title 21, Volume 2, Part 101, Section 101.81, *Health Claims: Soluble Fiber from Certain Foods and Risk of Coronary Heart Disease (CHD)*', Department of Health and Human Services. Food Labeling http://a257.g.akamaitech.net/7/257/2422/12feb20041500/edocket.access.gpo.gov/cfr_2004/aprqtr/pdf/21cfr101.81.pdf; accessed 12.5.2005).

Feldheim W and Wisker E (2000), 'Studies on the improvement of dietary fibre intake', *Deutsche Lebensm-Rundsch*, **96**, 327–330.

Flander L, Salmenkallio-Marttila M, Suortti T and Autio K (2002), 'Baking a functional oat bread', in AACC Annual Meeting, Montreal, Canada, 13–17 October 2002, St Paul, Minnesota, American Association of Cereal Chemists, pp. 138–139.

Forssell P, Cetin M, Wirtanen G and Mälkki Y (1990), 'Antioxidative effects of oat oil and its fractions', *Fat Sci Technol*, **8**, 319–321.

Forssell P, Kervinen R, Alkio M and Poutanen K (1992), 'Comparison of methods for separating polar lipids from oat oil', *Fat Sci Technol*, **9**, 355–358.

Gómez C, Navarro A, Garnier C, Horta A and Carbonell J V (1997), 'Physical and structural properties of barley (1→3),(1→4)-β-D-glucan – III. Formation of aggregates analyzed through its viscoelastic and flow behaviour', *Carbohydr Polym*, **34**, 141–148.

Hamad A M and Fields M L (1979), 'Evaluation of the potential and available lysine of germinated and fermented cereals', *J Food Sci*, **44**, 456–459.

Hammond E G (1983), 'Oat lipids', in Barnes P J (Ed.), *Lipids in Cereal Technology*, London, Academic Press Inc., pp. 331–352.

Heiniö R-L, Oksman-Caldentey K-M, Latva-Kala K, Lehtinen P and Poutanen K (2001), 'Effect of drying treatment conditions on sensory profile of germinated oat', *Cereal Chem*, **78**, 707–714.

Heiniö R-L, Lehtinen P, Oksman-Caldentey K-M and Poutanen K (2002), 'Differences between sensory profiles and development of rancidity during long-term storage of native and processed oat', *Cereal Chem*, **79**, 367–375.

Hill M J (1997), 'Cereals, cereal fibre and colorectal cancer risk: a review of the epidemiological literature', *Eur J Cancer Prev*, **6**, 219–225.

Högberg L, Laurin P, Fälth-Magnusson K, Grant C, Grodzinsky E, Jansson G, Ascher H, Browaldh L, Hammersjö J-Å, Lindberg E, Myrdal U and Stenhammar L (2004), 'Oats to children with newly diagnosed celiac disease: a randomised double blind study', *Gut*, **53**, 649–654.

Janatuinen E, Kemppainen T, Julkunen R, Kosma V-M, Mäki M, Heikkinen M and Uusitupa M (2002), 'No harm from five year ingestion of oats in coeliac disease', *Gut*, **50**, 332–335.

Jaskari J, Henriksson K, Nieminen A, Suortti T, Salovaara H and Poutanen K (1995), 'Effect of hydrothermal and enzymic treatments on the viscous behaviour of dry- and wet-milled oat brans', *Cereal Chem*, **72**, 625–631.

Joint Health Claims Initiative (2004), Generic health claim for oats and blood cholesterol, 06/05/04 (www.jhci.org.uk/approv/oats.htm; accessed 20.12.2005).

Kahlon T S (1989), 'Nutritional implications and uses of wheat and oat kernel oil'. *Cereal Foods World*, **34**, 872–875.

Karppinen S, Liukkonen K, Aura A, Forssell P and Poutanen K (2000), 'In vitro fermentation of polysaccharides of rye, wheat and oat brans and inulin by human faecal bacteria', *J Sci Food Agric*, **80**, 1469–1476.

Kaukovirta-Norja A, Oksman-Caldentey K-M, Heiniö R-L, Lehtinen P, Pihlava J-M and Poutanen K (2001), 'Germination increased bioactivity and stability of whole grain oats', in Liukkonen K, Kuokka A and Poutanen K (Eds), VTT Symposium 213, International symposium on *Whole Grain and Human Health,* Haikko Manor, 13–15 June 2001, Espoo, VTT, Technical Research Centre of Finland.

Keenan J M, Wenz J B, Myers S, Ripsin C and Huang Z Q (1991), 'Randomized, controlled, crossover trial of oat bran in hypercholesterolemic subjects', *J Fam Pract*, **33**, 600–608.

Kerckhoffs D A J M, Hornstra G and Mensink R P (2003), 'Cholesterol-lowering effect of β-glucan from oat bran in mildly hypercholesterolemic subjects may decrease when β-glucan is incorporated into bread and cookies', *Am J Clin Nutr*, **78**, 221–227.

Kestin M, Moss R, Clifton P M and Nestel P J (1990), 'Comparative effects of three cereal brans on plasma lipids, blood pressure, and glucose metabolism in mildly hypercholesterolemic men', *Am J Clin Nutr*, **52**, 661–666.

Lásztity R (1998), 'Oat grain – a wonderful reservoir of natural nutrients and biologically active substances', *Food Rev Int*, **14**, 99–119.

Leadbetter J, Ball M J and Mann J I (1991), 'Effects of increasing quantities of oat bran in hypercholesterolemic people', *Am J Clin Nutr*, **54**, 841–845.

Lehtinen P, Kiiliäinen K, Lehtomäki K and Laakso S (2003), 'Effect of heat treatment on lipid stability in processed oats', *J Cereal Sci*, **37**, 215–221.

Lindahl L, Ahldén I, Öste R and Sjöholm I (1997), 'Homogenous and stable cereal suspension and a method of making the same', US Patent 5686123.

Liukkonen K, Kaukovirta-Norja A and Laakso S (1993), 'Elimination of lipid hydrolysis in aqueous suspensions of oat flour', *J Cereal Sci*, **17**, 255–165.

Lockhart H B and Hurt H D (1986), 'Nutrition of oats', in Webster F H (Ed.), *Oats: Chemistry and Technology*, St Paul, Minnesota, American Association of Cereal Chemists, pp. 297–308.

Lovegrove J A, Clohessy A, Milon H and Williams C M (2000), 'Modest doses of β-glucan do not reduce concentrations of potentially atherogenic lipoproteins', *Am J Clin Nutr*, **72**, 49–55.

Lundin K, Nilsen E, Scott H, Løberg E, Gjøen A, Bratlie J, Skar V, Mendez E, Løvik A and Kett K (2003), 'Oats induced villous atrophy in coeliac disease', *Gut*, **52**, 1649–1652.

Lyly M, Salmenkallio-Marttila M, Suortti T, Autio K, Poutanen K and Lähteenmäki L (2004), 'The sensory characteristics and rheological properties of soups containing oat and barley β-glucan before and after freezing', *Lebensm Wiss Technol*, **37**, 749–761.

Lyly M, Roininen K, Honkapää K, Poutanen K and Lähteenmäki L (2007), 'Factors influencing consumers' willingness to use beverages and ready-to-eat frozen soups containing oat β-glucan in Finland, France and Sweden', *Food Qual Prefer*, **18**, 242–255.

Maier S M, Turner N D and Lupton J R (2000), 'Serum lipids in hypercholesterolemic men and women consuming oat bran and amaranth products', *Cereal Chem*, **77**, 297–302.

Mårtensson O, Björklund M, Mbou Malbo A, Dueñas-Chasco M, Irastorza A, Holst O, Norin E, Welling G, Öste R and Önning G (2005), 'Fermented, ropy, oat-based products reduce cholesterol levels and stimulate the bifidobacteria flora in humans', *Nutr Res*, **25**, 429–442.

Mattila P, Pihlava J-M and Hellström J (2005), 'Contents of phenolic acids, alkyl- and alkenylresorcinols, and avenanthramides in commercial grain products', *J Agric Food Chem*, **53**, 8290–8295.

McMullen M S (1991), 'Oats', in Lorenz K J and Kulp K (Eds), *Handbook of Cereal Science and Technology*, New York, Marcel Dekker Inc., pp. 199–232.

Mikola M (2004), 'Natural oat bran instant drink', in Peltonen-Sainio P and Topi-Hulmi M (Eds), Proceedings of the 7th International Oat Conference, Jokioinen, MTT Agrifood Research Finland, p. 99.

Miyagawa H, Ishihara A, Nishimoto T, Ueno T and Mayama S (1995), 'Induction of avenanthramides in oat leaves inoculated with crown rust fungus, *Puccinia coronata* f. sp. *avenae', Biosci Biotechnol Biochem*, **59**, 2305–2306.

Molteberg E L, Magnus E M, Bjorge J M and Nilsson A (1996), 'Sensory and chemical studies of lipid oxidation in raw and heat-treated oat flours', *Cereal Chem*, **73**, 579–587.

Naumann E, van Rees A, Önning G, Öste R, Wydra M and Mensink R (2006), 'β-glucan incorporated into a fruit drink effectively lowers serum LDL cholesterol concentrations', *Am J Clin Nutr*, **83**, 601–605.
Önning G, Åkesson B, Öste R and Lundquist L (1998), 'Effects of consumption of oat milk, soya milk, or cow's milk on plasma lipids and antioxidative capacity in healthy subjects', *Ann Nutr Metab*, **42**, 211–220.
Önning G, Wallmark A, Persson M, Åkesson B, Elmstahl S and Öste R (1999), 'Consumption of oat milk for 5 weeks lowers serum cholesterol and LDL cholesterol in free-living men with moderate hypercholesterolemia', *Ann Nutr Metab*, **43**, 301–309.
Peterson D M (1976), 'Protein concentration, concentration of protein fractions and amino acid balance in oats', *Crop Sci*, **16**, 663–666.
Peterson D M (2001), 'Oat antioxidants', *J Cereal Sci*, **33**, 115–129.
Peterson D, Hahn M and Emmons C (2002), 'Oat avenanthramides exhibit antioxidant activities in vitro', *Food Chem*, **79**, 473–478.
Picarelli A, Di Tola M, Sabbatella L, Gabrielli F, Di Cello T, Anania M, Mastracchio A, Silano, M and De Vincenzi M (2001), 'Immunologic evidence of no harmful effect of oats in celiac disease', *Am J Clin Nutr*, **74**, 137–140.
Pratt D E (1992), 'Natural antioxidants from plant material', in Huang M-T, Ho, C-T and Lee C Y (Eds), *Phenolic Compounds in Food and their Effects on Health. II Antioxidants and Cancer Prevention,* ACS Symposium Series 507, Washington DC, American Chemical Society, p. 54.
Ripsin C M, Keenan J M, Jacobes D R, Elmer P J, Wlech R R, Van Horn L, Liu K, Turnbull W H, Thye F, Kestin M, Hegsted M, Davidson D M, Davidson M H, Dugan D L, Denmark-Wahnefried W and Beling S (1992), 'Oat products and lipid lowering. A meta-analysis', *JAMA*, **267**, 3317–3325.
Robbins G S, Pomeranz Y and Briggle L W (1971), 'Amino acid composition of oat groats', *J Agric Food Chem*, **19**, 536–539.
Robitaille J, Fontaine-Bisson B, Couture P, Tchernol A and Vohl M-C (2005), 'Effect of an oat bran-rich supplement on the metabolic profile of overweight premenopausal women', *Ann Nutr Metab*, **49**, 141–148.
Saastamoinen M, Kumpulainen J and Nummela S (1989), 'Genetic and environmental variation of oil content and fatty acid composition of oats', *Cereal Chem*, **66**, 296–300.
Sahasrabudhe M R (1979) 'Lipid composition of oat (*Avena sativa* L.)', *J Am Oil Chem Soc*, **56**, 80–84.
Salovaara H (1997), 'A future food now: A yogurt-type snack from oat bran', in Poutanen K (Ed.), VTT Symposium 177, *Biotechnology in the Food Chain: New Tools and Applications for Future Foods,* Helsinki, Finland, 28–30 January 1997, Espoo, Technical Research Centre of Finland, p. 231.
Salovaara H and Kurka A-M (1993), 'Food product containing dietary fiber and method of making said product'. European Patent Specification, WO/91/17672.
Sjövall O, Lapveteläinen A, Johansson A and Kallio H (1997), 'Analysis of volatiles formed during oxidation of extruded oats', *J Agric Food Chem*, **45**, 4452–4455.
Sosulski F, Krygier K and Hogge L (1982), 'Free, esterified, and insoluble-bound phenolic acids. 3. Composition of phenolic acids in cereal and potato flours', *J Agric Food Chem*, **30**, 337–340.
Størsrud S, Olsson M, Lenner R, Nilsson L Å and Kilander A (2003a), 'Adult coeliac patients do tolerate large amounts of oats', *Eur J Clin Nutr*, **57**, 163–169.
Størsrud S, Hutlhén L and Lenner R (2003b), 'Beneficial effects of oats in the gluten-free diet of adults with special reference to nutrient status, symptoms and subjective experiences', *Br J Nutr*, **90**, 101–107.
Suortti T, Johansson L and Autio K (2000), 'Effect of heating and freezing on molecular weight of oat β-glucan', in AACC, Annual Meeting 2000, Abstract 332, St Paul, Minnesota, American Association of Cereal Chemists.

Tapola N, Karvonen H, Niskanen L, Mikola M and Sarkkinen E (2005), 'Glycemic responses of oat bran products in type 2 diabetes patients', *Nutr Metab Cardiovasc Dis*, **15**, 255–261.

Tappy L, Gugolz E and Wursch P (1996), 'Effects of breakfast cereals containing various amounts of β-glucan fibers on plasma glucose and insulin responses in NIDDM subjects', *Diabetes Care*, **19**, 831–834.

Törrönen R, Kansanen L, Uusitupa M, Hänninen O, Myllymäki O, Härkönen H and Mälkki Y (1992), 'Effects of an oat bran concentrate on serum lipids in free-living men with mild to moderate hypercholesterolemia', *Eur J Clin Nutr*, **46**, 621–627.

Tosh S, Brummer Y, Wood P J, Wang Q and Weisz J (2004), 'Evaluation of structure in the formation of gels by structurally diverse (1→3) (1→4)-β-D-glucans from four cereal and one lichen species', *Carbohydr Polym*, **57**, 249–259.

Tungland B C (2003), 'Fructooligosaccharides and other fructans: structures, occurrence, production, regulatory aspects, food applications, and nutritional health significance', in Egglestone G and Cote *G L, Oligosaccharides in Food and Agriculture,* ACS Symposium Series 849, Washington DC, American Chemical Society, pp. 135–152.

Urala N (2005), 'Functional Foods in Finland. Consumers' Views, Attitudes and Willingness to Use', VTT Publications 581, Espoo VTT, Technical Research Centre of Finland.

Urala N and Lähteenmäki L (2003), 'Reasons behind consumers' functional food choices', *Nutr Food Sci*, **33**, 148–158.

Urquhart A A, Altosaar I and Matlashewski G J (1983), 'Localization of lipase activity in oat grains and milled oat fractions', *Cereal Chem*, **60**, 181–183.

Urquhart A A, Brumell C A, Altosaar I, Matlashewski G J and Sahasrabudhe M R (1984), 'Lipase activity in oats during grain maturation and germination', *Cereal Chem*, **61**, 105–108.

Uusitupa M I, Ruuskanen E, Mäkinen E, Laitinen J, Toskala E, Kervinen K and Kesäniemi Y A (1992), 'A controlled study on the effect of β-glucan-rich oat bran on serum lipids in hypercholesterolemic subjects: relation to apolipoprotein E phenotype', *J Am Coll Nutr*, **11**, 651–659.

Velasquez M, Davies C, Marret R, Slavin J and Feirtag J (2000), 'Effect of oligosaccharides and fibre substitutes on short chain fatty acid production by human microflora', *Anaerobe*, **6**, 87–92.

Welch R W (1977), 'Fatty acid composition of grain from winter and spring sown oats, barley and wheat', *J Sci Food Agric*, **26**, 269–274.

Welch R W (1995), 'Oats in human nutrition and health', in Welch R W (Ed.), *The Oat Crop*, London, Chapman & Hall, pp. 432–479.

White P J and Armstrong L S (1986), 'Effect of selected oat sterols on the deterioration of heated soybean oil', *J Am Oil Chem Soc*, **63**, 525–529.

Whyte J L, McArthur R, Tomming D and Nestel P (1992), 'Oat bran lowers plasma cholesterol levels in mildly hypercholesterolemic men', *J Am Diet Assoc*, **92**, 446–449.

Wilhelmson A, Oksman-Caldentey K-M, Laitila A, Suortti T, Kaukovirta-Norja A and Poutanen K (2001), 'Development of a germination process for producing high β-glucan, whole grain food ingredients from oat', *Cereal Chem*, **78**, 715–720.

Wisker E, Daniel M, Rave G and Feldeim W (2000), 'Short chain fatty acids produced in vitro from fibre residues obtained from mixed diets containing different breads and in human faeces during ingestion diets', *Br J Nutr*, **84**, 31–37.

Wood P J (1986), 'Oat β-glucan: structure, location and properties', in Webster F H (Ed.), *Oats: Chemistry and Technology*, St Paul, Minnesota, American Association of Cereal Chemists, pp. 121–152

Wood P J (2001), 'Cereal β-glucans: structure, properties and health claims', in McClearly B V and Prosky L (Eds), *Advanced Dietary Fibre Technology,* London, Blackwell Science, pp. 315–327.

Wood P J (2002), 'Relationships between solution properties of cereal β-glucans and physiological effects – a review', *Trends Food Sci Technol*, **13**, 313–320.

Wood P J, Braaten J T, Scott F W, Riedel D and Poste L M (1990), 'Comparisons of viscous properties of oat and guar gum and the effects of these and oat bran on glycemic index', *J Agric Food Chem*, **38**, 753–757.

Wood P J, Weisz J and Mahn W (1991), 'Molecular characterization of cereal β-glucans. II. Size-exclusion chromatography for comparison of molecular weight. *Cereal Chem*, **68**, 530–536.

Wood P J, Braaten J T, Scott F W, Riedel K D, Wolynetz M S and Collins M W (1994a), 'Effect of dose and modification of viscous properties of oat gum on plasma glucose and insulin following an oral glucose load', *Br J Nutr*, **72**, 731–743.

Wood P J, Weisz J and Blackwell B A (1994b), 'Structural studies of (1→3) (1→4)-β-D-glucans by ^{13}C-NMR by rapid analysis of cellulose-like regions using high-performance anion-exchange chromatography of oligosaccharides released by lichenase', *Cereal Chem*, **71**, 301–307.

Wu Y V and Stringfellow A (1995), 'Enriched protein- and β-glucan fractions from high-protein oats by air classification', *Cereal Chem*, **72**, 132–134.

Yilmaz İ and Dağlioğlu O (2003). 'The effect of replacing fat with oat bran on fatty acid composition and physicochemical properties of meatballs', *Meat Sci*, **65**, 819–823.

Youngs V L (1986), 'Oat lipids and lipid-related enzymes', in Webster F H (Ed.), *Oats: Chemistry and Technology*, St Paul, Minnesota, American Association of Cereal Chemists, pp. 205–226.

Zhang D, Moore W R and Doehlert D C (1998), 'Effects of oat grain hydrothermal treatments on wheat-oat flour dough properties and breadbaking quality', *Cereal Chem*, **75**, 602–605.

Zhou M, Robards K, Glennie-Holmes M and Helliwell S (1998), 'Structure and pasting properties of oat starch', *Cereal Chem*, **75**, 273–281.

11

Rye bread and other rye products

A. Kamal-Eldin and P. Åman, Swedish University of Agricultural Sciences (SLU), Sweden; J.-X. Zhang, University of Umeå, Sweden; K.-E. Bach Knudsen, University of Aarhus, Denmark and K. Poutanen, VTT and University of Kuopio, Finland

11.1 Introduction

Among the cereal grains, rye (*Secale cereale*, L.) is quite similar to wheat despite important differences in its chemical composition and technological properties. Both grains have a husk that is normally removed by threshing. The outermost layers of the naked kernel include the pericarp and the testa, which surround the germ and endosperm (aleurone and starchy endosperm) (Shewry and Bechtel, 2001). Rye is mainly grown and consumed as food in northern Europe; the major producers being Russia, Poland, Germany, Belarus and Ukraine (FAOSTAT, 2006). The annual consumption of rye as food varies greatly between countries with >35 kg/capita in Poland, Belarus, and Estonia and 10–15 kg/capita in Finland, Denmark, Sweden and Germany compared with the average world consumption of 1 kg/capita. This makes rye an important source of dietary fiber (DF) in these countries, e.g. almost 40% of DF intake comes from rye foods in Finland and Denmark. The main rye foods include dark and sour breads, loaf bread with sifted rye flour, crisp bread and flakes for porridges and breakfast cereals. Currently, there is also a growing interest in including rye in convenience and snack foods.

Epidemiological studies suggest that consumption of whole grain cereals is associated with reduced incidences of chronic diseases, e.g. diabetes, cardiovascular disease, and certain cancers (Jacobs *et al.*, 1998a,b; Liu *et al.*, 2000a,b; Pereira *et al.*, 2002). DF and several bioactive components (e.g. lignans, phenolic acids, phytosterols, minerals, tocopherols and tocotrienols) concentrated in the germ and outer layers of the kernel (Hegedüs *et al.*, 1985; Nilsson *et al.*, 1997; Glitsø and Bach Knudsen, 1999) are presumably responsible for the health benefits of whole grain cereals. Consumer interest in the health aspects of whole grain foods is growing, but research has not yet verified specific physiological functions and the health relevance of different cereal products at the molecular level. Moreover, safety and acceptable sensory profiles are important prerequisites for future functional foods.

The rye grain is one potential candidate raw material for functional and healthy foods. In this chapter, we review the chemistry of the rye grain, its nutrients and bioactive compounds, rye food products and technologies involved in their production. The bioavailability of various bioactive compounds in rye is also described, and the health benefits of rye foods are briefly discussed.

11.2 Chemistry and properties of rye grain nutrients and bioactive compounds

Whole grain rye flour contains, as a percentage of dry matter: 56–70% starch, 8–13% proteins, 2–3% lipids, 2% ash and 15–21% total DF (of which *c.* 20% is soluble fiber) (Vinkx and Delcour, 1996). Rye contains higher levels of DF as compared with wheat (15–21% versus 11–13%). The main rye fibers are arabinoxylan (AX) (8–12%), β-glucan (1.3–2.2%) and cellulose (1–1.7%) (Hansen *et al.*, 2003). Rye also contains fructans (including fructooligosaccharides), and levels of 4.6–6.6% have been detected (Karppinen *et al.*, 2003). As fructans are currently considered as part of DF in certain countries, this further increases the DF content of rye.

Rye AX consists of a backbone of (1→4)-β-D-xylopyranosyl residues (X) mainly substituted with α-L-arabinofuranosyl residues (A) to varying degrees at the O-2 position (2mXyl), the O-3 position (3mXyl) or both positions (dXyl) (Fig. 11.1). AX also contains small amounts of ferulate residues bound to arabinose as esters at its O-5 position. Analysis has shown that low amounts of ferulic acid dimers (Fig. 11.2) were present in AX extracted from rye, indicating that some extractable AX exists in coupled form(s) in the walls of the endosperm. Extractable AX from rye gave stronger gels than AX from wheat after treatment with peroxidase/H_2O_2 due to cross-linking of AX chains through dimerization of ferulic acid substituents (Dervilly-Pinel *et al.*, 2001). Some ferulate residues in rye are linked to tyrosine suggesting also a covalent cross-link between AX and proteins that is induced by peroxidase (Piber and Koehler, 2005). The physico-chemical properties of AX are highly influenced by the extent of substitution and distribution of substituents along the xylan backbone (Fig. 11.1). The degree of substitution has been used as a classification criterion for rye AX; the xylan with an A:X ratio 0.5 was totally water extractable, whereas the xylan with an A:X ratio of 1.4 was only partly water extractable (Bengtsson and Åman, 1990). Rye AX has been classified based on extractability and structure into four distinguishable classes with partly overlapping extractabilities. The molecular weight, degree of A substitution and presence of other substituents influence the rheological properties and subsequently the technological and physiological functionality

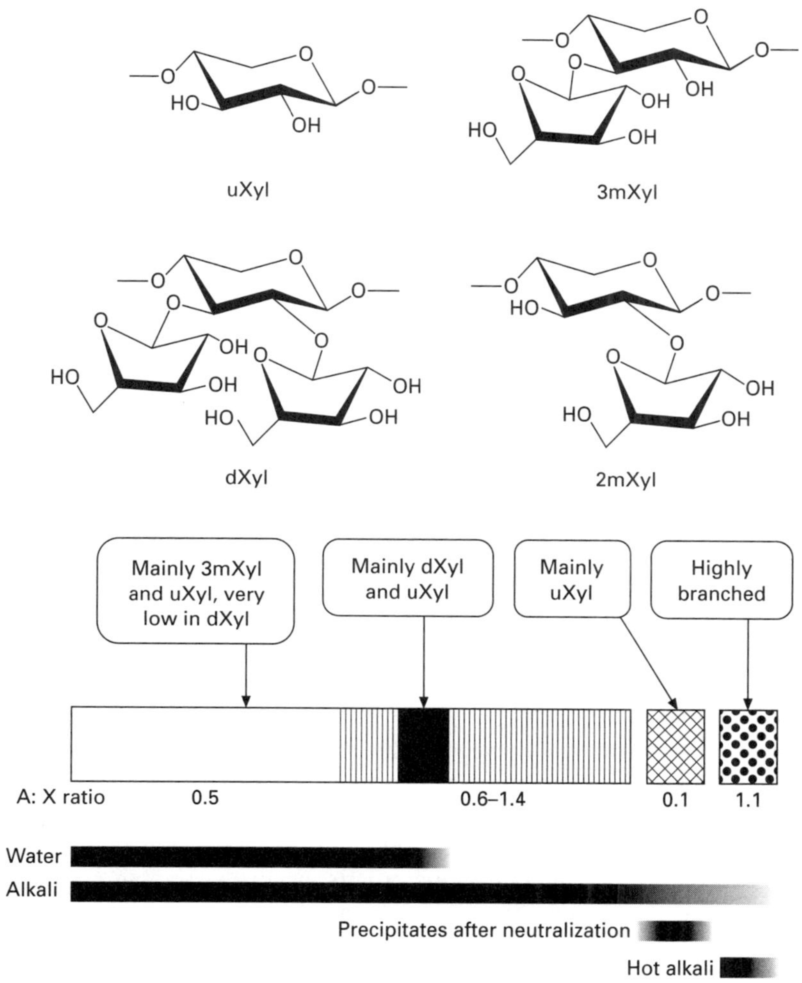

Fig. 11.1 Main structural units in rye arabinoxylan (AX) and a schematic representation of the extractability of rye AX classes. Adapted from Nilsson (1999).

of rye AX. The AX has been indicated to show a similar intrinsic viscosity to guar gum and to be more viscous than dextran and gum arabic (Mitchell, 1979). When comparing viscous properties of water-extractable AX of different rye varieties, a higher proportion of high molecular weight polymers was shown to be the reason behind the higher viscosity (Nilsson *et al.*, 2000).

The substitution pattern varies between the different parts of the kernel leading to different A:X ratios in different milling fractions. This ratio – for example 0.63 for whole grain rye flour, 0.75 for endosperm, 0.42 for the aleurone and 1.04 for the pericarp/testa – reflects different classes of AX in the different structures (Glitsø *et al.*, 1999). The different rye fractions differed accordingly also with respect to their functional characteristics, e.g. the water

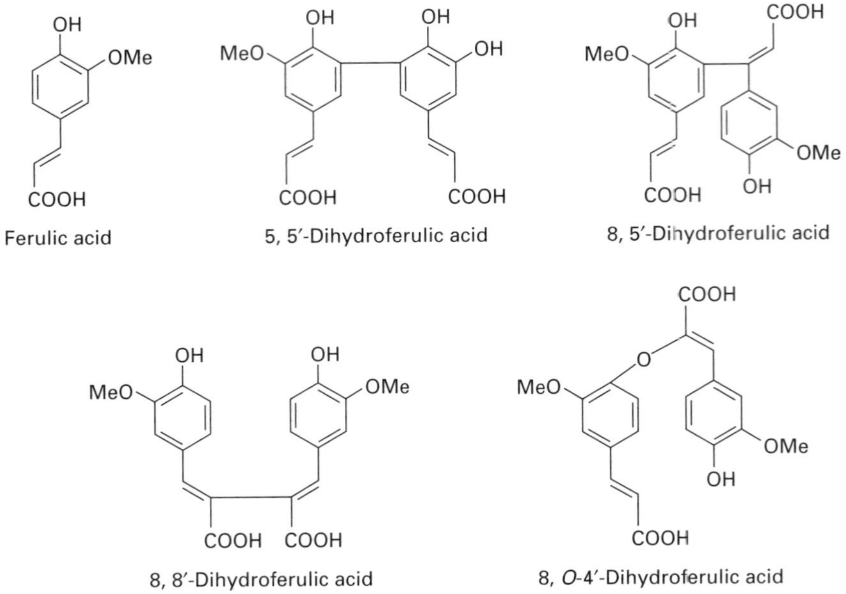

Fig. 11.2 Structures of ferulic acid and ferulic acid dimers in rye grain.

solubility of AX is *c.* 70% in the endosperm while almost null in the aleurone and pericarp/testa. Rye AXs have high water-binding and gelling capacities and are able to form highly viscous solutions, a property that correlates with the content of the highest molecular weight water-extractable AX and that has been linked to anti-nutritive effects in chicks and rats (Petterson and Åman, 1989; Vinkx and Delcour, 1996; Liu *et al.*, 2000a; Ragaee *et al.*, 2001). Unlike its unextractable counterparts, water-extractable AX imposes positive effects on dough structure and bread characteristics (Jankiewicz and Michniewicz, 1987; Kühn and Grosch, 1989; Seibel and Weipert, 2001).

Rye proteins, called secalins, are prolamins of different molecular weights including high molecular weight secalins, 75 kDa S-rich γ-secalins, S-poor ω-secalins and 40 kDa S-rich γ-secalins (Tatham and Shewry, 1995). The amount of starch in rye grain was found to be lower in non-adapted cultivars with high protein contents than in adapted hybrids (Hansen *et al.*, 2004). The starch is exclusively present in the starchy endosperm and rye starch has a lower gelatinization temperature than wheat starch. Both rye and wheat starches are of type A crystallinity but rye starch contains more A-type granules (up to 62.5 μm in diameter, 85–90%) and a smaller population of B-type granules (9.3 μm or less in diameter, 10–15%) than wheat starch. Moreover, rye starch has larger granules than wheat starch; it has both a lower proportion of B-type granules and the A-type granules are of larger particle size (Wasserman *et al.*, 2001; Verwimp *et al.*, 2004).

The lipids of rye grain are concentrated in the germ and are mainly made of unsaturated fatty acids. Oleic, linoleic and linolenic acids constitute about

81.6% of total fatty acids in the rye grain (Shewry and Bechtel, 2001). The major tocopherols and tocotrienols in the rye grain are α-tocopherol (10 mg/kg), α-tocotrienol (14 mg/kg), β-tocopherol (3 mg/kg), and β-tocotrienol (11 mg/kg) (Syväoja *et al.*, 1986). The amount of total vitamin E in rye meal, which mainly accounts for α-tocopherol content, is 1.6 mg tocopherol equivalent/100 g fresh weight. Rye has a high concentration of phytosterols/phytostanols (955 mg/kg), mainly concentrated in the bran (Piironen *et al.*, 2002). The major sterols/stanols in rye are β-sitosterol (*c.* 50%), β-sitostanol (*c.* 15%), campesterol (*c.* 15%), campestanol (*c.* 10%) and stigmasterol (<5%). The phenolic acids (1366 mg/kg), lignans (124 mg/kg) and alkylresorcinols (*c.* 930 mg/kg) of rye are also concentrated in the bran (Nilsson *et al.*, 1997; Glitsø *et al.*, 2000a; Bach Knudsen *et al.*, 2003; Liukkonen *et al.*, 2003; Mattila *et al.*, 2005). Alkylresorcinols are particularly interesting as possible biomarkers of whole grain rye and wheat products, and of their intake (Ross *et al.*, 2004a). Whole grain rye flour is a good source of minerals including (in mg/kg): phosphorus (*c.* 3800), magnesium (1190–1300), calcium (370–700), iron (30–90), copper (4–9), manganese (*c.* 29), zinc (24–40) and selenium (*c.* 0.22) (as reviewed by Liukkonen *et al.* (2006)). Rye grain also contains thiamin (vitamin B1), riboflavin (vitamin B2), niacin (vitamin B3), pantothenic acid (vitamin B5) and pyridoxine (vitamin B6) at levels comparable with wheat, as shown in Chapter 5. The level of folate (vitamin B9) in the rye grain is *c.* 0.7 mg/kg with formylfolates (5-formyltetrahydrofolate, 10-formyldihydrofolate and 10-formylfolic acid) and 5-methyltetrahydrofolate being the predominant vitamers (Liukkonen *et al.*, 2006). On the other hand, some of the compounds in whole grain cereals – such as phytic acid, polyphenols and digestive enzyme inhibitors – may have anti-nutritive properties (Slavin, 2003).

11.3 Process technologies and rye food products

Rye food products, mostly bread, are made of whole, steel cut, crushed or milled rye kernels that can also be malted (Liukkonen *et al.*, 2006). During milling, the kernel is ground and fractionated into different grades of flours and brans. In refined flours, the fiber- and nutrient-rich pericarp/testa and aleurone fractions and germ are separated from the starchy endosperm. Since minerals are concentrated in the outer parts of the rye kernel, the ash content of the flour is an indicator of the amount of the outer layers in the flour. Roller milling of rye grain can provide as many as 50 fractions of variable chemical composition and technological properties. The ash content in whole grain rye flour (14% DF), medium refined rye flour (9% DF) and light rye flour (7% DF) is about 2%, 0.9% and 0.8%, respectively.

Various types of soft and crisp rye breads are available in the marketplace in central, northern and eastern European countries, e.g. yeast-fermented, air-leavened breads and breads baked with sourdough (Meuser *et al.*, 1994;

Salovaara and Autio, 2001; Liukkonen *et al.*, 2006). Rye flours of different extraction rates as well as rye flours mixed with wheat flour in varying proportions are used for the production of rye bread (Bushuk, 2001). The DF content of rye bread is thus variable, but is typically in the range of 6–16%. In the milling process the distribution of both the health-relevant compounds and the flavor-intensive compounds is of interest. As is the case for DF and bioactive compounds, flavor attributes are unevenly distributed in the rye kernel. Between the mild-tasting endospermic part of rye grain and the bitter-tasting outermost bran fraction, a rye-like flavor without any obvious bitterness could be obtained (Heiniö *et al.*, 2003a). This fraction was also shown to contain significant levels of alkylresorcinols, lignans and phenolic acids.

The composition of rye products may influence the perceived flavor in two ways: by certain volatile compounds (such as aldehydes, ketones and alcohols), or by non-volatile compounds (such as sugars, amino acids and small peptides, free fatty acids, phenolic compounds and lipids). These compounds may be directly flavor-active but may also act indirectly as flavor precursors through reactions forming diverse new volatile compounds during processing (Heiniö, 2003, 2006). The volatiles in rye flour – e.g. malty-odor 2-methyl butanal, buttery-odor 2,3-butadione, green-fatty methional, and sweet, honey-like 2-phenylacetic acid – are flavor active. In addition, other flavor compounds are formed in biochemical reactions catalyzed by the starter culture during sourdough bread processing (e.g. malty 3-methylbutanal, vanillin, sweaty 3-methylbutanoic acid, fatty/waxy (E,E)-2,4-decadienal, and sour-pungent acetic acid (Kirchhoff and Schieberle, 2002).

The flavor of native, untreated rye grain is mild, and much of the flavor is formed during processing, as reviewed by Heiniö (2003, 2006). Sourdough fermentation is known to modify the flavor, but also texture, and nutritional value of rye breads and bran-rich breads (Salmenkallio-Marttila *et al.*, 2001; Brummer and Lorenz, 2003; Katina *et al.*, 2005, 2007). Traditional multiple-stage sourdough fermentation procedures provide the bread, especially the crumb, with desirable texture and its characteristic special flavor (Brümmer *et al.*, 1977; Hansen *et al.*, 1989). Fermentation of rye dough may also be accompanied by an increase in folate content due to folate synthesis by yeasts (Liukkonen *et al.*, 2003; Kariluoto *et al.*, 2006; Katina *et al.*, 2007). Sourdough fermentation may also assist in liberation of phenolic compounds from the cell wall matrix, with potential effects on bioavailability (Liukkonen *et al.*, 2003; Katina *et al.*, 2007).

Rye may also be germinated and malted for further uses in bread and various heat-processed foods (Heiniö *et al.*, 2003b; Liukkonen *et al.*, 2003). Germination is another bioprocess affecting the taste and flavor, and the levels of bioactive compounds (Liukkonen *et al.*, 2007). In heat processing, such as baking and extrusion cooking, further changes may modulate the physiological functionality both in terms of chemical composition and structural properties. In puffed rye, used in ready-to-eat breakfast cereals, the kernel is expanded along the minor axis yielding a light and crispy product. Puffed rye, similar to rice, is characterized by a porous matrix of numerous cavities that increased in size outwards and that are separated by very thin walls (Mariotti *et al.*, 2006).

Rye grain has the potential to add variety to the cereal foods available for human consumption, and to increase the level of DF and associated bioactive compounds. White wheat flour has, however, gained popularity, leading to a decline in the traditional use of rye as a food in many countries. Due to the increased consumer awareness of the healthiness of rye and whole grain foods, the cereal food industry is challenged to develop innovative healthy rye products with improved and varied sensory profiles. In recent years, the Nordic food industry has developed many new products such as convenient rye bread mixes with sour dough, hamburger buns with rye flour, crisp breads with tasty fillings, puddings and salads with rye prepared as boiled rice, and pasta with rye (Liukkonen *et al.*, 2006).

Rye bread, especially crisp bread, is subjected to high temperatures during baking and this leads to the formation of acrylamide, a neurotoxin and carcinogen (Friedman, 2003). Acrylamide in cereal products is mainly formed from asparagine and reducing sugars via Maillard reaction intermediates. Because of the nutritional importance of rye bread, it is important to find strategies to minimize its level of acrylamide. Research has shown that this can be partly achieved by lowering the temperature and time of baking (Mustafa *et al.*, 2005), and by incorporation of yeast fermentation which reduces the content of asparagine in the process (Fredriksson *et al.*, 2004).

11.4 Health effects of dietary fiber

As stated above, rye is a rich and versatile source of DF, the main constituents being AX, β-glucan, cellulose, Klasson lignin and also fructan if included. The health effects of DF are particularly studied with respect to interactions in the gastrointestinal tract. Several mechanisms for a beneficial effect of DF and associated phytochemicals have been proposed, including dilution and binding of toxic compounds, decreased transit time, stimulated production of short-chain fatty acids (SCFA), reduced pH and reduced formation of putative pro- and co-carcinogenic compounds such as secondary bile acids and toxic nitrogenous compounds (Cummings and Branch, 1982). The large intestine of humans – characterized as dark, warm, moist, anaerobic, and filled with food residues that flow at a relatively low speed (MacFarlane and Cummings, 1991) – provides a favorable environment for the growth of microorganisms. Its microbial ecosystem, thus, contains hundreds of species of anaerobic bacterial species each occupying a particular niche with numerous interrelationships with the other species. The physiological effects of DF in the large intestine depend on its fermentability, which is influenced by its chemical composition, solubility, physical form and the presence of lignin and other components (e.g. Cummings and Englyst, 1995).

The most important determinants of fermentation in the large intestine are the types and levels of non-digestible carbohydrates in the form of non-starch polysaccharides (NSP), resistant starch (RS), non-digestible oligosaccharides (NDO) and hydrogenated carbohydrates (polyols) (Fig. 11.3). When consuming cereal-based diets, the carbohydrate fraction accounts for 30–50% of the dry ileal solids (Englyst and Cummings, 1985; Pettersson *et al.*, 1996; Bach Knudsen and Canibe, 2000) and consists predominantly of NSP (AX, β-glucan and cellulose) whereas starch, RS, sugars and NDO represent only a minor part (Englyst and Cummings, 1985; Pettersson *et al.*, 1996; Bach Knudsen and Canibe, 2000). In the large intestine, the carbohydrates are exposed to the action of the hydrolytic and non-hydrolytic bacteria (MacFarlane and Cummings, 1991) that convert the carbohydrate monomers (pentoses and hexoses), through a variety of intermediates, to SCFA – mostly acetate, propionate and butyrate. Protein fermentation leads, in a similar way, to formation of the three main acids along with branched-chain fatty acids (MacFarlane *et al.*, 1986).

Although predominantly formed from only ten common monosaccharides, the NDO fraction is very diverse in terms of sugar residues, structural organization and physicochemical properties (Cummings *et al.*, 1997). All these factors will to a large extent influence the biochemical pathways that convert carbohydrates to SCFA during anaerobic fermentation and the

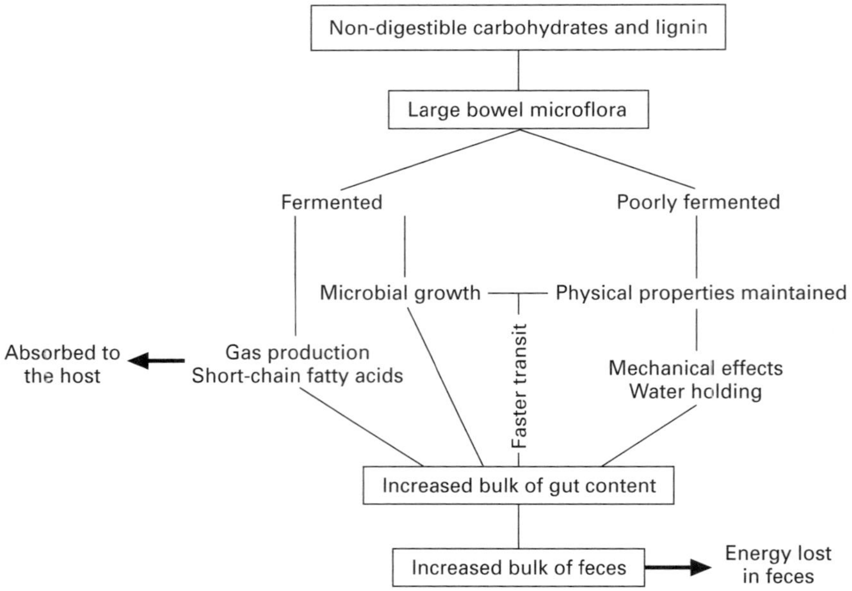

Fig. 11.3 Mechanisms of action of non-digestible carbohydrates and lignin in the large intestine.

carbohydrate residues will, consequently, have a profound influence on the molar proportions of the formed SCFA. For instance, enhanced butyrate production was found in pigs fed breads or rolls baked with wheat, oats and rye fractions (Bach Knudsen *et al.*, 1993, 2000, 2005), when rats were fed an aleurone-enriched wheat fraction (Mclntosh *et al.*, 2001), when humans were fed whole grain rye (Gråsten *et al.*, 2000, 2003; McIntosh *et al.*, 2003) as well as during *in vitro* fermentation of rye bran (Aura *et al.*, 2006). A common feature of all the DF sources that enhanced butyrate production was a high AX proportion of the NSP. Besides supplying the host with 5–7% of the energy requirement for maintenance (McNeil, 1984), SCFA are thought to have specific metabolic roles and health implications. Butyrate is the preferred fuel over glucose for the epithelial cells lining the large intestine (Roediger, 1980; Scheppach and Weiler, 2004) and with regulatory functions during cell proliferation and differentiation (Smith *et al.*, 1998; Williams *et al.*, 2003; Scheppach and Weiler, 2004), propionate has been proposed to modify hepatic cholesterol metabolism (Anderson, 1990), while acetate is used as fuel for muscle tissues (Bergman, 1990).

A consequence of an active fermentation is lowering of the luminal pH and modification of the metabolism of bile acids. In Finnish subjects fed low-fiber wheat bread, whole meal wheat bread or whole meal rye bread, the concentration of fecal free bile acids in the rye group was only 54–60% of that of the wheat group (Korpela *et al.*, 1992; Gråsten *et al.*, 2000). It has been shown that rye bread decreases the concentration of free secondary bile acids by changing the mode of their conjugation (Korpela *et al.*, 1992), which indicates an altered metabolic activity of the microflora. The likely cause is that the reduced pH inhibits the dehydroxylation of primary bile acids and their conversion to secondary bile acids, which are considered to be more carcinogenic and mutagenic.

An increased intake of DF will inevitably influence the bowel habitat not only because of the lowering of pH, and stimulation of microbial growth and SCFA production, but also because of mechanical action and water-holding properties (Fig. 11.3). The consequence is an increased content and bulk in the colon and feces, and a reduction of the transit time, which has also been shown for rye bread (Gråsten *et al.*, 2000; McIntosh *et al.*, 2003). The effect of the various DF sources on the stool weight, however, is tightly correlated with the type of polymers entering the large intestine. Thus, soluble DF polysaccharides such as β-glucan and soluble AX, which are partly degraded already in the small intestine, are extensively degraded in the large intestine in contrast to insoluble DF components such as cellulose, insoluble AX and lignin that are very resistant against microbial degradation (Bach Knudsen and Hansen, 1991; Glitsø *et al.*, 1998). Because of that, the bulking properties of DF from brans and whole grain products are substantially higher than those of refined flour. Rye bran, with whole grain rye, contains both rapidly fermentable, slowly fermentable and non-fermentable fractions (Karppinen *et al.*, 2000, 2001, 2003). Among the rapidly fermentable fractions are fructans (about 7% in rye bran) and the easily extractable AX, also shown to be bifidogenic (Crittenden *et al.*, 2002).

There is a general positive correlation between the intake of NSP and fecal bulking, which also influences the mean transit time (MTT) (Cummings *et al.*, 1992). Because of the higher proportion of DF from cereals, particularly as whole grain products, the intake of DF is roughly 50% higher in Scandinavia than in the United Kingdom. This has a profound influence on fecal bulking and the MTT has been reported to be ~20 h faster in Denmark and Finland than in the United Kingdom (Englyst *et al.*, 1982; Cummings *et al.*, 1992). A recent intervention study with low-DF wheat bread or a high-DF whole grain rye bread, confirmed the effect of DF on MTT (Gråsten *et al.*, 2000). Moreover, DF includes a wide range of phenolic and phytoestrogenic compounds associated with different beneficial health effects as discussed below.

11.5 Bioavailability and health effects of bioactive phenolic compounds in rye

Ferulic acid is the major phenolic acid in rye (*c.* 85%) and is found mainly in bound forms (Fig. 11.2); esterified predominantly to some of the arabinose residues of AX (Zhao *et al.*, 2003a,b). Esterase activities in the small intestine and the colon of humans are, thus, necessary to cleave the ester bonds and release the phenolic acids into the lumen before absorption and metabolism (Teuchy and van Sumere, 1971; Kroon *et al.*, 1997; Andreasen *et al.*, 2001a; Kern *et al.*, 2003a,b; Konishi and Shimizu, 2003). Released phenolic acids are absorbed via a Na^+-dependent saturable transport mechanism in the intestine (Ader *et al.*, 1996). Ferulic acid was recovered in both free and conjugated forms in the portal vein plasma, celiac arterial plasma, bile and urine of rats (Nardini *et al.*, 2002; Rondini *et al.*, 2002; Zhao *et al.*, 2003a,b, 2004). The major site of absorption of ferulic acid from rye bread seems to be the large intestine following fermentation of DF. Interestingly, the slow release of ferulic acid bound in bran was found to enhance its bioavailablity compared with the free form in the rat (Adam *et al.*, 2002; Rondini *et al.*, 2004). However, the total extent of digestion in the colon and absorption and urinary excretion of ferulic acid from cereal products in comparison with other foods is still unknown (Bourne and Rice-Evans, 1998; Kern *et al.*, 2003a; Harder *et al.*, 2004). Ferulic acid, its dimers and the other phenolic acids of cereal grains are potent antioxidants (Garcia-Conesa *et al.*, 1999; Andreasen *et al.*, 2001b) and may contribute to the health benefits of whole grain cereals, including rye.

Plant lignans are diphenolic compounds that are similar in structure to endogenous sex steroid hormones, and they are hypothesized to inhibit the development of cancer and cardiovascular disease. In particular, hormone-related cancer types, such as breast and prostate cancer, seem to be affected but also colon cancer (e.g. Adlercreutz, 2002). The known plant lignans in rye (Fig. 11.4) are secoisolariciresinol, matairesinol, pinoresinol, syringaresinol, medioresinol and lariciresinol (Heinonen *et al.*, 2001; Penalvo *et al.*, 2005). In cereals plant lignans are found in the outer fiber-rich layers (aleurone and pericarp/testa) (Nilsson *et al.*, 1997; Glitsø *et al.*, 2000b). Plant lignans are metabolized by the gut microbes to the mammalian lignans or enterolignans enterodiol and enterolactone (Borriello *et al.*, 1985), and were first detected in the urine of human subjects and monkeys (Setchell and Adlercreutz, 1979; Setchell *et al.*, 1980a,b). The microbial formed enterolignans are absorbed by passive diffusion from the intestinal tract to the portal vein (Bach Knudsen *et al.*, 2003), and then transported to the liver and other tissues in the body. Most of the absorbed lignans are conjugated in the liver

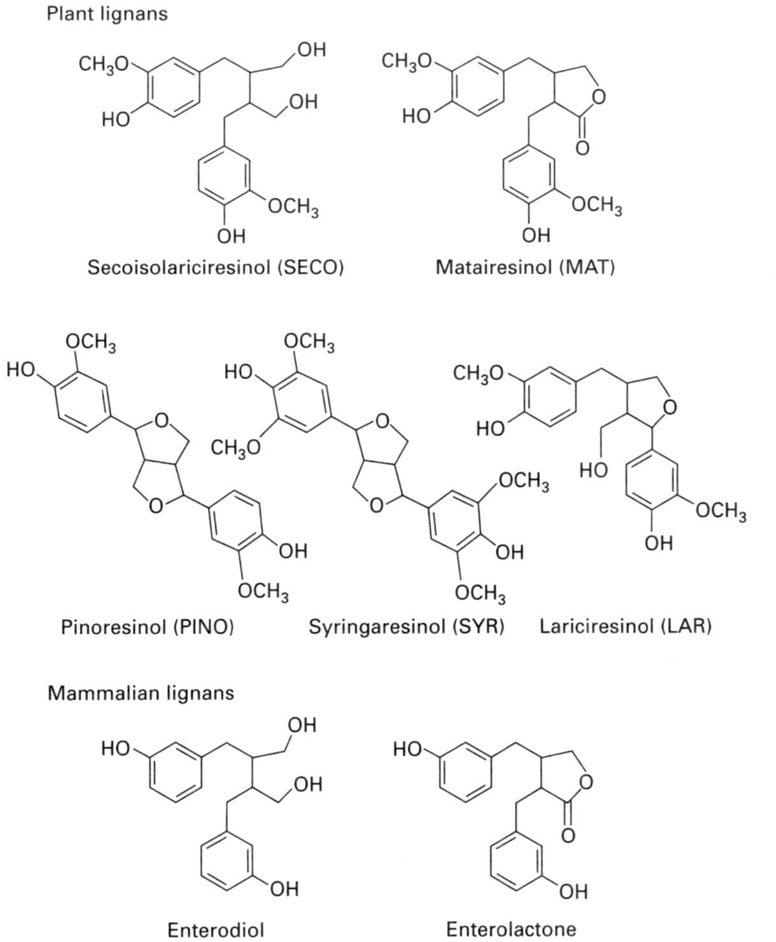

Fig. 11.4 Rye plant lignans and the mammalian lignans, enterodiol and enterolactone, produced by their microbial fermentation in the colon.

and recycled by the bile to the gut where they are deconjugated by bacteria (Axelson and Setchell, 1981). Lignans are not oxidized in the body but excreted via the urine or feces.

Although we know that consumption of whole grain diets and whole grain rye in particular leads to higher plasma and urine concentrations of enterolignans (Juntunen *et al.*, 2000; Jacobs *et al.*, 2002; McIntosh *et al.*, 2003), our understanding of the factors in cereal products that affect the bioavailability and pharmokinetics of these compounds in human subjects is currently limited. Therefore, we have to largely rely on *in vitro* studies or *in vivo* animal studies, e.g. rats and pigs, for our understanding of possible interactions. Results obtained with rats and pigs that are fed (a) breads baked with whole grain rye or various rye fractions (Glitsø *et al.*, 2000b), (b) high-fiber breads baked from whole grain rye with added rye bran or (c) high-fiber wheat bread baked from refined wheat flour and with added purified wheat fiber (Vitacel) (Bach Knudsen *et al.*, 2003) are shown in Table 11.1. In spite of the significant difference in the structures of DF and consequently the bioavailability of DF polysaccharides, e.g. the digestibility of AX varied from nil to 83% (Glitsø *et al.*, 1998; Bach Knudsen *et al.*, 2005), the bioavailability of plant lignans was remarkably similar in all diets with values in the range 47–69%. However, the bioavailability of the individual plant lignans varies, being 79–83% for secoisolariciresinol but only 10–19% for the sum of pinoresinol, syringaresinol and lariciresinol (Bach Knudsen *et al.*, 2003). This is essentially in agreement with *in vitro* data that also demonstrated a substantial variation in the conversion of plant to mammalian

Table 11.1 Dietary fibre and plant lignans of experimental diets and bioavailability of plant lignans in rats and pigs

	DF (g/kg DM)	PL* (mg/kg DM)	Bioavailability (% of intake)			
			SECO	MAT	ΣPINO, SYR and LAR	PL*
Rats[†]						
Whole rye	157	24	–	–	–	47
Pericarp/testa	177	77	–	–	–	61
Aleurone	177	63	–	–	–	60
Endosperm	93	6	–	–	–	69
Pigs[‡]						
High-fibre wheat	230	2.5	83	71	10	60
High-fibre rye	235	31	79	13	19	48

Abbreviations: DF, dietary fibre; PL, plant lignans; SECO, secoisolariciresinol; MAT, matairesinol; PINO, pinoresinol; SYR, syringaresinol; LAR, lariciresinol.

* Sum of secoisolariciresinol, matairesinol, pinoresinol, syringaresinol and lariciresinol.

† Calculated from the urinary excretion of enterolactone (ENL) as:

$$\text{Bioavailability of PL} = \left(1 - \frac{\text{Urinary ENL}}{\text{PL}}\right) \times 100$$

‡ For calculation see Bach Knudsen *et al.* (2003).

lignans, from 4% for syringaresinol to 101% for lariciresinol (Heinonen *et al.*, 2001). Recently it was found that rats are able to produce enterolactone from lignin (Begum *et al.* 2004) and it is likely that this also occurs in man.

The bioconversion of plant to mammalian lignans *in vitro* is far slower than is the case for the conversion of polysaccharides into butyrate (and other SCFA) (Fig. 11.5) (Glitsø *et al.*, 2000a). The same is reported for the formation of enterolactone from flaxseed plant lignans (Aura *et al.*, 2006). An *in vivo* study with pigs further concluded that there was no effect of fermentation pattern on the intestinal metabolism of lignans as plant lignans were liberated from the pericarp/testa in the large intestine, although the plant cell walls remained largely undegraded (Glitsø *et al.*, 2000b). In this aspect, the formation of enterolactone differs from the conversion of daidzein to equol. In the latter case, the amount of carbohydrates available for fermentation *in vitro* was found to influence the conversion significantly (Setchell and Cassidy, 1999).

An *in vivo* kinetic study with catheterized pigs virtually confirmed the *in vitro* data (Fig. 11.6). After a pulse dose with a high-DF rye diet (Bach Knudsen *et al.*, 2003, 2005), the concentration of butyrate starts to increase 4 h after feeding, whereas with enterolactone it was not until 6–7 h post-feeding that concentration differences in the portal vein were observed. It should also be noticed that, while the portal concentration of butyrate reached a plateau level after 8–9 h, the concentration of enterolactone continued to rise during the entire study period presumably reflecting the continuous supply to the portal vein of enterolactone undergoing enterohepatic circulation (Axelson and Setchell, 1981, Bach Knudsen *et al.*, 2003), as is the case with phenolic estrogens (Adlercreutz; and Martin, 1980).

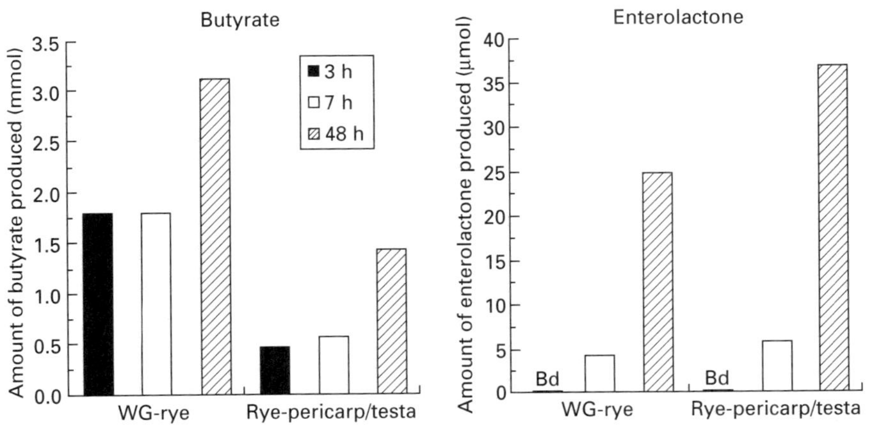

Fig. 11.5 Amounts of butyrate (mmol) and enterolactone (μmol) produced during a 48 h fermentation of whole grain rye flour (WG-rye) and a bran fraction (Rye-pericarp/testa). Data for butyrate are from Glitsø *et al.* (2000a). Bd, below detection limit.

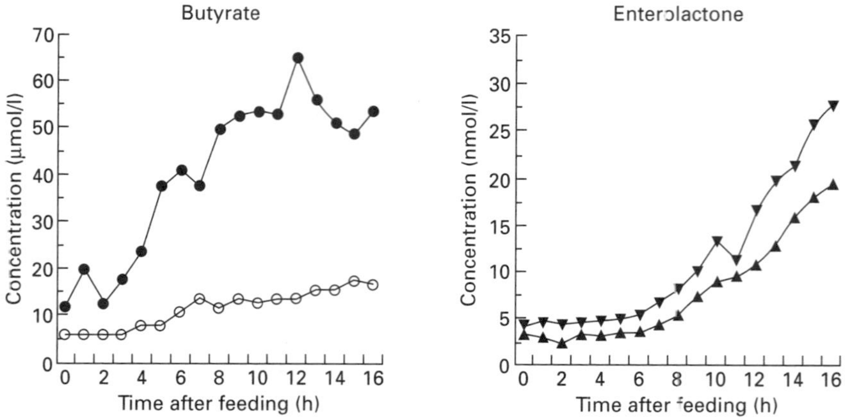

Fig. 11.6 Portal (● or ▼) and arterial (○ or ▲) blood concentrations of butyrate (μmol/l) and enterolactone (nmol/l) in pigs after the intake of a single dose of a high-fibre rye diet. Data from Bach Knudsen *et al.* (2003, 2005).

11.6 Health benefits of rye bread and other rye products

11.6.1 Diabetes and obesity

Epidemiological studies have shown that a high intake of whole grain products is associated with a decreased risk of diabetes (Liu, 2002, 2003; Murtaugh *et al.*, 2003). Weight gain has also been inversely associated with the intake of a whole grain high-fiber diet (Koh-Banerjee and Rimm, 2003; Liu *et al.*, 2000a, 2003). Whole grain foods were observed to have favorable effects on insulin sensitivity on overweight and obese adults (Pereira *et al.*, 2002). A recent study showed that changes in lifestyle of high-risk subjects by individualized counseling aimed at reducing weight, total intake of fat and intake of saturated fat and increasing intake of DF and level of physical activity may prevent type 2 diabetes mellitus (Tuomilehto *et al.*, 2001). Postprandial studies in healthy humans have repeatedly shown that rye bread induces lower insulin responses than white wheat bread, indicating an improved insulin economy (Leinonen *et al.*, 1999; Juntunen *et al.*, 2002, 2003a). Rye bread structure has been suggested to play an important role here. A number of intervention studies have shown that whole grain rye bread improves postprandial first-phase insulin secretion in the oral glucose tolerance test, indicating improved glucose metabolism, which would assist in lowering the risk of type 2 diabetes (Juntunen, 2003; Juntunen *et al.*, 2003b; Laaksonen *et al.*, 2005).

11.6.2 Cardiovascular disease

Whole grain high-fiber diets have a wide variety of beneficial health effects. Epidemiological studies have consistently shown that a high intake of whole grain products is associated with a decreased risk of cardiovascular diseases

(Anderson, 2003; Liu *et al.*, 2000b; Liu, 2002; Seal, 2006). It has previously been shown that rye breads containing rye bran and whole grain rye reduce serum total and low-density lipoprotein cholesterol (Zhang *et al.*, 1994; Leinonen *et al.*, 2000; Lundin *et al.*, 2004), which is known as one of the main risk factors of coronary heart disease. Whole grain rye contains a remarkable amount of soluble AX, which may have a similar cholesterol-lowering ability as the β-glucan in oats (FDA, 1997; Leinonen *et al.*, 2000). The effect of whole grain rye on other metabolic factors related to diabetes may further enhance the cardioprotective effect. In addition, some components in whole grain rye – such as phenolic acids, lignans and phytic acids – are suggested to play a protective role in cardiovascular diseases by acting as antioxidants (Slavin, 2003; Seal, 2006). High serum enterolactone levels in populations have been associated with reduced risk of acute coronary events (Vanharanta *et al.*, 1999).

11.6.3 Prostate cancer and related metabolic factors

The relationship between obesity incidence and mortality from prostate cancer related to different ages has been inconsistent for a long period. In a recent study by Stocks *et al.* (2006), however, strong support is given for the concept that the impact of insulin and factors associated with insulin on prostate cancer growth is age dependent with a protective association of c-peptide in younger cases of prostate cancer. It appears that metabolic and hormonal factors may act in a different way on tumor progression than on tumor initiation at different ages. In older men with prevalent prostate cancer, the results more clearly point to the direction of an inverse effect of insulin on tumor progression and death from prostate cancer (Hsing *et al.*, 2001; Lehrer *et al.*, 2002; Hammarsten and Högstedt, 2005). Type 2 diabetes mellitus is characterized by high plasma levels of glucose and insulin during the development of the disease as well as at diagnosis and is subsequently followed by a decline in insulin levels. Diabetes has consistently been shown to be associated with a decreased risk of prostate cancer in the later stages of the disease associated with low insulin concentrations but with an increased risk at the early stages of the disease with increased insulin levels (Giovannucci *et al.*, 1998; Tavani *et al.*, 2002; Rodriguez *et al.*, 2005). A number of experimental studies in animals and in humans have indicated that rye bran has an inhibitory effect on the growth of prostate cancer (Landström *et al.*, 1998; Bylund *et al.*, 2000, 2003) and that one active substance in the rye bran may be the lignans (Bylund *et al.*, 2005).

11.6.4 Colon cancer

On basis of observations of the lifestyle of western and rural African people, Burkitt (1973) hypothesized that increased large bowel cancer risk was partly due to lack of DF. Since then, the association between DF and the large bowel cancer risk has been intensively studied. Some recent epidemiological studies have supported this hypothesis (Bingham *et al.*, 2003; Peters *et al.*, 2003) while others have not (Fuchs *et al.*, 1999; Flood *et al.*, 2002). Rye bread has been shown to improve bowel function and to decrease the concentrations of some molecular biomarkers of increased colon cancer risk (Gråsten *et al.*, 2000; McIntosh *et al.*, 2003).

11.6.5 Gastric and duodenal ulceration and stomach cancer

Epidemiological and clinical studies showed that infection with *Helicobacter pylori*, a common pathogen infecting approximately half the world's population, may be of significance for a large variety of diseases ranging from superficial gastritis to gastric and duodenal ulceration and gastric cancer (Rothenbacher and Brenner, 2003; Israel and Peek, 2006; Leung *et al.*, 2006). The pathogenesis of severe gastric disorders caused by *H. pylori* is multi-factorial and involves complex interaction between the microbe and the gastric mucosa, e.g. the genotype of the *H. pylori* and various host factors (Blaser and Atherton, 2004; Mahdavi, 2004; Ahmed and Sechi, 2005). There has been a general agreement that whole grain cereals possibly decrease the risk of stomach cancer (American Institute for Cancer Research, 1997; Chatenoud *et al.*, 1999). The data are almost completely consistent from a number of case control studies. No studies have been performed so far to investigate the interaction between *H. pylori* infections and a possible protective effect of whole grains on stomach cancer.

11.7 Alkylresorcinols as biomarkers of rye intake

Alkylresorcinols (ARs; Fig. 11.7) are phenolic lipids present in a number of plants and bacteria (Kozubek and Tyman, 1999; Ross *et al.*, 2004a). Within food raw materials, ARs are only present in the outer parts of kernels of wheat and rye and in very low concentrations in barley (Tluscik, 1978; Ross *et al.*, 2004a). The length of the alkyl tail of ARs in cereals varies from 15 to 25 carbons. The dominant AR homologs have saturated 19 and 21 carbons in the side chain in wheat and 17, 19 and 21 carbons in rye. The variation in relative distribution of AR homologs within species of common wheat, durum wheat and rye is relatively small, despite a wider variation in their contents making them distinctly different (Chen *et al.*, 2004; Landberg *et al.*, 2006a). Thus, it is possible to distinguish between common wheat, durum wheat and rye on the basis of the ratio between the C17 and C21 homologs: about 0.01 for durum wheat, 0.1 for common wheat and 1.0 for rye. ARs have been shown to be stable during food processing – such as cooking, baking and pasta manufacturing – as well as resistant to oxidation (Kamal-Eldin *et al.*, 2000). The content of ARs in 45 Swedish wheat- and rye-containing food products correlated with that calculated from the average contents in the

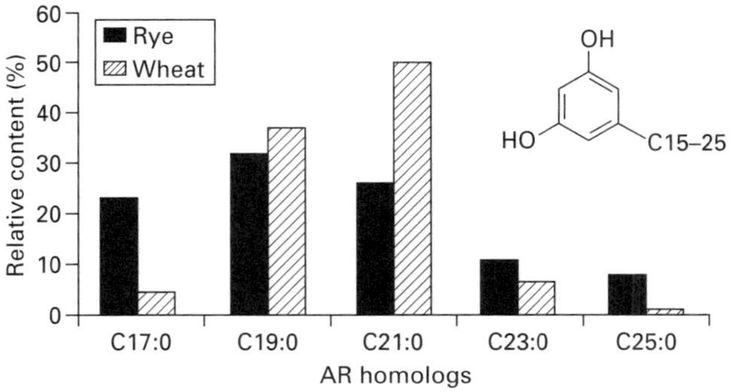

Fig. 11.7 Common structure of alkylresorcinols (ARs), potential biomarkers for whole grain rye and wheat intake, and the relative composition of AR homologs in rye and wheat. Data from Chen *et al.* (2004).

cereal ingredients ($R^2 = 0.91$). These studies showed clearly that ARs can be used as markers for whole grain and bran in foods containing wheat and rye and that the C17/C21 homolog ratio can be used to differentiate between durum wheat, common wheat and rye in foods (Chen *et al.*, 2004; Landberg *et al.*, 2006a).

About 60% of ingested ARs have been shown to disappear from the small intestine of ileostomy-operated humans, indicating that more than half the ingested ARs are absorbed (Ross *et al.*, 2003a). A rat study using a radiolabeled AR (C21:0) showed that about 50% of the radioactivity was recovered in feces in intact form and about 30% in urine in the form of metabolites (Ross *et al.*, 2003b). ARs have also been detected in the plasma (Linko *et al.*, 2002) and erythrocyte membranes (Linko and Adlercreutz, 2005) of people eating whole grain rye. Two probable urinary metabolites of ARs, 3-(3,5-dihydroxyphenyl)-1-propanoic and 1,3-dihydroxybenzoic acids, were detected in human urine after consumption of wheat bran (Ross *et al.*, 2004c). The metabolism of ARs probably occurs by shortening of the side chain by β-oxidation after ω-hydroxylation in accordance with data showing that ARs inhibit γ-tocopherol metabolism *in vitro* and in rats (Ross *et al.*, 2004b). The resorcinol part of the molecule is presumably not affected by phase 1 metabolic enzymes but is probably converted to glucuronide and sulfate conjugates for excretion in urine.

Human plasma pharmacokinetics and relative bioavailability of ARs were studied after a single intake of rye bran containing 190 mg of ARs (Landberg *et al.*, 2006b). The shapes of the plasma concentration–time curves of ARs were similar in the subjects ($n = 6$) with two average peaks at 3 and 6 h for all homologs after ingestion, with maximum average total AR concentration of 1300 and 3400 nmol/l for the two peaks. The relative elimination half-lives were similar, between 6.4 and 6.8 h, for the different AR homologs. The relative bioavailability of the different AR homologs increased with the number of carbon atoms in the side chain and that of C25:0 was more than five-fold that of C17:0. In conclusion, the relatively short terminal half-life suggests that ARs are rapidly eliminated from the central plasma compartment. The relatively short terminal half-life of ARs in plasma suggests that total plasma concentrations could give a short- to medium-term reflection of regular intake of whole grain from wheat or rye. Other compartments for ARs or their metabolites – such as erythrocytes, adipose tissue or urine – may provide a longer-term reflection of intake.

ARs have been suggested as a biomarker for intake of whole grain wheat- and rye-containing foods (Ross *et al.*, 2004a; Linko *et al.*, 2005). ARs seem to satisfy most of the important criteria for a useful dietary biomarker including: specificity for the food, presence in the food in amounts and forms that can be easily analyzed, consumption of these foods at levels that allow reasonable intake of the biomarker, reasonable absorption into the body, metabolism to one or a few major metabolites that can be easily analyzed in biological samples, a plausible dose–response relationship and a consistent response across studies (Wild *et al.*, 2001). The daily intake of ARs in the United Kingdom and Sweden was estimated as 12 mg/person/day in the United Kingdom and around 20 mg/person/day in Sweden (Ross *et al.*, 2005). Ninety-six percent of all Swedes consumed some ARs, compared with only 50% in the United Kingdom. For such a wide difference in intake, ARs may be good markers of diets rich or poor in whole grain wheat and rye foods.

11.8 Future trends

Rye is an interesting grain with many proven positive physiological effects and long traditions in food uses in Northern and Eastern Europe. In the Nordic countries, where knowledge of the positive health effects of rye has been well disseminated, consumption of rye is, after a long decline, again increasing. This development is also due to the continuous development of new types of rye breads and other foods meeting the needs of the modern consumer. One of the major challenges for the increase in the use of rye is the design of the sensory properties of foods containing rye. The strong flavor, dark color and hard structure typical of traditional rye foods are all challenges for food technologists as they develop new technologies for new food uses of rye. Controlled and optimized fermentation and use of enzyme technology for flavor and texture engineering are among the important tools being currently investigated.

Epidemiological studies mainly suggest that whole grain is a major part of a potential dietary prevention of several diet-related diseases. In order to investigate the relationship between whole grain cereal intake and health and disease, a dietary biomarker for whole grain cereals is urgently needed. If further research proves that ARs can provide a reliable marker of intake, this will open up new possibilities for studying the potential links between diets high in whole grain rye and wheat and health. Rye is an excellent source of dietary fiber and associated compounds, among which lignans, phenolic acids and alkylresocinols are of great interest. It will be important to learn more about the effects of processing on rye food structure, the physical properties of dietary fiber and the levels of bioactive compounds. Research into the physiological mechanisms of these compounds will yield criteria for development of new rye foods fully exploiting the health potential of rye.

11.9 Sources of further information and advice

In the Nordic countries, the Nordic Rye Group was established in 1994 and has performed research and published information about the health benefits of rye food. More information is available at the website http://rye.vtt.fi. The book edited by W. Bushuk, *Rye: Production, Chemistry and Technology*, and published by the American Association of Cereal Chemists in 2001, as well as the book chapters by Salovaara and Autio (2001) and Liukkonen *et al.* (2006), provide excellent references.

11.10 References

Adam A, Crespy V, Levrat-Verny M-A, Leenhart F, Leuillet M, Demigné C and Rémésy, C (2002), 'The bioavailability of ferulic acid is governed primarily by the food matrix rather than its metabolism in intestine and liver in rats', *J Nutr* **132**, 1962–1968.

Ader P, Grenacher B, Langguth P, Scharrer E and Wolffram S (1996), 'Cinnamate uptake by rat small intestine: transport kinetics and tranepithelial transfer', *Exp Physiol* **81**, 943–955.

Adlercreutz H (2002), 'Phyto-oestrogens and cancer', *Lancet Oncol* **3**, 364–373.

Adlercreutz H and Martin F (1980), 'Biliary excretion and intestinal metabolism of progestrone and estrogens in man', *J Steroid Biochem* **13**, 231–244.

Ahmed N and Sechi L A (2005), '*Helicobacter pylori* and gastroduodenal pathology: new threats of the old friend', *Ann Clin Microbiol Antimicrob* **4**, 1.

American Institute for Cancer Research (1997), '*Food Nutrition and the Prevention of Cancer: a Global Perspective*', Chapter 4.6.2.1, World Cancer Research Fund and Amercan Institute for Cancer Research, Washington DC, pp. 159–160.

Anderson J W (1990), 'Hypocholesterolemic effects of oat products', in Furda I and Brine C J (Eds), *New Developments in Dietary Fibre: Physiological, Physicochemical and Analytical Aspects*, Plenum Publishing Corp., New York, pp. 17–36.

Anderson J W (2003), 'Whole grains protect against atherosclerotic cardiovascular diseases', *Proc Nutr Soc* **62**, 135–142.

Andreasen M F, Kroon P A, Williamson G and Garcia-Conesa M-T (2001a), 'Esterase activity able to hydrolyze dietary antioxidant hydroxycinnamates is distributed along the intestine of mammals', *J Agric Food Chem* **49**, 5679–5684.

Andreasen M F, Landbo A-K, Christensen L P, Hansen Å and Meyer A S (2001b), 'Antioxidant effects of phenolic rye (*Secale cereale* L.) extracts, monomeric hydroxycinnamates and ferulic acid dehydrodimers on human low-density lipoproteins', *J Agric Food Chem* **49**, 4090–4096.

Aura A M, Oikarinen S, Mutanen M, Heinonen S M, Adlercreutz H C, Virtanen H and Poutanen K S (2006), 'Suitability of a batch in vitro fermentation model using human

faecal microbiota for prediction of conversion of flaxseed lignans to enterolactone with reference to an in vivo rat model', *Eur J Nutr* **45**, 45–51.
Axelson M and Setchell D R (1981), 'The excretion of lignans in rats – evidence for an intestinal source for this new group of compounds', *FEBS Letters* **123**, 337–343.
Bach Knudsen K E and Canibe N (2000), 'Breakdown of plant carbohydrates in the digestive tract of pigs fed on wheat or oat based rolls', *J Sci Food Agric* **80**, 1253–1261.
Bach Knudsen K E and Hansen I (1991), 'Gastrointestinal implications in pigs of wheat and oat fractions 1. Digestibility and bulking properties of polysaccharides and other major constituents', *Br J Nutr* **65**, 217–232.
Bach Knudsen K E, Jensen B B and Hansen I (1993), 'Oat bran but not oat gum enhanced butyrate production in the large intestine of pigs', *J Nutr* **123**, 1235–1247.
Bach Knudsen K E, Canibe N and Jørgensen H (2000), 'Quantification of the absorption of nutrients deriving from carbohydrate assimilation: Model experiment with catheterised pigs fed on wheat and oat based rolls', *Br J Nutr* **84**, 449–458.
Bach Knudsen K E, Serena A, Kjaer A K, Tetens I, Heinonen S M, Nurmi T and Adlercreutz H (2003), 'Rye bread in the diet of pigs enhances the formation of enterolactone and increases its levels in plasma, urine and feces', *J Nutr* **133**, 1368–1375.
Bach Knudsen K E, Serena A, Kjær A B K, Jørgensen H and Engberg R (2005), 'Rye bread enhances the production and plasma concentration of butyrate but not the plasma concentrations of glucose and insulin in pigs', *J Nutr* **135**, 1696–1704.
Begum A N, Nicolle C, Mila I, Lapierre C, Nagano K, Fukushima K, Heinonen S M, Adlercreutz H, Remesy C and Scalbert A (2004), 'Dietary lignins are precursors of mammalian lignans in rats', *J Nutr* **134**, 120–127.
Bengtsson S and Åman, P (1990), 'Isolation and chemical characterization of water-soluble arabinoxylans in rye grain', *Cereal Chem* **75**, 10–14.
Bergman E N (1990), 'Energy contributions of volatile fatty acids from the gastrointestinal tract in various species', *Physiol Rev* **70**, 567–590.
Bingham S A, Day N E, Luben R, Ferrari P, Slimani N, Norat T, Clavel-Chapelon F, Kesse E, Nieters A, Boeing H, Tjonneland A, Overvad K, Martinez C, Dorronsoro M, Gonzalez C A, Key T J, Trichopoulou A, Naska A, Vineis P, Tumino R, Krogh V, Bueno-de-Mesquita H B, Peeters P H, Berglund G, Hallmans G, Lund E, Skeie G, Kaaks R and Riboli E (2003), 'Dietary fibre in food and protection against colorectal cancer in the European Prospective Investigation into Cancer and Nutrition (EPIC): an observational study', *Lancet* **361**, 1496–1501.
Blaser M J and Atherton J C (2004), '*Helicobacter pylori* persistence: biology and disease', *J Clin Invest* **113**, 321–333.
Borriello S P, Setchell K D R, Axelson M and Lawson AM (1985), 'Production and metabolism of lignans by the human faecal flora', *J App Bacteriol* **58**, 37–43.
Bourne L C and Rice-Evans C (1998), 'Bioavailability of ferulic acid', *Biochem Biophys Res Commun* **253**, 222–227.
Brummer J-M and Lorenz K (2003), 'Preferments and sourdoughs in German breads', in Kulp K and Lorenz K (Eds), *Handbook of Dough Fermentations*, Marcel Dekker Inc., New York, pp. 247–267.
Brümmer J M, Stephan H and Morgenstern G (1977), 'On the use of acidifying agents in the preparation of rye and wheat bread', *Getreide Mehl Brot* **31**, 186–190 [in German].
Burkitt D P (1973), 'Epidemiology of large bowel disease: the role of fibre', *Proc Nutr Soc* **32**, 145–149.
Bushuk W (2001), 'History, world distribution, production, and marketing', in Bushuk W (Ed.), *Rye: Production, Chemistry and Technology*, American Association of Cereal Chemists, St Paul, MN, pp. 1–7.
Bylund A, Zhang J X, Bergh A, Damber J E, Widmark A, Johansson A, Adlercreutz H, Aman P, Shepherd M J and Hallmans G (2000), 'Rye bran and soy protein delay growth and increase apoptosis of human LNCaP prostate adenocarcinoma in nude mice', *Prostate* **42**, 304–314.

Bylund A, Lundin E, Zhang J X, Nordin A, Kaaks R, Stenman U-H, Aman P, Adlercreutz H, Nilsson T K, Hallmans G, Bergh A and Stattin P (2003), 'Randomised controlled short-term intervention pilot study on rye bran bread in prostate cancer', *Eur J Cancer Preven* **12**, 407–415.

Bylund A, Saarinen N, Zhang J-X, Bergh A, Widmark A, Johansson A, Lundin E, Adlercreutz H, Hallmans G, Stattin P and Mäkelä S (2005), 'Anticancer effects of a plant lignan 7-hydroxymatairesinol on a prostate cancer model *in vivo*', *Exp Biol Med* **230**, 217–223.

Chatenoud L, La-Vecchia C, Franceschi S, Tavani A, Jacobs D-R Jr, Parpinel M-T, Soler M and Negri E (1999), 'Refined-cereal intake and risk of selected cancers in Italy', *Am J Clin Nutr* **70**, 1107–1110.

Chen Y, Ross A, Åman P and Kamal-Eldin A (2004), 'Alkylresorcinols as markers of whole grain wheat and rye in cereal products', *J Agric Food Chem* **52**, 8242–8346.

Crittenden R, Karppinen S, Ojanen S, Tenkanen M, Fagerström R, Mättö J, Saarela M, Mattila-Sandholm T and Poutanen K (2002), '*In vitro* fermentation of cereal dietary fibre carbohydrates by probiotic and intestinal bacteria', *J Sci Food Agric* **82**, 781–789.

Cummings J H and Branch W J (1982), 'Postulated mechanisms whereby fibre may protect against large bowel cancer', in Vahouny G V and Kritchevsky D (Eds), *Dietary Fibre in Health and* Disease, Plenum Press, New York and London, pp. 313–325.

Cummings J H and Englyst H N (1995), 'Gastrointestinal effects of food carbohydrate', *Am J Clin Nutr* **61**, 938S–945S.

Cummings J H, Bingham S A, Heaton K W and Eastwood M A (1992), 'Fecal weight, colon cancer risk, and dietary intake of nonstarch polysaccharides (dietary fibre)', *Gastroenterology* **103**, 1783–1789.

Cummings J H, Roberfroid M B, Andersson H, Barth C, Ferro-Luzzi A, Ghoos Y, Gibney M, Hermensen K, James W P T, Korver O, Lairon D, Pascal G and Voragen A G S (1997), 'A new look at dietary carbohydrate: chemistry, physiology and health', *Eur J Clin Nutr* **51**, 417–423.

Dervilly-Pinel G, Rimsten L, Saulnier L, Andersson R and Åman P (2001), 'Water-extractable arabinoxylan from pearled flours of wheat, barley, rye and triticale. Evidence for the presence of ferulic acid dimers and their involvement in gel formation', *J Cereal Sci* **34**, 207–214.

Englyst H N and Cummings J H (1985), 'Digestion of polysaccharides of some cereal foods in the human small intestine', *Am J Clin Nutr* **42**, 778–787.

Englyst H N, Bingham S A, Wiggins H S, Southgate D A T, Seppänen R, Helms P, Anderson V, Day KC, Choolun R, Collinson E and Cummings J H (1982), 'Nonstarch polysaccharide consumption in four Scandinavian populations', *Nutr Cancer* **4**, 50–60.

FAOSTAT, http://faostat.fao.org, accessed October 2006.

FDA (Food and Drug Administration) (1997), '*FDA Allows Whole Oat Foods to Make Health Claim on Reducing The Risk of Heart Disease*', FDA Talk Paper, January. 21, 1997.

Flood A, Velie E M, Chaterjee N, Subar A F, Thompson F E, Lacey J V Jr., Schairer C, Troisi R and Schatzkin A (2002), 'Fruit and vegetable intakes and the risk of colorectal cancer in the Breast Cancer Detection Demonstration Project follow-up cohort', *Am J Clin Nutr* **75**, 936–943.

Fredriksson H, Tallving J, Rosén J and Åman P (2004), 'Fermentation reduces free asparagine in dough and acrylamide content in bread', *Cereal Chem* **81**, 650–653.

Friedman M (2003), 'Chemistry, biochemistry, and safety of acylamide. A review', *J Agric Food Chem* **51**, 4504–4526.

Fuchs C S, Giovannucci E L, Colditz G A, Hunter D J, Stampfer M J, Rosner B, Speizer F E and Willett W C (1999), 'Dietary fibre and the risk of colorectal cancer and adenoma in women', *N Engl J Med* **340**, 169–176.

Garcia-Conesa M T, Wilson P D, Plumb G W, Ralph J and Williamson G (1999), 'Antioxidant properties of 4,4′-dihydroxy-3,3′-dimethoxy-β,β′-bicinnamic acid (8-8′-diferulic acid, non-cyclic form)', *J Sci Food Agric* **79**, 379–384.
Giovannucci E, Rimm E B, Stampfer M J, Colditz G A and Willett W C (1998), 'Diabetes mellitus and risk of prostate cancer (United States)', *Cancer Causes Control* **9**, 3–9.
Glitsø L V and Bach Knudsen K E (1999), 'Milling of whole grain rye to obtain fractions with different dietary fibre characteristics', *J Cereal Sci* **29**, 89–97.
Glitsø L V, Brunsgaard G, Hojsgaard S, Sandstrom B and Bach Knudsen K E (1998), 'Intestinal degradation in pigs of rye dietary fibre with different structural characteristics', *Br J Nutr* **80**, 457–468.
Glitsø L V, Gruppen H, Schols H A, Hojsgaard S, Sandstrom B and Bach Knudsen K E (1999), 'Degradation of rye arabinoxylans in the large intestine of pigs', *J Sci Food Agric* **79**, 961–969.
Glitsø L V, Jensen B B and Bach Knudsen K E (2000a), 'In vitro fermentation of rye carbohydrates including arabinoxylans of different structure', *J Sci Food Agric* **80**, 1211–1218.
Glitsø L V, Mazur W M, Adlercreutz H, Wahala K, Makela T, Sandstrom B and Bach Knudsen K E (2000b), 'Intestinal metabolism of rye lignans in pigs', *Br J Nutr* **84**, 429–437.
Gråsten S, Juntunen K S, Poutanen K S, Gylling H K, Miettinen T A and Mykkänen H M (2000), 'Rye bread improves bowel function and decreases the concentrations of some compounds that are putative colon cancer markers in middle-aged women and men', *J Nutr* **130**, 2215–2221.
Gråsten S, Liukkonen K H, Chevatidis A, El-Nezami H, Poutanen K S and Mykkänen H M (2003), 'Effects of wheat pentosan and inulin on the metabolic activity of fecal microbiota and on bowel function in healthy humans', *Nutr Res* **23**, 1503–1514.
Hammarsten J and Högstedt B (2005), 'Hyperinsulinaemia: A prospective risk factor for lethal clinical prostate cancer', *Eur J Cancer* **41**, 2887–2895.
Hansen Å, Lund B and Lewis M J (1989), 'Flavour of sourdough rye bread crumb', *Lebensm Wiss Technol* **22**, 141–144.
Hansen H B, Rasmussen C V, Bach Knudsen K E and Hansen Å (2003), 'Effect of genotype and harvest year on content and composition of dietary fibre in rye (*Secale cereale* L.) grain', *J Sci Food Agric* **83**, 76–85.
Hansen H B, Møller B, Andersen S B, Jørgensen J R and Hansen Å. (2004), 'Grain characteristics, chemical composition, and functional properties of rye (*Secale cereale* L.) as influenced by genotype and harvest year', *J Agric Food Chem* **52**, 2282–2291.
Harder H, Tetens I, Let M B and Meyer A S (2004), 'Rye bran bread intake elevates urinary excretion of ferulic acid in humans, but does not affect the susceptibility of LDL to oxidation *ex vivo*', *Eur J Nutr* **43**, 230–236.
Hegedüs H, Pedersen B and Eggum B O (1985), 'The influence of milling on the nutritive value of flour from cereal grains. 7. Vitamins and tryptophan', *Quality Plant Foods Human Nutr* **35**, 175–180.
Heiniö R (2003), 'Influence of processing on the flavour formation of oat and rye', VTT Publications 494, Espoo, Finland
Heiniö R-L (2006), 'Sensory attributes of bakery products', in Hui Y H, Corke H, De Leyn I, Nip W-K and Cross N (Eds.), *Bakery Products: Science and Technology*, Blackwell Publishing, Ames, IO, pp. 285–298.
Heiniö R-L, Liukkonen K-H, Katina K, Myllymäki O and Poutanen K (2003a), 'Milling fractionation of rye produces different sensory profiles of both flour and bread', *Food Sci Technol* **36**, 577–583.
Heiniö R-L, Katina K, Wilhelmson A, Myllymäki O, Rajamäki T, Latva-Kala K, Liukkonen K-H and Poutanen K (2003b), 'Relationship between sensory perception and flavour-active volatile compounds of germinated, sourdough fermented and native rye following the extrusion process', *Food Sci Technol* **36**, 533–545.

Heinonen S, Nurmi T, Liukkonen K H, Poutanen K S, Wähälä K, Deyama T, Nishibe S and Adlercreutz H (2001), '*In vitro* metabolism of plant lignans: new precursors of mammalian lignans enterolactone and enterodiol', *J Agric Food Chem* **49**, 3178–3186.
Hsing A W, Chua S Jr, Gao Y T, Gentzschein E, Chang L, Deng J and Stanczyk F Z (2001), 'Prostate cancer risk and serum levels of insulin and leptin: a population-based study', *J Natl Cancer Inst* **93**, 783–789.
Israel D A and Peek R M Jr (2006), 'The role of persistence in *Helicobacter pylori* pathogenesis', *Curr Opin Gastroenterol* **22**, 3–7.
Jacobs D R, Marquart L, Slavin J and Kushi L H (1998a), 'Whole-grain intake and cancer: an expanded review and meta-analysis', *Nutr Cancer* **30**, 85–96.
Jacobs D R, Meyer K A, Kushi L H and Folsom A R (1998b), 'Whole-grain intake may reduce the risk of ischemic heart disease death in postmenopausal women: the Iowa women's health study', *Am J Clin Nutr* **68**, 248–257.
Jacobs D R Jr, Pereira M A, Stumpf K, Pins J J and Adlercreutz H (2002), 'Whole grain food intake elevates serum enterolactone', *Br J Nutr* **88**, 111–116.
Jankiewicz M and Michniewicz J (1987), 'The effect of soluble pentosans isolated from rye grain on staling of bread', *Food Chem* **25**, 241–249.
Juntunen K (2003), 'Rye bread and glucose and insulin in healthy adults', PhD Thesis, Kuopio University Publications D, Medical Sciences 313, Kuopio, Finland.
Juntunen K S, Mazur W M, Liukkonen K H, Uehara M, Poutanen K S, Adlercreutz H C T and Mykkänen H M (2000), 'Consumption of wholemeal rye bread increases serum concentrations and urinary excretion of enterolactone compared with consumption of white wheat bread in healthy Finnish men and women', *Br J Nutr* **84**, 839–846.
Juntunen K, Niskanen L, Liukonen K H, Poutanen K, Holst J and Mykkänen H (2002), 'Postprandial glucose, insulin and incretin responses to grain products in healthy subjects', *Am J Clin Nutr* **75**, 254–262.
Juntunen K, Laaksonen D E, Autio K, Niskanen L, Holst J, Savolainen K E, Liukonen K H, Poutanen K and Mykkänen H (2003a), 'Structural differences between rye and wheat breads but not total fibre content may explain the lower postprandial insulin response to rye bread', *Am J Clin Nutr* **78**, 957–964.
Juntunen K, Laaksonen D E, Poutanen K, Niskanen L and Mykkänen H (2003b), 'High fibre rye bread and insulin secretion and sensitivity in healthy post-menopausal women', *Am J Clin Nutr* **77**, 385–391.
Kamal-Eldin A, Pouru A, Eliasson C and Åman P (2000), 'Alkylresorcinols as antioxidants: hydrogen donation and peroxyl radical-scavenging effects', *J Sci Food Agric* **81**, 353–356.
Kariluoto S, Aittamaa M, Korhola M, Salovaara H, Vahteristo L and Piironen V (2006), 'Effects of yeasts and bacteria on the levels of folates in rye sourdoughs', *Int J Food Microbiol* **106**, 137–143.
Karppinen S, Liukkonen K, Aura A-M, Forssell P and Poutanen K (2000), '*In vitro* fermentation of polysaccharides of rye, wheat and oat brans and inulin by human faecal bacteria', *J Sci Food Agric* **80**, 1469–1476.
Karppinen S, Kiiliäinen K, Liukkonen K, Forssell P and Poutanen K (2001), 'Extraction and *in vitro* fermentation of rye bran fractions', *J Cereal Sci* **34**, 269–278.
Karppinen S, Myllymäki O, Forssell P and Poutanen K (2003), 'Fructan content of rye and rye products', *Cereal Chem* **80**, 168–171.
Katina K, Arendt E, Liukkonen K-H, Autio K, Flander L and Poutanen K (2005), 'Potential of sourdough for healthier cereal product', *Trends Food Sci Technol* **16**, 104–112.
Katina K, Laitila A, Juvonen R, Liukkonen K-H, Kariluoto S, Piironen V, Landberg R, Åman P and Poutanen K (2007), 'Bran fermentation as a means to enhance technological properties and bioactivity of rye', *Int J Food Microbiol* **24**, 175–186.
Kern S M, Bennett R N, Mellon F A, Kroon P A and Garcia-Conesa, M T (2003a), 'Absorption of hydroxycinnamates in humans after high-bran cereal consumption', *J Agric Food Chem* **51**, 6050–6055.

Kern S M, Bennett R N, Needs P W, Mellon F A, Kroon P A and Garcia-Conesa M T (2003b), 'Characterisation of metabolites of hydroxycinnamates in the in vitro model of human small intestinal epithelium Caco-2 cells', *J Agric Food Chem* **51**, 7884–7891.

Kirchhoff E and Schieberle P (2002), 'Quantitation of odor-active compounds in rye flour and rye sourdough using stable isotope dilution assay', *J Agric Food Chem* **50**, 5378–5385.

Koh-Banerjee P and Rimm E B (2003), 'Whole grain consumption and weight gain: a review of the epidemiological evidence, potential mechanisms and opportunities for future research', *Proc Nutr Soc* **62**, 25–29.

Konishi Y and Shimizu M (2003), 'Transepithelial transport of ferulic acid by monocarboxylic acid transporter in Caco-2 cell monolayers', *Biosci Biotechnol Biochem* **67**, 856–862.

Korpela J T, Korpela R and Adlercreutz H (1992), 'Fecal bile acid metabolic pattern after administration of different types of bread', *Gastroenterol* **103**, 1246–1253.

Kozubek A and Tyman J H P (1999), 'Resorcinolic lipids, the natural non-isoprenoid phenolic amphiphiles and their biological activity', *Chem Rev* **99**, 1–25.

Kroon P A, Faulds C B, Ryden P, Robertson J A and Williamson G (1997), 'Release of covalently bound ferulic acid from fibre in the human colon', *J Agric Food Chem* **45**, 661–667.

Kühn M C and Grosch W (1989), 'Baking functionality of reconstituted rye flours having different nonstarchy polysaccharide and starch contents', *Cereal Chem* **66**, 149–153.

Laaksonen D E, Toppinen L K, Juntunen K S, Autio K, Poutanen K S, Niskanen L K and Mykkänen H M (2005), 'Dietary carbohydrate modification enhances insulin secretion in individuals with the metabolic syndrome', *Am J Clin Nutr* **82**, 1218–1227.

Landberg R, Kamal-Eldin A, Andersson R and Åman P (2006a), 'Alkylresorcinol content and homologue composition in durum wheat (*Triticum durum*) kernels and pasta', *J Agric Food Chem* **54**, 3012–3014.

Landberg R, Linko A-M, Kamal-Eldin A, Vessby B, Adlercreutz H and Åman P (2006b), 'Human plasma kinetics and relative bioavailability of alkylresorcinols after intake of rye bran', *J Nutr* **136**, 2760–2765.

Landström M, Zhang J X, Hallmans G, Åman P, Berg A, Damber J E, Mazur W, Wähälä K and Adlercreutz H (1998), 'Inhibitory effects of soy and rye diets on the development of Dunning R3327 prostate adenocarcinoma in rats', *Prostate* **36**, 151–161.

Lehrer S, Diamond E J, Stagger S, Stone N N and Stock R G (2002), 'Increased serum insulin associated with increased risk of prostate cancer recurrence', *Prostate* **50**, 1–3.

Leinonen K, Liukonen K-H, Poutanen K, Uusitupa M and Mykkänen H (1999), 'Rye bread decreases postprandial insulin response but does not alter glucose response in healthy Finnish subjects', *Eur J Clin Nutr* **53**, 262–267.

Leinonen K, Poutanen K and Mykkänen H (2000), 'Rye bread decreases serum total and LDL cholesterol', *J Nutr* **130**, 164–170.

Leung W K, Ng E K, Lam C C, Chan K F, Chan W Y, Auyeung A C, Wu J C, Ching J Y, Lau J Y and Sung J J (2006), '*Helicobacter pylori* infection in 1st degree relatives of Chinese gastric cancer patients', *Scand J Gastroenterol* **41**, 274–279.

Linko A M and Adlercreutz H (2005), 'Whole grain rye and wheat alkylresorcinols are incorporated into human erythrocyte membranes', *Br J Nutr* **93**, 11–13.

Linko A, Parikka K, Wahala K and Adlercreutz H (2002), 'Gas chromatographic-mass spectrometric method for the determination of alkylresorcinols in human plasma', *Anal Biochem* **308**, 307–313.

Linko A M, Juntunen K S, Mykkanen K S and Adlercreutz H (2005), 'Whole-grain rye bread consumption by women correlates with plasma alkylresorcinols and increases their concentration compared with low-fibre wheat bread', *J Nutr* **135**, 580–583.

Liu S (2002), 'Intake of refined carbohydrates and whole grain foods in relation to risk of type 2 diabetes mellitus and coronary heart disease', *J Am Coll Nutr* **21**, 298–306.

Liu S (2003), 'Wholegrain foods, dietary fibre, and type 2 diabetes: searching for a kernel of truth', *Am J Clin Nutr* **77**, 527–529.

Liu S, Manson J E, Stampfer M J, Hu F, Giovannucci E, Colditz G, Hennekens C H and Willett W C (2000a), 'A prospective study of whole-grain intake and risk of type 2 diabetes mellitus in US women', *Am J Public Health* **90**, 409–1415.

Liu S, Manson J E, Stampfer M J, Rexrode K M, Hu F B, Rimm E B and Willett W C (2000b), 'Whole grain consumption and risk of ischemic stroke in women: A prospective study', *JAMA* **284**, 1534–1540.

Liu S, Willett W C, Manson J E, Hu F B, Rosner B and Colditz G (2003), 'Relation between changes in intakes of dietary fibre and grain products and changes in weight and development of obesity among middle-aged women', *Am J Clin Nutr* **78**, 920–927.

Liukkonen K-H, Katina K, Wilhelmsson A, Myllymäki O, Lampi A, Kariluoto S, Piironen V, Heinonen S, Nurmi T, Adlercreutz H, Peltoketo A, Pihlava J, Hietaniemi V and Poutanen K (2003), 'Session: Health effects of whole grains. process-induced changes on bioactive compounds in whole grain rye', *Proc Nutr Soc* **62**, 117–122.

Liukkonen K-H, Heiniö R-L, Salmenkallio-Marttila M, Autio K, Katina K and Poutanen K (2006), 'Rye', in Hui Y H, Corke H, De Leyn I, Nip W-K, and Cross N (Eds), *Bakery Products: Science and Technology*, Blackwell Publishing, Ames, Iowa, USA, pp. 109–122.

Liukkonen K-H, Katina K, Kaukovirta-Norja A, Lampi A-M, Kariluoto S, Piironen V, Heinonen S-M, Adlrecreutz H, Nurmi A, Pihlava J-M and Poutanen K, (2007), 'Influence of germination conditions on the bioactivity of rye', in Marquardt L, Reicks M, Jacobs D, McIntosh G H and Poutanen K (Eds), *Whole Grains and Health*, Blackwell Publishing, Ames, Iowa, USA.

Lundin E A, Zhang J X, Lairon D, Tidehag P, Åman P, Adlercreutz H and Hallmans G (2004), 'Effects of meal frequency and high-fibre rye-bread diet on glucose and lipid metabolism and ileal excretion of energy and sterols in ileostomy subjects', *Eur J Clin Nutr* **58**, 1410–1419.

MacFarlane G T and Cummings J H (1991), 'The colonic flora, fermentation, and large bowel digestion function', in Phillips S F, Pemberton J H, and Shorter R G (Eds), *The Large Intestine: Physiology, Pathophysiology and Disease*, Raven Press Ltd, New York, pp. 51–92.

MacFarlane G T, Cummings J H and Allison C (1986), 'Protein degradation by human intestine bacteria', *J Gen Micro* **132**, 1647–1656.

Mahdavi J (2004), *'Helicobacter pylori Adhension and Patho-Adaptation: The Role of BabA And SabA Adhesins in Persistent Infection and Chronic Inflammation'*, Umeå University Medical Dissertation No. 83, Umeå, Sweden.

Mariotti M, Alamprese C, Pagani M A and Lucisano M (2006), 'Effect of puffing on ultrastructure and physical characteristics of cereal grains and flours', *J Cereal Sci* **43**, 47–56.

Mattila P, Pihlava J-M and Hellström J (2005), 'Contents of phenolic acids, alkyl- and alkenylresorcinols and avenanthramides in commercial grain products', *J Agric Food Chem* **53**, 8290–8295.

McIntosh G H, Royle P J and Pointing G (2001), 'Wheat aleurone flour increases ceacal β-glucuronidase activity and butyrate concentration and reduces colon adenoma burden in azoxymethane-treated rats', *J Nutr* **131**, 127–131.

McIntosh G H, Noakes M, Royle P J and Foster P R (2003), 'Whole-grain rye and wheat foods and markers of bowel health in overweight middle-aged men', *Am J Clin Nutr* **77**, 967–974.

McNeil N I (1984), 'The contribution of the large intestine to energy supplies in man', *Am J Clin Nutr* **39**, 338–342.

Meuser, F, Brümmer, J M and Seibel W (1994), 'Bread varieties in Central Europe', *Cereal Foods World* **39**, 222–230.

Mitchell J R (1979), 'Rheology of polysaccharide solutions and gels', in Blanshard J M C and Mitchell J R (Eds), *Polysaccharides in Food*, Butterworth, London, pp. 51–72.

Murtaugh M A, Jacobs D R Jr, Jacob B, Steffen L M and Marquart L (2003), 'Epidemiological support for the protection of whole grains against diabetes', *Proc Nutr Soc* **62**, 143–149.

Mustafa A, Andersson R, Rosén J, Kamal-Eldin A, and Åman P (2005), 'Factors influencing acrylamide content and color in rye crisp bread', *J Agric Food Chem* **53**, 5985–5989.

Nardini M, Cirillo E, Natella F and Scaccini C (2002), 'Absorption of phenolic acids in humans after coffee consumption', *J Agric Food Chem* **50**, 5735–5741.

Nilsson M (1999), *'The Dietary Fibre Complex of Rye Grain, with Emphasis on Arabinoxylan'*, PhD Thesis, Agaria 142, Swedish University of Agricultural Sciences.

Nilsson M, Åman P, Häkönen H, Hallmans G, Bach Knudsen K E, Mazur W and Adlercreutz H (1997), 'Content of nutrients and lignans in roller milled fractions of rye', *J Sci Food Agric* **73**, 143–148.

Nilsson M, Andersson R, Andersson R E, Autio K and Åman P (2000), 'Heterogeneity in a water-extractable rye arabinoxylan with a low degree of disubstitution', *Carbohydr Polym* **41**, 397–405.

Penalvo J L, Haajanen K M, Botting N and Adlercreutz H (2005), 'Quantification of lignans in food using isotope dilution gas chromatography/mass spectrometry', *J Agric Food Chem* **53**, 9342–9347.

Pereira M A, Jacobs D R Jr, Pins J J, Raatz S K, Gross M D, Slavin J L and Seaquist E R (2002), 'Effect of whole grains on insulin sensitivity in overweight hyperinsulinemic adults', *Am J Clin Nutr* **75**, 848–855.

Peters U, Sinha R, Chatterjee N, Subar A F, Ziegler R G, Kulldorff M, Bresalier R, Weissfeld J L, Flood A, Schatzkin A and Hayes R B (2003), 'Dietary fibre and colorectal adenoma in a colorectal cancer early detection programme', *Lancet* **361**, 1491–1495.

Petterson D and Åman P (1989), 'Enzyme supplementation of a poultry diet containing rye and wheat', *Br J Nutr* **62**, 139–149.

Pettersson D, Åman P, Bach Knudsen K E, Lundin E, Zhang J-X, Hallmans G and Härkönen H (1996), 'Intake of rye bread by ileostomists increases ileal excretion of fibre polysaccharide components and organic acids but does not increase plasma or urine lignans and isoflavonoids', *J Nutr* **126**, 1594–1600.

Piber M and Koehler P (2005), 'Identification of dehydro-ferulic acid-tyrosine in rye and wheat: evidence for a covalent cross-link between arabinoxylans and proteins', *J Agric Food Chem* **53**, 5276–5284.

Piironen V, Toivo J and Lampi A M (2002), 'Plant sterols in cereals and cereal products', *Cereal Chem* **79**, 148–154.

Ragaee S M, Campbell G L, Scoles G J, Mcleod J G and Tyler R T (2001), 'Studies on rye (*Secale cereale* L.) lines exhibiting a range of extract viscosities. 1. Composition, molecular weight distribution of water extracts, and biochemical characteristics of purified water-extractable arabinoxylan', *J Agric Food Chem* **49**, 2437–2445.

Rodriguez C, Patel A V, Mondul A M, Jacobs E J, Thun M J and Calle E E (2005), 'Diabetes and risk of prostate cancer in a prospective cohort of US men', *Am J Epidemiol* **161**, 147–152.

Roediger W E W (1980), 'The colonic epithelium in ulcerative colitis: an energy-deficiency disease?', *Lancet* **2**, 712–715.

Rondini L, Peyrat-Maillard M-N, Marsset-Baglieri A and Berset C (2002), 'Sulfated ferulic acid is the main *in vivo* metabolite found after short-term ingestion of ferulic acid in rats', *J Agric Food Chem* **50**, 3037–3041.

Rondini L, Peyrat-Maillard M N, Marsset-Baglieri A, Fromentin G, Durant P, Tomé D, Prost M and Berset C (2004), 'Bound ferulic acid from bran is more bioavailable than the free compound in rat', *J Agric Food Chem* **52**, 4338–4343.

Ross A B, Kamal-Eldin A, Lundin E, Zhang J-X, Hallmans G and Åman P (2003a), 'Cereal alkylresorcinols are absorbed by humans', *J Nutr* **133**, 2222–2224.

Ross A B, Shepard M J, Bach Knudsen K E, Glitsø L V, Bowey E, Philips J, Rowland I, Gou Z-X, Massy D J R, Åman P and Kamal-Eldin A (2003b), 'Absorption of dietary alkylresorcinols in ileum-canulated pigs and rats', *Br J Nutr* **90**, 787–794.

Ross A, Kamal-Eldin A and Åman P (2004a), 'Dietary alkylresorcinols: Absorption, bioactivities, and possible use as biomarkers of wholegrain wheat- and rye-rich foods', *Nutr Rev* **62**, 81–95.

Ross A, Åman P and Kamal-Eldin A (2004b), 'Identification of cereal alkylresorcinol metabolites in human urine – potential biomarkers of wholegrain wheat and rye intake', *J Chromatogr B* **809**, 125–130.

Ross A B, Chen Y, Frank J, Swanson J E, Parker R S, Kozubeck A, Lundh T, Vessby B, Åman P and Kamal-Eldin A (2004c), 'Cereal alkylresorcinols elevate γ-tocopherol levels in rats and inhibit γ-tocopherol metabolism *in vitro*', *J Nutr* **134**, 506–510.

Ross A, Becker W, Chen Y, Kamal-Eldin A and Åman P (2005), 'Intake of alkylresorcinols from wheat and rye in the United Kingdom and Sweden', *Br J Nutr* **94**, 496–499.

Rothenbacher D and Brenner H (2003), 'Burden of *Helicobacter pylori* and *H. Pylori*-related diseases in developed countries: recent developments and future implications', *Microbes Infect* **5**, 693–703.

Salmenkallio-Marttila M, Katina K and Autio K (2001), 'Effect of bran fermentation on quality and microstructure of high-fibre wheat bread', *Cereal Chem* **78**, 429–435.

Salovaara H and Autio K (2001), 'Rye and triticale', in Dendy S A V and Dobraszczyk B J (Eds), *Cereals and Cereal Products, Chemistry and Technology*, Aspen Publishers, Gaithersburg, Maryland, USA, pp. 391–410.

Scheppach W and Weiler F (2004), 'The butyrate story: old wine in new bottles?', *Curr Opin Clin Nutr Metab Care* **7**, 563–567.

Seal C J (2006) 'Whole grains and CVD risk', *Proc Nutr Soc* **65**, 24–34.

Seibel W and Weipert D (2001) 'Bread baking and other food uses around the world', in Bushuk W (Ed.), *Rye: Production, Chemistry and Technology*, American Association of Cereal Chemists, St Paul, Minnesota, USA, pp. 147–211.

Setchell K D and Adlercreutz H (1979), 'The excretion of two new phenolic compounds (180/442 and 180/410) during the human menstrual cycle and in pregnancy', *J Steroid Biochem* **11**, xv–xvi.

Setchell K D and Cassidy A (1999), 'Dietary isoflavones: biological effects and relevance to human health', *J Nutr* **129**, 758S–767S .

Setchell K D, Bull R and Adlercreutz H (1980a), 'Steroid excretion during the reproductive cycle and in pregnanacy of the vervet monkey (*Cercopithecus aethiopus pygerythrus*)', *J Steroid Biochem* **12**, 375–384.

Setchell K D, Lawson A M, Mitchell F L, Adlercreutz H, Kirk D N and Axelson M (1980b), 'Lignans in man and in animal species', *Nature* **287**, 740–742.

Shewry P R and Bechtel D B (2001), 'Morphology and chemistry of rye grain', in Bushuk W (Ed.), *Rye: Production, Chemistry and Technology*, American Association of Cereal Chemists, St Paul, Minnesota, USA, pp. 69–127.

Slavin J (2003), 'Why whole grains are protective: biological mechanisms', *Proc Nutr Soc* **62**, 129–134.

Smith J G, Yokoyama W H and German J B (1998), 'Butyric acid from the diet: Action at the level of gene expression', *Critical Rev Food Sci* **38**, 259–297.

Stocks T, Lukanova A, Rinaldi S, Biessy C, Lindahl B, Hallmans G, Kaaks R and Stattin P (2006), 'Insulin resistance is inversely related to prostate cancer: a prospective study in Northern Sweden', Abstract in Nordic Research Seminar on Biobanking, Malmö, Sweden, 26–28 August 2006.

Syväoja E-L, Piironen V, Varo P, Koivistoinen P and Salminen K (1986), 'Tocopherols and tocotrienols in Finnish foods: Oils and fats', *J Am Oil Chem Soc* **63**, 328–329.

Tatham A S and Shewry P R (1995), 'The S-poor prolamins of wheat, barley and rye', *J Cereal Sci* **22**, 1–16.

Tavani A, Gallus S, Bosetti C, Tzonou A, Lagiou P, Negri E, Trichopoulos D and La Vecchia C (2002), 'Diabetes and the risk of prostate cancer', *Eur J Cancer Prev* **11**, 125–128.

Teuchy H and Van Sumere C F (1971), 'Metabolism of [1-^{14}C] phenylalanine, [3-^{14}C] cinnamic acid and [2-^{14}C] ferulic acid in the rat', *Arch Int Physiol Biochem* **79**, 589–618.

Tluscik F (1978), 'Localization of alkylresorcinols in rye and wheat caryopses', *Acta Societatis Botanicorum Poloniae* **47**, 211–218.

Tuomilehto J, Lindstrom J, Eriksson J G, Valle T T, Hamalainen H, Ilanne-Parikka P, Keinanen-Kiukaanniemi S, Laakso M, Louheranta A, Rastas M, Salminen V and Uusitupa M (2001), 'Prevention of type 2 diabetes mellitus by changes in lifestyle among subjects with impaired glucose tolerance', *N Engl J Med* **344**, 1343–1350.

Vanharanta M, Voutilainen S, Lakka T, van der Lee M, Adlercreutz H and Salonen J T (1999), 'Risk of acute coronary events according to serum concentrations of enterolactone: A prospective population-based case-control study', *Lancet* **354**, 2112–2115.

Verwimp T, Vandeputte G E, Marrant K and Delcour J A (2004), 'Isolation and characterisation of rye starch', *J Cereal Sci* **39**, 85–90.

Vinkx C J A and Delcour J A (1996) 'Rye (*Secale cereale* L) arabinoxylans: A critical review', *J Cereal Sci* **24**, 1–14.

Wasserman L A, Eiges N S, Koltysheva G K, Andreev N R, Karpov V G and Yuryev V P (2001), 'The application of different physical approaches for the description of structural features in wheat and rye starches. A DSC study', *Starch/Stärke* **53**, 629–634.

Wild C, Andersson C, O'Brien N, Wilson L, and Woods J (2001), 'A critical evaluation of the application of biomarkers in epidemiological studies on diet and health', *Br J Nutr* **86**, S37–S53.

Williams E A, Coxhead J M and Mathers J C (2003), 'Anti-cancer effects of butyrate: use of micro-array technology to investigate mechanisms', *Proc Nutr Soc* **62**, 107–115.

Zhang J X, Lundin E, Hallmans G, Adlercreutz H, Andersson H, Bosaeus I, Åman P, Stenling R and Dahlgren S (1994), 'Effect of rye bran on excretion of bile acids, cholesterol, nitrogen and fat in human subjects with ileostomies', *Am J Clin Nutr* **59**, 389–394.

Zhao Z, Egashira Y and Sanada H (2003a), 'Digestion and absorption of ferulic acid sugar esters in rat gastrointestinal tract', *J Agric Food Chem* **51**, 5534–5539

Zhao Z, Egashira Y and Sanada H (2003b), 'Ferulic acid sugar esters are recovered in rat plasma and urine mainly as the sulfoglucuronide of ferulic acid', *J Nutr* **133**, 1355–1361.

Zhao Z, Egashira Y and Sanada H (2004), 'Ferulic acid is quickly absorbed from rat stomach as the free form and then conjugated mainly in liver', *J Nutr* **134**, 3083–3088.

12

Functional barley products

A. A. H. Andersson and P. Åman, Swedish University of Agricultural Sciences, Sweden

12.1 Introduction

Barley is one of the oldest cultivated cereal grains and is in some parts of the world consumed in large quantities, for example, in Tibet and Morocco. In the Western world barley is mostly used for feed, malt and alcohol production. There is, however, a large interest in increasing the food consumption of barley in Western countries also, mainly due to the high content of dietary fibre, especially partly soluble β-glucan which can lower serum cholesterol levels in humans. The endosperm cell walls in barley are also more resistant to degradation than in wheat and maize, and thereby the rate of release of nutrients in the upper gastrointestinal tract may be reduced. This chapter deals with barley for food, and includes a section on the history of barley foods as well as the use of barley for foods today in the West Asia–North Africa regions. Research on some different products made in the Western world is also discussed in this section. Section 12.3 deals with the health benefits of barley, focusing on the cholesterol-lowering and the insulin- and glucose-lowering effects. Section 12.4 describes some problems encountered in the development of functional barley products, and Section 12.5 the manufacturing technology for barley. In the last section some future trends for barley are outlined. The use of barley for malt, beer and whisky will not be dealt with in this chapter.

12.2 The range of barley products

12.2.1 History of barley foods

The location and age of the earliest remains of barley are the subject of much discussion. Wendorf *et al.* (1979) have reported remains from Aswan in

Egypt from as long as 18 000 years ago. The Fertile Crescent of the Near East – in present-day Israel, northern Syria, southern Turkey, eastern Iraq and western Iran – is, however, the most likely area of barley origin, where barley cultivation has been archeologically traced back to 7000 BC (Evans, 1968). Harlan (1968) has reported remains from Iran dating from about 7900 BC, although the remains were few and cultivation uncertain. At that time only two-rowed barleys were found. About 6000 BC six-rowed and naked characters appeared and barley had become perhaps the most important cultivated plant of the region (Harlan, 1968). Other areas where barley was grown at an early date were Ethiopia, China and Korea. In Scandinavia, barley was cultivated as a cereal crop during the Bronze Age, about 2000 BC (Munck, 1981).

Barley has been used as a food grain for a long time and biblical references suggest that barley was used as a food in Old Testament times (Newman & McGuire, 1985). Indian literature from about 1500 BC describes the organoleptic properties of barley as astringent, cold, rough, light and sweet (Grivetti, 1991). In Greece, around 800 BC, barley was eaten as *alphita* which was roasted, ground barley mixed with water and eaten with oil and other condiments (Beaven, 1947). Barley was also the staple food of the Roman and Greek gladiators, who were called 'hordearii' because of barley's Latin name *Hordeum* (Kent, 1983). In Finland, the earliest of all breads was *Rieska*, which was unleavened barley bread (Bhatty, 1993a), and in England bread from barley and rye was the main food in the diet of peasants and country folk in the 15th century (Newman & McGuire, 1985). It was also a staple food in the rural areas of Denmark as late as the beginning of the 19th century. The diet at the beginning of the 1900s in Lunnede on the island of Funen in Denmark could be exemplified as follows (Munck, 1981):

Morning:	Porridge of barley grits cooked in milk or beer.
Noon:	Meat broth with abraded barley.
Evening:	Porridge of barley cooked in sufficient amounts to provide for next day's breakfast as well.

In Scandinavia pearled barley and grits are traditionally used in soups, stews and porridge. In Dalarna in Sweden, peameal, barkmeal and oats were blended into barley meal for baking a thin bread cake (Munck, 1981). Today, however, the common thin crisp bread (*tunnbröd*) is mostly made of wheat instead of barley.

12.2.2 Barley foods today

Very little barley is used in human foods today except in the West Asia–North Africa region, particularly Morocco, where per capita consumption of barley is 39 kg/year (FAOSTAT, 2006), and in certain mountain areas of the Himalayas where the consumption can be very high. In Tibet, for example, naked barley is still the staple food, and the per capita consumption is reported to be as much as 155 kg/year (Tashi, 2005). Many different traditional dishes are made from barley in these regions. In Morocco, green immature barley grains are used to make *azenbou*, which is steamed, dried and fried grains. The grains are thereafter milled to flour or semolina, which can be mixed with hot or cold water, oil, eggs or buttermilk to make different dishes (Amri *et al.*, 2005). Grains that have been stored for over 3 years, and have a special odor, are called *aballagh*; these are also used to make flour and semolina. Common barley flour and semolina is used for making different types of foods, for example, bread and couscous. Dishes from whole barley grain are also made, for example, *tijmirout*, which is made by roasting the whole mature spikes of barley. Whole grains are also often consumed like popcorn (Amri *et al.*, 2005). In Tunisia, barley bread called *kisra*, and barley soup called *fric*, are made from whole cracked barley (El Felah & Medimagh, 2005). In Tibet, the main product made from hull-less barley is *tsangpa*, which is roasted barley flour (Tashi, 2005). *Tsangpa* can be added to tea, skimmed milk or cold water and consumed as a beverage. It can also be mixed with sugar and eaten directly, or used in different cakes. Most commonly it is mixed with a little tea and kneaded into a dough-like ball called *ba*, which can be cooked in different ways. Porridge, soup, snack foods and beverages are also made from barley in Tibet. Other traditional barley dishes are made in Egypt, Libya, Eritrea, Ethiopia, Yemen, Iran, Nepal, Ecuador and Peru (Grando & Gomez MacPherson, 2005).

The consumption of barley has, however, decreased in most countries. For example in Korea, per capita consumption of barley has decreased from 64 kg/year in 1961 to only 6.5 kg/year in 2004 because of increased consumption of rice and wheat (FAOSTAT, 2006). In Denmark, Italy and the United Kingdom the consumption of barley is only 0.3 kg per person per year; this is also the level of consumption in Sweden.

12.2.3 Barley muffins

Barley flour can replace part of or all of the wheat flour in muffins with good results. Using 70% barley flour Berglund *et al.* (1992) made blueberry muffins that did not differ from wheat flour muffins in appearance, texture, flavour, sweetness and aftertaste. The volume of the barley muffins, however, was lower. Hudson *et al.* (1992) compared high-fibre muffins prepared with oat bran (100%), barley bran (40%) and rice bran (60%), with a commercial oat bran mix muffin. They found that the high-fibre muffins were rated acceptable or better in overall quality by four different taste panels. Muffins have also been successfully prepared using as much as 100% barley flour, although cultivar differences influenced volume and density (Newman *et al.*, 1990). Taste panels scored the barley muffins equal or higher in overall acceptability than control wheat muffins. Muffins made from hull-less and/or waxy types of barley were most tender, and muffins made from the flour of covered barley with normal starch scored the best in height measurements. The greatest volume was produced in muffins made from hull-less barley with waxy starch and short awns. Muffins made with a barley milling fraction (shorts) replacing 23% of the flour, were similar to standard muffins made of wheat in overall acceptability (Newman *et al.*, 1998).

12.2.4 Yeast-leavened bread made of barley

In yeast-leavened bread, barley flour is less desirable because of its lack of gluten and its greater water retention due to soluble fibre (Newman & Newman, 1991). Loaf volume and crumb texture is negatively affected by the addition of barley flour. According to Bhatty (1986a) no more than 10% of barley flour could be incorporated into white pan bread without affecting loaf volume. Briggs (1978), however, reported that barley flour can be mixed with wheat flour up to 30% and still produce loaves of acceptable volume. Dhingra and Jood (2004) reported that 15% barley flour could be added to produce acceptable breads, while 20% addition resulted in breads with a compact texture of the bread crumb. Berglund *et al.* (1992) developed a formula for whole wheat bread containing 26% barley flour. There was no difference in sensory scores for texture, flavour and overall acceptability compared with whole wheat bread. However, the colour score was significantly lower and the barley bread was denser than the whole wheat bread. Trogh *et al.* (2004) showed that baking with 40% hull-less barley flour decreased the volume compared with wheat breads. This decrease was, however, overcome to a large extent by the use of endoxylanases, which transformed detrimental water-unextractable arabinoxylans into solubilized arabinoxylans. Barley flour can also be used in sour dough breads to improve taste (Marklinder *et al.*, 1996). The loaf volume was, however, not increased by the addition of barley sour dough, which is the case for rye sour dough due to the increased ability of the pentosans (arabinoxylan and arabinogalactan peptides) in rye to hold water (Spicher & Stephan, 1992). The levels of pentosans in barley are lower than in rye, which may explain why the loaf volume was not improved for barley as it was for rye (Marklinder *et al.*, 1996). Different fractions of barley can also be incorporated into wheat bread. Newman *et al.* (1998) used 26% barley shorts, this decreased the loaf volume but did not affect the sensoric properties of the bread. Barley milling fractions enriched in β-glucan (Cavallero *et al.*, 2002) and purified barley β-glucan fractions (Symons & Brennan, 2004) have also been tested in wheat breads to study the physiological effects of barley β-glucan.

12.2.5 Barley pasta

Barley flour can also be used in pasta. Melland *et al.* (1984) reported acceptable spaghetti prepared with barley. Since product colour was the major drawback, the barley flour was bleached to obtain a whiter product. Oriental noodle products, traditionally made from buckwheat that produces a normal dark colour, are possible uses for barley flour (Newman & Newman, 1991). Noodles made with 75% barley flour and 25% semolina flour were given lower scores for appearance and texture than wheat noodles, but similar scores for flavour (Berglund *et al.*, 1992). Asian noodles prepared with 25% hull-less barley fibre-rich fractions were also given lower scores than wheat noodles for appearance and colour (Izydorczyk *et al.*, 2005). On the other hand, cooking time was reduced, cooked texture was improved due to increased firmness and resistance to compression, and nutritional quality was improved due to increased levels of total dietary fibre and β-glucan.

12.2.6 Cookies and biscuits

Barley flour and/or flakes can be incorporated into cookies and biscuits. Chocolate chip cookies prepared with 50% barley flour and raisin cookies with 36% barley flakes received similar ratings for texture, flavour and overall appeal to the control cookies made of wheat and oat bran, respectively (Berglund *et al.*, 1992). The appearance, however, was rated lower for barley cookies and the surface of the chocolate chip cookies was more 'cracked' and 'grainy', while the raisin cookies spread more than oat bran cookies. Newman *et al.* (1992) prepared cookies with 23–28% barley flour and found no difference in spread area from control cookies. They also prepared biscuits with 23–28% barley flour: these were depressed in height and had the lowest score for sensory quality of the products made (bread, muffins, biscuits and cookies). Berglund *et al.* (1992) found that the volume of biscuits decreased as the content of barley increased. There was no significant difference in preference for flavour or overall acceptability when 25% and 35% barley biscuits were compared with wheat flour biscuits. The panelists preferred, however, the appearance and texture of the wheat biscuits compared with the barley biscuits. Cookies and biscuits made of 50% hull-less barley shorts were not significantly different from wheat flour cookies and biscuits in sensory evaluation (Newman *et al.*, 1998). The height of the barley biscuits and the thickness of the barley cookies, however, were lower than those of the wheat biscuits and cookies.

12.2.7 Other products

Carrot–spice bars and granola bars have been prepared with 100% barley flour and 100% barley flakes or 32% extruded crisp barley, respectively (Berglund *et al.*, 1992). The barley carrot–spice bars were similar to wheat carrot–spice bars in terms of appearance, texture and flavour. The colour values were lower, indicating a darker colour for barley spice bars than for wheat spice bars. The appearance, flavour and texture for barley granola bars were scored higher than for control granola bars made of oatmeal.

Traditional Swedish barley products such as porridge and puddings have been prepared with barley flour and flakes from different barley genotypes (Sundberg & Falk, 1994; Sundberg *et al.* 1994). The type of barley had a significant effect on the juiciness and grain consistency of the puddings, according to a sensory taste panel (Sundberg *et al.*, 1994) (see also Section 12.5, Manufacturing technology). Tortillas have been prepared with barley flour from different barley genotypes, and these were liked slightly more or better than wheat tortillas by 90% of the panelists (Ames *et al.*, 2006). Tortillas made from waxy barley genotypes were easier to roll and broke less than tortillas made from other genotypes. Breakfast sausages with barley β-glucan as a fat replacer have also been prepared with good results (Morin *et al.*, 2002).

The examples above show that wheat can be successfully substituted by barley flour/flakes in many food products in amounts ranging from 25 to 100%. In many cases, food products prepared with barley are accepted equally as well as products prepared using the original formulae containing wheat or oats. Figure 12.1 shows some examples of barley products that can be found in Sweden today.

12.3 Health benefits of barley

12.3.1 Nutrients

The barley grain has a healthy nutrient profile. It has a high content of dietary fibre, especially β-glucan, and also of vitamins and minerals. There

Fig. 12.1 Some examples of Swedish barley products. From left to right in the front: flour, flakes, grits, grains and two types of traditional Swedish sausages made with barley grits. At the back: traditional Swedish thin crisp bread (*tunnbröd*) made of barley flour.

is a large variation in the content of the different chemical constituents of barley due to genotype and environment, and also due to interactions between the two. The major components are starch, dietary fibre and protein, with contents of 47–67, 13–28 and 9–22% of dry matter, respectively (Åman *et al.*, 1985; Åman & Newman, 1986; Oscarsson *et al.*, 1996; Andersson *et al*, 1999a). Fat, ash and low molecular weight sugars are minor components constituting about 2–4, 2–3 and 1–7% of dry matter, respectively (Åman *et al.*, 1985; Andersson *et al.*, 1999a). Figure 12.2 shows a comparison of the chemical composition of different types of barley. The major dietary fibres in barley are β-glucan and arabinoxylan, which have been reported to constitute 3–11% and 4–7% of dry matter, respectively (Åman & Graham, 1987; Miller & Fulcher, 1994; Oscarsson *et al.*, 1996; Andersson *et al.*, 1999a). There is a large difference in dietary fibre content between different types of barleys. In hull-less barleys the content is lower than in covered barleys, due to the lack of a husk which has a high content of cellulose (30%), arabinoxylan (35%) and Klason lignin (20%) (Salomonsson *et al.*, 1980). Hull-less barley is, however, higher in dietary fibre content than pearled covered barley, indicating that the content in the endosperm cell walls is higher in hull-less barley (Heen *et al.*, 1991) or that part of the bran is removed during the pearling process (see Section 12.5, Manufacturing technology). The content of β-glucan is also very different in different types of barley. Barleys with waxy and high-amylose starches have a higher content of β-glucan than barleys with normal starch (Ullrich, 1986; Oscarsson *et al.*, 1996; Andersson *et al.*, 1999a). Extreme types of barley, such as Prowashonupana, contain as much as 15% β-glucan (Åman & Newman, 1986; Andersson *et al.*, 1999b;

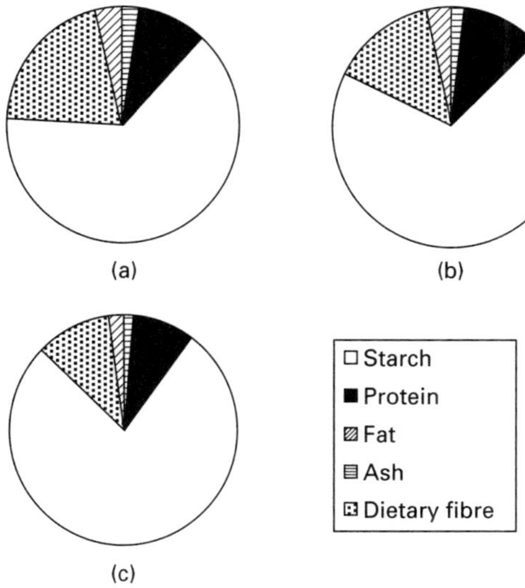

Fig. 12.2 Chemical composition of (a) hulled, (b) hull-less and (c) pearled hulled barley. Compiled from Heen *et al.* (1991).

Munck *et al.*, 2004). These types have a shrunken endosperm with thick cell walls and a very low content of starch.

12.3.2 The cholesterol-lowering effect

Whole grain barley and barley-containing products are today allowed to claim that they reduce the risk of coronary heart disease (FDA, 2005; Åman, 2006; SNF, 2007). An example of the health claim that may be used is: 'Soluble fibre from foods such as barley, as part of a diet low in saturated fat and cholesterol, may reduce the risk of coronary heart disease'. To qualify for the health claim, the barley-containing foods must provide at least 0.75 g of soluble fibre per serving of the food. The reduced risk of coronary heart disease is linked to the cholesterol-lowering effect of barley, which is probably caused by the soluble dietary fibres. In barley the major component of the soluble fibre fraction is β-glucan, but other cell wall components such as arabinoxylan may also be partly soluble (Bhatty, 1992a). The mechanism by which soluble fibres, and especially β-glucan, lowers serum cholesterol is not completely understood, but it is generally believed that soluble fibres increase the viscosity in the small intestine leading to increased excretion of bile acids and cholesterol to the large intestine. This results in a reduction in bile acids returning to the liver, which means that new bile acids have to be synthesized by using cholesterol from the blood and thereby the serum cholesterol levels are decreased. Viscosity is influenced by, among other things, the concentration and molecular weight of the β-glucan molecules, and these two factors are therefore important for the nutritional effects of barley (Wood *et al.*, 1991a).

The cholesterol-lowering effect of barley β-glucan has been shown in several animal studies. Chickens fed a barley diet had lower plasma total cholesterol and low-density lipoprotein (LDL) cholesterol than chickens fed a corn diet (Newman *et al.*, 1991; Wang *et al.*, 1992). Supplementation of the barley diet with β-glucanase resulted in higher plasma total cholesterol and LDL cholesterol concentrations and also higher digestibility of lipids than in diets without β-glucanase, suggesting that β-glucan is responsible for the cholesterol-lowering effect. The effect of viscosity on the cholesterol-lowering effect was shown by Fadel *et al.* (1987) by feeding chickens with two different barley diets with similar total β-glucan contents (5%), but with different proportions of soluble β-glucan. The diet with a higher proportion of soluble β-glucan, with a higher molecular weight resulting in higher viscosity, caused significant reductions in both total (–16%) and LDL (–30%) cholesterol, which were reversed by supplementing the diet with β-glucanase. The other diet had no effect on serum cholesterol. Another barley, a genotype grown in Arizona, with a β-glucan content of 10–11% gave reductions of as much as 26% in total serum cholesterol and 52% in LDL-cholesterol (Newman *et al.*, 1987). This barley genotype caused even higher viscosity in the intestine, due to its higher molecular weight of β-glucan.

The cholesterol metabolism of chickens is similar to that of humans in that the liver is the major site of cholesterol synthesis, hydroxymethylglutaryl-CoA reductase is the rate-limiting enzyme and it is feedback inhibited by cholesterol feeding (Wang *et al.*, 1992). The cholesterol-lowering effect of barley has also been shown in human clinical studies. Newman *et al.* (1989a) studied 14 volunteer men, 35 years or older, who were given a barley or a 75% wheat/25% wheat bran diet for 28 days. Subjects who consumed wheat had significantly increased serum total cholesterol and LDL cholesterol when compared with pre-treatment levels. Subjects who had average pre-treatment levels had no significant effects of the barley diet, but for those subjects who had higher pre-treatment levels, total and LDL cholesterol levels were reduced. It is known that the reduction in cholesterol levels in response to treatment, increases with increasing pre-treatment levels of the subjects (Ripsin *et al.*, 1992). In another human clinical study barley was compared with oats (Newman *et al.*, 1989b). The same barley genotype was used and 22 men and women were given 56 g of breakfast cereals, one muffin and one flatbread, made of the same amount of barley or oats, daily for 6 weeks. In both the oats and the barley groups, there was a lowering of serum total and LDL cholesterol. All subjects except three experienced reductions in cholesterol from initial levels to final levels. The three non-responders had the lowest initial levels, which suggests that they were not as likely to experience cholesterol reduction by moderate dietary treatment. In a study by Behall *et al.* (2004), where 25 subjects were given whole grain foods containing 0, 3 or 6 g barley β-glucan per day, the reductions in total and LDL cholesterol were as much as 10% and 17%, respectively after eating the high-β-glucan diet. Bourdon *et al.* (1999) showed that barley pasta naturally rich in β-glucan and pasta enriched with β-glucan reduced total serum cholesterol levels, while no significant change in cholesterol levels occurred after eating wheat pasta.

Most clinical studies that have shown a cholesterol-lowering effect of β-glucan have used bran, porridge, pasta, muffins or breakfast cereals, and only the β-glucan content, but not the molecular weight, has been taken into account (Braaten *et al.*, 1994; Kerckhoffs *et al.*, 2002). However, when β-glucan has been consumed in breads or as purified fractions, there has been no, or a less pronounced, effect on serum cholesterol in moderately hypercholesterolaemic humans (Keogh *et al.*, 2003; Frank *et al.*, 2004). In yeast-leavened breads the molecular weight of β-glucan has been shown to be low, due to degradation of the molecule during the baking process (Åman *et al*, 2004; Andersson *et al.*, 2004; Frank *et al.*, 2004). Purified β-glucan fractions may also have a low molecular weight (e.g. Glucagel™ with a molecular weight of around 14 000–56 000 g/mol according to Morgan and Ofman (1998)). Oat bran, porridge, muffins and breakfast cereals, on the other hand, are foods that have been shown to have high β-glucan molecular weight (Åman *et al.*, 2004), and therefore it seems likely that high molecular weight is important for the cholesterol-lowering effect of β-glucan. In an abstract of a so far unpublished study with high and low molecular weight β-glucan extracts that were given to hyperlipidaemic subjects at a 5 g/day or a 3 g/day dose for 6 weeks, the LDL cholesterol levels fell by 15% in the 5 g of high molecular weight β-glucan per day group, by 13% in the 5 g of low molecular weight β-glucan per day group and by 9% in both of the 3 g of β-glucan per day groups (Pins *et al.*, 2005; Pins and Kaur, 2006). The 5 g/day dose was thus more effective than the 3 g/day dose, but molecular weight did not influence the findings. The efficacy of low molecular weight β-glucan is still in question and has to be tested further (Pins & Kaur, 2006).

Several animal studies have shown that the cholesterol-lowering effect of β-glucan was lost after enzymatic treatment with β-glucanases which resulted in an extensive hydrolysis of the β-glucan (Bengtsson *et al.*, 1990). It has also been shown in a human study that bile acid excretion was lower after enzymatic treatment with oat β-glucan (Lia *et al.*, 1996). These results suggest that a high molecular weight is needed for a physiological effect of β-glucan. It may also be important to have low enzyme activity in the raw material, to avoid enzymatic hydrolysis of β-glucan, resulting in low molecular weights.

12.3.3 Glucose- and insulin-lowering effect

There is currently a global increase in the prevalence of obesity and the metabolic syndrome, a condition characterized by a cluster of metabolic dysfunctions that may include abdominal obesity, atherogenic dyslipidaemia, small LDL particles, low high-density lipoprotein (HDL) cholesterol levels, elevated blood pressure, insulin resistance and glucose intolerance (NCEP, 2002). Recent epidemiological studies suggest that diets rich in whole grain cereals can protect against disorders related to the so-called insulin resistance syndrome (Östman *et al.*, 2006). In addition, accumulating observational data indicate that foods with low glycaemic index (GI) have a protective potential, and a low GI diet has been related to reduced risk of the insulin resistance syndrome (McKeown *et al.*, 2004) and type 2 diabetes (Salmeron *et al.*, 1997a,b).

It has been shown that several food factors may be responsible for differences in glucose and insulin responses. These factors include, among others, the content of soluble fibre and food structure (Granfeldt *et al.*, 1994; Östman *et al.*, 2006). As discussed above, whole grain barley is known to be rich in dietary fibre and soluble fibre (mainly β-glucan), and to have resistant endosperm cell walls. Resistant cell walls in barley foods will reduce the release of nutrients in the upper gastrointestinal tract. Several studies have shown that bread or porridge with inclusion of high-fibre barley or β-glucan isolated from barley can lower the GI as well as the insulinaemic index compared with reference products (Granfeldt *et al.*, 1994; Liljeberg *et al.*, 1996; Cavallero *et al.*, 2002; Östman *et al.*, 2006). The reduction in GI seems to be correlated to the β-glucan content in the food and the fluidity index of enzymatic digesta. Bread or porridge supplemented with barley flour with normal dietary fibre content, however, does not always seem to result in a significant reduction in GI compared with the control (Liljeberg *et al.*, 1996). High-fibre barley foods had higher satiety scores when compared with white wheat bread (Granfeldt *et al.*, 1994) and improved glucose tolerance at the second meal (Liljeberg *et al.*, 1999).

12.4 Particular difficulties in the development of functional barley products

One of the main problems in the development of functional barley products is that the hulls of covered barley have to be removed for food consumption. Most of the barley harvested in the Western world today is covered or hulled (Gaines *et al.*, 1985). In hulled barley the flowering glumes are fused and adhere strongly to the seed with a cementing substance (Bhatty, 1986a). The hull constitutes about 10% of the grain dry weight (Munck, 1981) and consists mainly of cellulose, hemicellulose, lignin and smaller quantities of protein (Palmer & Bathgate, 1976). Since the hull is prickly and inedible, it has to be removed by abrasion or pearling, which also removes part of the bran, resulting in loss of essential amino acids and vitamins, other bioactive components and dietary fibre. In hull-less or naked barley, fusion of the flowering glumes does not occur (Bhatty, 1986a) and the hull falls off during threshing (Newman & Newman, 1991). The resulting hull-less barley is especially suitable as a food grain, since it requires no pearling to remove the hull, thus retaining the desirable nutrients, and is regarded as a whole grain cereal. In some countries hull-less barley is already used for food products, and in, for example, Tibet most of the cultivated barley varieties used for human consumption are hull-less (Tashi, 2005).

Another problem in the development of functional barley products is that β-glucan is degraded during food processing due to the activity of endogenous β-glucanases. A high molecular weight of the β-glucan is probably important for the health effects of barley (see Section 12.3, Health benefits). In a study made by Åman *et al.* (2004), it was shown that the molecular weight of β-glucan in several different oat products was lower or much lower than in the raw material. These products were bread loaves, crisp breads, teacakes, pancake batter, fried pancakes, a yogurt-like product, a fermented oat soup, apple juice and fresh pasta. In porridge made of rolled oats, extruded oats, a breakfast cereal product, macaroni and muffins, on the other hand, the β-glucan molecular weight was high. This indicates that no β-glucanase activity was present or that any endogenous enzymes present had only a very limited time for hydrolysis during the food processing (Åman *et al.*, 2004).

In yeast-leavened barley breads the molecular weight of β-glucan has also been shown to be low (Andersson *et al.*, 2004; Trogh *et al.*, 2004, 2005). In a study made by Andersson *et al.*, (2004) barley breads were made with 40% hull-less barley flour and 60% wheat flour to investigate the effect on the molecular weight of β-glucan of flour type (sifted or wholemeal), water content, yeast addition, mixing and fermentation time of the dough and oven-baking. The results showed that the molecular weight of β-glucan decreased with increasing mixing and fermentation time, and that the distribution of β-glucan was polymodal. These results indicate that β-glucan was degraded by endogenous β-glucanases in the flour mix. The molecular weight was not affected by flour type, yeast addition, water content or oven-baking. To investigate the breakdown of β-glucan in more detail, a fermentation experiment was performed with fermentation times from 0 to 90 min. The results showed that β-glucan was degraded rapidly and extensively with a decrease in molecular weight from 120×10^4 to 36×10^4 g/mol. The distribution was polymodal with three different populations, which did not change during fermentation, but which moved to lower molecular weights (Fig. 12.3). These findings indicate that the enzymes responsible for hydrolysis did not act randomly on the β-glucan, or that certain regions of the polysaccharides were not available for hydrolysis (Andersson *et al.*, 2004).

Trogh *et al.* (2004, 2005) also showed that β-glucan in breads made of 40% hull-less barley and 60% wheat was degraded by mixing and fermentation of the dough. The extractability of β-glucan was also affected by the bread-making process, with higher extractability after mixing and lower extractability after proofing and oven-baking. The lower extractability may be due to association of the partly degraded β-glucan molecules with each other and/

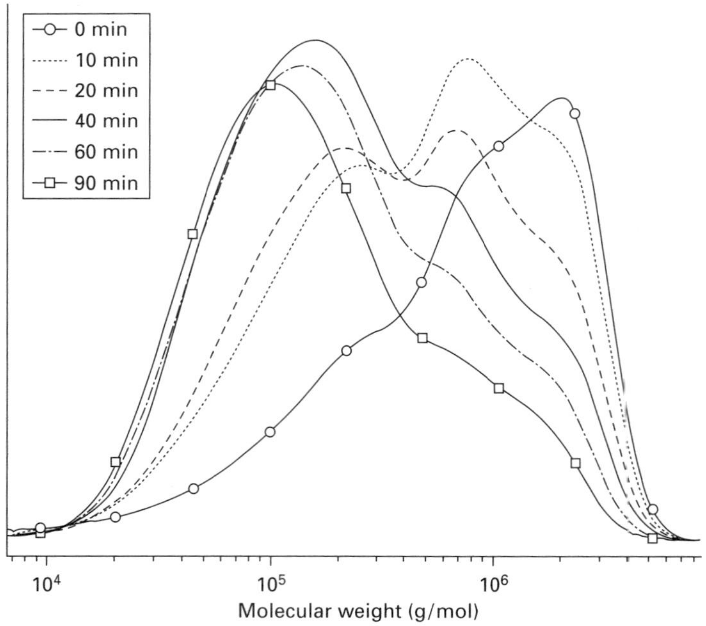

Fig. 12.3 Molecular weight distribution of β-glucan in doughs made of sifted barley flour blends fermented for the times shown in the key. Reprinted from Andersson *et al.* (2004), with permission from Elsevier.

or with other components to form unextractable aggregates (Trogh *et al.*, 2004). The addition of endoxylanases to the flour to improve the baking performance, for example, loaf volume, and to increase soluble arabinoxylan content in the bread, did not affect the extractability or molecular weight of β-glucan. The effect of the endoxylanases was influenced by endoxylanse inhibitors that have been found in both wheat and barley flour (Trogh *et al.*, 2005). To overcome the impact of endoxylanase inhibition, uninhibited endoxylanases were developed, and in bread-making experiments, it was shown that the use of uninhibited endoxylanases gave improved bread quality.

12.5 Manufacturing technology

As mentioned above, most of the barley harvested today is covered with a hull (palea and lemma), which constitutes about 10% of the grain dry weight (Munck, 1981). The hull is undesirable for human consumption but generally has beneficial or at least no adverse effects for malt and animal feed production. Manufacturing technology therefore needs to be employed to remove the hulls before the covered grain can be used as a raw material for food production.

Pot and pearled barley are prepared by gradual removal of hull, bran and germ by abrasive action in a mill (Bhatty, 1993b). When about 10% of the outer parts of the grain are removed, pot barley is produced. Further abrasion results in the removal of the seed coat, aleurone, germ and part of the starchy endosperm, leaving behind the central part of the starchy endosperm. This pearled barley is generally obtained at a yield of 60–70% of the covered kernels. Barley suitable for pearling should be uniform, white, medium hard and thin-hulled, as well as having a high kernel weight. In one pearling process, the cleaned barley kernels are conditioned to about 15% moisture, blocked, aspirated, sized by sifting, groat cutted, pearled, graded and polished. Pearling rates of 30–40% were found to be most desirable since at this rate the content of insoluble dietary fibre was found to be minimal, indicating that most of the hulls were removed, and the content of β-glucan from the starchy endosperm cell walls was maximal (Pedersen *et al.*, 1989).

Barley flakes that have been pre-damped, steam-cooked, flaked and dried in hot air (Sundberg *et al.*, 1994) can be used in porridges, breakfast cereals and as a substitute for rice. A common problem, when used as rice substitute, is that the dishes become dry and compact due to continuous water absorption after preparation of the food. Rolled barley flakes were prepared from three different batches of barley grain by three different techniques. Both the batch of barley and the process had a significant effect on the chemical composition and viscosity of slurries of the flakes. Puddings were prepared from the flakes and mechanical consistency, juiciness and grain consistency were graded in both newly prepared and re-heated puddings evaluated by a sensory test panel. The batch of barley had no effect on mechanical consistency but had significant effects on juiciness and grain consistency. The type of processing had a significant effect on all three parameters for both newly prepared and re-heated puddings. The problem of dry and compact dishes could at least partly be overcome by using malted barley when producing the flakes.

Barley flour is generally produced by roller milling of pot or pearled barley. The flour produced may not be regarded as whole grain since at least part of the outer parts of the kernel and germ are removed. Bran and sifted flour are rarely produced and are generally not found on the food market. Barley bran is brittle and shatters regardless of kernel conditioning, which makes milling of barley into sifted flour difficult and yields are low compared with wheat milling (Bhatty 1992a, 1993a; Sundberg & Åman, 1994). The milling performance of North European hull-less barleys has recently been studied. Four samples of barley were milled in a Bühler MLU 202 laboratory mill and individual and combined milling fractions were characterized (Andersson *et al.*, 2003; Trogh *et al.*, 2005). The best milling performance was obtained when the samples were conditioned to 14.3% moisture. Yields were 37–48% for straight run flours, 47–56% for shorts and 5–8% for bran. The chemical composition of the milling fractions varied depending on barley type (Table 12.1).

Dry and wet processes have been used for the enrichment of β-glucan in different products prepared from de-hulled or naked barley and oats. Dry processes include milling and sieving procedures resulting in different types

Table 12.1 Yield and content of ash, protein, starch, β-glucan and arabinoxylan (% of dry matter) in milling fractions of different hull-less barleys

	Yield	Ash	Protein	Starch	β-Glucan	Arabinoxylan
SW 1290 (L)[a]						
SR flour*	47.8	0.8	8.2	79.3	2.1	1.5
Shorts	47.4	3.0	14.2	54.8	4.6	7.5
Bran	4.7	4.5	15.0	40.5	3.6	10.5
SW 1290 (H)[b]						
SR flour	44.7	0.9	9.4	75.8	1.6	1.5
Shorts	48.7	3.3	14.9	48.2	4.2	7.7
Bran	6.6	1.9	15.4	34.8	3.0	11.8
SW 1291 (H)						
SR flour	36.9	0.9	9.0	77.0	2.0	1.2
Shorts	56.0	2.3	14.2	51.8	5.8	6.1
Bran	7.1	3.3	15.9	43.9	4.7	8.1
SW 8775 (H)						
SR flour	38.7	1.0	7.7	76.1	2.0	1.3
Shorts	53.7	2.9	12.7	51.2	5.8	6.9
Bran	7.6	4.3	12.2	38.1	4.2	10.4

* Straight run white flour. Compiled from Andersson *et al.*, 2003.
(a) Grown in Landskrona in Sweden.
(b) Grown in Haga in Sweden.

of brans. The content of β-glucan is highest in the sub-aleurone layer of oat groats, which is enriched in oat bran (Wood *et al.*, 1991b). The β-glucan content in different oat brans varies but can be as high as over 8% (Luhaloo *et al.*, 1998). Barley does not contain a sub-aleurone layer with a thick β-glucan-rich cell wall and therefore the content of β-glucan in barley bran is lower. Bran has been produced from naked barley at a yield of about 30% and with a β-glucan content of 6.5% (Bhatty, 1992b). Fractions with higher contents of β-glucan can be obtained by air classification (Sundberg & Åman, 1995; Andersson *et al.*, 2000). Wet processes for the extraction of β-glucan generally include three steps (Brennan & Cleary, 2005). First, an inactivation of endogenous β-glucan degrading enzymes – which otherwise will degrade the polymers resulting in changes in physiochemical and functional properties – is performed, generally with extensive treatment in aqueous ethanol (Irakli *et al.*, 2004). Extraction of β-glucan can be performed with sequential water extractions, sodium carbonate or sodium hydroxide, with variable yields and purity. Saulnier *et al.* (1994) used hot water extraction in the presence of thermostable α-amylase to minimize the contamination of starch. The extracted β-glucan can be precipitated in aqueous ethanol or purified from protein by a change in pH. A cost-effective method for extraction of relatively pure (89–94%), partially degraded and gelling β-glucan has been developed (Morgan & Ofman, 1998). The isolation procedure includes hot water extraction and recovery of the partly degraded β-glucan as a gel by freezing and thawing of the extract. The preparation (Glucagel™) is commercially available.

12.6 Future trends

An increased variety in the cereal food market is desirable and barley could help to fulfil this wish. This will, however, require that high-yielding naked cultivars and new barley products and foods with desirable sensory profiles are developed. Naked barley is preferred in areas where barley is used as a staple food and is today also available in the Western world (Bhatty, 1986b). High-yielding naked barley with high amylose or waxy starch has also been developed and can be used in products where designed starch properties are needed (Oscarsson *et al.*, 1997; Andersson *et al.*, 1999a,c 2001). Since barley β-glucan is a valuable hydrocolloid, cultivars with high β-glucan content are of special interest to be used in functional foods or as a starting material for enrichment of β-glucan in fractions or extracts. As mentioned earlier, Prowashonupana is one cultivar with high β-glucan content and several other barleys with 15–17% β-glucan have been identified (Åman & Newman, 1986; Andersson *et al.*, 1999b; Munck *et al.*, 2004). Growing conditions will also influence the β-glucan and starch content in the barley (Oscarsson *et al.*, 1998). Whole grain barley flour from naked barley can be used in whole grain foods which are known to possess health properties. Barley can also be a valuable component in low glycaemic foods. Of special importance is the stable structure of barley endosperm, the high content of soluble fibre (mainly β-glucan) and the high amylose content in high-amylose cultivars.

Health claims can today be used on barley foods in Sweden, the United States and the United Kingdom under certain conditions. Of special interest are claims regarding high content of dietary fibre in barley foods and reduced risk for constipation, high content of soluble fibre (mainly β-glucan in barley) and reduced risk for cardiovascular diseases, and high content of whole grain barley and reduced risk for heart disease. These relatively new claims will probably increase the interest from the food industry in developing new tasty and healthy barley foods.

12.7 References

Åman, P (2006), Cholesterol-lowering effects of barley dietary fibre in humans. Scientific support for a generic health claim, *Scand J Food Nutr*, **50**, 173–176.

Åman P and Graham H (1987), Analysis of total and insoluble mixed linked (1-3),(1-4)-β-D-glucans in barley and oats, *J Agric Food Chem*, **35**, 704–709.

Åman P and Newman C W (1986), Chemical composition of some different types of barley grown in Montana, USA, *J Cereal Sci*, **4**, 133–141.

Åman P, Hesselman K and Tilly A-C (1985), The variation in chemical composition of Swedish barleys, *J Cereal Sci*, **3**, 73–77.

Åman P, Rimsten L and Andersson R (2004), Molecular weight distribution of β-glucan in oat-based foods, *Cereal Chem*, **81**, 356–360.

Ames N, Rhymer C, Rossnagel B, Therrien M, Ryland D, Dua S and Ross K (2006), Utilization of diverse hulless barley properties to maximize food product quality, *Cereal Foods World*, **51** (1), 23–28.

Amri A, Ouammou L and Nassif F (2005), Barley-based food in southern Morocco, in: Grando S and Gomez MacPerson H (Eds), *Food Barley: Importance, Uses and Local Knowledge*. Proceedings of the International Workshop on Food Barley Improvement 14–17 January 2002, Hammamet, Tunisia, ICARDA, Aleppo, Syria, pp. 22–28.

Andersson A A M, Elfverson C, Andersson R, Regner S and Åman P (1999a), Chemical and physical characteristics of different barley samples, *J Sci Food Agric*, **79**, 979–986.

Andersson A A M, Andersson R, Autio K and Åman P (1999b), Chemical composition and microstructure of two naked barleys, *J Cereal Sci*, **30**, 183–191.

Andersson L, Fredriksson H, Oscarsson Bergh M, Andersson R and Åman P (1999c), Characterization of starch from inner and peripherial parts of normal and waxy barley kernels, *J Cereal Sci*, **30**, 165–171.

Andersson A A M, Andersson R and Åman P (2000), Air classification of barley flours, *Cereal Chem*, **77**, 463–467.

Andersson A A M, Andersson R and Åman P (2001), Starch and by-products from a laboratory-scale barley starch isolation procedure, *Cereal Chem*, **78**, 507–513.

Andersson A A M, Courtin C M, Delcour J, Fredriksson H, Schofield J D, Trough I, Tsiami A A and Åman P (2003), Milling performance of north European hull-less barleys and characterization of resultant millstreams, *Cereal Chem*, **80**, 667–673.

Andersson A A M, Armö E, Grangeon E, Fredriksson H, Andersson R and Åman P (2004), Molecular weight and structure units of (1→3, 1→4)-β-glucans in dough and bread made from hull-less barley milling fractions, *J Cereal Sci*, **40**, 195–204.

Beaven E S (1947), The origin and antiquity of barley, in *Barley, Fifty Years of Observation and Experiment,* Edinburgh, The Riverside Press, pp. 7–13.

Behall K M, Scholfield D J and Hallfrisch J (2004), Diets containing barley significantly reduce lipids in midly hypercholesterolemic men and women, *Am J Clin Nutr*, **80**, 1185–1193.

Bengtsson S, Åman P, Graham H, Newman C W and Newman R K (1990), Chemical studies on mixed-linked β-glucans in hull-less barley cultivars giving different hypocholesterolaemic responses in chicken, *J Sci Food Agric*, **52**, 435–445.

Berglund P T, Fastnaught C E and Holm E T (1992), Food uses of waxy hull-less barley, *Cereal Foods World*, **37**, 707–714.

Bhatty R S (1986a), Physiochemical and functional (breadmaking) properties of hull-less barley fractions, *Cereal Chem*, **63**, 31–35.

Bhatty R S (1986b), The potential of hull-less barley – A review, *Cereal Chem*, **63**, 97–103.

Bhatty R S (1992a), β-Glucan content and viscosities of barleys and their roller-milled flour and bran products, *Cereal Chem*, **69**, 469–471.

Bhatty R S (1992b), Dietary and nutritional aspects of barley in human foods, in Munck L (Ed.), *International Barley Genetics Symposium VI,* Vol. 2, Copenhagen, Muncksgaard International Publishers Ltd, pp. 913–923.

Bhatty R S (1993a), Nonmalting uses of barley, in MacGregor A W and Bhatty R S (Eds), *Barley – Chemistry and Technology*, St Paul, Minnesota, AACC Inc., pp. 353–417.

Bhatty, R S (1993b), Physiochemical properties of roller-milled barley bran and flour, *Cereal Chem*, **70**, 397–402.

Bourdon I, Yokoyama W, Davis P, Hudson C, Backus R, Richter D, Knuckles B and Schneeman B O (1999), Postprandial lipid, glucose, insulin and cholecystokinin response in men fed barley pasta enriched with β-glucan, *Am J Clin Nutr*, **69**, 55–63.

Braaten J T, Wood P J, Scott F W, Wolynetz M S, Lowe M K, Bradley-White P and Collins M W (1994), Oat beta-glucan reduces blood-cholesterol concentrations in hypercholesterolemic subjects, *Eur J Clin Nutr*, **48**, 465–474.

Brennan C S and Cleary L J (2005), The potential use of cereal (1→3, 1→4)-β-D-glucans as functional food ingredients, *J Cereal Sci*, **42**, 1–13.

Briggs D E (1978), *Barley*, London, Chapman & Hall.

Cavallero A, Empili S, Brighent F and Stanca A M (2002), High (1→3) (1→4)-β-D-glucan fractions in bread making and their effect on human glycaemic response, *J Cereal Sci*, **36**, 59–66.

Dhingra S and Jood S (2004), Effect of flour blending on functional, baking and organoleptic characteristics of bread, *Int J Food Sci*, **39**, 213–222.

El Felah M and Medimagh S (2005), Food barley in Tunisia, in Grando S and Gomez MacPerson H (Eds), *Food Barley: Importance, Uses and Local Knowledge*, Proceedings of the International Workshop on Food Barley Improvement, 14–17 January 2002, Hammamet, Tunisia. Aleppo, Syria, ICARDA, pp. 29–35.

Evans R D (1968), Knossos Neolithic, part II, *Annu Br School Archaeol Athens*, **63**, 239–276.

Fadel J G, Newman R K, Newman C W and Barnes A E (1987), Hypocholesterolemic effects of β-glucans in different barley diets fed to broiler chickens, *Nutr Rep Int*, **35**, 1049–1058.

FAOSTAT, (2006), http://faostat.fao.org/.

FDA (Food and Drug Administration) (2005), FDA allows barley products to claim reduction in risk of coronary heart disease, *Federal Register*, **70**, 76150–76162.

Frank J, Sundberg S, Kamal-Eldin A, Vessby B and Åman P (2004), Yeast-leavened oat breads with high or low molecular weight β-glucan do not differ in their effects on blood concentrations of lipids, insulin or glucose in humans, *J Nutr*, **134**, 1384–1388.

Gaines R L, Bechtel D B and Pomeranz Y (1985), A microscopic study on the development of a layer in barley that causes hull-caryopsis adherence, *Cereal Chem*, **62**, 35–40.

Grando S and Gomez MacPherson H (Eds) (2005), *Food Barley: Importance, Uses and Local Knowledge*, Proceedings of the International Workshop on Food Barley Improvement, 14–17 January 2002, Hammamet, Tunisia. Aleppo, Syria, ICARDA.

Granfeldt Y, Liljeberg H, Drews A, Newman R and Björck I (1994), Glucose and insulin responses to barley products: influence of food structure and amylase–amylopectin ratio, *Am J Clin Nutr*, **59**, 1075–1082.

Grivetti L E (1991), Nutrition past – nutrition today. Prescientific origins of nutrition and dietetics, *Nutr Today*, **26**, 13–24.

Harlan J R (1968), On the origin of barley, in *Barley: Origin, Botany, Culture, Winter Hardiness, Genetics, Utilization, Pests,* Agricultural Handbook No. 388, Washington, DC, USDA, pp. 9–29.

Heen A, Eide O R and Frølich W (1991), Hull-less barley, agronomical traits and nutritional quality, Proceedings of the Sixth International Barley Genetics Symposium, 22–27 July, Helsingborg, Sweden, pp. 446–448.

Hudson C A, Chiu M M and Knuckles B E (1992), Development and characteristics of high-fibre muffins with oat bran, rice bran, or barley fibre fractions, *Cereal Foods World*, **37**, 374–378.

Irakli M, Biliaderis C G, Izydorczyk M S and Papadoyannis I N (2004), Isolation, structural features and rheological properties of water-extractable beta-glucans from different Greek barley cultivars. *J Sci Food Agric*, **84**, 1170–1178.

Izydorczyk M S, Lagassé S L, Hatcher D W, Dexter J E and Rossnagel B G (2005), The enrichment of Asian noodles with fibre-rich fractions derived from roller milling of hull-less barley, *J Sci Food Agric*, **85**, 2094–2104.

Kent N L (1983), Barley, in *Technology of cereals*, Oxford, Pergamon Press, pp. 154–164.

Keogh G F, Cooper G J S, Mulvey T B, McArdle B H, Coles G D, Monre J A and Poppitt S D (2003), Randomized controlled crossover study of the effect of a highly beta-glucan enriched barley on cardiovascular disease risk factors in mildly hypocholesterolemic men, *Am J Clin Nutr*, **78**, 71.

Kerckhoffs D A J M, Brouns F, Hornstra G and Mensink R P (2002), Effects on the human serum lipoprotein profile of β-glucan, soy protein and isoflavones, plant sterols and stanols, garlic and tocotrienols, *J Nutr*, **132**, 2492–2505.

Lia Å, Hallmans G, Sandberg A S, Sundberg B, Åman P and Andersson H (1996), Oat β-glucan increases bile acid excretion and a fibre-rich barley fractions increases cholesterol excretion in ileostomy subjects, *Am J Clin Nutr*, **62**, 1245–1251.

Liljeberg H G M, Granfeldt Y E and Björck I M E (1996), Products based on a high fibre barley genotype, but not on common barley or oats, lower postprandial glucose and insulin responses in healthy human, *J Nutr*, **125**, 1503–1511.

Liljeberg H G M, Åkerberg A K E and Björck I M E (1999), Effect of the glycemic index and content of indigestible carbohydrates of cereal-based breakfast meals on glucose tolerance at lunch in healthy subjects, *Am J Clin Nutr*, **69**, 647–655.

Luhaloo M, Mårtensson A-C, Andersson R and Åman P (1998), Compositional analysis and viscosity measurements of commercial oat bran, *J Sci Food Agric*, **76**, 141–148.

Marklinder I, Haglund Å and Johansson L (1996), Effects of flour from different barley varieties on barley sour dough bread, *Food Qual Pref*, **7**(3/4), 275–284.

McKeown N M, Meigs J B, Liu S, Saltzman, E, Wilson P W and Jacques P F (2004), Carbohydrate nutrition, insulin resistance, and the prevalence of the metabolic syndrome in the Framingham offspring cohort. *Diabetes Care*, **27**, 538–546.

Melland R, Newman R K, McGuire C F and Eslick R F (1984), The effects of bleach treatment on pasta made from a series of barley genotypes, *Cereal Res Comm*, **12**, 201–207.

Miller S S and Fulcher R G (1994), Distribution of (1→3),(1→4)-β-D-glucan in kernels of oats and barley using microspectrofluorometry, *Cereal Chem*, **71**, 64–68.

Morgan K R and Ofman D J (1998), Glucagel, a gelling β-glucan from barley, *Cereal Chem*, **75**, 879–881.

Morin L A, Temelli F and McMullen L (2002), Physical and sensory characteristics of reduced-fat breakfast sausages formulated with barley β-glucan, *J Food Sci*, **67**(6), 2391–2396.

Munck L (1981), Barley for food, feed and industry, in Pomeranz Y and Munck L (Eds), *Cereals. A Renewable Resource, Theory and Practice*, St Paul, MN, AACC, pp. 427–459.

Munck L, Møller B, Lacobsen S and Søndergaard I (2004), Near infrared spectra indicate specific mutant endosperm genes and reveal a new mechanism for substituting starch with (1→3, 1→4)-β-glucan in barley. *J Cereal Sci*, **40**, 213–222.

NCEP (National Cholesterol Education Program) (2002). Third Report of the Expert Panel on Detection, Evaluation, and Treatment of High Blood Cholesterol in Adults (Adult Treatment Panel III) Final Report, Circulation 1, Bethesda, MD, NCEP.

Newman C W and McGuire C F (1985), Nutritional quality of barley, in Rasmusson D C (Ed.), *Barley, Agronomy Monograph No. 26,* Madison, WI, ASA-CSSA-SSSA, pp. 404–456.

Newman R K and Newman C W (1991), Barley as a food grain, *Cereal Foods World*, **36**, 800–805.

Newman R K, Newman C W, Fadel J and Graham H (1987), Nutritional implications of β-glucans in barley, in *Barley Genetics V*, Proceedings of the Fifth International Barley Genetics Symposium, October 1986, Okayama Japan, p. 773.

Newman R K, Lewis S E, Newman C W, Boik R J and Ramage R T (1989a), Hypocholesterolemic effect of barley food on healthy men, *Nutr Rep Int*, **39**, 749.

Newman R K, Newman C W and Graham H (1989b), The hypocholesterolemic function of barley β-glucans, *Cereal Foods World*, **34**(10), 883–886.

Newman R K, McGuire C F and Newman C W (1990), Composition and muffin-baking characteristics of flours from barley cultivars, *Cereal Foods World*, **35**, 563–566.

Newman R K, Newman C W, Hofer P J and Barnes A E (1991), Growth and lipid metabolism as affected by feeding of hull-less barleys with and without supplemental β-glucanase, *Plant Food Hum Nutr*, **41**, 371–380.

Newman R K, Ore K C and Newman C W (1992), Fibre enrichment of baked products with a barley milling fraction, in *Barley for Food and Malt*, ICC/SCF International Symposium, Uppsala, Sweden, pp. 242–244.

Newman R K, Ore K C, Abbott J and Newman C W (1998), Fibre enrichment of baked products with a barley milling fraction, *Cereal Foods World*, **43**(1), 23–25.

Oscarsson M, Andersson R and Salomonsson A C (1996), Chemical composition of barley samples focusing on dietary fibre components, *J Cereal Sci*, **24**, 161–170.

Oscarsson M, Parkkonen T, Autio K and Åman P (1997), Composition and microstructure of waxy, normal and high amylose barley samples, *J Cereal Sci*, **26**, 259–264.

Oscarsson M, Andersson R, Åman P, Olofsson S and Jonsson A (1998), Effects of cultivar, nitrogen fertilization rate and environment on yield and grain quality of barley, *J Sci Food Agric*, **78**, 359–366.

Östman E, Rossi E, Larsson H, Brighenti F and Björck I (2006), Glucose and insulin response in healthy men to barley bread with different levels of (1→3, 1→4)-β-glucans; predictions using fluidity measurements of in vitro enzyme digests, *J Cereal Sci*, **43**, 230–235.

Palmer G H and Bathgate G N (1976), Malting and brewing: III. Anatomy of the barley grain, in Pomeranz Y (Ed.), *Advances in Cereal Science and Technology*, St Paul, MN, AACC, pp. 238–245.

Pedersen B, Bach Knudsen K E and Eggum B (1989), Nutritive value of barley products with emphasis on the effect of milling, *World Rev Nutr Diet*, 60, 1–91.

Pins J J and Kaur H (2006), A review of the effects of barley β-glucan on cardiovascular and diabetic risk, *Cereal Foods World*, **51**, 8–11.

Pins J, Keenan J M, Curry L L, Goulson M J and Kolberg L W (2005), Extracted barley beta-glucan improves CVD risk factors and other biomarkers in a population of generally healthy hypercholesterolemic men and women, *Prev Control*, **1**, 131.

Ripsin C, Keenan J, Jacobs D, Elmer P, Welch R, Van Horn L, Liv K, Turnbull W H, Tnye F W, Kestin M, *et al.* (1992), Oat products and lipid lowering: a meta-analysis, *JAMA*, **267**, 3317–3321.

Salmeron Jascherio A, Rimm E B, Colditz G A, Spiegelman D, Jenkins D J, Stampfer M J, Wing A L and Willett W C (1997a), Dietary fibre, glycemic load, and risk of NIDDM in men, *Diabetes Care*, **20**, 545–550.

Salmeron J, Manson J E, Stampfer M J, Colditz G A, Wing A L and Willett W C (1997b), Dietary fibre, glycemic load, and risk of non-insulin-dependent diabetes mellitus in women, *JAMA*, **277**, 472–477.

Salomonsson A C, Theander O and Åman P (1980), Composition of normal and high-lysine barleys, *Swedish J Agric Res*, **10**, 11–16.

Saulnier L, Gevaudan S and Thibault J F (1994), Extraction and partial characterization of β-glucan from the endosperms of two barley cultivars, *J Cereal Sci*, **19**, 171–178.

SNF (Swedish Nutrition Foundation) (2007), http://www.snf.ideon.se/snf/en/.

Spicher G and Stephan H (1992), *Handbuch Sauerteig. Biologie, Biochemie, Technologie*, Hamburg, Behr's verlag.

Sundberg B and Åman P (1994), Fractionation of different types of barley by roller milling and sieving, *J Cereal Sci*, **19**, 179–184.

Sundberg B and Åman P (1995), Enrichment of mixed-linked (1→3), (1→4)-β-D-glucans from a high-fibre barley-milling stream by air classification and stack-sieving, *J Cereal Sci*, **21**, 205–208.

Sundberg B and Falk H (1994), Composition and properties of bread and porridge prepared from different types of barley flour, *Am J Clin Nutr*, **59** (suppl.), 780S.

Sundberg B, Abrahamsson L and Åman P (1994), Quality of rolled barley flakes as affected by batch of grain and processing technique, *Plant Foods Hum Nutr*, **45**, 145–154.

Symons L J and Brennan C S (2004), The influence of (1→3) (1→4)-beta-D-glucan-rich fractions from barley on the physicochemical properties and in vitro reducing sugar release of white wheat breads, *J Foods Sci*, **69**, C463–C467.

Tashi N (2005), Food preparation from hull-less barley in Tibet, in Grando S and Gomez MacPerson H (Eds), *Food Barley: Importance, Uses and Local Knowledge*, Proceedings of the International Workshop on Food Barley Improvement, 14–17 January 2002, Hammamet, Tunisia. Aleppo, Syria, ICARDA, pp. 115–120.

Trogh I, Courtin C M, Andersson A A M, Åman P, Sørensen J F and Delcour J (2004), The combined use of hull-less barley flour and xylanase as a strategy for wheat/hull-less barley flour breads with increased arabinoxylan and (1→3, 1→4)-β-D-glucan levels, *J Cereal Sci*, **40**, 257–267.

Trogh I, Courtin C, Goesaert H, Delcour J, Andersson A A M, Åman P, Fredriksson H, Pyle D L and Sørensen J F (2005), From hull-less barley and wheat to soluble dietary fibre-enriched bread, *Cereal Foods World*, **50**, 253–260.

Ullrich S E, Clancy J A, Eslick R F and Lance R C M (1986), β-Glucan content and viscosity of extracts from way barley, *J Cereal Sci*, **4**, 279–275.

Wang L, Newman R K, Newman C W and Hofer P J (1992), Barley β-glucans alter intestinal viscosity and reduce plasma cholesterol concentrations in chickens. *J Nutr*, **122**, 2292–2297.

Wendorf F, Schild R, Hadidi N E, Close A E, Kobusiewicz M, Wieckowska H, Issawi B and Haas H (1979), Use of barley in the Egyptian Late Paleolithic, *Science*, **205**, 1341–1347.

Wood P J, Weisz J and Blackwell B A (1991a), Molecular characterization of cereal β-D-glucans, structural analysis of oat β-D-glucan and rapid structural evaluation of β-D-glucan from different sources by high-performance liquid chromatography of oligosachharides released by lichenase, *Cereal Chem*, **68**, 31–39.

Wood P J, Weisz J and Fedec P (1991b), Potential for β-glucan enrichment in brans derived from oat (*Avena sativa* L) cultivars of different (1→3), (1→4)-β-glucan concentrations, *Cereal Chem*, **68**, 48–51

13

Products containing other speciality grains: sorghum, the millets and pseudocereals

J. R. N. Taylor and M. N. Emmambux, University of Pretoria, South Africa

13.1 Introduction

This chapter deals with starchy grains that, although relatively minor in terms of world grain production, are major constituents of a wide range of healthy, grain-based foods and beverages, and have considerable potential for broader use. The grains that will be discussed are sorghum, the millets and pseudocereals. Millets comprise several different small-seeded cereal grain species that are cultivated in both tropical and temperate regions of the world. The name comes from the French word '*mil*' for 1000, implying that there are thousands of grains per handful. The choice of millets described in detail here – viz. pearl millet, finger millet and tef (also written as teff) – is somewhat arbitrary. These particular millets, like sorghum, are tropical grasses and each has some interesting healthy nutritional attributes and is the basis of many traditional foods. Pseudocereals can be defined as starchy food grains excluding those currently classified as cereals, legumes, oilseeds and nuts (Fletcher, 2004). Here the choice of which to discuss is much more obvious since three pseudocereals predominate, viz. grain amaranth (*Amaranthus* spp.), buckwheat (*Fagopyrum* spp.) and quinoa (*Chenopodium quinoa* (Willd)). With regard to their healthy aspects, a common feature of all these grains is that they do not contain gluten or gluten-like proteins. Therefore they are suitable for coeliacs and people with wheat intolerance.

For each of the grains, brief information about their botany, history and cultivation will be given. Then the specific and unique potential health benefits in terms of the nutrients in each of these grains will be described in some detail, as well any potential negative effects on health. It is important to note that the nutrient and antinutrient compositions described relate to the whole grain and not to the final foods. Critical in this respect is that, when grains are processed into foods, invariably part of the grain is removed. Often the pericarp (commonly referred to as the bran) and part of the germ are removed during milling. Kent (1983) presents data that show the effect of this on the nutritive value of wheat flour. For example, the vitamin B_1 content is reduced to 85% of that of the whole grain content if 15% of the grain is removed during milling (typical of brown bread flour) and to only 25% if 25% of the grain is removed (typical of white bread flour). Other micro- and macronutrients are affected in the same way, except for starch which increases. It is because whole grains are so nutritionally superior to refined grain products that the United States Department of Agriculture (USDA) is recommending the daily consumption of three servings of whole grain products for American adults (USDA, 2005). Processing of grains into foods also affects their nutritional value. For example, browning due to the Maillard reaction takes place when grain foods are baked. This reduces the essential amino acid content of the food.

This chapter will also discuss changing consumer attitudes towards these speciality food grains. The important traditional food and beverage products that are made from the grains will be described, together with the associated processing technologies. Then, modern/novel foods and beverages that are or can be produced from the grains will be described, together with the associated technologies. Attention will be given to processing technologies such as malting and fermentation that can improve grain food nutritional value. Lastly, other reliable sources of information about these grains will be given.

13.2 The grains

13.2.1 Sorghum

Description, history and production

Sorghum (*Sorghum bicolor* (L.) Moench) – also commonly known as milo (USA and the Middle East), jowar (India), sorgo (French speaking countries), mtama (East Africa), mabela (Southern Africa) and kaoliang (China) – originated in Africa and was probably first domesticated some 5000 years ago in the broad-leaved savanna belt which stretches from Lake Chad in the west to Sudan in the east (De Wet and Harlan, 1971). It first reached India some 2000 years ago and then China (National Research Council, 1996). Sorghum was introduced into the USA in the middle of the nineteenth century. It is now also an important crop in Mexico, Central America and Australia. Food and Agriculture Organization (FAO) statistics show an annual world production of 54.2–59.9 million tonnes over the years 2001–2005 (FAO, 2006). This makes sorghum the world's fifth most important cereal crop after maize, wheat, rice and barley. The major production regions in descending order are the USA, West Africa (Nigeria, Burkina Faso, Mali and Niger), Mexico, the Sudan, East Africa (Ethiopia and Tanzania), Australia and Southern Africa. Sorghum flourishes in hot, arid climates with rainfall between 500 and 700 mm. It is uniquely drought resistant and can also resist waterlogging. In the developing world, open pollinating varieties of sorghum are generally cultivated. Although traditional landraces are still widely used, these are gradually being replaced by modern improved varieties with higher yield potential. The traditional landraces may be up to 3 m high; whereas the modern varieties are invariably rather shorter (Fig. 13.1). In the developed world, hybrid dwarf cultivars are the norm; these are high yielding with high-input, mechanised agriculture.

Physical characteristics

Sorghum grains are more or less spherical but flattened on the germ side. They have a length of some 4 mm. The 1000 kernel weight is around 25–35 g, about the same as that of wheat. Sorghum grains vary in colour from almost white to almost black, with shades of red and brown being common. The pigments responsible for the colour of sorghum grain have important nutritional implications. Sorghum is a naked kernel, i.e. unlike barley, oats and rice, and like wheat it does not have a husk (hull). The kernel, like all cereal kernels, consists of three main parts: the pericarp (the branny outer layer which is sometimes erroneously referred to as a hull), the germ (the living part which on germination develops into the new plant) and the endosperm (the starch-rich storage tissue) (Fig. 13.2). Sorghum is notable in that, in a minority of varieties, it contains condensed tannins in a thick testa

Fig. 13.1 Red improved variety sorghum being cultivated in Ethiopia.

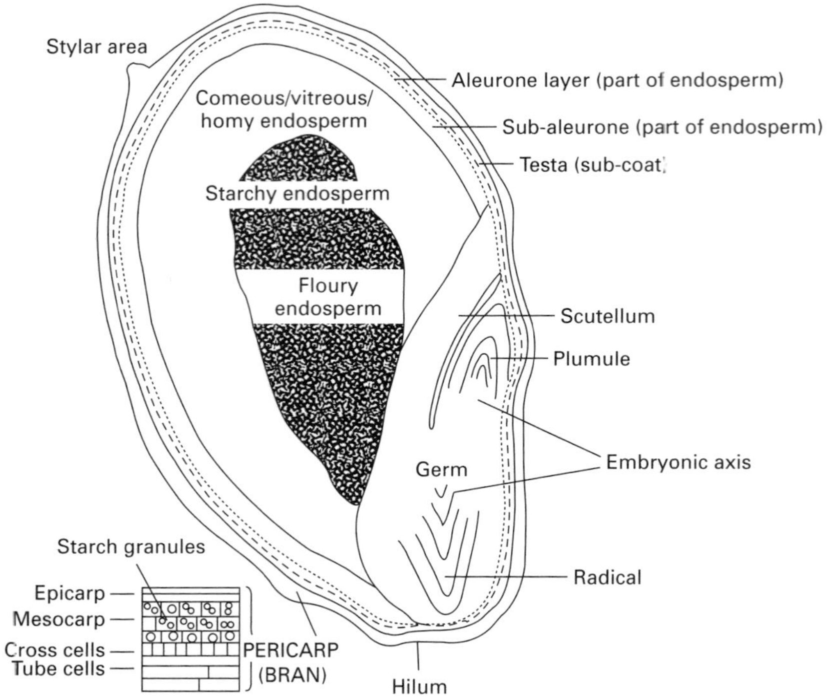

Fig. 13.2 Schematic longitudinal section through a sorghum grain showing cereal kernel component parts. Redrawn from Rooney (1973).

layer under the pericarp. The tannins in these tannin sorghum varieties (often referred to as brown, high-tannin, bitter, bird-proof or bird-resistant) are of considerable nutritional interest. It needs to be clearly pointed out, however, that the overall colour of a sorghum grain has very little bearing on whether or not it is a tannin type. Most red sorghums do not contain tannins and some white sorghums do.

Nutrients

In terms of its general nutritional composition, sorghum is not exceptional. Its lipid content around 3.4% is relatively high for cereals (Table 13.1), on account of its proportionally large germ, around 8% of the kernel. As with most cereals the major fatty acids are unsaturated, linoleic acid (C18:22) (32–50%) and oleic acid (C18:1) (31–41%) predominate. Quantitatively the most important saturated fatty acid is palmitic acid (C16:0) (12–14%) (Serna-Saldivar and Rooney, 1995).

The protein content of sorghum, generally around 11% (Table 13.1) is highly variable but generally rather higher than that of maize and lower than of wheat. As for virtually all cereals, the essential amino acid composition is rather poor and the first limiting amino acid is lysine. The lysine content of sorghum is only around 2.0 g/100 g protein and data suggest that sorghum

Table 13.1 Nutrient composition of various grains expressed on a 12% moisture basis. Data from the USDA food composition database, except where indicated (USAID, 1999)

Nutrient composition (per 100 g grain)	Sorghum	Pearl millet[1]	Finger millet[1]	Tef[1]	Buckwheat	Amaranth	Quinoa	Wheat (hard red spring)
Energy (kJ)	1374	1443	1396	1389	1399	1505	1518	1389
Protein (g)	11.6	11.5	7.3	9.5	12.9	13.9	12.7	15.9
Lysine (g)	0.24	0.36	0.19	0.22	0.66	0.72	0.72	0.41
(Lysine (g/100 g protein))	2.0	3.1	2.6	2.3	5.1	5.2	5.6	2.6
Lipid (g)	3.4	4.7	1.3	2.0	3.3	6.3	5.6	1.9
Ash (g)	1.6	2.3	2.6	2.9	2.1	2.9	2.8	1.9
Carbohydrate (by difference) (g)	77.0	71.6	74.0	77.2	69.7	63.6	66.8	68.6
Dietary fibre (g)	6.3 (white sorghum 9.1–11.5 (other types)	9.7[2]	3.2 (crude fibre)	2.0 (crude fibre)	9.8	8.9	6.1	12.3
Ca (mg)	29	36	358	157	18	147	58	25
Fe (mg)	4.5	9.6	9.9	5.7	2.1	7.3	9.0	3.6
Mg (mg)	135[1]	111	140	168	225	256	204	125
P (mg)	296	332	250	374	338	437	398	335
K (mg)	361	498	314	397	449	352	718	343
Na (mg)	6	15	49	46	1	2	20	2
Zn (mg)	1.4[2]	2.0	1.5	2.0	2.4	3.1	3.2	2.8
Cu (mg)	1.7[1]	0.5	0.5	0.7	1.1	0.8	0.8	0.4
Mn (mg)	ND[3]	0.8	1.9	6.3	1.3	2.2	2.2	4.1

Table 13.1 Continued

Nutrient composition (per 100 g grain)	Sorghum	Pearl millet[1]	Finger millet[1]	Tef[1]	Buckwheat	Amaranth	Quinoa	Wheat (hard red spring)
Se (μg)	ND	ND	ND	ND	8.1	ND	ND	71.3
Vitamin C (mg)	0	ND	1.0?[4]	ND	0	4.0	0	0
Thiamin (mg)	0.24	0.30	0.24	0.30	0.10	0.08	0.19	0.51
Riboflavin (mg)	0.15	0.19	0.11	0.18	0.41	0.20	0.38	0.11
Niacin (mg)	3.0	2.5	1.0	2.5	6.8	1.24	2.8	5.8
Vitamin B_6 (mg)	0.48[1] 0.18[2]	ND	ND	0.38[2]	0.21	0.21	0.13	0.34
Folate (μg)	84[2]	ND	ND	84[2]	29	47	44	43
Vitamin A (μg retinol equivalents)	20[1] 10[2]	22	6	8	0	0	0	3
Vitamin E (α-tocopherol) (mg)	ND	ND	ND	ND	ND	0	ND	1.02

[1]From National Research Council (1996).
[2]From Schakel *et al.* (2004).
[3]No data.
[4]Data questionable.

has somewhat lower lysine amino acid scores than other major cereals and the millets (Klopfenstein and Hoseney, 1995). However, the range of values is very wide and probably the real differences are marginal. Of interest with respect to lysine is that certain Ethiopian varieties of sorghum have significantly higher levels of lysine, in the range 3.1–3.3 g/100 g protein.

An apparently unique feature of sorghum protein is that its digestibility is substantially reduced when the flour is wet cooked to make a food product such as a porridge or bread (Duodu *et al.*, 2003). This seems to be primarily due to cross-linking by disulphide bonding of the major proteins in sorghum grain, which are called kafirins. A reduction in protein digestibility is clearly disadvantageous for people whose protein intake is limited and improving the digestibility of sorghum protein is a continuing, important research topic. However, for people whose protein intake is adequate the reduction in digestibility may be advantageous. It has been shown *in vitro* that the rate and extent of digestion of the starch in sorghum, particularly cooked sorghum, is low (Ezeogu *et al.*, 2005). This work and other studies suggest that this is related to the cross-linking of the kafirin storage proteins, which seem to limit the access of the amylase digestive enzymes to the starch. This *in vitro* work is supported by animal feeding studies, mainly using raw grain, which have consistently shown that sorghum has a lower metabolisable energy content than maize. The difference cannot be explained by differences in proximate composition. The advantage of slower starch breakdown is that sorghum seems to be a more sustained source of energy than, for example, maize. African people who eat both sorghum and maize frequently refer to sorghum 'filling the stomach' more than maize and giving one energy all through the day. For the same reason, many doctors in Africa recommend that people who are at risk from type II diabetes consume sorghum in preference to maize on account of its slower starch digestibility.

The dietary fibre in sorghum seems to be mainly insoluble (water-unextractable) glucuronoarabinoxylans (Verbruggen *et al.*, 1993). Thus, it has benefits in terms of preventing constipation, but probably does not have the cholesterol-lowering effect of soluble dietary fibre.

With regard to micronutrients, sorghum, like other cereal grains, is an important source of B vitamins (Table 13.1). These are concentrated in the aleurone layer and the germ; hence, their content can be adversely affected by milling. The niacin in sorghum, as in maize and probably in the millets as well, is bound to carbohydrate rendering it partially unavailable (Magboul and Bender, 1982).

A very important nutritional feature of sorghum grain is that many varieties contain substantial quantities of phenolic compounds that have strong antioxidant activity (Awika and Rooney, 2004). The phenolic compounds can be categorised into three groups: simple phenolic acids, monomeric polyphenolic flavonoids and complex polymeric polyphenolic condensed tannins (also called proanthocyanidins or procyanidins). Phenolic acids have the lowest antioxidant activity and condensed tannins have the highest. Grain of white tan plant-type sorghums essentially just has phenolic acids. Flavonoids such as anthocyanin are present in substantial quantity in coloured sorghum grains as they are pigment compounds. Condensed tannins are only present in tannin sorghums. Table 13.2 shows the relationship between sorghum type, phenolic content and antioxidant activity. Many health benefits have been claimed in respect of consumption of phenolics including prevention of cardiovascular disease and cancer (reviewed by Awika and Rooney, 2004). Unfortunately, much of the evidence is currently only circumstantial.

However, there is clear evidence that tannin sorghums have somewhat lower nutritional value than tannin-free sorghums and other non-tannin-containing cereals. There are numerous animal studies that show lower weight gain for animals fed tannin sorghums (Hancock, 2001). The mechanisms involved are complex. Tannins form complexes with food proteins and carbohydrates. In the case of proteins, this is referred to as tanning, as in leather tanning. Tannins also bind with and inactivate digestive enzymes such as proteases and amylases. In fact, tannins seem to enhance the slow digestibility of sorghum foods as discussed above. Thus, what applies to sorghum in general applies even more strongly to tannin sorghums.

Until recently, there has been an emphasis on the negative aspects of these antinutritional effects of the tannins. Certainly, for people whose diet is nutritionally deficient, it would be better for them to eat tannin-free sorghum than tannin sorghum. However, for people who are obese and at risk of type II diabetes, tannin sorghums appear to be a potentially useful food for fighting obesity and its associated diseases. It has also been suggested that tannin sorghums are actually toxic and can cause oesophageal cancer (Morton, 1970). In fact, there is no evidence to support this. In areas where sorghum and millet are consumed there is actually low incidence of oesophageal cancer (Van Rensburg, 1981). Furthermore, it appears that the proline-rich protein in human saliva, which strongly binds tannins, exists specifically to protect against an adverse effects of tannins (Baxter *et al.*, 1997).

Sorghum wax, which is on the pericarp surface of the grain, may also have some unique health properties. Of specific interest are the primary long-chained alcohols in the wax, which are called policosanols. These represent

Table 13.2 Relationship between sorghum type, phenolic content and antioxidant activity. Adapted from Awika and Rooney (2004)

Sorghum type	Phenolic content (mg gallic acid equivalent/g)	Oxygen radical absorbance capacity (ORAC)	2,2′-Azinobis (3-ethyl-benzothiaziline-6-sulphonic acid) activity (ABTS)
White sorghum	0.8	22	6
Red sorghum	6.6	140	53
Black sorghum	6.4	220	57
Tannin sorghum	19.8	870	226

approximately 37–44% of sorghum wax (Hwang *et al.*, 2004), which translates to about 80 mg policosanols/100 g sorghum grain. Policosanols are reportedly effective in lowering the amount of low-density lipoprotein (LDL) cholesterol and raising the amount of high-density lipoprotein (HDL) cholesterol, thus improving the LDL/HDL ratio (Hargrove *et al.*, 2004). It is been proposed that policosanols are a promising resource for the prevention and therapy of cardiovascular disease (Varady *et al.*, 2003).

13.2.2 Pearl millet

Description, history and production

Pearl millet (*Pennisetum glaucum* (L.) R. Br.) is also commonly known as bulrush millet on account of the shape of the grain head (Fig. 13.3). The plant is up to 2 m tall. Other common names are bajra (India) and babala (Southern Africa) and *mil à chandelles* (North and West Africa). Pearl millet, like sorghum, also originated in Africa south of the Sahara and was first domesticated some 5000 or more years ago, and was introduced into India about 3000 years ago (House, 1995). Pearl millet is the most widely cultivated millet. The FAO does not distinguish between the different millets in its production statistics but gives a total annual production figure of 33.8–37.3 million tonnes over the years 2001–2005 for millets (FAO, 2006). It is estimated that pearl millet accounts for slightly more than half of millet production

Fig. 13.3 Pearl millet being cultivated in Kenya.

(ICRISAT and FAO, 1996). The major production regions are India, Nigeria and the Sahelian countries of North and West Africa, and the drier parts of East and Southern Africa. Pearl millet is probably the most hardy grain crop. It is cultivated in areas of very low rainfall (300–500 mm/year) and very high temperatures (day time temperatures above 30 °C). Currently, pearl millet is cultivated almost exclusively by smallholder farmers and is of critical importance to food security in arid, tropical regions. However, large-scale production is now beginning, notably in Queensland in Australia.

Physical characteristics

Pearl millet grains are small, up to 2 mm in length, and the 1000 kernel weight is typically approximately only 8 g, in other words about one-quarter the size of a wheat grain. The grains are ovoid to tear shaped and often pearly white in colour (hence the name). Although the grain colour can also be yellow, brown or purple, individual grains are often not uniformly coloured. The pearl millet grain has the typical structure of a cereal grain. Like sorghum, it is a naked kernel. An important difference, however, is that the germ is proportionally rather large, about 18% of the whole kernel. This has important implications in terms of its content of various nutrients.

Nutrients and antinutrients

Whole pearl millet grain has a high energy content for a cereal grain, approximately 1443 kJ/100 g, This is due to the high fat content of the grain (approximately 4.7%) (Table 13.1), resulting from its large germ. Similarly the protein content is relatively high, 11.5% or higher, for the same reason. The large germ also means that pearl millet has a proportionally high content of the albumin- and globulin-type proteins (25–26%) which are rich in essential amino acids, and a lower proportion of the poor nutritional quality prolamin proteins (31–44%) (Serna-Saldivar and Rooney, 1995). These factors have a positive impact on the essential amino acid content of pearl millet grain. It has a high lysine content for a cereal grain (2.8–3.2 g/100 g protein). However, it should be noted that even this level by no means meets the requirements of infants, or for that matter school-age children.

The lipids of pearl millet, like those of other cereals are rich in unsaturated fatty acids. The major fatty acids are: linoleic acid (C18:2), typically 43–45% of the total; oleic acid (C18:1), 26–27% of the total; and the saturated fatty acid palmitic acid (C16:0), 20–21% of the total (Serna-Saldivar and Rooney, 1995). Because of its high fat content, pearl millet is a good source of tocopherols (vitamin E), located mainly in the germ.

The dietary fibre content of pearl millet grain is around 9.7% (Table 13.1) and, as with sorghum, is mainly of the insoluble arabinoxylan type, substituted with uronic acid (Hadimani *et al.*, 2001).

Like sorghum, pearl millet can contain flavonoid phenolics in the pericarp, which are responsible for brown/grey pigmentation of the grain (Reichert, 1979) and have antioxidant activity. There is evidence that some of these can act as antinutrients, specifically as goitrogens. The causal agents seem to be the C-glycosyl flavones: vitexin, glucosyl vitexin and glucosyl orientin (Birzer and Klopfenstein, 1988). Other phenolic compounds – such as phloroglucinol, resorcinol and p-hydroxybenzoic acid – could also be involved. These compounds apparently inhibit the deiodination of the hormone thyroxine to its more active form triiodothyronine. As they are concentrated in the pericarp, they are considerably reduced when the grain is 'dehulled' during milling (see below). In fact, the nutritional significance of the goitrogens in pearl millet should not be overstated. Although some rural people who consume pearl millet as a staple have been found to suffer from goitre, for example in the Sudan, it is probable that this is because their diet is very restricted and hence deficient in iodine.

Perhaps of more significance is that the C-glycosyl flavones appear to be responsible for the characteristic 'mousy' flavour of damp pearl millet flour. The sometimes slightly 'mousy' or 'mouse-droppings'-like flavour of pearl millet foods, such as porridges, can be rather off-putting to those not familiar with the food. Unlike sorghum, no pearl millet varieties appear to contain tannins.

13.2.3 Finger millet

Description, history and production

Finger millet (*Eleucine coracana* (L.) Gaertn.) gets its name from the heads of grain that resemble the fingers of a hand (Fig. 13.4). The plant is quite short, about 1 m. Finger millet is also commonly known as wimbi (East Africa), rapoko (Southern Africa) and ragi (India). It seems to have been domesticated in East Africa at least 3000 years ago (House, 1995). At about the same time it was introduced into India. Some 3–4.5 million tonnes of finger millet are produced annually (ICRISAT and FAO, 1996), nearly half in Africa, mainly East Africa, and half in India. Finger millet is a locally highly popular food fetching a high price relative to grains such as maize. It is popular on account of its taste and also because of its outstanding storage qualities. Regarding cultivation, finger millet requires somewhat more water than sorghum or pearl millet but thrives at relatively high altitude, up to 2000 m (Obilana and Manyasa, 2002).

Physical characteristics

Finger millet grains are very small, just 1.2–1.8 mm in diameter, and are essentially spherical. The grains range in colour from almost white to brown, with red being common. In terms of structure, finger millet grain is unusual in that it is a utricle. This means that the pericarp is not completely fused to the testa/seed coat and can be easily removed by rubbing, particularly after soaking. In fact, a so-called 'mini-millet mill' has been developed in India for finger millet which wets the grain in a water mixer then employs a plate mill action to rub the pericarp off (National Research Council, 1996). Like

Fig. 13.4 Finger millet being cultivated in Kenya.

sorghum, some finger millet varieties contain tannins (Ramachandra *et al.*, 1977). As with sorghum, the tannins seem to be located in the testa layer.

Nutrients and antinutrients

Perhaps the most significant nutritional characteristic of finger millet is its low fat content, only about 1.3% (Table 13.1). This undoubtedly contributes to the good storage qualities of the grain. In terms of fatty acid composition, finger millet seems to be unusual for cereals in that nearly half of the fatty acid content is oleic acid (C18:1), the monounsaturated fatty acid. The protein content of finger millet is low, but it has been reported that it is particularly rich in the essential amino acid methionine (National Research Council, 1996). This is not supported by data presented by Serna-Saldivar and Rooney (1995), which show a normal methionine content. Table 13.1 indicates that the overall mineral (ash) content of finger millet is high and that some particular minerals such as calcium and sodium are very high. However, these data are very probably misleading. A problem particular to grains cultivated by smallholder farmers is that they are often threshed on the ground, which means that they can be contaminated with soil, raising the apparent mineral content significantly. The situation is exacerbated with small grains such as finger millet where the effect is proportionally greater.

Finger millet is a rich source of phenolics. It contains flavone-type flavonoids (Hilu *et al.*, 1978), which are undoubtedly responsible for the red/brown colour of many varieties. Among the types reported are orientin and vitexin, the same compounds said to be goitrogens in pearl millet and believed to be responsible for its 'mousy' flavour. Finger millet foods are also strongly flavoured so it is possible that these compounds are also responsible. An association between goitre and finger millet consumption has not been reported. As stated previously, some finger millet varieties also contain tannins; levels up to 3.5% have been found (Ramachandra *et al.*, 1977). These levels are somewhat lower than those reported for tannin sorghums. Notwithstanding this, finger millet is a potent source of antioxidants. Sripriya *et al.* (1996) reported that a red/brown variety had higher radical scavenging activity than wheat, rice or other millet species, with the activity corresponding to its phenolic content.

13.2.4 Tef

Description, history and production

Tef (*Eragrostis tef* (Zucc.) Trotter) is closely related to finger millet, both belonging to the subfamily Chloridoideae. Like finger millet, it is a relatively short plant (up to 1 m), but the grains are on an open head (Fig. 13.5), similar to oats. Tef originated in Ethiopia and is a major cereal in that country with

Fig. 13.5 Tef being cultivated in Ethiopia.

a production of about 2 million tonnes, about 20% of the cereal crop (Bultosa and Taylor, 2004). It is widely grown in South Africa, but mainly as a forage crop. There is some production for food use in the USA (National Research Council, 1996). Tef is uniquely well adapted to cultivation at high altitudes, up to 3000 m, but the best conditions are 1800–2100 m (Bultosa and Taylor, 2004). Tef cultivation in Ethiopia is almost exclusively manual. A problem with the tef plant is that it tends to lodge because the stalks are thin, making mechanical harvesting difficult.

Physical characteristics
The grain is tiny and oval shaped, some 0.9–1.7 mm in length. It has a 1000 kernel weight of only around 2 g. The grains range widely in colour from white to red and brown. The white grains are preferred for food use. Despite its colour, brown tef does not seem to contain tannins (G. Bultosa and J.R.N. Taylor, unpublished data, 2004). In terms of grain structure, tef is conventional like sorghum and pearl millet, in spite of its much closer taxonomic relationship to finger millet.

Nutrients and antinutrients
On the face of it, the nutrient composition of tef appears quite ordinary (Table 13.1). It has a low fat content which, as with finger millet, undoubtedly is a major factor in its good storage qualities. The fatty acid composition of the lipids is normal for cereals with linoleic acid (C18:2) being the highest at around 42%, followed by oleic acid (C18:1) and palmitic acid (C16:0) (Hundera, 1998). With regard to protein composition, there is evidence that tef is much richer in the higher quality and more digestible albumin and globulin proteins, and correspondingly poorer in the low-quality prolamin proteins, compared with other cereals. A range of only 3–15% of prolamin expressed as a proportion of total protein has been found (Bekele, 1995). However, this is not really borne out by the amino acid composition of tef, which shows lysine to be rather low (Table 13.1) and glutamine, alanine, leucine and proline to be high (Bekele, 1995), clearly indicating a high proportion of prolamins. With regard to micronutrients, it is perhaps of significance that tef appears to contain a high level of folate (Table 13.1). It has been suggested that tef is particularly rich in iron (National Research Council, 1996). However, this is not borne out by the data (Table 13.1) and the reported high level of iron was probably an artefact caused by soil contamination of the grain.

Perhaps the major issue regarding tef's nutritional value is the fact that, despite it being gluten-free, it can make a high-quality leavened flatbread. This pancake-like flatbread, injera, is the staple food of Ethiopia. The reason for tef's good leavened bread making quality is not known. It seems to be related to the fact that its starch occurs in the form of small granules, only 2–6 μm in diameter (Bultosa and Taylor, 2004), like those in rice, which is often used as a tef substitute in injera making.

13.2.5 Quinoa

Description, history and production

Quinoa is dicotyledonous plant (Fig. 13.6), unlike cereals which are monocotyledonous. It belongs to the Chenopodiaceae family and is native to the Andean region of South America. The name quinoa was given by the Spanish, and the common names used in the Andean region are: quinua, kuina, parca in Ecuador, Peru and Bolivia; jopa, jupha, juira, aara, callapi, vocali in Bolivia; quinhua in Chile; and suba and pasca in Columbia (Valencia-Chamorro, 2004). Quinoa was one of the staple foods for the Inca Empire. The quinoa plant is well adapted to a range of altitudes, from sea level in the south of Chile to 3500–4000 m around Lake Titicaca at the Peru/Boliva border (Valencia-Chamorro, 2003), depending on the ecotypes. It is frost- and drought-resistant and can be cultivated in areas with a rainfall of just 200–400 mm (Weber, 1978). The potential of quinoa has been highlighted as a new crop considered for NASA's controlled ecological life support system, because of its high nutritional composition, harvest index, canopy structure and life cycle duration (Schlick and Bubenheim, 1996). According to the FAO (2006), the major quinoa-producing countries are Bolivia, Peru and Ecuador, but it is gaining popularity worldwide. Since 1970 the cultivation area and the production of quinoa have steadily increased from 30 269 ha and 18 768 tonnes respectively, in 1970–1974 to about 67 277 ha and 51 152 tonnes in 2005 (FAO, 2006).

Fig. 13.6 Quinoa harvested in Peru (courtesy of Mrs N. N. Dersley, University of Pretoria).

Physical characteristics

The fruit of quinoa is an achene. It is disc-shaped and ranges from 1–3 mm in diameter (Varriano-Marston and De Francisco, 1984) with a seed weight of 250–500 seeds/gram (Valencia-Chamorro, 2004). The grain structure of quinoa is complex. Basically it consists of a pericarp, a seed coat which is beneath the pericarp, the embryo with two cotyledons and radicle seen as coiled around the perisperm (Fig. 13.7) (Prego *et al.*, 1998). The main starch reserve is in the perisperm as the endosperm is reduced to one to two cell layers in the micropylar region (Prego *et al.*, 1998). Protein and lipid bodies are abundant in the endosperm and the embryo.

Nutrients

The protein content of quinoa is about 12.7% (Table 13.1). The major proteins are the globulin and albumin types, with the major protein group being 11S globulin, known as the chenopodin (Brinegar and Goundan, 1993). This protein is rich in the essential amino acid lysine and low in proline. The 2S globulin-type proteins have also been isolated in quinoa and are rich in other essential amino acids: cysteine, histidine and arginine (Brinegar *et al.*, 1996). The protein quality of quinoa is similar to casein. The protein efficiency

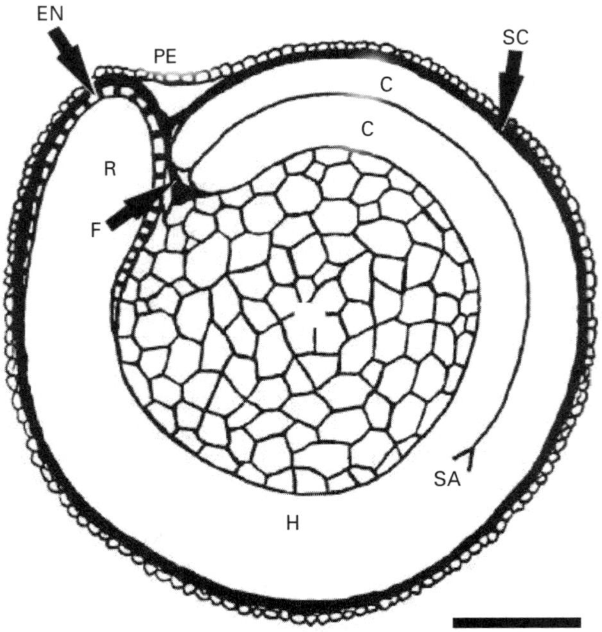

Fig. 13.7 *Chenopodium quinoa*: median longitudinal section of the grain. Pericarp (PE) covers the seed. The embryo consists of a hypocotyl radicle axis (H) and two cotyledons (C). Endosperm (EN) is present in the micropylar region. F, funicle; P, perisperm; R, radicle; SA, shoot apex. Scale bar is 500 µm. From Prego *et al.* (1998); published with permission of Oxford University Press.

ratio (PER) of quinoa has been reported as 2.7 (Ranhotra *et al.*, 1993) and 3.1 (Gross *et al.*, 1989), with an apparent *in vivo* protein digestibility of 84% (Ranhotra *et al.*, 1993; Gross *et al.*, 1989).

The fat content of quinoa (5.6%) is high compared with most cereals (Table 13.1) and is concentrated in the germ (Prego *et al.*, 1998). Quinoa contains a very high proportion of polar lipids (25%) and free fatty acid (19%) (Przybylski *et al.*, 1994). Triglycerides are the major fraction of the neutral lipids. Quinoa is rich in oleic (monounsaturated) and linoleic and linolenic (polyunsaturated) fatty acids with values of about 23, 55 and 7%, respectively.

Starch is the major carbohydrate component of quinoa and is mostly located in the perisperm cells. Starch occurs as small granules of 1.0–2.5 μm diameter in a compound granule of 6.4–32 μm diameter (Atwell *et al.*, 1983). *In vitro* starch digestibility of precooked quinoa was reported to be 77% (Ruales and Nair, 1994). The dietary fibre content is high compared with other cereals, with over 80% being of the insoluble type (Ruales *et al.*, 2002). Quinoa is rich in micronutrients, for example calcium, phosphorus, iron, zinc, magnesium, vitamin E, vitamin C, folic acid and thiamin.

The chemical composition of quinoa shows its potential as a major source of nutrients for both adults and children. *In vivo* studies showed that the level of insulin-like growth factor-1 (IGF-1) in the plasma of undernourished children increased significantly when their diet was supplemented with a portion of 2 × 100 g of quinoa food product for 15 days compared with undernourished children without the supplementation (Ruales *et al.*, 2002). Despite being a nutrient-dense grain, quinoa can be used as an appetite-controlling food for the developed world, as it has been shown to have a higher satiety index than common cereal grains (Berti *et al.*, 2005).

Antinutrients

The major antinutrients in quinoa are saponins and phytic acid. Phytic acid is a common antinutrient present in all grains, but saponins are quite unique to quinoa, being concentrated in the pericarp seed coat Layer (Chauhan *et al.*, 1992). The saponins in quinoa are secondary metabolites of triterpenoid aglycones. Six triterpenoid saponins have been isolated from quinoa seeds (Dini *et al.*, 2001) and lately, twelve triterpene saponins have been characterised in debittered seed of quinoa (Zhu *et al.*, 2002). They are antinutrients because of the bitter taste they impart and because they can be membranolytic against cells of the small intestine *in vivo* (Gee *et al.*, 1993). Saponins can also bind iron and reduce its absorption (Southon, 1988). Sweet genotypes of quinoa contain 0.02–0.04% saponins compared with bitter genotypes with levels of 0.47–1.13% (Mastebroek *et al.*, 2000). Saponins are traditionally removed from quinoa by washing with water until little or no foam is observed. Alternatively, saponins can be reduced by up to 99% by abrasive dehulling (Reichert *et al.*, 1986). Although saponins are antinutrients, they may have possible health benefits in terms of reducing serum cholesterol levels (Oakenfull and Sidhu, 1990).

13.2.6 Grain amaranth

Description, history and production

Amaranth belongs to the genus *Amaranthus* and the major types can be divided into grain amaranth, wild vegetable amaranth, ornamental amaranth and weed amaranth. Here we will only discuss grain amaranth. The species of grain amaranth are *A. caudatus*, *A. cruentus* and *A. hypochondriacus*. Grain amaranth is believed to have originated in Central and South America (Saunders and Becker, 1984). It was an important staple food for the ancient Mayan and Aztec cultures; it was prohibited during the Spanish occupation. Since the 1970s, there has been a renaissance in the production and processing of amaranth because of its potential as a highly nutritional food grain (Berghofer and Schoenlechner, 2002). Nowadays, grain amaranth is grown in Central and Latin America, China and Russia. Amaranth can be grown in marginal land with poor soil conditions. It is quite drought resistant and grows from sea level to 3500 m. Unfortunately, there are no FAO statistics on grain amaranth production. However, the major producers are China and Russia with 300000 and 100000 ha cultivated, respectively (Cai *et al.*, 2004a). Worldwide, yields seem to vary from 2 to 5.5 tonnes/ha.

Physical description

Grain amaranth seeds (Fig. 13.8) are very small with diameters of 1.2–1.4 mm and weights ranging from 0.56 to 0.78 g/1000 grains for the species *A. caudatus* and *A. cruentus* (Gamel *et al.*, 2004). Like quinoa, amaranth seeds have an embryo which encloses the starch-rich perisperm (Berghofer and Schoenlechner, 2002). The embryo is more than 25% of the grain weight. The seed coat is smooth and thin, unlike quinoa. The starch granules are small, about 1 μm in diameter, and located in the perisperm (Nakamura *et al.*, 1998). The protein-containing protein bodies are 2–5 μm in diameter and are mainly found in the embryo. They contain globulin- and albumin-2-type proteins. The albumin-2 fraction comprises five polypeptides with molecular masses of 56, 35, 31, 25 and 22 kDa.

Nutrients

Amaranth is a good source of protein and lipids (Table 13.1). The protein content can range from 11.7–18.4% (Berghofer and Schoenlechner, 2002). Like quinoa, amaranth is also rich in the limiting essential amino acid lysine. *A. caudatus* grain is reported to have only a slightly lower PER of 2.59 and apparent protein digestibility of 73% compared with casein with values of 3.1 and 87.5%, respectively (Gross *et al.*, 1989). Gamel *et al.* (2004) reported a higher *in vivo* protein digestibility ranging from about 85 to 86% for raw and popped *A. caudatus* and *A. cruentus*. The protein in amaranth is mainly globulins and albumins. Like legumes, both the 7S and 11S globulins are the storage globulins (Marcone, 1999). Interestingly, the globulin of amaranth has very good foaming properties (Fidantsi and Doxastakis, 2001).

Fig. 13.8 Grain amaranth (centre) for sale in Puorto Maldonado, Peru (courtesy of Mrs N. N. Dersley, University of Pretoria).

The lipids are concentrated in the germ and seed coat in amaranth (Betschart *et al.*, 1981). They are composed mainly of palmitic (C16:0), palmitoleic (C16:1), oleic (C18:1) and linoleic acid (C18:2). The unsaturated fatty acid level is three times higher than the saturated fatty acid level (Escudero *et al.*, 2004). The unsaponifiable fraction of amaranth lipids is mainly squalene. This is a highly unsaturated open-chain triterpene phytosterol. Squalene is a biosynthetic precursor to steroids. It has a high commercial functional value in cosmetics and as a computer disc lubricant (Cai *et al.*, 2004a). Squalene content ranges from 2.26 to 5.94% in *A. cruentus* oil (Berganza *et al.*, 2003). The squalene content depends on agronomic practices and environmental conditions. Amaranth is also a rich dietary source of other phytosterols, for example β-sitosterols, campesterols and stigmasterols (Marcone *et al.*, 2004). These phytosterols can reduce blood cholesterol level (Kritchevsky and Chen, 2005). Plate and Arêas (2002) demonstrated reduction in LDL and cholesterol when hypercholesterolaemic rabbits were fed with extruded *A. caudatus* grains. However, Berger *et al.* (2003) did not find any cholesterol-lowering properties when *A. cruentus* flakes, and crude and refined oils were fed to hamsters. This discrepancy may be because of species, products and processing conditions.

Starch is the major carbohydrate in amaranth. *In vitro* starch digestibility of *A. hypochondriacus* was 63.5%, (Yanez *et al.*, 1986). *A. cruentus* starch caused higher glycaemic and insulinaemic responses when consumed as an extruded snack compared with white bread (Guerra-Matias and Arêas, 2005). This suggests that amaranth is a good source of energy for athletes, but not for diabetics.

Antinutrients
Berghofer and Schoenlechner (2002) reviewed the presence in grain amaranth of the antinutrients phytic acid, phenolic compounds, protease inhibitors and saponins. These can interfere with digestion and absorption of nutrients. However, the polyphenols in amaranth can have antioxidant properties and have been associated with a reduction in blood cholesterol levels in rats fed a cholesterol-containing diet (Czerwinski *et al.*, 2004).

13.2.7 Buckwheat

Description, history and production
Buckwheat belongs to the family Polygonaceae and is in the genus *Fagopyrum*. It is believed to have originated in China and is widely cultivated as a minor crop around the world. There are 15 species of the genus *Fagopyrum*, and the two that are widely cultivated are *F. esculentum* (Moench), known as common buckwheat and *F. tataricum* (L.), known as tartary wheat (Cai *et al.*, 2004b). Buckwheat favours warm and moist climates but can be adapted to most types of soil conditions. The yields for common and tartary buckwheat are, respectively, 527 kg/ha and 1294 kg/ha (Cai *et al.*, 2004b). Common buckwheat can be milled into flour for food and feed. Tartary buckwheat has limited use in low-grade animal feed rations and is not commonly used in human food because it is bitter (Oplinger *et al.*, 1989). The major buckwheat producing countries are China and Russia. According to the FAO (2006), world production of buckwheat in 1971–1975 ranged from 3.1–3.5 million tonnes and the cultivated area was 3.7–4.2 million ha. The production of buckwheat decreased from years 1971–1975 to 2001–2005. The world production of buckwheat in 2001–2005 was in the range 1.8–2.6 million tonnes with a cultivated area of 2.1–3.1 million ha.

Physical characteristics
Buckwheat grain is an achene. It is 4–9 mm long and can weigh 15–35 mg (Fig. 13.9). The grain consists of a hull (or pericarp), spermoderm, endosperm and embryo. The hull represents 17–26% of grain weight in common buckwheat and 30–35% in tartary wheat (Mazza and Oomah, 2005). The grain has a very compact external layer and a sinusoidal-shaped embryo, which extends throughout the cross-section (Mariotti *et al.*, 2006). The endosperm consists of an aleurone layer and a subaleurone endosperm. The endosperm is rich in starch granules surrounded by a protein matrix. The starch granules are 1–7 μm and of regular shape. The protein is stored in protein bodies, which are 1–3 μm in diameter (Gao *et al.*, 2002).

Fig. 13.9 Buckwheat grain.

Nutrients

Buckwheat is a good source of energy, protein, vitamins and minerals (Table 13.1). The protein is rich in the essential amino acid lysine. The biological value (percentage of absorbed nitrogen retained in the body) of buckwheat protein is 92.3%, compared with 81.5% for defatted dried milk and 62.5% for wheat (Rayas-Duarte *et al.*, 1998). The major storage protein in common buckwheat is of the globulin class, accounting for almost 70% of the total. The major globulins are a 13S legumin-like protein and an 8S vicilin-like protein (Radović *et al.*, 1996). Albumins account for about 25% of total protein and are mainly 2S type (Radović *et al.*, 1999). Contrarily, the major storage protein in tartary buckwheat is albumin (44%), followed by glutelin, prolamin and globulin (Guo and Yao, 2006). The globulin accounts for only about 8% of total protein. The digestibility of buckwheat protein is somewhat low. The apparent *in vivo* protein digestibility of high-protein buckwheat flour (36.5% protein) was reported to be 74% compared with 96.1% for casein (Tomotake *et al.*, 2006). The low digestibility has been ascribed to an endogenous protease inhibitor and tannins (Ikeda *et al.*, 1991). However, the low digestibility has been found to have some beneficial effects in reducing cholesterol, obesity and gallstone formation (Tomotake *et al.*, 2006).

Other functional components in buckwheat are starch, phenols and rutin. Boiled buckwheat groats have a high content of resistant starch (4.4% dry basis) and have a very slow rate of starch digestion (Skrabanja *et al.*, 2001). Adding buckwheat with wheat bread significantly reduced the rate of starch digestion compared with whole wheat bread. The consumption of buckwheat and whole wheat composite flour bread reduced postprandial blood glucose and insulin response compared with whole wheat bread alone. This indicates the potential of buckwheat to lower the glycaemic index of foods.

Although, as stated, phenolic compounds can decrease protein digestibility, they can be a potential source of antioxidants. The total phenolic contents of white and dark buckwheat flour are about 1.79 and 10.74 mg/g gallic acid equivalents, respectively (Şensoy *et al.*, 2006). Buckwheat contains higher level of phenolics than oats and barley (Holasova *et al.*, 2002). Acetone extract of buckwheat contained 34 mg/g catechin equivalent of phenolic compounds and showed similar radical scavenging ability compared with the synthetic antioxidants, butylated hydroxyanisole, butylated hydroxytoluene and tertiary butylhydroquinone (Sun and Ho, 2005). Buckwheat hulls contain the phenolic antioxidants quercin, hyperic, rutin, protocatecholic acid and 2,4-dihydroxybenzaldehyde (Watanabe *et al.*, 1997). Rutin is considered to be the most important flavonoid in buckwheat (Biacs *et al.*, 2002).

Because of the above-mentioned health benefits, buckwheat has been proposed as a functional food. The major functional compounds in buckwheat are considered to be flavonoids, flavones, phytosterols, phagopyrins and thiamin-binding proteins (Krkošková and Mrázová, 2005). Buckwheat can also be considered as a prebiotic as it was found to increase the probiotic bacteria in the intestine of rats (Préstamo *et al.*, 2003).

Antinutrients

As mentioned, the major antinutrients in buckwheat are endogenous protease inhibitor and polyphenolic compounds (Ikeda *et al.*, 1991). Although not considered as an antinutrient, the presence of potentially allergenic proteins in buckwheat is critical to some hypersensitive individuals. The proteins responsible for the allergenic reaction have been identified and are low molecular mass proteins of 15 and 22 kDa (Morita *et al.*, 2006). These proteins are mostly located in the outer part of the grain (Morita *et al.*, 2006), and in the embryo, but not in the endosperm (Ličen and Kreft, 2005). Thus, dehulling to remove the outer grain parts, including the embryo, can produce less allergenic buckwheat primary products.

13.3 Consumer attitudes

Even today, these grains play an important role in the culture of the people who consume them traditionally. For example, traditional African beer made from sorghum is invariably consumed by urbanised black South Africans at events such as weddings and funerals. This is despite the fact that many of the people will only consume beverages such as lager beer the rest of the time.

The grains also remain highly respected by people for their nutritional value. This aspect has been capitalised upon. A most striking example concerns that of a lactic bacteria-soured sorghum soft porridge called ting, a traditional African breakfast food in South Africa and Botswana. A company in South Africa has produced an instant ting-type product by dry precooking the sorghum and replacing the lactic acid fermentation by adding fruit acid to provide characteristic sourness. The product, which is called Morvite (Fig. 13.10), alone accounts for some 20% of all sorghum flour sales in South

Fig. 13.10 Morvite, an instant sour sorghum porridge from South Africa.

Africa. Interestingly, Morvite is purchased and consumed by persons of all cultural groups in South Africa. Development of new products from these traditional grains seems to be a general trend. In the Sahel region of West Africa, a wide range of processed traditional cereal products, based on sorghum and millets, but with the convenience level of rice are becoming available in urban areas (Vitale and Sanders, 2005), with the number of processors of these grains increasing rapidly. Similarly, in Latin America, there are many commercial products based on amaranth and quinoa (see Section 13.5).

13.4 Traditional food products and processing technologies

Despite the fact that these grains do not contain gluten, a very wide range of different types of food products have been produced from them. The food products can be divided into dry cooked and wet cooked. Dry cooked products are generally whole grain and are consumed as snack foods.

13.4.1 Dry cooked products

Whole grains may be roasted. The preparation process for roasting generally involves harvesting the heads of the plants at the dough stage. The heads are covered with hot coals or ashes for a few minutes to roast the grains (Murty and Kumar, 1995). The heads are then threshed to remove the grains, which in turn are winnowed. In Ethiopia, high-lysine types of sorghum are used, which in theory should mean that the product has high protein quality. However, it is likely that the lysine content in the product will be considerably reduced during the roasting process, as Maillard browning-type reactions occur between lysine and reducing sugars with heat (Damodaran, 1996). Other dried products made from amaranth are alegria (sweet meat) in Mexico and laddoos (sweet meat) in India.

These grains may also be popped like popcorn. Most other types of grains pop to some extent but the degree of popping is quite variable. Hard endosperm types definitely pop better. Popped sorghum is popular in Ethiopia where it is called kollo. A similar product is consumed in Nigeria.

13.4.2 Wet cooked products

The preparation of most wet cooked grain products – for example, most rice products, flatbreads and most porridges – invariably involves first removing the bran (the pericarp and at least part of the germ) from the grain to produce a product that is essentially starchy endosperm. The bran removal process is generally done by a process of gradually abrading (rubbing) off the outer layers of the grain (Taylor and Dewar, 2001). The process is generally referred to as decortication or dehulling, although in fact the term dehulling is incorrect since all the grains considered here are naked, i.e. they do not have a hull. As explained in the Introduction, refining brings about significant changes in the nutrient composition of the grain. The extent of these changes is directly dependent on the extent of removal of the grain outer layers.

Rice-type products

Rice-type products are generally prepared from decorticated grain. Sorghum and millet rice-type products are widely consumed across Africa. An example of a whole sorghum grain rice-type product is Lesasaoka/Lehata (Botswana) and examples of decorticated sorghum grain rice-type products are Mosutlhane (Botswana), Nifro (Ethiopia) and Balila/Pearl dura (Sudan). Generally sorghum requires a longer cooking time than rice (Murty and Kumar, 1995) and at least one change in cooking water is often required (J.R.N. Taylor, personal observation, 2007). Thus, water-soluble nutrients such as B vitamins and minerals will be lost from the grain during the cooking process. Sorghum varieties with a high proportion of vitreous endosperm are preferable for rice-type product making, as they should have lower micronutrient losses compared with the softer endosperm types. Quinoa seed is traditionally used as boiled grain; in soups, salads, casseroles, chilli and stew in Latin America.

Couscous

Couscous consists of particles of cereal endosperm meal agglomerated together and cooked by steaming. Although couscous is generally made from wheat semolina, decorticated pearl millet and sorghum are widely used in West Africa (Galiba *et al.*, 1987). Figure 13.11 shows the pearl millet/sorghum

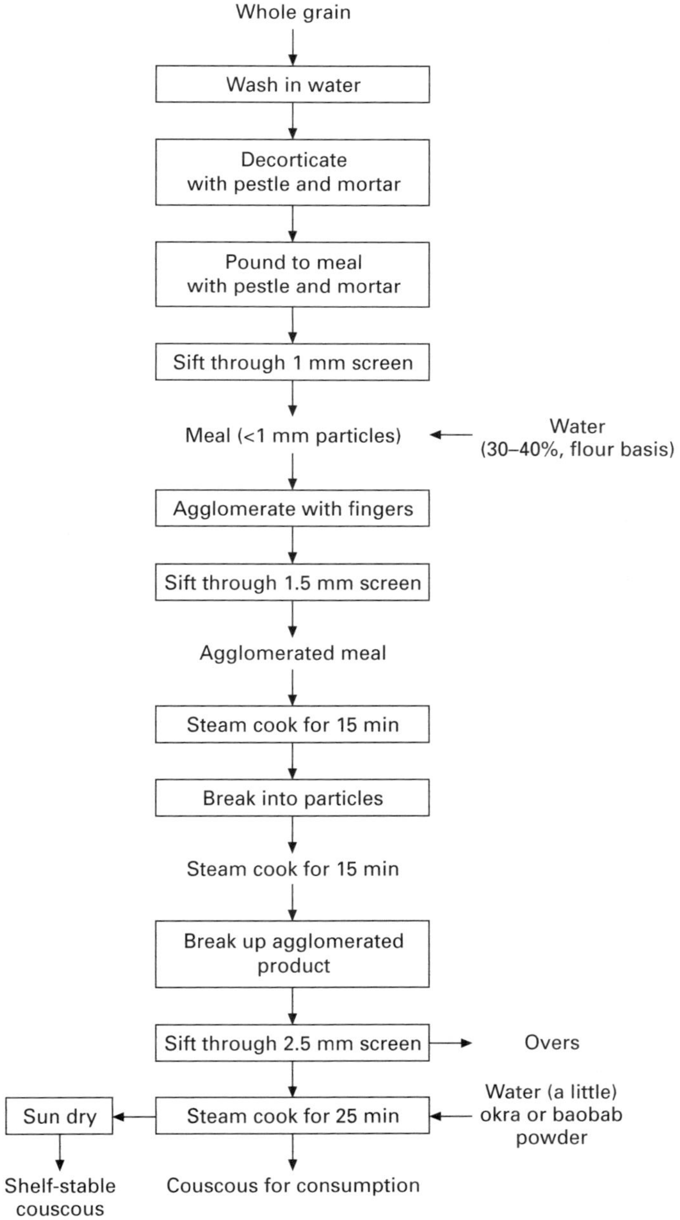

Fig. 13.11 Process for making pearl millet/sorghum couscous. Adapted from Galiba *et al.* (1987), and Murty and Kumar (1995).

couscous preparation process used. An example of a commercial pearl millet couscous product from Senegal is shown in Fig. 13.12. As explained in Section 13.2.1, the process of steaming is liable to considerably reduce protein and starch digestibility in sorghum couscous due to cross-linking of the kafirin storage proteins.

Flatbreads
These pancake-like products are staples in northeast Africa (Ethiopia, Eritrea and Sudan). They are made from a variety of different cereals, especially tef, sorghum and finger millet. A feature of many of these products is that they undergo a mixed lactic acid bacteria and yeast fermentation (Gashe *et al.*, 1982) that gives them a somewhat leavened texture and an acidic flavour. The nutritional effects lactic acid fermentation are discussed later in Section 13.4.3 on fermentation. Probably the two most well-known flatbreads are injera and kisra. Injera from Ethiopia and Eritrea is a large (approximately 50 cm diameter), spongy textured pancake about 5 mm thick. It has a honeycomb-like appearance, very similar to an English crumpet, which is

Fig. 13.12 A commercial pearl millet couscous product from Senegal.

created by the escaping carbon dioxide. Figure 13.13 shows a typical process used for preparing injera in Ethiopia. For reasons not fully understood, the keeping qualities of tef injera are superior to those of injera made from other cereals such as sorghum (Yetneberk *et al.*, 2004). Kisra, which is from the Sudan, in contrast is a flexible thin wafer (1–1.5 mm thick); with neither holes nor a spongy texture (Badi *et al.*, 1988).

Non-fermented flatbread products include chapatti from amaranth in the Himalayas, and pakora – a crisp fried bread-like snack – and the unleavened bread chillare in India.

Tortillas

Tortillas are soft pancakes about 150 mm in diameter and 5 mm thick, traditionally made from maize. They are traditional to Mexico and Central America. A feature of their method of preparation is that the maize is steeped

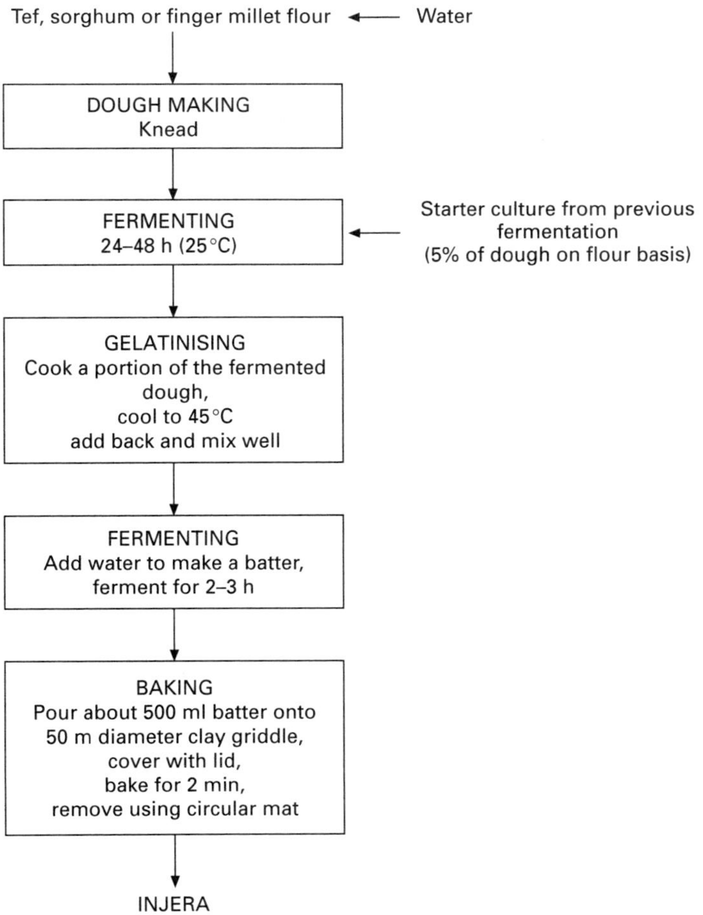

Fig. 13.13 Process for making injera. Adapted from Yetneberk *et al.* (2004).

in a solution of lime or the leachate of wood ash (Rooney, 1993). This process is known as nixtamalisation, an Aztec word for the process. An important nutritional feature of the lime-cooking process is that it frees the niacin in the grain, which is otherwise largely unavailable due to its occurrence as the bound form nyacitin (Magboul and Bender, 1982). Although maize is the grain of choice for tortillas, in parts of Central America, like Honduras, sorghum is used alone or in combination with maize (Rooney and Waniska, 2001).

Dumplings

Dumplings are balls of dough that are cooked by boiling in water or steaming. In Africa, there are both non-fermented and fermented dumpling products, which may be made from sorghum, maize or millets. Obviously, as described above, this wet cooking process will affect the protein and starch digestibility of sorghum dumplings. Figure 13.14 shows a traditional South African process for preparing sorghum dumplings.

Porridges

Porridges are the most common type of food product make from these non-gluten-containing grains. They are the most important staple foods in sub-Saharan Africa. Figure 13.15 shows a selection from Kenya of commercial finger millet and sorghum flours for making porridges. Porridges are served as the main staple in the evening, as breakfast porridge and they are very widely used as weaning foods. Making porridge involves cooking the meal or flour in boiling water to gelatinise the starch. Gelatinising the starch gives the porridge its desirable texture and removes the floury taste of the starch.

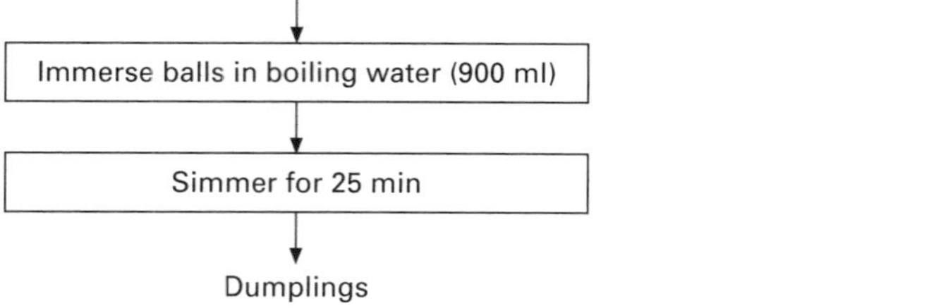

Fig. 13.14 Process for making sorghum dumplings used by the Pedi people of South Africa. Compiled from Quin (1959).

Fig. 13.15 Kenyan commercial finger millet and sorghum flours for making porridges.

There is a vast range of different types of porridges. They vary according to grain used and processing method. Porridges are also highly variable in terms of consistency and taste. Consistency is primarily a consequence of the solids content of the porridge. It is also a result of temperature of serving and whether the starch has been partially hydrolysed. For example, the solid Nigerian 'porridge' eko/kafa/agidi is actually thin ogi porridge that has been allowed to cool and gel. As described under 'Malting' in Section 13.4.3, the starch in the porridge can be partially hydrolysed by alpha-amylase to reduce porridge viscosity or even to liquefy it. The traditional source of the alpha-amylase is normally malted sorghum (Mtebe *et al.*, 1993). For the sake of convenience, porridges will be divided here into thick and thin. This is fairly arbitrary, but a rough guide is that thick porridges can be eaten with the hand, whereas thin porridges necessitate the use of a spoon. Porridges also vary greatly with respect to taste: they may be neutral, as for example ugali in East Africa; they may be fermented by lactic acid bacteria, like ting in Southern Africa; they may be acidified, for example, with tamarind, like kunun tsamiya in Nigeria; or they may be made alkaline with wood ash, like tô in Mali. Obviously, all these variables, and particularly the solids content, will affect the nutritional value of the porridge.

Thick porridges are normally in the range 20–30% solids but, as mentioned previously, other factors like temperature of serving and whether the starch has been hydrolysed affect consistency. Figure 13.16 shows a process for

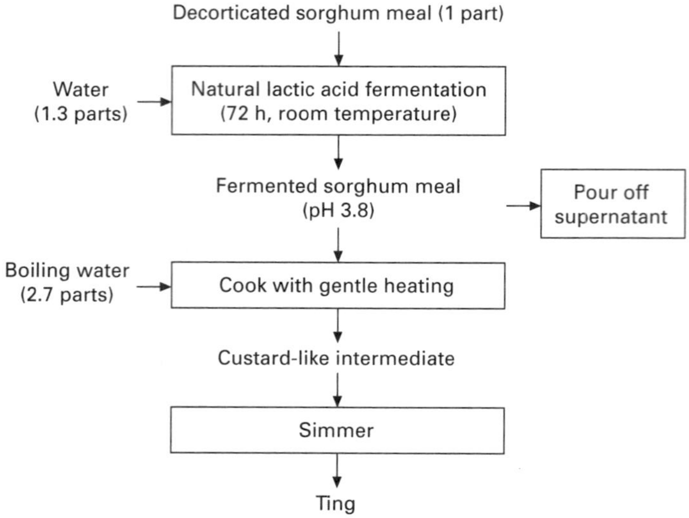

Fig. 13.16 Process for making ting, a fermented thick sorghum porridge from Southern Africa. Adapted from Taylor (2004).

preparing ting, a fermented thick sorghum porridge widely consumed in Botswana and South Africa. Thin porridges are normally in the range 10–20% solids. Figure 13.17 shows a process for preparing ogi, a fermented thin porridge often made from maize, sorghum or pearl millet and very widely consumed in Nigeria (Onyekwere *et al.*, 1989).

Buckwheat porridges are also made in China. In Korea, a jelly-like food product, known as memilmuk is a common traditional buckwheat food.

Non-alcoholic gruels and beverages

The difference between a gruel/beverage and a porridge is also to some extent arbitrary. It again depends primarily on solids content, but factors such as serving temperature and degree to which the starch is hydrolysed also have a bearing. Normally, gruels/beverages contain less than 10% solids. For example, commercial South African mageu maize gruel contains approximately 8% solids; this is the same as is normal for ogi (5–8% solids), which is regarded as a thin porridge. A practical definition is that a gruel/beverage is pourable and is generally consumed as a beverage. Gruels can be distinguished from beverages in that the former contain a substantial proportion of gelatinised starch. Around the world, many different types of gruels and beverages are prepared from all these grains. For example, in Namibia oshikundu is a very popular fermented pearl millet beverage. In Zimbabwe, a fermented gruel product mahewu, prepared from sorghum or maize, is widely consumed. Figure 13.18 shows the traditional process for preparing sorghum mahewu in Zimbabwe (Bvochora *et al.*, 1999). Sorghum malt is used as an ingredient to provide fermentable sugars and to partially hydrolyse the starch to thin the paste into a gruel. Another example of a non-alcoholic

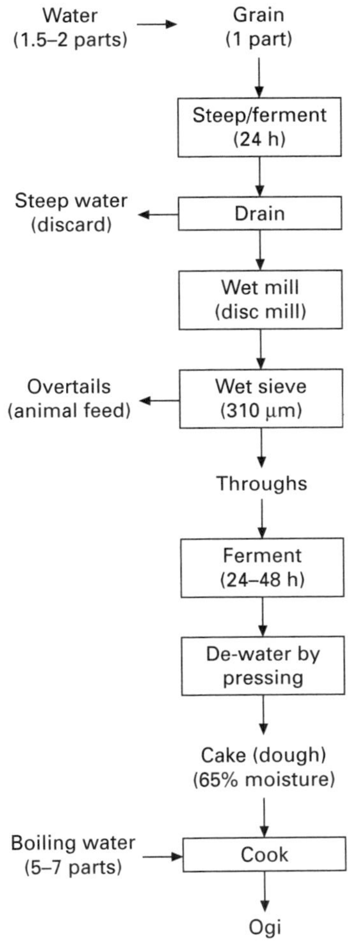

Fig. 13.17 Process for making ogi, a fermented thin porridge from Nigeria. Adapted from Taylor (2004).

drink is atole, a beverage from amaranth that is commonly consumed in Mexico.

Traditional beers

In Africa, the traditional beers are made from sorghum, pearl millet, finger millet and today, increasingly, maize. The beers are characteristically cloudy in the case of the beers of West Africa commonly known as dolo and pito, or opaque as in the beers of East, Central and Southern Africa, often referred to as sorghum beer, opaque beer and by the brand name of Chibuku. The cloudiness and opaqueness are due to the fact that they are coarsely filtered so that yeast cells, pieces of cereal and especially starch are still present. Another characteristic of these traditional beers is that, in addition to the yeast alcoholic fermentation, they undergo lactic acid bacteria fermentation.

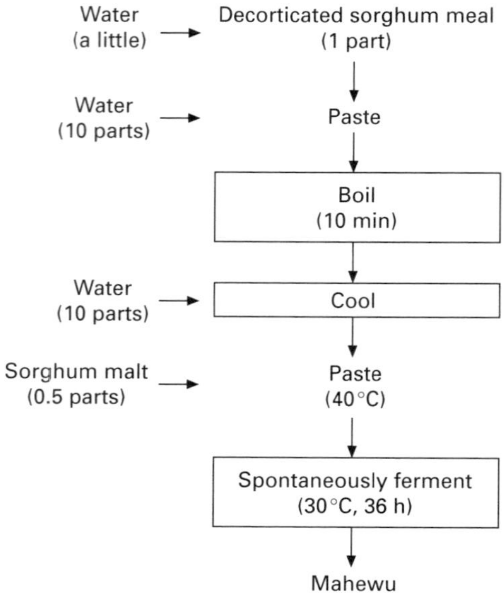

Fig. 13.18 Process for making sorghum mahewu. Compiled from Bvochora *et al.* (1999).

This gives them a characteristic sour taste. Unlike beers of European origin they are not hopped. Traditional African beers are also not pasteurised, but are often consumed in an active state of fermentation. When these beers are brewed in the traditional way using just malt, as shown in the flow chart (Fig. 13.19), or malt and whole grain, they have high nutritional value (Table 13.3). Unusually for beer, they are a source of complex carbohydrates in the form of starch and dietary fibre and, because they contain yeast cells, they are also a source of B vitamins. The Pedi people of South Africa actually refer to their opaque beer as food. When the host serves the beer to his guest, he says 'Mother, may we eat' (Quin, 1959). Figure 13.20 shows commercially brewed opaque beer together with traditional woven-grass beer strainers and serving vessels.

13.4.3 Traditional grain processing technologies

As described above, the traditional grain processing technologies of malting and fermentation are used for both alcoholic and non-alcoholic beverages. These technologies have important nutritional effects.

Malting

Malting involves the sprouting (germination) of grain in moist air, under controlled environmental conditions. The malt is generally dried to produce a shelf-stable product. The main objective of malting is to modify the grain structure and composition. This is achieved primarily by mobilising the

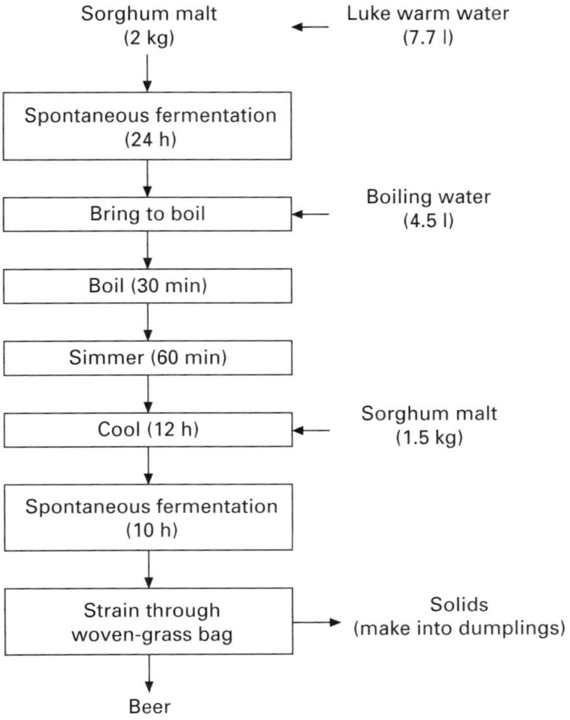

Fig. 13.19 Process of making opaque sorghum beer used by the Pedi people of South Africa. Compiled from Quin (1959).

malt's endogenous enzymes to hydrolyse the biopolymer molecules. For example, the amylases hydrolyse starch to produce fermentable sugars and the proteases break down protein to produce free amino acids. In addition, many micronutrients are made more available through malting. Although malting is primarily used for the beer-making process, malt can be used as food or added to other grain foods to improve their nutritional value. The nutritional benefits of malting in various grain products are shown in Table 13.4. The benefits can be summarised as follows:

1 Overall improvement in essential amino acid composition, especially an increase in total lysine in lysine-deficient grains.
2 General improvement in protein and starch digestibility, possibly because of structural modification increasing accessibility to enzymes. However, further heat treatment, like boiling, can reduce digestibility (Hugo, 2002).
3 Increase in availability of vitamins and minerals.
4 Increase in nutrient density.
5 Decrease in levels of antinutritional factors like phytate and tannins.

Not only is malt itself a nutrient-rich food, it can be added to unmalted grain foods to improve their nutrient density and palatability. In this application,

Table 13.3 Nutrient content of traditional African opaque sorghum beers brewed with sorghum malt and whole sorghum grain adjunct. Data from Van Heerden (1988)

Nutrient	Content	Contribution to recommended dietary allowance (RDA) (%/l)
Energy	1695 kJ/l	13
Ethanol	32.0 g/l	Not applicable
Starch	32.8 g/l	Not applicable
Other carbohydrates	7.1 g/l	Not applicable
Protein	5.2 g/l	9
Crude fibre	1.1 g/l	Not applicable
Thiamin	0.96 mg/l	69
Riboflavin	0.47 mg/l	29
Nicotinic acid	5.58 mg/l	31
Zn	1.94 mg/l	13
Cu	0.27 mg/l	9
Mn	1.83 mg/l	46
Fe	3.44 mg/l	34
Ca	52 mg/l	6
Mg	178 mg/l	51
K	407 mg/l	11
Na	18 mg/l	1
P	305 mg/l	38

Fig. 13.20 Commercial opaque sorghum beer, traditional serving vessels and grass beer strainer.

Table 13.4 The effects of malting on the nutritive value of various grains and their food products

Grain	Food product	Nutritional effects of malting	Reference
Sorghum	Malt flour	An average 68–87% decrease in phytic acid content in four cultivar	Mahgoub and Elhag (1998)
Red and white sorghum varieties	Malt flour	Increase in *in vitro* protein digestibility by 10.6–17.2%	Hugo (2002)
Red sorghum and pearl millet	Malt flour	Increase nutrient density Decrease phytate content	Traoré *et al.* (2004)
White sorghum, pearl millet and finger millet	Malt flour	Increase in dietary fibre Increase in lysine content except for sorghum Increase in riboflavin, niacin and ascorbic acid	Malleshi and Klopfenstein (1998)
Pearl millet	Malt flour	Decrease in phenolics by 50% and phytic acid by five-fold Increase in riboflavin, thiamin, ascorbic acid, vitamin A, carotene and tocopherol content	Opoku *et al.* (1981)
Pearl millet	Malt flour	Increase in *in vitro* protein digestibility from 68.8% to 95.5% and from 58.5% to over 90% Increase in lysine for a specific variety A 10-fold decrease in phytic acid	Pelembe (2001)
Pearl millet	Malt flour	A reduction of 38–48% in polyphenols and 46–50% decrease in phytic acid	Sehgal and Kawatra (1998)
Brown and white finger millet	Malt flour	*In vivo* studies with rats showed an increase in food intake, body weight gain and protein efficiency ratio for diets with malted flours of both brown and white finger millet compared with unmalted flour	Roa (1994)

Table 13.4 Continued

Grain	Food product	Nutritional effects of malting	Reference
Finger millet	Malt flour	Increase in antioxidant activity coefficient of malted millet as a result of an increase in free phenolic acids	Rao and Muralikrishna (2002)
Finger millet	Malt flour	A three-fold increase in water-extractable, non-starch polysaccharides	Rao and Muralikrishna (2004)
Finger millet	Malt flour	Increase in lysine	Mbithi-Mwikya *et al.* (2000)
Buckwheat	Sprouts	Increase in linoleic acid by 52% and an overall increase in unsaturated fatty acid by 83%	Lim *et al.* (2004)
		Increase in lysine content by almost eight-fold	
		Increase in rutin and quercitin by almost 35- and 65-fold, respectively	
		Increase in vitamin B (thiamin and pyridoxine) and ascorbic acid	
Quinoa	Sprouts	Reduction of phytate and its degradation products by 35–39%	Valencia *et al.* (1999)
		Increase in iron solubility by two to four-fold	

malt is referred to as power flour or amylase-rich flour. Power flour is particularly useful for porridge-based infant weaning foods and for the elderly and infirm. To make porridge with high nutrient density, it must have a high concentration of solids. Hence, such porridges are very highly viscous, stiff and unpalatable for infants. Adding water as a diluent facilitates ingestion by infants (Lorri and Svanberg, 1995), but the porridges then have very low nutrient density. This has been identified as a contributing factor to malnutrition among children in Africa (Mosha and Svanberg, 1990). Power flour malt addition to low nutrient density porridges increases the nutrient density, because it is highly nutritive and because it decreases the porridge viscosity, enabling porridge with higher solids content (i.e. higher nutrient density), but of palatable viscosity, to be produced (Wambugu, 1999; Thaoge *et al.*, 2003). This action of Power Flour malt is due to the alpha-amylase enzyme in the malt, which depolymerises starch molecules, thereby rapidly reducing the viscosity of the porridge.

One possible danger with malted sorghum is the formation of high amounts of dhurrin in roots and shoots. Dhurrin is a cyanogenic glucoside, and yields hydrogen cyanide upon hydrolysis. However, drying at 30°C and above (Aniche, 1990), removal of shoots and roots (Traoré *et al.*, 2004) and the traditional fermentation processes (Ahmed *et al.*, 1996) can reduce the cyanide content to a safe level. Another potential problem with malt is the presence of mycotoxins due to fungal contamination. Lefyedi *et al.* (2005) found the presence of four toxigenic species of filamentous fungi in sorghum malt produced by an outdoor malting process. Thus, hygienic malting is essential.

Fermentation

Fermentation is a microbial metabolic process, usually anaerobic. During fermentation, fermentable nutrients from food are used as substrate by microorganisms for growth and reproduction. Food is also chemically and physically modified by microbial enzymes, and metabolic processes change its nutritional and functional properties. Fermentation can be by yeast (*Saccharomyces cerevisiae*) to produce alcoholic beverages and by lactic acid bacteria to produce non-alcoholic foods and beverages.

Traditional African beer production from sorghum and millet involves the use of malting, followed by both lactic acid bacterial and yeast alcoholic fermentations. The beer can be very nutritious and is thus considered to be a food as well as a drink. One litre of traditional sorghum beer can contribute about 1695 kJ, 13% of an adult's recommended dietary allowance (RDA) (Van Heerden, 1989). It is a good source of starch and dietary fibre (Van Heerden, 1989). Another potential health benefit, shown in the traditional sorghum beer from Ghana, is the presence of a probiotic strain of *S. cerevisiae* (Kühle *et al.*, 2005). This strain can survive the pH 2.5 and bile environment, and expresses anti-inflammatory cytokines that are inhibitory to *Escherichia coli*.

Traditional lactic acid fermentation of cereal foods is not a well-controlled process (Taylor and Belton, 2002) and the fermented food may contain some alcohol depending on the growth conditions of the yeast. The end-product will vary depending on many intrinsic factors and extrinsic factors. Intrinsic factors include pH, moisture, redox potential and chemical composition of the raw material. The major extrinsic factor is temperature. The potential nutritional benefits of lactic acid fermentation of food products are presented in Table 13.5. In summary, the nutritional benefits of lactic acid fermentation are:

1 Improvement in protein and starch digestibility. Although the mechanism is not fully understood, it appears that this improvement is related to grain structural changes, making these components accessible to enzymes (Taylor and Taylor, 2002).
2 Reduction in porridge viscosity, depending on the type of microorganism involved. If the microorganisms produce amylases, there is the possibility of viscosity reduction because of starch hydrolysis (Westby and Gallat, 1991).
3 Increase in the amount of B vitamins, especially thiamin.
4 Increase in the bioavailability of minerals.

The most important beneficial effect of lactic acid fermentation in relation to health is the potential production of a microbiologically safe product. During lactic acid fermentation, the production of lactic acid substantially decreases the pH which appears to inhibit growth of Gram-negative intestinal pathogenic bacteria (Svanberg *et al.*, 1992; Thaoge *et al.*, 2003). This is of particular significance in developing countries where many households do not have access to potable/clean water and refrigeration. In addition, some traditional fermented grain foods such as koko, a fermented millet porridge from Ghana, have been shown to have potential as probiotics (Lei and Jakobsen, 2004).

13.5 Modern foods and processing technologies

13.5.1 Ready-to-eat snacks and breakfast cereals

Ready-to-eat (RTE) foods are invariably dry products. To produce them, grains or grits are subjected to a cooking process that gelatinises the starch and then expands the product, normally through the vapourisation of steam. The expansion process gives RTE foods their characteristic crispy/crunchy texture. The technologies for producing RTE snacks and breakfast cereals are often the same. After processing, savoury flavourings are added to snacks and sweet flavourings may be added to breakfast cereals.

Probably the most popular technology for producing RTE foods is extrusion cooking. Extrusion cooking involves a screw turning in a barrel that has small hole known as a die at the end. The screw transports meal through the barrel and subjects it to pressure, shear and heat. The meal melts and turns

Table 13.5 The effect of lactic acid fermentation on the nutritive value of various grains and their food products

Grain	Food product	Nutritional effects of fermentation	Reference
Sudanese sorghum	Flour from traditional fermentation	A 57–60% decrease in phytic acid	Mahgoub and Elhag (1998)
Sorghum (white tannin-free)	Lactic acid fermented and dried flour	Increase in protein content Increase in *in vitro* protein digestibility from 35.3% to 52.7%	Hugo (2002)
Sorghum	Extruded fermented/plain flour composites	Significant decrease in viscosity Increase in lysine content Increase in *in vitro* protein digestibility	Wambugu (1999)
Sorghum	Lactic acid fermented sorghum	Increase in thiamin content	Kazanas and Fields (1981)
Finger millet	Germinated grain followed by lactic acid fermentation	A 10-fold increase in free amino acids A 60% decrease in phytate content Potential increase in zinc bioavailability	Sripriya *et al.* and Chandra (1997)
Buckwheat	Fermented flour for soba noodle in Japan	Decrease of about 70–90% in phytic acid levels Over 75-fold increase in lysine Removal of allergenic protein by enzymatic hydrolysis	Handoyo *et al.* (2006)
Quinoa	Fermented flours from low and saponin varieties	Decrease of phytate and its degradation products by 82–98% Increase in iron solubility by three to five-fold	Valencia *et al.* (1999)

into a plastic material. As this material emerges through the die, the superheated water within it vapourises expanding the product, producing, for example, the characteristic worm-like corn curl snack. Extrusion-cooked RTE starchy foods are invariably made from decorticated flour or meal. The type of flour is to a large extent immaterial as it is the starch that is the most important constituent. However, this means that the nutrient composition of the extrusion-cooked RTE foods is often restricted to essentially starch and relatively low quality protein. Pelembe *et al.* (2002) improved the protein quality of an extrusion-cooked instant sorghum porridge by co-extruding the sorghum flour with cowpea flour. Combining cowpea with sorghum in a 50:50 ratio increased the protein content from 8.5% to 16% so that 100 g of the flour would meet 28% of an adult's protein RDA. Figure 13.21 shows the extrusion-cooking process used. As an alternative way of improving the nutritional value of extruded RTE foods, researchers at Texas A&M University successfully produced expanded whole sorghum grain snacks using extrusion cooking (Professor L W Rooney, Texas A&M University, personal communication, 2005) by using a short-barrel single-screw extruder. The corn curl-like snacks, which have name Bongos, have a firm crunchy texture and nutty flavour.

A technology often used to produce expanded whole grain RTE foods is gun puffing, as commonly used in the processing of puffed wheat. In gun puffing, whole cereal grain and a little water is put into a steep pressure cooker (the gun) and the quick-release lid, operated by a system of levers and cams, is closed (Fast, 2000). The vessel is heated by direct flame and a pressure in excess of 14 bar may be reached. The lid is then automatically opened and the pressure released instantaneously. This causes the superheated water to turn to steam which puffs the grain. At the same time, the grain is blown out

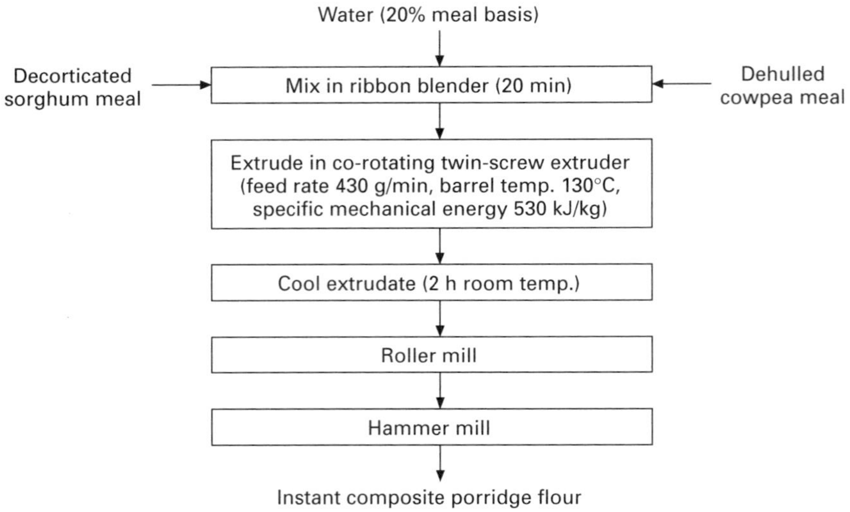

Fig. 13.21 Process for making a composite sorghum–cowpea instant porridge by extrusion cooking. Compiled from Pelembe *et al.* (2002).

of the gun. Invariably the puffing process results in fracturing of the grain layers so that some, but not all of the pericarp and germ may be lost. In South Africa, instant sorghum porridge flour is produced using gun puffing. Figure 13.22 shows a gun-puffed millet breakfast cereal from the USA.

Quinoa is also used as instant breakfast cereal and instant food in the form of popped quinoa (Fig. 13.23). Instant breakfast cereals, popped grains, instant infant food and other snacks are modern food products made from amaranth (Fig. 13.24). Composites of quinoa and amaranth are also common modern products found in Latin America (Fig. 13.25).

13.5.2 Baked products (breads, pasta, cakes and biscuits)

Wheat is unique in that when its flour is mixed with a limited amount of water it forms a dough with both viscous flow and elastic (viscoelastic) properties. These viscoelastic properties enable the dough to hold the gas produced by yeast or baking powder and expand to give a leavened product with a light airy texture. These properties of wheat dough are attributed to its gluten proteins. No other grain has such proteins. Hence, the production of baked goods from other grains that have similar texture to those from wheat is a challenge. Invariably some form of hydrocolloid (water-holding biopolymer) is added. The hydrocolloid increases the viscosity of the aqueous phase of the dough, thereby improving its gas-holding properties. It also acts

Fig. 13.22 Gun-puffed millet breakfast cereal from the USA.

Fig. 13.23 Quinoa products form Peru (from left to right: quinoa flour, instant breakfast cereal and popped quinoa) (courtesy of Mrs N. N. Dersley, University of Pretoria).

Fig. 13.24 Amaranth products form Peru (from left to right: an extruded product, biscuits with amaranth and wheat flour, and popped amaranth) (courtesy of Mrs N. N. Dersley, University of Pretoria).

as a binder, helping to prevent the product from crumbling. Industrially produced hydrocolloids, such as modified cellulose (Hart *et al.*, 1970) and the bacterial polysaccharide xanthan gum (Schober *et al.*, 2005), have been investigated for this purpose and are used in commercial practice. For home production and for developing countries where such products are prohibitively

Fig. 13.25 Modern cereal bars (left) and snacks made with quinoa and amaranth from Peru (courtesy of Mrs N. N. Dersley, University of Pretoria).

expensive, starches are a good alternative. In a seminal piece of research, Mr Morton Satin of the FAO used gelatinised cassava (tapioca or manioc) starch to improve the quality of breads made from sorghum and non-wheat flours (Satin, 1988). Cassava starch, probably because it has a high proportion of the branched-chain amylopectin starch polymer, seems uniquely suitable as an additive in non-wheat breadmaking as it produces a gel with far greater cohesion than starches from other sources (De Ruiter, 1978). Figure 13.26 shows a flow chart for producing sorghum and cassava bread from Nigeria using precooked cassava starch as the hydrocolloid additive. A characteristic of these breads is, that to optimise the quality, they are made with a high proportion of water. Hence, a batter rather than a dough is produced and the process is known as the batter or custard process. The breads have a rigid crumb structure and more closely resemble cake than bread. Because of this, the process can also be used to produce non-wheat flour cakes. In the case of biscuits (cookies), since they are not leavened, in theory they should be more easily made with non-wheat flours. In practice, however, a hydrocolloid binder is required, otherwise they crumble easily. A further problem with cookies made from grains that have both a vitreous and floury endosperm, like sorghum (Fig. 13.2) and pearl millet, is that the vitreous endosperm flour tends to give the biscuits a gritty texture. This problem can be alleviated by precooking the flour by extrusion cooking and sifting out the coarse vitreous endosperm particles (Munck, 1995).

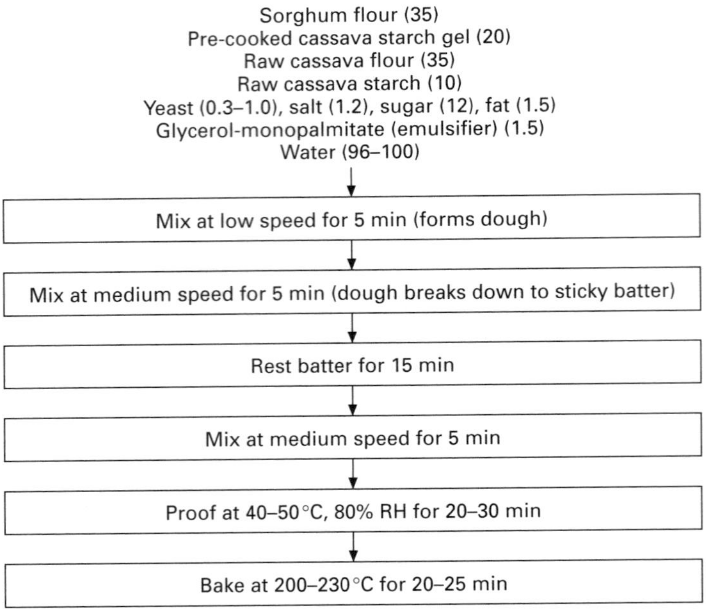

Fig. 13.26 Process for making sorghum–cassava bread using precooked cassava starch as a hydrocolloid. R H, relative humidity. Ingredients other than flours and starches are expressed as a percentage of total flour plus starch. Compiled from Olatunji *et al.* (1992).

Milled quinoa flour (Fig. 13.23) has also been substituted for wheat flour in modern food products; for example, bread, noodles, pasta and sweet biscuits. Modern buckwheat food products include dried noodles, instant noodles, spaghetti, macaroni and pasta (Fig. 13.27).

13.5.3 Lager and stout beers and non-alcoholic malt beverages

Because sorghum can be cultivated economically in the semi-arid tropics of Africa and barley cannot, commercial brewing of sorghum-based lager and stout beers and non-alcoholic beverages has grown rapidly. In the USA, commercial sorghum lager beer brewing is also taking place to cater for the coeliac market. Figure 13.28 shows a selection of sorghum-based beers and non-alcoholic malt beverages from Nigeria.

Due to the fact that sorghum differs in several important respects from barley, substantial changes in brewing equipment and process have been required. In barley brewing, the barley hulls are used as a filter bed to separate the wort (unfermented beer) from the spent grain in a vessel called a lauter tun. Sorghum does not have a hull and this is has necessitated the use of a piece of equipment called a mash filter. This is basically a series of fine plastic screens through which the mash is pumped through. The wort passes through the filter screens and the spent grain accumulates in the form of a

Fig. 13.27 Buckwheat pasta from Australia.

Fig. 13.28 Sorghum-based lagers, stouts and non-alcoholic malt beverages from Nigeria.

'cake' between them. Mash filters have proven to be very popular and are now replacing lauter tuns in some barley breweries. Because the grain or malt can be milled to a fine flour, it yields more extract (greater solubilisation of the cereal).

Sorghum also differs biochemically from barley in two important respects. Its starch gelatinises at a rather higher temperature range (approximately 68–78 °C) and sorghum malt has much lower levels of beta-amylase (the maltose-producing enzyme) than barley malt (Taylor, 1992). These two factors make it problematic to achieve the simultaneous gelatinisation and hydrolysis of sorghum starch during the mashing process, as takes place with barley malt. Hence, when brewing with sorghum, sorghum malt or unmalted sorghum grain, it is first cooked to gelatinise the starch, then hydrolysed with amylase enzymes (Fig. 13.29). Thus the sorghum functions primarily as an adjunct, a source of starch in the brewing process and not as a source of enzymes. In Africa, two different sorghum brewing processes are used. In Nigeria, sorghum brewing is done using sorghum grain and malt which is cooked then mashed with barley malt and/or commercial enzymes, whereas in East and Central Africa sorghum brewing simply involves mashing cooked whole sorghum grain with commercial enzymes. Surprisingly, the latter product does not differ substantially in taste from beer made with malt. It may be expected that brewing beer using whole grain, rather than refined grits (endosperm) has a positive effect on its nutritional value. This, however, does not yet appear to have been studied.

In Nigeria, several non-alcoholic beverages based on sorghum are also produced. The process of their manufacture is similar to brewing, except that the wort is not hopped or fermented. The products come into two types, a carbonated beverage called Malta and Maltina (Fig. 13.28), and a dried malt powder that can be made into a cold or hot beverage, of which Milo is an example (Fig. 13.28). These beverages are highly nutritious as not only are they rich in simple sugars, they also contain B vitamins and minerals.

Other grains have been used experimentally to brew beers, including pearl millet (Eneje *et al.*, 2001), amaranth (Berghofer and Schoenlechner, 2002) and buckwheat (Biacs *et al.*, 2002). In all cases, the results were promising.

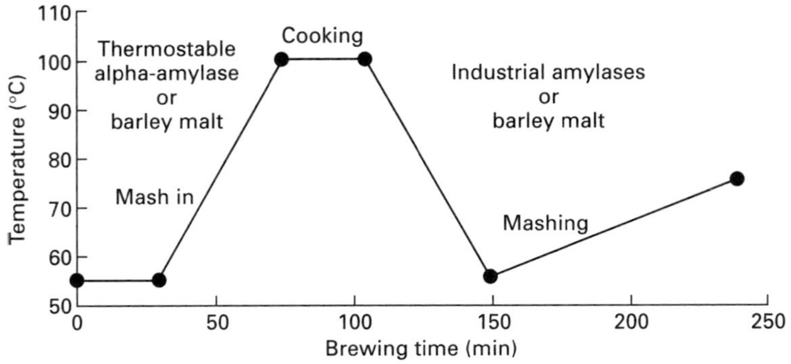

Fig. 13.29 Mashing process for brewing lager and stout beers with sorghum malt or unmalted sorghum grain.

13.6 Conclusions

As described, consumer attitudes to these grains are very positive in the countries where they are traditional foodstuffs and there is considerable new product development activity. Obviously, a major issue is whether they will become popular foodstuffs worldwide. Important negative aspects of the grains are that, with the exception of sorghum, firstly their grain is very small and secondly modern high-yielding varieties are only now being developed. Thus, they are not ideal for modern mechanised agriculture. Probably the biggest opportunity is the rapidly growing market for organic foods as the profit margins are potentially much higher, and thus efficiency of production is less important.

13.7 Sources of further information and advice

Abdel-Aal E and Wood P (2005), *Speciality Grains for Food and Feed*, St Paul, MN, American Association of Cereal Chemists.

Bavec F and Bavec M (2007), *Organic Production and Uses of Alternative Crops*, Boca Raton, FL, CRC Taylor and Francis.

Belton P S and Taylor J R N (2002), *Pseudocereals and Less Common Cereals: Grain Properties and Utilization Potential*, Berlin, Springer.

Center for New Crops & Plants Products, Purdue University, www.hort.purdue.edu/newcrop.

Dendy D A V (1995), *Sorghum and Millets: Chemistry and Technology*, St Paul, MN, American Association of Cereal Chemists.

Food and Agriculture Organization (1995), *Sorghum and Millets in Human Nutrition*, Rome, FAO (available at www.fao.org).

National Research Council (1989), *Lost Crops of the Incas*, Washington, DC, National Academies Press (available at http://darwin.nap.edu/openbook).

National Research Council (1996), *Lost Crops of Africa: Volume I: Grains*, Washington, DC, National Academies Press (available at www.nap.edu/books). As noted in the text, some of the nutritional information should be treated with caution.

Wrigley C, Corke H and Walker C E (2004), *Encyclopedia of Grain Science*, Amsterdam, Elsevier.

13.8 References

Ahmed S B, Mahgoub S A and Badikar B E (2006), 'Changes in tannin and cyanide contents and diastatic activity during germination and the effect of traditional processing on cyanide content of sorghum cultivars', *Food Chem*, **25**, 159–162.

Aniche G N (1990), 'Studies on the effect of germination and drying conditions on the cyanide content of sorghum sprouts', *J Food Sci Technol*, **27**, 202–204.

Atwell W A, Patrick B M, Johnson L A and Glass R W (1983), 'Characterization of quinoa starch', *Cereal Chem*, **60**, 9–11.

Awika J M and Rooney L W (2004), 'Sorghum phytochemicals and their potential impact on human health', *Phytochemistry*, **65**, 1199–1221.

Badi S M, Bureng P L and Monowar L Y (1988), 'Commercial production: A breakthrough in kisra technology', in *ICC 4th Quadrennial Symposium on Sorghum and Millets*, Lausanne, Switzerland, ICC, pp. 31–45.

Baxter N J, Lilley T H, Haslam E and Williamson M P (1997), 'Multiple interactions between polyphenols and a salivary proline-rich protein repeat result in complexation and precipitation', *Biochemistry*, **36**, 5566–5577.

Bekele E (1995), 'Essential and non-essential amino acid in a free state and in the major protein fractions of teff seeds', *SINET: Ethiop J Sci*, **18**, 79–99.

Berganza B E, Moran A W Rodriguez G M, Coto, N M, Santamaria M and Bressani R (2003), 'Effect of variety and location on the total fat, fatty acids and squalene content of Amaranth', *Plant Foods Hum Nutr*, **58**, 1–6.

Berger A, Monnard I, Dionisi F, Gumy D, Hayes K C and Lambelet P (2003), 'Cholesterol-lowering properties of amaranth flakes, crude and refined oils in hamsters', *Food Chem*, **81**, 119–124.

Berghofer E and Schoenlechner R (2002), 'Grain amaranth', in Belton P S and Taylor J R N (eds), *Pseudocereals and Less Common Cereals: Grain Properties and Utilization Potential*, Berlin, Springer, pp. 219–260.

Berti C, Riso P, Brusamolino A and Porrini M (2005), 'Effect on appetite control of minor and pseudocereal products', *Brit J Nutr*, **94**, 850–858.

Betschart A A, Wood I D, Shepherd A D and Saunders R M (1981), '*Amaranthus cruentus*: milling characteristics, distribution of nutrients within seed components, and the effects of temperature on nutritional quality', *J Food Sci*, **46**, 1181–1187.

Biacs P, Aubrecht E, Léder I and Lojos J (2002), 'Buckwheat', in Belton P S and Taylor J R N (eds), *Pseudocereals and Less Common Cereals: Grain Properties and Utilization Potential*, Berlin, Springer, pp. 123–151.

Birzer D M and Klopfenstein C F (1988), 'The pearl millet goitrogens', *Cereal Foods World*, **33**, 229–231.

Brinegar C and Goundan S (1993), 'Isolation and characterisation of chenopodin, the 11S seed storage protein of quinoa (*Chenopodium quinoa*)', *J Agric Food Chem*, **41**, 182–185.

Brinegar C, Sine B and Nwokocha L (1996), 'High cysteine 2S seed storage proteins from quinoa (*Chenopodium quinoa*)', *J Agric Food Chem*, **44**, 1621–1623.

Bultosa G and Taylor J R N (2004), 'Teff', in Wrigley C, Corke H and Walker C E (eds), *Encyclopedia of Grain Science*, Amsterdam, Elsevier, pp. 281–290.

Bvochora J M, Reed J D, Read, J S and Zvauya R (1999), 'Effect of fermentation processes on proanthocyanidins in sorghum during preparation of *Mahewu*, a non-alcoholic beverage', *Process Biochem*, **15**, 21–25.

Cai Y Z, Corke H and Wu H X (2004a), 'Amaranth', in Wrigley C, Corke H and Walker C (eds), *Encyclopedia of Grain Science*, Vol. 1, Oxford, Elsevier, pp. 1–10

Cai Y Z, Corke H and Li W D (2004b), 'Buckwheat', in Wrigley C, Corke H and Walker C (eds), *Encyclopedia of Grain Science*, Vol. 1 Oxford, Elsevier, pp. 120–128.

Chauhan G S, Eskin N A M and Tkachuk R (1992), 'Nutrients and antinutrients in quinoa seed', *Cereal Chem*, **69**, 85–88.

Czerwinski J, Bartnikowska E, Leontowicz H, Lange E, Leontowicz M, Katrich E, Trakhtenberg S and Gorinstein S (2004), 'Oat (*Avena sativa* L.) and amaranth (*Amaranthus hypochondriacus*) meals positively affect the plasma lipid profile in rats fed cholesterol containing diets', *J Nutr Biochem*, **15**, 622–629.

Damodaran S (1996), 'Amino acids, peptides, and proteins', in, Fennema O R (ed.), *Food Chemistry*, 3rd Edn, New York, Marcel Dekker, pp. 321–429.

De Ruiter D (1978), 'Composite flours', in Pomeranz Y (ed.), *Advances in Cereal Science and Technology*, Vol. II, St Paul, MN, American Association of Cereal Chemists, pp. 349–385.

De Wet J M J and Harlan J (1971), 'The origin and domestication of *Sorghum bicolor*', *Econ Bot*, **25**, 128–135.

Dini I, Tenore G C, Schettino O and Dini A (2001), 'New oleanane saponins in *Chenopodium quinoa*', *J Agric Food Chem*, **49**, 3976–3981.

Duodu K G, Taylor J RN, Belton P S and Hamaker B R (2003), 'Mini review: Factors affecting sorghum protein digestibility', *J Cereal Sci*, **38**, 117–131.

Eneje LO, Obiekezie SO, Aloh CU and Agu R C (2001), 'Effect of mashing procedures on millet (*Pennisetum maiwa*) malt wort properties', *Process Biochem*, **36**, 723–727.

Escudero N L, De Arellano M L, Luco J M, Gimenez M S and Mucciarelli S I (2004), 'Comparison of the chemical composition and nutritional value of *Amaranthus cruentus* flours and its protein concentrate', *Plant Foods Hum Nutr*, **59**, 12–21.

Ezeogu L I, Duodu K G and Taylor J R N (2005), 'Effects of endosperm texture and cooking conditions on the *in vitro* starch digestibility of sorghum and maize flours', *J Cereal Sci*, **42**, 33–44.

FAO (Food and Agriculture Organization) (2006), http://faostat.fao.org/, accessed July 2006.

Fast R B (2000), 'Manufacturing technology of ready-to-eat cereals', in Fast R B and Caldwell E F (eds), *Breakfast Cereals and How They are Made*, St Paul, MN, American Association of Cereal Chemists, pp. 17–55.

Fidantsi A and Doxastakis G (2001), 'Emulsifying and foaming properties of amaranth seed protein isolates', *Colloids Surfaces*, **21**, 119–124.

Fletcher R J (2004), 'Pseudocereals, overview', in Wrigley G, Corke H and Walker C E (eds), *Encyclopedia of Grain Science*, Oxford, Elsevier, pp. 488–493.

Galiba M, Rooney L W, Waniska R D and Miller F R (1987), 'The preparation of sorghum and millet couscous in West Africa', *Cereal Foods World*, **32**, 878–884.

Gamel T H, Linssen J P, Alink G M, Mosallem A S and Shekib L A (2004), 'Nutritional study of raw and popped seed proteins of *Amaranthus cruentus* L.', *J Sci Food Agric*, **84**, 1153–1158.

Gao X Q, Wang X L and Xi X Y (2002), 'The ultrastructures of storage protein accumulation in aleurone layer and cotyledons of *Fagopyrum esculentum*', *Shi Yan Sheng Wu Xue Bao*, **35**, 283–288.

Gashe B A, Girma M and Bisrat A (1982). Tef fermentation. 1. The role of microorganisms in fermentation and their effect on the nitrogen content of tef. *SINET: Ethiopian J Sci*, **5**, 69–76.

Gee J M, Price K R, Ridout C L, Wortley G M, Hurrell R F and Johnson I T (1993), 'Saponins of quinoa (*Chenopodium quinoa*): Effects of processing on their abundance in quinoa products and their biological effects on intestinal mucosal tissue', *J Sci Food Agric*, **63**, 201–209.

Gross R, Koch F, Malaga I, De Miranda A F, Schoeneberger H and Trugo L C (1989), 'Chemical composition and protein quality of some Andean food sources', *Food Chem*, **34**, 25–34.

Guerra-Matias A C and Arêas J A G (2005), 'Glycemic and insulinemic responses in women consuming extruded amaranth (*Amaranthus cruentus* L)', *Nutr Res*, **25**, 815–822.

Guo X and Yao H (2006), 'Fractionation and characterization of tartary buckwheat flour proteins', *Food Chem*, **98**, 90–94.

Hadimani N A, Muralikrisna G, Tharanathan R N and Malleshi N G (2001), 'Nature of carbohydrates and proteins in three pearl millet varieties varying in processing characteristics and kernel texture', *J Cereal Sci*, **33**, 17–25.

Hancock J D (2001), 'Value of sorghum and sorghum coproducts in diets for livestock', in Wayne Smith C and Frederiksen R A (eds), *Sorghum – Origin, History, Technology, and Production*, New York, John Wiley & Sons, pp. 731–748.

Handoyo T, Maeda T, Urisu A, Adachi T and Morita N (2006), 'Hypoallergenic buckwheat flour preparation by *Rhizopus oligosporus* and its application to soya noodle', *Food Res Int*, **39**, 598–605.

Hargrove J L, Greenspan P and Hartle D K (2004), 'Nutritional significance and metabolism of very long chain fatty alcohols and acids from dietary waxes', *Exp Biol Med*, **229**, 215–226.

Hart M R, Graham R P, Gee M and Morgan I A Jr (1970), 'Bread from sorghum and barley flours', *J Food Sci*, **35**, 661–665.

Hilu K W, De Wet J M J and Seigler D (1978), 'Flavonoid patterns and systematics in *Eleucine*', *Biochem Systematics Ecol*, **6**, 247–249.

Holasova M, Fiedlerova V, Smrcinova H, Orsak M, Lachman J and Vavreinova S (2002), 'Buckwheat – the source of antioxidant activity in functional food', *Food Res Int*, **35**, 207–211.

House L R (1995), 'Sorghum and millets: History, taxonomy, and distribution' in Dendy D A V (ed.), *Sorghum and Millets: Chemistry and Technology*, St Paul, MN, American Association of Cereal Chemists, pp. 1–10.

Hugo L F (2002), *Malted and fermented sorghum as ingredients in composite breads*, PhD Thesis, Pretoria, University of Pretoria.

Hundera F (1998), 'Variations of morpho-agronomic characteristics and grain chemical composition of released varieties of tef [*Eragrostis tef* (Zucc.) Trotter]', *J Genet Breed*, **52**, 307–311.

Hwang, K T, Weller C L, Cuppett S L and Hanna M A (2004), 'Policosanol contents and composition of grain sorghum kernels and dried distillers grains', *Cereal Chem*, **81**, 345–349.

ICRISAT and FAO (1996), *The World Sorghum and Millet Economies: Facts, Trends and Outlook*, Patancheru, India, ICRISAT and Rome, FAO.

Ikeda K, Sakaguchi T, Kusano T and Yasumoto K (1991), 'Endogenous factors affecting protein digestibility in buckwheat', *Cereal Chem*, **68**, 424–427.

Kazanas N and Fields M L (1981), 'Nutritional improvement of sorghum by fermentation', *J Food Sci*, **46**, 819–821.

Kent N L (1983), *Technology of Cereals*, 3rd Edn, Oxford, Pergamon.

Klopfenstein C F and Hoseney R C (1995), 'Nutritional properties of sorghum and the millets' in Dendy D A V (ed.), *Sorghum and Millets: Chemistry and Technology*, St Paul, MN, American Association of Cereal Chemists, pp. 125–168.

Kritchevsky D and Chen S C (2005), 'Phytosterols – health benefits and potential concerns: a review', *Nutr Res*, **25**, 415–428.

Krkošková B and Mrázová Z (2005), 'Prophylactic components of buckwheat', *Food Res Int*, **38**, 561–568.

Kühle A V der A, Skovgaard K and Jepersen L (2005), '*In vitro* screening of probiotic properties of *Sacharomyces cerevisciae* var boulardii and food-borne *Saccharomyces cerevisciae* strains', *Int J Food Microbiol*, **101**, 29–39.

Lei V and Jakobsen M (2004), Microbiological characterization and probiotic potential of Koko and Koko sour water, African spontaneously fermented millet porridge and drink. *J Appl Microbiol*, **96**, 384–397.

Lefyedi M L, Marais G J, Dutton M F and Taylor J R N (2005), 'The microbial contamination, toxicity and quality of turned and unturned outdoor floor malted sorghum', *J Inst Brew*, **111**, 190–196.

Ličen M and Kreft I (2005) 'Buckwheat (*Fagopyrum esculentum* Moench) low molecular weight seed proteins restricted to the embryo and are not detectable in the endosperm. *Plant Physiol Biochem*, **43**, 862–865.

Lim S L, Kim S K and Park C H (2004), 'Introduction and nutritional evaluation of buckwheat sprouts as a new vegetable', *Food Res Int*, **37**, 319–327.

Lorri W and Svanberg U (1995), 'An overview of the use of fermented foods for child feeding in Tanzania', *Ecol Food Nutr*, **34**, 65–81.

Magboul B I and Bender D A (1982), 'The nature of niacin in sorghum', *Proc Nutr Soc*, **41**: 50A.

Mahgoub S E O and Elhag S A (1998), 'Effect of milling, soaking, malting, heat treatment and fermentation on phytate level of four Sudanese sorghum cultivars', *Food Chem*, **61**, 77–80.

Malleshi N G and Klopfenstein C F (1998), 'Nutrient composition, amino acid and vitamin contents of malted sorghum, pearl millet, finger millet and their rootlets', *Int J Food Sci Nutr*, **49**, 412–422.

Marcone M F (1999) 'Evidence confirming the existence of a 7S globulin like storage protein in *A. hypochondriacus* seed', *Food Chem*, **65**, 533–542.

Marcone M F, Kakuda Y and Yada R Y (2004), 'Amaranth as a rich dietary source of β-sitosterols and other phytosterols', *Plant Foods Human Nutr*, **58**, 207–211.

Mariotti M, Alamprese C, Pagani M A and Lucisano M (2006), 'Effect of puffing on utrastructure and physical characteristics of cereal grains and flours', *J Cereal Sci*, **43**, 47–56.

Mastebroek H D, Limburg H, Gilles T and Marvin H J P (2000), 'Occurrence of sapogenins in leaves and seeds of quinoa (*Chenopodium quinoa* Willd)', *J Sci Food Agric*, **80**, 152–156.

Mazza G and Oomah B D (2005), 'Buckwheat as a food and feed', in Abdel-Aaal E and Wood P (eds), *Speciality Grains for Food and Feed*, St Paul, MN, American Association of Cereal Chemists, pp. 375–393.

Mbithi-Mwikya S, Ooghe W, Van Camp J, Ngudi D and Huyghebaert A (2000), 'Amino acid profiles after sprouting, autoclaving, and lactic acid fermentation of finger millet (*Eleusine coracana*) and kidney beans (*Phaseaolus vulgaris* L.)', *J Agric Food Chem*, **48**, 3081–3085.

Morita N, Maeda T, Sai R, Miyake K, Yoshioka H, Urisu A and Adachi T (2006), 'Studies on distribution of protein and allergen in graded flours prepared from whole buckwheat grains', *Food Res Int*, **39**, 782–790.

Morton J F (1970), 'Tentative correlations of plant usage and esophageal cancer zone', *Econ Bot*, **24**, 217–226.

Mosha A C and Svanberg U (1990), 'The acceptance and intake of bulk reduced weaning foods: the Luganga village study', *Food Nutr Bull*, **12**, 69–74.

Mtebe K, Ndabikunze B K, Bangu N T A and Mwemezi E (1993), 'Effect of germination on the energy density of togwa', *Int J Food Sci Nutr*, **44**, 175–180.

Munck L (1995), 'New milling technologies and products: Whole plant utilization by milling and separation of the botanical and chemical components', in Dendy D A V (ed.), *Sorghum and Millets: Chemistry and Technology*, St Paul, MN, American Association of Cereal Chemists, pp. 223–281.

Murty D S and Kumar K A (1995), 'Traditional uses of sorghum and millets', in Dendy D A V (ed.), *Sorghum and Millets: Chemistry and Technology*, St Paul, MN, American Association of Cereal Chemists, pp. 185–222.

Nakamura R, Konishi Y, Kojima A and Nakatani N (1998), 'Globulin and albumin-2 associated with protein bodies in *Amaranthus cruentus* seeds', *Biosci Biotech Biochem*, **62**, 1231–1233.

National Research Council (1996), *Lost Crops of Africa: Volume I: Grains*, Washington, DC, National Academies Press.

Oakenfull D and Sidhu G (1990), 'Could saponins be a useful treatment for hypercholesterolaemia?', *Eur J Clin Nutr*, **44**, 79–88.

Obilana A B and Manyasa E (2002), 'Millets', in Belton P S and Taylor J R N (eds), *Pseudocereals and Less Common Cereals*, Berlin, Springer, pp. 177–217.

Olatunji O, Koleoso O A and Oniwinde A B (1992), 'Production of bread from composite flours', in *5th Quadrennial Symposium on Sorghum and Millets*, Schwachet, Austria, ICC, pp. 45–54.

Onyekwere O O, Akinrele I A and Koleoso O A (1989) 'Industrialization of ogi fermentation', in Steinkraus K H (ed.), *Industrialization of Indigenous Fermented Foods*, New York, Marcel Dekker, pp. 329–362.

Oplinger E S, Oelke E A, Brinkmen M A and Kelling K A (1989), 'Buckwheat', in *Alternative Field Crops Manual*, http:/www.hort.purdue.edu/newcrop/afcm/buckwheat.html, accessed June 2006.

Opoku A R, Ohenhen S O and Ejiofor N (1981), 'Nutrient composition of millet (*Pennisetum typhoides*) grains and malt', *J Agric Food Chem*, **29**, 1247–1248.

Pelembe L A M (2001), *Pearl millet malting: factors affecting product quality*, PhD Thesis, Pretoria, University of Pretoria.

Pelembe L A M, Erasmus C and Taylor J R N (2002), 'Development of a protein-rich sorghum-cowpea instant porridge by extrusion cooking process', *Lebensm Wiss Technol*, **35**, 120–127.

Plate A Y A and Arêas J A G (2002), 'Cholesterol-lowering effect of extruded amaranth (*Amaranthus caudatus* L.) in hypercholesterolemic rabbits', *Food Chem*, **76**, 1–6.

Prego I, Maldonado S and Otegui M (1998), 'Seed structure and localization of reserves in *Chenopodium quinoa*', *Ann Bot*, **82**, 481–488.

Prěstamo G, Pedrazuela A, Peňas E, Lasunción M A and Arroyo G (2003), 'Role of buckwheat diet on rats as prebiotic and healthy food', *Nutr Res*, **23**, 803–814.

Przybylski. R, Chauhan, G S and Eskin N A M (1994), 'Characterisation of quinoa (*Chenopodium quinoa*) lipids', *Food Chem*, **51**, 187–192.

Quin P J (1959), *Foods and Feeding Habits of the Pedi*, Johannesburg, Witwatersrand University Press.

Radović R S, Maksimović R V, Brkljacic M J, Gasic I E V and Savic P A (1999), '2S albumin from buckwheat (*Fagopyrum esculentum* Moench) seeds', *J Agric Food Chem*, **47**, 1497–1470.

Radović S R, Maksimović V R and Varkonji-Gašić E I (1996), 'Characterization of buckwheat seed storage proteins', *J Agric Food Chem*, **44**, 972–974.

Ramachandra G, Virupaksha T K and Shadaksharaswamy M (1997), 'Relationship between tannin levels and *in vitro* digestibility in finger millet (*Eleucine coracana* Gaertn.)', *J Agric Food Chem*, **25**, 1101–1104.

Ranhotra G S, Gelroth J A, Glaser B K, Lorenz K J and Johnson D L (1993), 'Composition and protein nutritional quality of quinoa', *Cereal Chem*, **70**, 303–305.

Rao M V S S T S and Muralikrishna G (2002), 'Evaluation of antioxidant properties of free and bound phenolic acids from native and malted finger millet (Ragi, *Eleusine coracana* Indaf-15)', *J Agric Food Chem*, **50**, 889–892.

Rao P U (1994), 'Evaluation of protein quality of brown and white ragi (*Eleusine coracana*) before and after malting', *Food Chem*, **51**, 433–436.

Rao R S P and Muralikrishna G (2004), 'Non starch polysaccharide–phenolic acid complexes from native and germinated cereals and millet', *Food Chem*, **84**, 527–531.

Rayas-Duarte P, Majewska K and Doetkott C (1998), 'Effect of extrusion process parameters on the quality of buckwheat flour mixes', *Cereal Chem*, **75**, 338–345.

Reichert R D (1979), 'The pH-sensitive pigments in pearl millet', *Cereal Chem*, **56**, 291–294.

Reichert R D, Tatarynovich J T and Tyler R T (1986), 'Abrasive dehulling of Quinoa (*Chenopodium quinoa*); Effect on saponin content as determined by an adapted hemolytic assay', *Cereal Chem*, **63**, 471–475.

Rooney L W (1973), 'A review of the physical properties, composition and structure of sorghum grain in relation to utilization', in Pomeranz Y (ed.), *Industrial Uses of Cereals*, St Paul, MN, American Association of Cereal Chemists, pp. 316–342.

Rooney L W (1993), 'Chemistry, technology and quality of products from maize nixtamalization', in Taylor J R N, Randall P G and Viljoen J H (eds), *Cereal Science and Technology: Impact on a Changing Africa,* Pretoria, South Africa, CSIR, pp. 541–555.

Rooney L W and Waniska R D (2001), 'Sorghum food and industrial utilization', in Wayne Smith C and Fredericksen R A (eds), *Sorghum: Origin, History, Technology, and Production,* New York, John Wiley & Sons, pp. 689–729.
Ruales J and Nair B M (1994), 'Effect of processing on *in vitro* digestibility of protein and starch in quinoa seeds', *Int J Food Sci Tech*, **29**, 449–456.
Ruales J, Grijalva Y D, Lopez-Jaramillo P and Nair B M (2002), 'The nutritional quality of an infant food from quinoa and its effect on the plasma level of insulin–like growth factor-1 (IGF-1) in undernourished children', *Int J Food Sci. Nutr*, **53**, 143–154.
Satin M (1988), 'Bread without wheat', *New Scientist*, 28 April, 56–59.
Saunders R M and Becker R (1984), '*Amaranthus*: A potential food and feed resource', in Pomeranz Y (ed.), *Advances in Cereal Science and Technology*, Vol. VI, St Paul, MN, American Association of Cereal Chemists, pp. 357–396.
Schakel S F, Van Heel N and Harnack J (2004), 'Grain composition table', in Wrigley C, Corke H and Walker C E (eds), *Encyclopedia of Grain Science*, Vol. 3, Appendix 1, Amsterdam, Elsevier.
Schlick G and Bubenheim D L (1996), Quinoa: candidate crop for NASA's controlled ecological life support systems, in Janick J (ed.), *Progress in New Crops*, Arlington, VA, ASHS Press, pp. 632–640.
Schober T J, Messerschmidt M, Bean S R, Park S-H and Arendt E K (2005), 'Gluten-free bread from sorghum: Quality differences among hybrids', *Cereal Chem*, **82**, 394–404.
Sehgal A S and Kawatra A (1998), 'Reduction of polyphenols and phytic acid content of pearl millet grains by malting and blanching', *Plant Foods Hum Nutr*, **53**, 93–98.
Şensoy I, Rosen R T, Ho C T and Karwe M V (2006), 'Effect of processing on buckwheat phenolics and antioxidant activity', *Food Chem*, **99**, 388–393.
Serna-Saldivar S and Rooney L W (1995), 'Structure and chemistry of sorghum and millets', in Dendy D A V (ed.), *Sorghum and Millets: Chemistry and Technology*, St Paul, MN, American Association of Cereal Chemists, pp. 69–124.
Skrabanja V, Liljeberg-Elmstahl H G M, Kreft I and Björk I M E (2001), 'Nutritional properties of starch in buckwheat products: studies *in vitro* and *in vivo*', *J Agric Food Chem*, **49**, 490–496.
Southon S, Wright A J A, Price K R, Fairweather-Tait S J and Fenwick G R (1988), 'The effect of three types of saponin on iron and zinc absorption from a single meal in rat', *Brit J Nutr*, **59**, 389–396.
Sripriya G, Antony U and Chandra T S (1997), 'Changes in carbohydrate, free amino acids, organic acids, phytate and HCl extractability of minerals during germination and fermentation of finger millet (*Eleusine coracana*)', *Food Chem*, **58**, 345–350.
Sripriya G, Chandrasekharan K, Murty V S and Chandra T S (1996), 'ESR spectroscopic studies on free radical quenching action of finger millet (*Eleucine coracana*)', *Food Chem*, **57**, 537–540.
Sun T and Ho C T (2005), 'Antioxidant activities of buckwheat extracts', *Food Chem*, **90**, 743–749.
Svanberg U, Sjögren E, Lorri W, Svennerholm A-M and Kaijser B (1992), 'Inhibited growth of common enteropathogenic bacteria in lactic-fermented cereal gruels', *World J Microbiol Biotech*, **8**, 601–606.
Taylor J and Taylor J R N (2002), 'Alleviation of the adverse effect of cooking on sorghum protein digestibility through fermentation in traditional African porridges', *Int J Food Sci Tech*, **37**, 129–137.
Taylor J R N (1992), 'Mashing with malted grain sorghum', *J Am Soc Brew Chem*, **50**, 13–18.
Taylor J R N (2004), 'Fermentation: foods and nonalcoholic beverages', in Wrigley C, Corke H and Walker C E (eds), *Encyclopedia of Grain Science*, Vol. 1, Amsterdam, Elsevier, pp. 380–390.
Taylor J R N and Belton P S (2002), 'Sorghum', in Belton P S and Taylor J R N (eds), *Pseudocereals and Less Common Cereals*, Berlin, Springer, pp. 20–91.

Taylor J R N and Dewar J (2001), 'Developments in sorghum food technologies', in Taylor S (ed.), *Advances in Food and Nutrition Research,* Vol. 43, San Diego, Academic Press, pp. 217–264.

Thaoge M L, Adams M R, Sibara M M, Watson T G, Taylor J R N and Goyvaerts E M (2003), 'Production of improved porridges from pearl millet using a lactic acid fermentation step and addition of sorghum malt to reduce viscosity of porridges with high protein, energy and solids (30%) content', *World J Microbiol Biotech*, **19**, 305–310.

Tomotake H, Yamamoto N, Yanaka N, Ohinata H, Yamazaki R, Kayashita J and Kato N (2006), 'High protein buckwheat flour suppresses hypercholesterolemia in rats and gallstone formation in mice by hypercholesterolemic diet and body fat in rats because of its low protein digestibility', *Basic Nutr Invest*, **22**, 166–173.

Traoré T, Mouquet C, Icard-Vernière C, Traoré A S and Trèche S (2004), 'Changes in nutrient composition, phytate and cyanide contents and α-amylase activity during cereal malting in small production units in Ouagadougou (*Burkina faso*)', *Food Chem*, **88**, 105–114.

USAID (US Agency for International Development) (1999), *Commodities Reference Guide*, www.usaid.gov, accessed February 2006.

USDA (United States Department of Agriculture) and DHHS (Department of Health and Human Services) (2005), *Dietary Guidelines for Americans 2005*, www.health.gov/dietaryguidelines/dga2005/document/, accessed July 2006.

Valencia S, Svanberg U, Sandberg A S and Ruales J (1999), 'Processing of quinoa (*Chenopodium quinoa*, Willd): Effects on *in vitro* iron availability and phytate hydrolysis', *Int J Food Sci Nutr*, **50**, 203–211.

Valencia-Chamorro S A (2003), 'Quinoa', in Caballero B, Trugo L and Finglas P (eds), *Encyclopedia of Food Sciences and Nutrition*, Oxford, Elsevier, pp. 4895–4902.

Valencia-Chamorro S A (2004), 'Quinoa', in Wrigley C, Corke H and Walker C (eds), *Encyclopedia of Grain Science*, Vol. 3, Amsterdam, Elsevier, pp. 1–8.

Van Heerden I V (1988), 'In defence of sorghum beer', *S Afr Food Rev*, **15** (1 Suppl.), 129–130, 133.

Van Heerden I V (1989), The nutritive content of African beers brewed with maize grits or sorghum adjunct, *J Inst Brew*, **95**, 17–20.

Van Rensburg S J (1981), 'Epidemiological and dietary evidence for a specific nutritional predisposition to esophageal cancer', *J Nat Cancer Inst*, **67**, 243–251.

Varady K A, Wang Y and Jones, P J H (2003), 'Role of policosanols in the prevention and treatment of cardiovascular disease', *Nutr Rev*, **61**, 376–383.

Varriano-Marston E and De Francisco A (1984), 'Ultrastructure of quinoa fruit (*Chenopodium quinoa* Willd)', *Food Microstruct*, **3**, 165–173.

Verbruggen M A, Beldman G, Voragen A G J and Hollemans M (1993), 'Water-unextractable cell wall material from sorghum: isolation and characterization', *J Cereal Sci*, **17**, 71–82.

Vitale J D and Sanders J H (2005), 'New markets and technological change for the traditional cereals in semiarid sub-Saharan Africa: the Malian case', *Agric Econ*, **32**, 111–129.

Wambugu S M (1999), *Malted and fermented sorghum as functional ingredients in extruded instant weaning porridges*, MSc Thesis, Pretoria, University of Pretoria.

Watanabe M, Ohshita Y and Tsushida, T (1997), 'Antioxidant compounds from buckwheat (*Fagopyrum esculentum* Moench) hulls', *J Agric Food Chem*, **45**, 1039–1044.

Weber E J (1978), 'The Inca's ancient answer to food shortage', *Nature*, **272**, 486.

Westby A and Gallat S (1991), 'The effect of fermentation on the viscosity of sorghum porridges', *Trop Sci*, **31**, 131–139.

Yanez G A, Messinger J K and Walker C E (1986), '*Amaranthus hypochondriacus*: starch isolation and partial characterization', *Cereal Chem*, **63**, 273–276.

Yetneberk S, De Kock H L, Rooney L W and Taylor J R N (2004), 'Effects of sorghum cultivar on injera quality', *Cereal Chem*, **81**, 314–327.

Zhu N, Sheng S, Sang S, Jhoo J W, Bai N, Karwe M V, Rosen R T and Ho C T (2002), 'Triterpene saponins from debittered quinoa (*Chenopodium quinoa*) seeds, *J Agric Food Chem*, **50**, 865–867.

14

Vitamin and mineral fortification of bread

C. M. Rosell, Instituto de Agroquímica y Tecnología de Alimentos (IATA-CSIC), Spain

14.1 Introduction: vitamin and mineral fortification of bread

A varied and balanced diet provides a healthy person with an adequate intake of all the nutrients they require. Inadequate intake may result from an unbalanced diet, one poor in specific nutrients – for instance, if a nutrient is removed during processing, handling or storage – or when physiological conditions increase dietary requirements. Nowadays, the typical diet includes fewer natural and raw foods and more over-refined and processed foods, where certain valuable ingredients have been destroyed or discarded. This situation is magnified in the case of cereals, because they are not consumed directly as grains, the grain is generally processed before consumption (Conner, 1941). Processing substantially reduces nutrient content, particularly B-vitamins and iron which are mainly concentrated in the outer layers and germ.

Cereal fortification methods have been developed to restore the nutrients removed during milling and to improve the nutrient intake level of specific populations (Tobey and Cathcart, 1941). Fortification is included in policy instruments for improving the nutrient intake of a population, along with education by health professionals and recommendations for the use of supplements (Serra-Majem, 2001). In general, fortification is a way of providing different nutrients using staple foods as vehicles, with the aim of supplementing the dietary intake of those substances, in order to improve health and prevent nutritional deficiencies. Fortification has the advantage of requiring fewer changes in consumer behaviours and food habits than other policy instruments, although nutrition education and social marketing cannot be ignored.

Fortification can be defined as the addition of one or more vitamins and/

or minerals to a food, whether or not it is usually contained in the food, in order to:

- prevent or correct a demonstrated deficiency of one or more vitamins and/or minerals in the population or specific population groups; and/or
- improve the nutritional status of the population and dietary intakes of vitamins or minerals due to changes in dietary habits.

This addition must be based on generally accepted scientific knowledge regarding the role of vitamins and minerals in health.

Some factors should be taken into account prior to adding any nutrient to foods.

- The nutrient added should be present at a level that will not result in either an excessive or an insignificant intake considering the amount provided from other dietary sources. The maximum safe amounts to be added should be set taking into account the levels established by generally accepted scientific data and bearing in mind the diverse degree of sensitivity shown by the different consumer groups.
- Nutrients should not be added to fresh products – including meat, fish, fruits and vegetables – nor to beverages containing more than 1.2% by volume of alcohol.
- The nutrient added should be biologically available from the food and should not interfere with the metabolism of any other nutrient.
- The nutrient added should be sufficiently stable during packaging, storage, distribution and end-use, and should not affect either the sensory characteristics or the shelf-life of the food.
- Technology and processing facilities should be available to permit the addition of the nutrient, along with methods to measure and control the levels of nutrient added.
- Fortification should not be used as a tool to substitute for the natural nutrient sources, which would lead to a change in dietary patterns and thus jeopardise good dietary practices.

Enrichment is a term used interchangeably with fortification (Williams, 1941). The term 'enriched flour or bread' emerged in the 1940s, but there was little public interest or economic incentive to encourage this practice. However, the situation changed dramatically during World War II, when Britain began to manufacture only enriched flour and started a public campaign to improve the nation's health during wartime. Similar actions were taken in the US, but they were only effective from 1942, when the US army decided to buy only enriched flour. Initially, the enrichment consisted of the addition of the three major B-vitamins (thiamine, riboflavin and niacin) and iron. Another mineral, calcium, was permitted later as an optional ingredient. This was adopted despite the fact that scientific information concerning the real nutritional improvement was scarce. Later, numerous scientific studies were conducted to verify the utility of these early measures.

Strictly speaking, enrichment should be equivalent to restoration, which means adding vitamins and minerals to a food that have been lost during good manufacturing practice, or during normal storage and handling procedures. In the conversion of cereal grains to flour, 50–90% of at least 20 nutrients – including thiamin, riboflavin, niacin, pyridoxine, folate, panthothenate, biotin, vitamin E, calcium, copper, iron, potassium, magnesium, manganese, phosphorus, zinc, chromium, fluorine, molybdenum and selenium – are lost during the milling process (Fig. 14.1). In 1941, flours were enriched with iron and the vitamins thiamin, riboflavin and niacin, restoring them to the levels found in the whole grain. In 1955, the first cereal fortified with iron and vitamins to a level above that found in the whole grain was introduced. Since fortification is a broader term, it will be used here to refer to the addition of nutrients.

14.2 Range of vitamins and minerals added to bread and the health benefits

Vitamins and minerals are essential micronutrients for human development and metabolism. The recommended daily allowance or daily intake of the different vitamins and minerals has been established (Table 14.1) (European Union, 1993). However, current diets very often do not provide the necessary amounts of these micronutrients and they must be obtained through the consumption of either fortified food or specific supplements.

One of the basic criteria concerning food fortification is the selection of the right vehicle, which should be a food that is commonly eaten by the

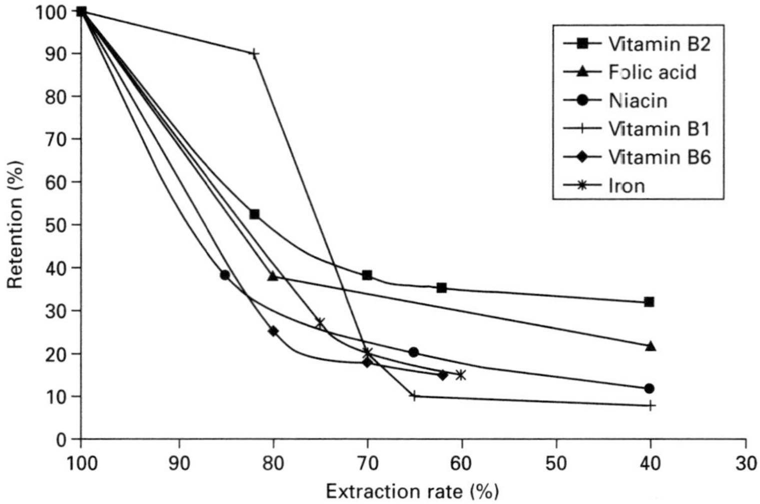

Fig. 14.1 Changes in micronutrient content of wheat grain during the milling process.

Table 14.1 Dietary recommendations for adults aged 19–50 years pertaining to mineral and vitamin intake

		RDA/AI	
Nutrient	Unit	Male	Female
Vitamin A	mg	900	700
Vitamin D	mg	5	5
Vitamin E	mg	15	15
Vitamin K	mg	120	90
Vitamin B1	mg	1.2	1.1
Vitamin B2	mg	1.3	1.1
Vitamin B6	mg	1.3	1.3
Vitamin B12	mg	2.4	2.4
Niacin	NE	16	14
Folate	DFE	400	400
Pantothenic acid	mg	5	5
Biotin	mg	30	30
Vitamin C	mg	90	75
Calcium	mg	1000	1000
Phosphorus	mg	700	700
Magnesium	mg	420	320
Fluoride	mg	4	3
Iron	mg	8	18
Zinc	mg	11	8
Copper	mg	900	900
Molybdenum	mg	45	45
Iodine	mg	150	150
Selenium	mg	55	55

RDA, Recommended daily allowance; AI, adequate intake; NE, niacin equivalent; DFE, dietary folate equivalent. Source: European Union (1993).

target population group, is economically affordable and available all year long (Baurenfiend and DeRitter, 1991). Flour, and by extension bread, is one of the best vehicles for food fortification because it meets these requirements.

Fortification of cereal products with iron and B-vitamins has been practised throughout the developing world for a number of years (Johnson, 2000). Fortification with multiple micronutrients is becoming more common, as deficiencies tend to manifest concurrently in at-risk populations. Several Latin American countries have extensive experience fortifying corn flour with iron, thiamin, riboflavin and niacin, and more recently with folic acid and vitamin A. Cereal flour fortification with B-vitamins, iron and sometimes vitamin A is extensive in several Middle East countries and is being explored in Asia and Africa (Nutriview, 2001). The Flour Fortification Initiative estimates that around 20% of the white flour produced in the world is currently fortified with at least iron and folic acid. Their goal is to raise this to 70% by the end of 2008. This will require launching fortification programmes in most wheat-eating countries, such as China, Turkey, Egypt, Ukraine and Russia.

14.2.1 Range of vitamins added to bread and health implications

B-vitamins (thiamine (B1), riboflavin (B2), and niacin (B3) have long been added to cereal flours in developed countries to ensure adequate intake of those nutrients through, for instance, the consumption of fortified bread (Fig. 14.2). Pyridoxine or vitamin B6 is also added to wheat flour, although bioavailability decreases by 5–10% when it is used in whole wheat bread (Leklem *et al.*, 1980). Vitamin B12-fortified bread is a good source of this vitamin, with even the elderly showing efficient absorption of the vitamin (Russell *et al.*, 2001).

Bread has also been fortified with vitamin D using cholecalciferol, showing good dispersion in the breads and stability. This fortified bread increases serum 25-hydroxyvitamin D concentration as effectively as other cholecalciferol supplements, without affecting serum intact parathyroid hormone concentration or urinary calcium excretion (Natri *et al.*, 2006). Therefore, fortified bread is a safe and feasible way to improve vitamin D nutrition (Mocanu *et al.*, 2005).

Folic acid is another B-vitamin added to cereals, due to its great importance for public health. Folate derivatives are essential for all cells as biochemical cofactors and serve as acceptors and donors of single-carbon units in a wide variety of reactions involved in amino acid and nucleotide metabolism. Folic acid deficiency can cause reductions in serum and erythrocyte folate and megaloblastic changes in the bone marrow, and anaemia. There is also increasing evidence demonstrating an association between folate deficiency

Fig. 14.2 Commercial bread fortified with multivitamins.

and carcinogenesis in certain epithelial tissues. Folate deficiency is probably not the causal factor, but may act as a 'co-carcinogen' with other risk factors (Kim, 1999).

Human requirements for folate vary depending on physiological condition, increasing during pregnancy, lactation and infancy. In 1991, the British Medical Research Council sponsored controlled trials that indicated that folic acid could reduce the risk for neural tube defects, including spina bifida, anencephaly and encephalocele. Neural tube defects cause serious birth defects that are important causes of infant mortality and disability. Spina bifida is an inclusive name for various conditions associated with the lack of closure of the spine, which causes permanent damage to the spinal cord and spinal nerves, resulting in varying degrees of paralysis. Anencephaly is a lethal malformation characterised by the absence of the cranial vault and cerebral hemispheres, whereas encephalocele is a defect in the skull that results in herniation of the meninges and brain tissue (Anon., 1991). Each year, nearly 400 000 infants worldwide are born with these conditions. In 1992, a randomised controlled trial conducted in Hungary reported the protective effect of folic acid-containing multivitamins against neural tube defects (Czeizel and Dudas, 1992). In the last decade, some studies confirming the role of folates in congenital malformations (neuronal tube defect) and the development of chronic diseases in elderly people, such as Alzheimer's disease, have recommended the addition of folic acid to cereal-based foods.

In 1996, the US Food and Drug Administration (FDA, 1996) concluded that 1000 μg folate/day is the safe upper limit of folate intake for the general population. The fortification level was established at 140 μg folic acid/100 g for enriched cereal grain products; for breads that level provides 10% of the daily requirement per serving. This level of fortification was chosen to help women of reproductive age increase their folic acid intake by an average of 140 μg folic acid daily. The effectiveness of this fortification programme in lowering homocysteine concentration and its relationship with a reduction in the incidence of cardiovascular disease has been investigated in elderly people (Koehler *et al.*, 1997). Research suggests that folate may play a role in reducing the risk of vascular disease. Although the folate status of many elderly people is adequate according to traditional, haematologic criteria, some elderly people have elevated blood concentrations of the metabolite homocysteine, which indicates subclinical deficiency of folate (and to a lesser extent of vitamins B6 and B12). Mean folate intake would increase by 16.5% if bread were fortified. Such fortification could help some individuals to lower serum homocysteine concentration and consequently reduce their cardiovascular disease risk (Koehler *et al.*, 1997).

Despite the widespread availability of fortified foods, relatively little is known about the impact of fortification programmes due to the limited number of rigorous assessments that have been carried out or documented. Some available data highlight the potential effectiveness of a properly implemented fortification programme. A survey conducted in the US to examine the possible health impact of niacin fortification of flour and bread found that fortification played an important role in the decline of pellagra-attributed mortality in the 1930s and 1940s, and its eventual elimination (Park *et al.*, 2000). Since 1996, when folic acid fortification of cereal products was authorised in the US, the incidence of neural tube birth defects has decreased by 19%, and fortification, along with educational outreach and consumer awareness, has contributed to that decrease (Caudill *et al.*, 2001; Honein *et al.*, 2001; Dietrich *et al.*, 2005). Since 2000, when the Chilean Ministry of Health mandated the fortification of wheat flour with folic acid, the concentration of blood folate in healthy women of reproductive age has increased significantly (Hertrampf *et al.*, 2003).

The effectiveness of iron fortification programmes in combating anaemia has been less clear. In fact, Sweden ceased fortifying its bread in 1994 because of a lack of perceived benefit. In Venezuela, fortification of wheat and corn with iron, B-vitamins and vitamin A was initiated in 1993, and a preliminary survey conducted among a small group of children showed that the prevalence of iron deficiency and anaemia had decreased significantly ($p < 0.05$) (Layrisse *et al.*, 1996).

14.2.2 Iron fortification and health benefits

Iron is highly localised in the human body in dopaminergic–peptidergic regions, such as the globus pallidus, substantia nigra, red nucleus thalamus, caudate nucleus and nucleus accumbens (Youdim and Ben-Shachar, 1987; Sandstead, 2000). Iron deficiency induces a decrease of both iron concentration in the brain and the number of dopamine receptors, and as a consequence learning is impaired. Some studies have revealed that iron deficiency during brain development results in residual abnormalities (Roncagliolo *et al.*, 1998).

Iron deficiency and the anaemia that results is a major public health problem in developing countries, affecting an estimated 2 billion people. Iron deficiency reduces work capacity and productivity. It is especially dangerous in children and pregnant women, where it increases the risk of premature and low birth weight babies, and also the risk of the woman's death during childbirth. Children with iron-deficiency anaemia have poor growth and physical development, impaired cognitive abilities and reduced resistance to diseases. Iron supplementation has been adopted in some developed countries to reduce the prevalence of iron-deficiency anaemia.

In planning a fortification strategy, a number of factors should be considered when determining the optimal level of iron fortification, including the prevalence of iron deficiency, the nature of the diet, the distribution of cereal foods and the bioavailability of the added iron. The minimum recommended amount to restore the iron naturally present in the whole grain product is 25 ppm iron for white flour, using ferrous sulphate or ferrous fumarate. This would give an iron level in the enriched flour of about 35 ppm, or equivalent to the original level found in whole wheat flour. Higher iron levels may be necessary in countries where iron deficiency is prevalent and where consumption levels of wheat-based products is low (Ranum *et al.*, 2001).

In 1996, the World Health Organization Eastern Mediterranean Region Office began recommending the use of ferrous sulphate at 30 ppm, or reduced iron at 60 ppm, along with 1.5 ppm folic acid in flour fortification in the Middle East. In 1999, a multidisciplinary task force of international iron experts established some guidelines on iron fortification. They recommended the use of ferrous sulphate or ferrous fumarate where possible, and, if elemental iron powder was to be used, the electrolytic form was recommended at twice the level of iron used for iron salts (Ranum *et al.*, 2001). Some studies confirmed previous findings that none of the elemental iron powders match the bioavailability of ferrous sulphate, which remains the preferred source of iron for flour fortification (Hurrell *et al.*, 2002). Iron enrichment of flour is generally regulated in most countries by specifying a minimum iron level in the enriched flour that includes both the added and native iron, rather than specifying the exact quantity of iron fortification to add.

14.2.3 Fortification with other minerals

Zinc, magnesium and calcium have also been added to wheat flour (Whittaker and Dunkel, 1995). Adequate intake of calcium during childhood and adolescence is crucial for attaining peak bone mass and attenuating bone loss and predisposition to osteoporosis in later years (Ulrich *et al.*, 1996). In addition, calcium has a protective effect against hypertension (Jorde and Bonna, 2000) and colorectal cancer (Parodi, 2001; Hambly *et al.*, 2002). The recommended intakes of calcium are 800–1000 mg/day in the European Union (European Union, 1993) and 1300 mg/day in the US (Institute of Medicine, 1997). Fortification of bread with calcium has contributed to increasing the calcium intake of the population. The bioavailability of calcium from fortified bread has been compared with that of a milk-based diet, showing that enriched bread can serve as a good source of bioavailable calcium, although special care should be taken with high-fat-content diets, because fat may interfere with calcium absorption (Juma *et al.*, 1999). In addition, it has been demonstrated that the consumption of bread fortified with 2.5% elemental calcium is an effective way to ameliorate hyperphosphataemia without inducing hypercalcaemia (Babarykin *et al.*, 2004). Therefore, calcium-fortified bread can be consumed as a treatment for uraemic hyperphosphataemia.

Calcium-fortified breads – fortified with calcium carbonate, calcium sulphate, calcium citrate or calcium lactate – are a good source of bioavailable calcium (Ranhotra *et al.*, 1997). For successful fortification, the calcium fortificant must not impart off-flavours, especially important as many calcium salts have intrinsic characteristic flavours (Flynn and Cashman, 1999). To avoid possible changes in the sensory profile of breads, calcium salts must be added at levels below their detection thresholds. The sensory thresholds of calcium carbonate, calcium sulphate and tricalcium dicitrate have been determined in pita bread loaves or pocket-type flat bread, and are defined as 2724 mg/100 g for calcium sulphate, 1984 mg/100 g for calcium carbonate and 2132 mg/100 g for tricalcium dicitrate. Calcium-fortified pita bread was similar in taste to regular pita bread (Ziadeh *et al.*, 2005). Despite the different calcium sources that can be used for bread fortification, calcium carbonate remains the preferred source due to its reasonable price, its calcium content and its small influence on product quality (Ranhotra *et al.*, 2000).

The addition of zinc and calcium was shown to improve dough expansion during breadmaking without having any adverse effects on the quality of the resulting bread, which retains between 87 and 93% of the added calcium and zinc (Wang and Tang, 1995).

Bread is used as a carrier in magnesium-fortified breads (Ranhotra *et al.*, 1980; Ranhotra and Winterringer, 1982; Salovaara, 1982; Whittaker and Dunkel, 1995). Different salts of magnesium – including dioxide, hydroxide and carbonate – have been used, although they induce a volume reduction and a deterioration of the flavour (Ranhotra *et al.*, 1976). Other magnesium sources include chloride, phosphate and sulphate, but they result in similar effects.

Selenium has been added to sourdough bread, showing that bread could be a good source of dietary selenium (Bryszewska *et al.*, 2005). An alternative method for fortifying bread with selenium is to use selenium-enriched yeast (Rumi *et al.*, 1994).

14.3 Consumer attitude towards fortified bread

Some nutritionists were against fortification when it was introduced in the US in 1941, but their attitude rapidly changed to a limited recommendation under proper control. Consumers wanted to know why bread and cereals were fortified with minerals and vitamins. Initially, bread fortification was followed by extravagant claims from the manufacturers, which threatened consumer confidence. Later, various information and education campaigns were launched to reassure consumers that fortification of flour and bread was being carried out for their benefit and not for that of the industry (Mitchell, 1941).

Today, consumers are becoming more positive about functional foods, and foods that have been fortified with a substance already present in the conventional product. Some population segments are willing to pay more for food products if they think there is a health benefit (Poulsen, 1999). The marketing of fortified foods should emphasise the convenience of obtaining vitamins and minerals through the daily diet in a natural way. Food industry marketing bodies perceive that consumers want foods that are convenient, fresh (less processing and less packaging), natural (without preservatives, or so-called 'clean label'), without a negative association (high fat, high salt, high sugar), and healthy. The industry perception is that consumers look for foods that not only cause no harm but that also act as a remedy for, or decrease the incidence of, diseases such as heart disease and osteoporosis, and help reduce fatigue and memory loss. Fortified foods fall in the category of foods that provide those benefits.

The fortification of foods with micronutrients is an economical, flexible and socially accepted method for improving processed foods without changing the physical or sensory properties, or the food consumption practices of the consumers (Lofti *et al.*, 1996; Darnton-Hill, 1998). Although cereal fortification can be an effective means of improving micronutrient status, there are some barriers to its general implementation. There is no broad consensus about the extent to which food fortification should be practised. Arguments against fortification include those that suggest that manufacturers could use fortification as a promotional tool, which could result in excessive intake of certain nutrients and risk consumer health.

The success of fortification programmes always depends on good control, and consequently they should be set up, regulated and enforced by national government. In addition, the long-term sustainability of a food fortification programme requires campaigns convincing consumers and policy makers of the need and benefits. Programmes that fail are constrained by a lack of consumer demand due to poor awareness of the benefits of consuming fortified foods. Fortification should require minimal consumer involvement and minimal change in dietary habits. An uncontrolled food fortification could result in excessive intakes of certain nutrients creating nutrient imbalances, consequently representing a risk to the health of consumers. Recent research suggests that, while fortification programmes may be necessary in developing countries, in the developed world they may be posing problems with implications that are not yet fully understood.

Manufacturers either add nutrients voluntarily or because it is compulsory under national rules. In general, vitamins and minerals are added to conform with the legislation of the country in which the product is being sold. The Economic and Social Department of the Food and Agriculture Organization (FAO) has reviewed the status of legislation concerning wheat flour fortification (Table 14.2). There are currently 14 countries with legislation or regulations that mandate the fortification of wheat flour, whereas in others – such as Finland, New Zealand and Norway – the addition of nutrients to wheat flour is prohibited by law. In the US, the FDA approves the addition of four B vitamins (thiamine, niacin, riboflavin and folic acid) and iron to wheat flour, to obtain 'enriched flour', and this practice is regulated by Federal law. Nevertheless, the following states have opted not to require this enrichment: Alaska, Delaware, District of Columbia, Illinois, Iowa, Maryland, Minnesota, Missouri, Nevada, New Mexico, New York, North Carolina, Oklahoma, Pennsylvania, Rhode Island, Tennessee, Vermont, Virginia and Wisconsin.

Table 14.2 Summary of legislation pertaining to fortification of wheat flour

Country	Law status	B vitamins	Vitamin E	Folic acid	Iron	Zinc	Magnesium	Calcium
Australia	Voluntary	B1, B2, B3, B6	+	+	+	+	+	
Canada	Mandatory	B1, B2, B3			+			
	Voluntary	B5, B6		+			+	+
Chile	Mandatory	B1, B2, B3			+			
Costa Rica	Mandatory	B1, B2, B3			+			
Dominican Republic	Mandatory	B1, B2, B3			+			
Ecuador	Mandatory	B1, B2, B3		+	+			
El Salvador	Mandatory	B1, B2, B3			+			
Finland	Prohibited							
Guatemala	Mandatory	B1, B2, B3		+	+			
Honduras	Mandatory	B1, B2, B3			+			
Hungary	Voluntary							
Malta	Voluntary	B1, B2, B3			+			
New Zealand	Prohibited							
Nigeria	Mandatory	B1, B2, B3			+			
Norway	Prohibited							
Panama	Mandatory	B1, B2, B3			+			
Philippines	Voluntary	+			+			
Saudi Arabia	Mandatory	B1, B2, B3			+			
Sweden	Voluntary	B1, B2, B3, B6			+			
Switzerland	Voluntary	B1, B2, B3			+			
UK	Mandatory	B1, B3			+			
Uruguay	—	B1, B3			+			+
USA	Mandatory	B1, B2, B3		+	+			
Venezuela	Mandatory	B1, B2, B3			+			

Source: FAO legislation pertaining to food fortification (FAO, 2002).

In 1996, the US Government made fortification of cereals and grains with folic acid mandatory at a level of 140 μg/100 g grain and this came into force in 1998 (Buttriss, 2004), which suggests a significant increase in the folic acid content of cereal-based products compared with the value before fortification (Table 14.3). A similar approach was taken in Canada in 1998, defining the level of 150 μg folic acid/100 g grain. Since 2000, fortification of wheat flour with 220 μg folic acid/100 g grain has been required in Chile. In 2003, 38 countries introduced or agreed to introduce folic acid fortification of flour, but no countries from the European Union were among them (Wald, 2004). In 2000, the United Kingdom Advisory Committee on Medical Aspects of Food and Nutrition Policy recommended mandatory folic acid fortification at a concentration of 240 μg/100 g grain, but in 2002 the Board of the United Kingdom's Food Standards Agency decided against such action. A similar decision was reached by the Dutch Health Council (Postma *et al.*, 2002).

In the United Kingdom, it is a legal requirement that the nutrients removed with the bran during the milling must be replaced in all types of flour, with exception of wholemeal, and that white and brown flour must have added thiamine, niacin, iron and calcium. Since 1997, the Organización Panamericana de la Salud (OPS) has supported the addition of iron to wheat flour due to its low cost and extensive consumption. Recently, several countries from South and Central America have implemented iron supplementation measures to reduce the incidence of iron-deficiency anaemia. Fifty-two countries now have iron-fortification standards and fortify some or all of their flour. Nine of those countries specify the use of ferrous sulphate, nine specify electrolytic iron powder and one specifies ferrous fumarate. In the other 33, there are no regulations concerning the type of iron to be added, which often results in the use of elemental iron powders for price reasons and due to their low reactivity (Ranum, 2006).

Table 14.3 Folate values for different cereal-based products

Enriched grain products	Value before fortification (g/100 g)	Value after fortification (g/100 g)
Flour	26	154
Cornmeal, grits	48	187
Farina	24	173
Rice	8	231
Macaroni	18	231
Noodles	29	231
Bread	34	95
Rolls	30	95
Buns	27	95

Source: USDA Nutrient Database (http://www.nal.usda.gov/fnic/foodcomp/ search).

14.4 Particular difficulties in the development of vitamin- and mineral-fortified breads

Three different requirements should be considered when selecting the source of added nutrients: the chemical compound should be stable during packaging, storage, distribution and end-use; it should not modify the sensory characteristics (appearance, texture, taste and aroma); and it should ensure the bioavailability of the nutrient. Different chemical compounds are added as a source of vitamins and minerals, depending on their intended use (Tables 14.4 and 14.5).

Nutrient stability is very important in the case of vitamin fortification. Vitamins are sensitive to heat, oxidising and reducing agents, light and other physical and chemical stresses. For instance, vitamin A is stable when added to flour, but the combination of high humidity and temperature adversely affects its stability. This vitamin is labile at high temperature, and unstable in the presence of oxygen and light, due to its susceptibility to oxidation. Therefore, when vitamin A is added, protection against vitamin loss is needed, in the form of suitable packaging materials and appropriate conditions of storage. Encapsulation of this vitamin in a more hydrophilic coat is a common practice aimed at obtaining a more water-dispersible product. Folic acid,

Table 14.4 Compounds added to cereals as a source of vitamins

Vitamin	Chemical compound
Vitamin A	Retinol
	Retinyl acetate
	Retinyl palmitate
	Beta-carotene
Vitamin B1	Thiamin hydrochloride
	Thiamin monohydrate
Vitamin B2	Riboflavin
	Riboflavin 5′-phosphate
Niacin (B3)	Nicotinic acid
	Nicotinamide
Pantothenic acid (B5)	D-Pantothenate, calcium
	D-Pantothenate, sodium
Vitamin B6	Pyridoxine hydrochloride
	Pyridoxine 5′-phosphate
	Pyridoxine dipalmitate
Vitamin B12	Cyanocobalamin
	Hydroxocobalamin
Vitamin C	Ascorbic acid
	Sodium ascorbate
	Calcium ascorbate
	Ascorbyl 6-palmitate
Vitamin E	Tocopherol
	Tocopheryl acetate
	Tocopheryl acid succinate
Folic acid	Pteroylmonoglutamic acid

Table 14.5 Compounds added to cereals as a source of minerals

Mineral	Chemical compound
Calcium	Calcium carbonate Calcium chloride Calcium salts of citric acid Calcium gluconate Calcium lactate Calcium hydroxide
Copper	Cupric carbonate Cupric citrate Cupric gluconate Cupric sulphate
Iodine	Sodium and potassium iodide Sodium and potassium iodate
Iron	Ferrous carbonate Ferrous citrate Ferric ammonium citrate Ferrous gluconate Ferrous fumarate Ferrous lactate Ferric diphosphate Elemental iron (carbonyl + electrolytic + hydrogen reduced) Hydroxocobalamin
Magnesium	Magnesium acetate Magnesium carbonate Magnesium chloride Magnesium salts of citric acid Magnesium oxide Magnesium sulphate
Manganese	Manganese carbonate Manganese sulphate Manganese chloride Manganese citrate
Phosphorus	Calcium salts of orthophosphoric acid
Zinc	Zinc acetate Zinc chloride Zinc oxide Zinc sulphate Zinc carbonate

unlike iron, is a food additive rather than a GRAS (generally recognised as safe) substance. The folate used as a food fortificant is the highly bioavailable oxidised monoglutamate form of folic acid.

Adding B-vitamins to flour is difficult due to the low stability of these compounds during storage. Higher effectiveness is obtained when B-vitamins are added at the bakery rather than at the mill. Comparing the effectiveness of adding thiamine, riboflavin and niacin at the mill or to the dough at the bakery shows that the content of vitamins decreases by 33.3–58.1% when the wheat flour is vitaminised at the mill, in contrast to a 17.0–38.7% reduction when the vitamins are added to the dough (Stepanova *et al.*, 1988). Nevertheless, the content of thiamine and niacin in the resulting bread prepared from flour vitaminised at the mill matches the standard intake recommendation.

The bioavailability of added iron varies widely depending on the iron compound used for fortification. Selecting the most convenient iron compound requires special consideration of the physical and chemical properties of the food to be fortified and the iron form used (Anon., 1955). Selection is based on organoleptic considerations, bioavailability, cost and safety. Iron powders (carbonyl iron, electrolytic iron and reduced iron) are extensively used in flour fortification due to their low cost and because they do not promote a rancid taste and do not interfere with bread volume (Germani *et al.*, 2001). However, they are poorly absorbed and their absorption is highly dependent on dietary composition (phytates, calcium and polyphenols decrease the bioavailability of iron) (Tetens *et al.*, 2005). In general, ascorbic acid enhances the bioavailability of elemental iron powders, whereas the baking process reduces their bioavailability from bread (Yeung *et al.*, 2005). Ferrous sulphate is also used because of its high bioavailability and low cost. It can be used in bakery flour and other types of low-extraction wheat flours. The main concern about ferrous sulphate has been its effect in promoting oxidative rancidity in flour, particularly when flour has a high fat content and is stored for prolonged periods at high temperature and humidity. The ferrous sulphate should be of a fine particle size and dehydrated. Large particle size or hydrated ferrous sulphate can cause colour and spotting problems, yielding unpleasant colours and flavours due to reactions with other components of the food matrix, and so are not recommended (Ranum *et al.*, 2001; Kajishima *et al.*, 2003). A comparison of breads fortified with different elemental iron powders indicates that bioavailability was comparable with ferrous sulphate, thus they provide a good alternative (Yeung *et al.*, 2005).

Ferrous fumarate is another good choice because its bioavailability is similar to that of ferrous sulphate. It is insoluble in water and therefore causes fewer organoleptic problems than the more soluble ferrous sulphate. However, it is more expensive than ferrous sulphate.

Sodium iron ethylene diamine tetraacetic acid (NaFeEDTA) is a chelated form of iron that has superior absorption when used in high phytic acid foods or diets. In China, NaFeEDTA has been used in a voluntary fortification programme, revealing that its effectiveness in reducing the prevalence of iron deficiency and anaemia is higher than that of ferrous sulphate. The main disadvantages of NaFeEDTA are its high cost and the fact that it does not respond to the iron spot test that millers use to assess fortification levels. In addition, a comparison test regarding the effect of ferrous sulphate and NaFeEDTA on bread quality showed that NaFeEDTA negatively affected appearance, suggesting the advantage of using ferrous sulphate despite the unpleasant taste conferred to bread (Ilyas *et al.*, 1996). Furthermore, NaFeEDTA may have a negative effect on the metabolism of other minerals, such as calcium and zinc, although contradictory results have been reported regarding NaFeEDTA and zinc absorption (Davidsson *et al.*, 1994). Recently Lopez *et al.* (2002) have reported the sensorial acceptance (taste, odour, texture and overall acceptability) of breads with 30 mg of added ferrous sulphate/100 g of flour.

The technological and sensorial characteristics of sandwich bread fortified with three different forms of iron (reduced iron, ferric pyrophosphate and microencapsulated ferrous sulphate) at 4.2 mg iron/100 g of flour have been compared with a standard formulation (Table 14.6) (Nabeshima *et al.*, 2005). Breads containing ferric pyrophosphate gave softer crumbs; despite some differences in sensorial properties, iron-fortified breads were accepted.

The effect of the fermentation and baking processes on the stability of added folic acid and some endogenous folates has been determined for rye and wheat breads. Breads produced using flour enriched with 0.2 mg folic acid/100 g resulted in total losses ranging from 12 to 21%, depending on the baking process (Gujska and Majewska, 2005).

In 2001, an analytical study of different cereal grain products revealed that the amounts of iron and folic acid were considerably higher than indicated on the label (Whittaker *et al.*, 2001). However, a more recent analysis conducted in Alabama, US indicated that the amount of folic acid in white bread decreased significantly from 2001 to 2003, whereas in whole wheat bread values were approximately constant (Johnston and Tamura, 2004).

14.5 Development and manufacturing technology of vitamin- and mineral-fortified breads

A high proportion of vitamins and minerals, mainly located in the outer layers of the grain, is lost during milling. These vitamins and minerals can

Table 14.6 Effect of different iron sources on the characteristics of iron-fortified white breads

			Iron source		
		Control	Ferric pyrophosphate	Reduced iron	Ferrous sulphate[a]
Iron (mg/100 g)		0	4.37	4.7	5.11
Specific volume (cm^3/g)		3.36	3.18	3.03	3.54
Sensory characteristics	Maximum value				
Crumb structure	10	7.5	7.0	7.0	6.0
Texture	10	7.5	9.0	8.0	6.5
Aroma	10	9.0	7.5	7.5	7.5
Taste	15	13.0	13.0	13.0	13.0
Global acceptance	100	71.7	70.6	69.1	70.8

[a] Microencapsulated ferrous sulphate.
Source: Nabeshima *et al.* (2005).

be replaced by using cereals grown in enriched soil, directly adding the micronutrients, modifying the milling process or using organic material as a mineral source. Breads have been produced using wheat grown in soil enriched with selenium. The resulting breads have been marketed with the claim of 'healthy bread builds immunity', due to the effect of selenium on the immune system (Fig. 14.3).

The preferred point for fortifying bakery products is at the milling stage (with the exception of wholemeal flour), rather than at the breadmaking stage, because the addition can be better controlled. However, a study carried out comparing the efficiency of fortification at the mill or at the bakery, showed that the vitamin content decreased by 33.3–58.1% when wheat flour was fortified at the mill, versus a 17.0–38.7% reduction observed when the vitamins were added to bread dough (Stepanova *et al.*, 1988).

Fortification of milled, refined cereals is a convenient way to deliver iron and other micronutrients to a general population whose diets are deficient in those micronutrients. Iron should be included in cereal fortification or enrichment programmes in countries where iron-deficiency anaemia is prevalent.

Fortification of flour is usually carried out at the mill, where a mixture of the required nutrients is added to the flour before packaging (Fig. 14.4). The addition of B-vitamins, iron and calcium is a common practice in some developed countries (Table 14.2). However, the nutrient added must have good stability during storage.

Iron fortification has faced several problems regarding bioavailability of the iron fortificant, as well as its cost and its stability in foods. Another challenge has been that the benefits of a national iron fortification programme aimed towards at-risk population groups, including children, has not been

Fig. 14.3 Bread marketed as selenium-enriched bread.

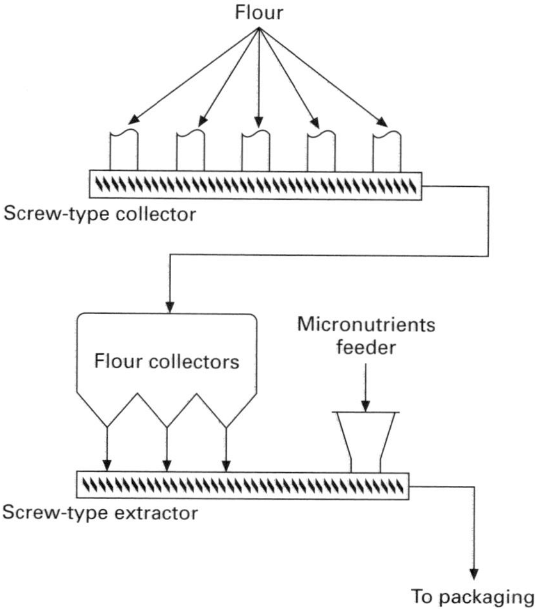

Fig. 14.4 Addition of micronutrients at the mill before packaging. Source: Nutraview, http://www.nutrivit.org/vic/staple/index.htm.

clearly demonstrated. The choice of which form of iron to use represents a compromise between avoiding detrimental effects on colour and shelf-life, maximising bioavailability and minimising cost. The maximum level of iron that can be added to milled cereals without causing quality problems in the final food product will vary with the type of product and should be tested. Iron addition levels up to 40 ppm from ferrous sulphate and 60 ppm from elemental iron powders have been successfully used in wheat flour (Ranum *et al.*, 2001).

Some countries, for example, Canada, which enriches all flour at the mill, prefer the use of elemental iron powders. The US uses both ferrous sulphate and elemental iron powder, with the latter being the most common. Venezuela uses a combination of both ferrous fumarate and elemental iron powder in corn flour, and ferrous fumarate in wheat flour.

In most cases, iron is added along with other micronutrients such as vitamins, in a premix formulated to meet the specified enrichment standards. The types of vitamins in the premix determine its shelf-life, because iron fortificants are usually more stable than the vitamins. In Turkey, where bread is a staple food and provides about 50% of daily energy intake, vitamins C, B1, B2, B6 and B12, folic acid, niacin and minerals – including iron, zinc and calcium – are added to wheat flours via ultra-microcapsules. Fortified bread formulations are optimised according to organoleptic properties (Karadzhov and Iserliyska, 2003; Loker, 2004).

Fortification during milling is based on controlled removal of the outer layer of the grain. Since the minerals are located in the outer layers, higher extraction yields will provide nutritionally richer wheat flours, although with worse rheological properties. However, it is possible to improve those properties by grinding and separating. Grinding is done using break rolls, sizing rolls and reduction rolls. The smaller particles of ground material obtained after grinding are called break middlings, and are a mixture of pure endosperm, endosperm attached to particles of bran and some small particles of bran. One way of fortifying the resulting wheat flour with minerals consists of mixing between 5 and 50% middlings (from the first and second breaks) with wheat flour (straight, patent, break or clear), obtaining a mineral enrichment of up to 20% with respect to the initial flour content of calcium, zinc, iron, manganese and phosphorus, without changing the fermentation and baking characteristics (Maldonado, 2000).

Another way to fortify wheat flour with minerals is by adding minerals indirectly, for instance, by blending small amounts of seaweed extracts like *Laminaria japonica* into the wheat flour, which acts as a source of iodine (Miki *et al.*, 1993a,b). Selenium-enriched bread is also available.

14.6 Drawbacks of fortification

With increasing fortification of foods, public health authorities are concerned that excess intakes of specific nutrients may result in toxic manifestations or adverse effects (Suojanen *et al.*, 2006). In the European Union, a first draft of harmonised EU legislation on food fortification aimed at preventing excessive intake of certain micronutrients is being prepared.

14.6.1 Excessive iron intake

Several observations have led researchers to examine the possible association between high iron intake and coronary heart disease. It appears that men with high concentrations of serum ferritin or high iron intake have an increased risk of heart attack or myocardial infarction. Conversely, lower rates of heart disease have been detected among people living in developing countries; this is possibly due to low meat (and iron) intake, high-fibre diets that inhibit iron absorption and gastrointestinal parasite concentrations that result in gastrointestinal blood (and iron) loss. Other studies have shown that serum ferritin is the strongest indicator of the presence and progression of carotid artery disease (Kiechl *et al.*, 1997), and this association was most evident in current or former smokers and in subjects with diabetes, raising the question of whether ferritin may adversely affect ischaemic heart disease risk in combination with other risk factors. However, further research is needed in order to provide convincing support for an association between high body iron stores and increased risk for coronary heart disease (Institute of Medicine, Food and Nutrition Board, 2001).

Excess intake of iron can also result in iron overload, which is a relatively common disorder of iron metabolism. A genetic form of iron overloading known as hereditary haemochromatosis affects 1 in 400 individuals of Northern European descent (Dadone *et al.*, 1982; Edwards *et al.*, 1988). Individuals with hereditary haemochromatosis are at increased risk for liver cancer. Some studies, based on data from the National Health and Nutrition Examination Survey (NHANES I and II), found an association between excess body iron stores and an increased risk of cancer (Stevens *et al.*, 1988). Stevens *et al.* (1994) reported a significant association between colon cancer and iron intake.

There are at least three possible mechanisms for the role of iron in carcinogenesis: (1) production of free oxygen radicals; (2) promotion of the growth of transformed cells; (3) action as an essential cancer cell growth nutrient (Nelson, 1992; Siegers *et al.*, 1992; Weinberg, 1994). There is inconclusive evidence that iron status is associated with the incidence of cancer in those who do not have hereditary haemochromatosis. Identification of definitive relationships between increased iron intake, increased iron bioavailability, and high-meat and high-fat diets, will require more research (Whittaker *et al.*, 2001). Nonetheless, because of the possible associations of increasing iron status with heart disease and several types of cancer, it may be important to reduce dietary iron, or at least to have additional cereals and breads available without iron fortification to allow individuals with an elevated iron status more choices.

Some researchers have raised concerns about the effects of iron fortification and supplementation on the absorption of other nutrients, such as zinc, calcium and copper. Research studies have shown that supplemental iron may decrease the absorption of these nutrients, but generally only when the supplement is taken on an empty stomach. Absorption of these nutrients is generally not affected when supplementary iron is taken with food.

14.6.2 Excessive folic acid intake

Concern has been raised in some studies about the potential negative effect of folic acid supplementation on zinc absorption (Milne *et al.*, 1984; Butterworth and Tamura, 1989). The long-term effect of folic acid on zinc status *per se* cannot be evaluated effectively in humans due to the lack of reliable biomarkers for zinc status, but recent reports based on absorption studies showed that fortification of white bread with the currently used concentrations of folic acid did not appear to influence zinc absorption (Hansen *et al.*, 2001). However, it would be relevant to monitor the effect of higher doses of folic acid in order to confirm that there is no association between folic acid and zinc absorption.

The ability of folate to mask the anaemia of vitamin B12 deficiency is the most recognised adverse effect of high folic acid intake (Lindenbaum *et al.*, 1988). An excess of folic acid intake together with an inadequate intake of vitamin B12 can mask the development of the megaloblastic anaemia of vitamin B12 deficiency, delaying diagnosis and treatment, and exacerbating the progression of neurological complications. There is also some indication that high folic acid intake can have adverse effects on individuals treated with anticonvulsants and methotrexate (Institute of Medicine, National Research Council, 1998).

The fortification of bread with calcium could result in health risks for some population groups, for instance, those persons in the uppermost calcium intake decile. Regular excessive intake of calcium (over 2500 mg/day) has been found to be harmful, especially if it is accompanied by a high intake of vitamin D. Excessive intake of calcium can result in hypercalcaemia and calciuria (Weawer and Heaney, 1999). High levels of calcium in urine can result in renal damage and the development of kidney stones. In addition, there is increasing evidence connecting high calcium intakes with prostate cancer.

14.7 Future trends

Currently, the terms 'functional foods' and 'nutraceuticals' seem to be dominating the food market. Bread is an excellent product in which to incorporate nutraceuticals or 'pharmafoods' due to its characteristics and wide consumption. One of the latest enrichments has been the addition of omega-3 polyunsaturated fatty acids to improve essential fatty acid intake. In Europe, consumption of bread enriched with omega-3 polyunsaturated fatty acids is steadily increasing because Europeans recognise the healthy component of such products. Therefore, the near future for nutrition could include extending the use of breads as vehicles for different micronutrients.

From the information provided in this chapter some conclusions can be drawn: (1) the necessity of standardising dietary survey methods around the world to facilitate comparisons and (2) the need for special attention to be paid to diverse dietary habits when general recommendations are made, in order to prevent overdoses. The addition of minerals and vitamins to flour and bread is considered an effective approach for maintaining a safe and nutritionally adequate micronutrient supply. However, monitoring is necessary to assure consumers that systematically fortified bread consumption does not produce micronutrient imbalances and/or overdoses. Monitoring and evaluation are critical aspects in food fortification that have not been properly assessed. A good monitoring system requires specific mechanisms for: (1) identifying the nutrition patterns of the population in order to identify nutrient deficiencies, (2) identifying the level of purchase and consumption of fortified foods; (3) assessing the intake levels of the nutrient of interest and the percentage of contribution of the fortified food to that intake; (4) assessing the impact of fortification on public health; and (5) determining safety problems associated with food fortification.

14.8 Sources of further information and advice

- FAO (Food and Agriculture Organization) (2002), Economic and Social Department. Food Fortification: Technology and Quality Control (FAO Food and Nutrition Paper 60). This website includes the report of an FAO technical meeting held in 1995 related to Food Fortification. http://www.fao.org/docrep/W2840E/w2840E00.htm.
- Hurrell R (1999), *The Mineral Fortification of Foods.* Surrey, UK, Leatherhead Food International Ltd.
- Institute of Medicine of the National Academies, Washington DC. http://www.iom.edu/CMS/3708.aspx.
- Nutriview, Vitamin Information Centre – Micro Nutrient Intervention Programs. http://www.nutrivit.org/vic/staple/index.htm.
- Ranum PM (2000), 'Cereal enrichment and nutrient labelling', in: Kulp K and Ponte JG (Eds), *Handbook of Cereal Science and Technology.* New York, Marcel Dekker Inc., pp. 697–704.
- Rosell CM (2003), 'The nutritional enhancement of wheat flour', in: Cauvain SP (Ed.), *Bread Making: Improving Quality.* Cambridge, Woodhead Publishing Limited, pp. 253–269.
- Rosell CM (2004), 'Fortification of grain-based foods', in: Wrigley C, Corke H, Walker CE (Eds), *Encyclopedia of Grain Science.* Oxford, UK, Elsevier Ltd, pp. 399–405.
- SCF (Scientific Committee on Food) (2003), Opinions of tolerable upper intake levels for vitamins and minerals (various). http://ec.europa.eu.food/fs/sc/scf/out80_en.html.

14.9 References

Anon. (1955), 'Utilization of iron compounds in enriched bread', *Nutr Rev*, **13**, 165–166.

Anon. (1991), 'Folic acid and neural tube defects', *Lancet*, **338**, 153–154.

Babarykin D, Adamsone I, Amerika D, Spudass A, Moisejev V, Berzina N, Michule L, Rozental R (2004), 'Calcium-enriched bread for treatment of uremic hyperphosphatemia', *J Renal Nutr*, **14**, 149–156.

Baurenfiend JC, DeRitter E (1991), 'Foods considered for nutrient addition: Cereal grain products', in Baurenfiend JC and LaChance PA (Eds), *Nutrient Additions to Food.* Connecticut, Food and Nutrition Press, pp. 143–209.

Bryszewska MA, Ambroziak W, Diowksz A, Baxter MJ, Laangford NJ, Lewis DJ (2005), 'Changes in the chemical form of selenium observed during the manufacture of a selenium-enriched sourdough bread for use in a human nutrition study', *Food Addit Contam*, **22**, 135–140.

Butterworth CE, Tamura T (1989), 'Folic acid safety and toxicity: a brief review', *Am J Clin Nutr*, **50**, 353–358.

Buttriss J (2004), 'Strategies to increase folate/folic acid intake in women: an overview', *Nutr Bull*, **29**, 234–244.

Caudill MA, Le T, Moonie SA, Torabian Esfahani S, Cogger EA (2001), 'Folate status in women of childbearing age residing in Southern California after folic acid fortification', *J Am College Nutr*, **20**, 129–134.

Conner RT (1941), 'Vitamin and mineral restoration and fortification from the point of the manufacturer', *Ind Eng Chem*, **33**, 711–713.
Czeizel AE, Dudas I (1992), 'Prevention of neural tube defects: results of the periconceptional vitamin supplementation', *N Engl J Med*, **327**, 1832–1835.
Dadone MM, Kushner JP, Edwards CQ, Bishop DT, Skolnick MH (1982), 'Hereditary hemochromatosis: analysis of laboratory expression of the disease by genotype in 18 pedigrees', *Am J Clin Pathol*, **78**, 196–207.
Darnton-Hill I (1998), 'Overview: rationale and elements of a successful food-fortification programme', *Food Nutr Bull*, **19**, 92–100.
Davidsson L, Kastenmayer P, Hurrell, RF (1994), 'Sodium iron EDTA (NaFe(III)EDTA) as a food fortificant. The effect on the absorption and retention of zinc and calcium in women', *Am J Clin Nutr*, **60**, 231–237.
Dietrich M, Brown CJ, Block G (2005), 'The effect of folate fortification of cereal-grain products on blood folate status, dietary folate intake, and dietary folate sources among adult non-supplement users in the US', *J Am Coll Nutr*, **24**, 266–274.
Edwards CQ, Griffen LM, Goldgar D, Drummond C, Skolnick MH, Kushner JP (1988), 'Prevalence of hemochromatosis among 11,065 presumably healthy blood donors', *N Engl J Med*, **318**, 1355–1362.
European Union (1993). Nutrient and Energy Intakes for the European Community. Reports of the Scientific Committee for Food (Thirty–first Series). Commission of the European Communities, Directorate-General Industry, Luxembourg.
FAO (Food and Agriculture Organization) (2002), 'Food fortification: technology and quality control', Economic and Social Department, FAO Food and Nutrition Paper 60, http://www.fao.org/docrep/W2840E/w2840E00.htm.
Flynn A, Cashman K (1999), 'Calcium', in: Hurrell R (Ed.), *The Mineral Fortification of Foods*, Surrey, UK, Leatherhead Food International, pp. 18–53.
Food and Drug Administration (1996), 'Food additives: folic acid, potassium iodide and kelp, final rule', *Fed Regist*, **38**, 20725.
Germani R, Aschieri JLR, Silva FT, Torrezan R, Lins e Silva K, Netto AG, Nutri MR (2001), *Manual de Fortificação de Farinha de Trigocom Ferro*, Rio de Janeiro, Embrapa Agroindustria de Alimentos.
Gujska E, Majewska K (2005), 'Effect of baking process on added folic acid and endogenous folates stability in wheat and rye breads', *Plant Foods Hum Nutr*, **60**, 37–42.
Hambly RJ, Saunders M, Rijken PJ, Rowland IR (2002), 'Influence of dietary components associated with high or low risk of colon cancer on apoptosis in the rat colon', *Food Chem Toxicol*, **40**, 801–808.
Hansen M, Samman S, Madsen LT, Jensen M, Sorensen SS, Sandstrom B (2001), 'Folic acid enrichment of bread does not appear to affect zinc absorption in young women', *Am J Clin Nutr*, **74**, 125–129.
Hertrampf E, Cortes F, Erickson JD, Cayazzo M, Freire W, Bailey LB, Howson C, Kauwell GPA, Pfeiffer C (2003), 'Consumption of folic acid fortified bread improves folate status in women of reproductive age in Chile' *J Nutr*, **133**, 3166–3169.
Honein MA, Paulozzi LJ, Mathews TJ, Ericksonn JD, Wong LYC (2001), 'Impact of folic acid fortification in the US food supply on the occurrence of neural tube defects', *J Am Med Assoc*, **285**, 2981–2986.
Hurrell RF, Bothwell T, Cook JD, Dary O, Davidsson L, Fairweather-Tait S, Hallberg L, Lynch S, Rosado J, Walter T, Whittaker P (2002), 'The usefulness of elemental iron for cereal flour fortification: A SUSTAIN Task Force report', *Nutr Rev*, **60**, 391–406.
Ilyas M, Khalil J, Ayub M, Khan S, Akhtar S (1996), 'The effect of iron fortification on the quality of fortified bread', *Sarhad J Agric*, **12**, 117–175.
Institute of Medicine, National Research Council (1998), *Dietary Reference Intakes: Folate, Other B Vitamins and Choline*. Washington, DC, National Academies Press.
Institute of Medicine, Food and Nutrition Board (2001), *Dietary Reference Intakes for Vitamin A, Vitamin K, Arsenic, Boron, Chromium, Copper, Iodine, Iron, Manganese,*

Molybdenum, Nickel, Silicon, Vanadium and Zinc. Washington, DC, National Academies Press, 2001.

Institute of Medicine (1997), *Dietary Reference Intakes for Calcium, Phosphorus, Magnesium, Vitamin D and Fluoride*. Washington, DC, National Academies Press.

Johnson Q (2000), 'Fortifying flour to improve world health', *World Grain*, November 7, http://www.worldgrain.com.

Johnston KE, Tamura T (2004), 'Folate content in commercial white and whole wheat sandwich breads', *J Agric Food Chem*, **52**, 6338–6340.

Jorde R, Bonna KH (2000), 'Calcium from dairy products, vitamin D intake, and blood pressure: the Tromso study', *Am J Clin Nutr*, **71**, 1530–1535.

Juma S, Sohn E, Arjmandi B H (1999), 'Calcium enriched bread supports skeletal growth of young rats', *Nutr Res*, **19**, 389–399.

Kajishima S, Pumar M, Germani R (2003), 'Efeito de adicao de diferentes sais de cálcio nas características da massa e na elaboracao de pao frances', *Cienc Tecnol Aliment*, **23**, 222–225.

Karadzhov G, Iserliyska D (2003), 'Sensory quality of minerally fortified bread', *Eur Food Res Technol*, **216**, 274–276.

Kiechl S, Willeit J, Egger G, Poewe W, Oberhollenzer F (1997), 'Body iron stores and the risk of carotid atherosclerosis: prospective results from the Bruneck study', *Circulation*, **96**, 3300–3307.

Kim YI (1999), 'Folate and carcinogenesis: Evidence, mechanisms, and implications', *J Nutr Biochem*, **10**, 66–88.

Koehler KM, PareoTubbeh SL, Romero LJ, Baumgartner RN, Garry PJ (1997), 'Folate nutrition and older adults: Challenges and opportunities', *J Am Dietetic Assoc*, **97**, 167–173.

Layrisse M, Chavez JF, Mendez-Castellano H, Bosch V, Tropper E, Bastardo B, Gonzalez E (1996), 'Early response to the effect of iron fortification in the Venezuelan population', *Am J Clin Nutr*, **64**, 903–907.

Leklem JE, Miller LT, Perera AD, Pefers DE (1980), 'Bioavailability of vitamin B-6 from wheat bread in humans', *J Nutr*, **110**, 1819–1828.

Lindenbaum J, Healton EB, Sabage DG, Brust JC, Garrett T, Podell ER, Marcell PD, Stabler SP, Allen RH (1988), 'Neuropsychiatric disorders caused by cobalamin deficiency in the absence of anaemia or macrocytosis', *N Engl J Med*, **318**, 1720–1728.

Lofti M, Mannar MGV, Merx RJHM, Naber-van den Heuvel P (1996), *Micronutrient Fortification of Foods: Current Practices, Research and Opportunities*. Ottawa, Ontario, The Micronutrient Initiative/International Agriculture Centre.

Loker GB (2004), 'Fortification of Turkish traditional bread with vitamins – minerals and evaluation in vulnerable group diets', *Asia Pac J Clin Nutr*, **13**, S160.

Lopez RD, Brown KH, Guinard JX (2002), 'Sensory trial to assess the acceptability of zinc fortificants added to iron fortified wheat products', *J Food Sci*, **67**, 461–465.

Maldonado A (2000), 'Mineral enhanced bakery products', US Patent 6126982.

Miki T, Fukatsu M, Harada S (1993a), 'Processed food made of iodine-enriched wheat flour', European Patent Application, EP 0 394 904 A2.

Miki T, Fukatsu M, Harada S (1993b), 'Processed food made of iodine-enriched wheat flour', US Patent 5232728.

Milne DB, Canfield WK, Mahalko JR, Sandstead HH (1984), 'Effect of oral folic acid supplements on zinc, copper, and iron absorption and excretion', *Am J Clin Nutr*, **39**, 535–539.

Mitchell HS (1941), 'What the consumer should know about fortified foods', *Ind Eng Chem*, **33**, 716–717.

Mocanu V, Stitt PA, Costan A R, Zbranca E, Luca V, Vieth R (2005), 'Long term efficacy and safety of high vitamin D intakes as fortified bread', *FASEB J*, **19**, A59.

Nabeshima EH, Ormenese R, Montenegro FM, Toda E, Sadahira MS (2005), 'Technological and sensorial properties of breads fortified with iron', *Cienc Tecnol Aliment*, **25**, 506–511.

Natri AM, Salo P, Vikstedt T, Palssa A, Huttunen M, Karkkainen MUM, Salovaara H, Piironen V, Jakobsen J, Lamberg-Allardt CJ (2006), 'Bread fortified with cholecalciferol increases the serum 25-hydroxyvitamin D concentration in women as effectively as cholecalciferol supplement', *J Nutr*, **136**, 123–127.

Nelson RL (1992), 'Dietary iron and colorectal cancer risk', *Free Rad Biol Med*, **12**, 161–168.

Nutriview (2005), 'Vitamin Information Centre – Micro Nutrient Intervention Programs', http://www.nutrivit.org/vic/staple/index.htm.

Park YK, Sempos CT, Barton CN, Vanderveen JE, Yetley EA (2000), 'Effectiveness of food fortification in the US: the case of pellagra', *Am J Public Health*, **90**, 727–738.

Parodi PW (2001), 'An assessment of the evidence linking calcium and vitamin D to colon cancer prevention', *Austr J Dairy Technol*, **56**, 38–58.

Postma MJ, Londeman J, Veenstra M, de Walle HEK, de Jong-van den Berg LTW (2002), 'Cost effectiveness of periconceptional supplementation of folic acid', *Pharmacy World Sci*, **24**, 8–11.

Poulsen J (1999), 'Danish consumers' attitudes towards functional foods', MAPP Working Papers **62**, http://www.asb.dk/centres/mapp.aspx.

Ranhotra GS, Loewe RJ, Lehman TA, Hepburn FN (1976), 'Effect of various magnesium sources on breadmaking characteristics of wheat flour', *J Food Sci*, **41**, 952–954.

Ranhotra GS, Lee C, Gelroth J (1980), 'Expanded cereal fortification: bioavailability and functionality (flavour) of magnesium in bread', *J Food Sci*, **45**, 915–917.

Ranhotra GS, Winterringer GL (1982), 'The use of magnesium powder in fortified bread', *Cereal Chem*, **59**, 446–448.

Ranhotra GS, Gelroth JA, Leinen SD, Schneller FE (1997), 'Bioavailability of calcium in breads fortified with different calcium sources', *Cereal Chem*, **74**, 361–363.

Ranhotra GS, Gelroth JA, Leinen SD (2000), 'Utilization of calcium in breads highly fortified with calcium as calcium carbonate or as dairy calcium', *Cereal Chem*, **77**, 293–296.

Ranum P (2006), 'Update on iron fortification of milled cereals', *Cereal Foods World*, **51**, 87–88.

Ranum P, Lynch S, Bothwell T, Hallberg L, Hurrell R, Whittaker P, Rosado J L, Davidsson L, Mannar V, Walter T, Fairweather-Tait S, Dary O, Rivera J, Bravo L, Reddy M, Garcia-Casal MN, Hertrampf E, Parvanta I (2001), *Guidelines for Iron Fortification of Cereal Food Staples*. Washington, DC, SUSTAIN, www.sustaintech.org.

Roncagliolo M, Garrido M, Walter T, Peirano P, Lozoff B (1998), 'Evidence of altered central nervous system development in infants with iron deficiency anemia at 6 mo: delayed maturation of auditory brainstem responses', *Am J Clin Nutr*, **68**, 683–690.

Rumi G, Imre L, Sulle C, Lassune MZ, Sarudi I, Kelemen J (1994), 'Selenium supplementation with bread', *Orv Hetil*, **135**, 2371–2372.

Russell RM, Baik H, Kehayias JJ (2001), 'Older men and women efficiently absorb vitamin B-12 from milk and fortified bread', *J Nutr*, **131**, 291–293.

Salovaara H (1982), 'Sensory limitations to replacement of sodium with potassium and magnesium in bread', *Cereal Chem*, **59**, 427–430.

Sandstead HH (2000), 'Causes of iron and zinc deficiences and their effects on brain', *J Nutr*, **130**, 347S–349S.

Serra-Majem L (2001), 'Vitamin and mineral intakes in European children. Is food fortification needed', *Public Health Nutr*, **4**(1A), 101–107.

Siegers CP, Bumann D, Trepkau HD, Schadwinkel B, Baretton G (1992), 'Influence of dietary iron overload on cell proliferation and intestinal tumorigenesis in mice', *Cancer Lett*, **65**, 245–249.

Stepanova EN, Shatnyuk LN, Verzhinskaya MF, Kostyleva MG, Bukina NA, Spirichev VB, Shukhnow AF, Kosteltseva NN, Bistrova AI, Emtseva IB (1988), 'Effect of method of vitaminization of high grade wheat flour bread on its content of thiamine, riboflavin and niacin', *Voprosy-Pitaniya*, **2**, 67–71.

Stevens RG, Jones DY, Micozzi MS, Taylor PR (1988), 'Body iron stores and the risk of cancer', *N Engl J Med*, **319**, 1047–1052.

Stevens RG, Graubard BI, Micozzi MS, Neriishi K, Blumberg BS (1994), 'Moderate elevation of body iron level and increased risk of cancer occurrence and death', *Int J Cancer*, **56**, 364–369.

Suojanen A, Raulio S, Ovaskainen ML (2006), 'Liberal fortification of foods: the risks. A study relating to Finland', *J Epidemiol Community Health*, **56**, 259–264.

Tetens I, Larsen TM, Kristensen MB, Hels O, Jensen M, Morberg CM, Thomsen A D, Hojgaard L, Henriksen M (2005), 'The importance of dietary composition for efficacy of iron absorption measured in a whole diet that includes rye bread fortified with ferrous fumerate: a radioisotope study in young women', *Br J Nutr*, **94**, 720–726.

Tobey JA, Cathcart WH (1941), 'Fortification and restoration in the baking and dairy industries', *Ind Eng Chem*, **33**, 714–716.

Ulrich CM, Georgiou CC, Snow-Harter CM, Gillis DE (1996), 'Bone mineral density in mother–daughter pairs: relations to lifetime exercise, lifetime milk consumption and calcium supplements', *Am J Clin Nutr*, **63**, 72–79.

Wald NJ (2004), 'Folic acid and the prevention of neural tube defects', *New Engl J Med*, **35**, 101–103.

Wang S, Tang YR (1995), 'Fortification of Ca and Zn in wheat flour', *J Zhengzhou Grain College*, **16**, 26–34.

Weaver CM, Heaney RP (1999), 'Calcium', in: Shils Mçe and Young VR (Eds), *Modern Nutrition in Health and Disease*. Baltimore, MD, Williams and Wilkins.

Weinberg ED (1994), 'Association of iron with colorectal cancer', *BioMetals*, **7**, 211–216.

Whittaker P, Dunkel V (1995), 'Iron, magnesium, and zinc fortification of food', in: Berthon G (Ed.), *Handbook of Metal-Ligand Interactions in Biological Fluids*, Vol 1. New York, Marcel Dekker, Inc.

Whittaker P, Tufaro PR, Rader JI (2001), 'Iron and folate in fortified cereals and folate in breakfast cereals', *J Am College Nutr*, **20**, 247–254.

Williams RR (1941), 'Fortification and restoration of processed foods', *Ind Eng Chem*, **33**, 718–720.

Yeung CK, Miller DD, Cheng Z, Glahn RP (2005), 'Bioavailability of elemental iron powders in bread assessed with an in vitro digestion/Caco–2 cell culture model', *J Food Sci*, **70**, S199–203.

Youdim MB, Ben-Shachar D (1987), 'Minimal brain damage induced by early iron deficiency: modified dopaminergic neurotransmission', *Isr J Med Sci*, **23**, 19–25.

Ziadeh G, Shadarevian S, Malek A, Khalil J, Haddad T, Haddad J, Toufeili I (2005), 'Determination of sensory thresholds of selected calcium salts and formulation of calcium-fortified pocket-type flat bread', *J Food Sci*, **70**, 548–552.

15

Omega-3-enriched bread

C. Hall III and M. C. Tulbek, North Dakota State University, USA

15.1 Introduction

Omega-3 lipids are a group of lipids that have a double bond between the third and fourth carbon from the terminal methyl group (Fig. 15.1). The most common plant omega-3 lipid is α-linolenic acid (ALA), whereas eicosapentaenoic acid (EPA) and docosahexaenoic acid (DHA) are the primary omega-3 lipids in fish and algae. However, other omega-3 lipids such as docosapentaenoic acid (DPA) and stearidonic acid have also been observed in fish and plant oils. The consumption of fish, as a source of omega-3 lipids, has been promoted for over 20 years by health professionals. However, a number of factors such as availability and cost of fish, and the lack of fish consumption by vegetarians may have contributed to the less than expected levels of consumption of fish. Finding alternative methods for delivering omega-3 lipids that are low cost is likely to increase the consumption of these lipids.

The addition of omega-3 oil to food systems will inherently make the product more susceptible to oxidation. Thus, one of the biggest challenges in the development of omega-3 fortified products is to maintain the shelf stability of the products. Omega-3 fortified products exist on the market; however, only products that are stored cold have the omega-3 lipid added directly to the product formulae. The application of omega-3 lipids to bread systems becomes more complicated because the oil can affect the production of the bread and the baking operation can promote the instability of the oil. Strategies have been developed that allow for the fortification of omega-3 lipids in breads.

This chapter will highlight the fortification of bread and bakery products as an alternative method for delivering omega-3 to humans. The chapter will

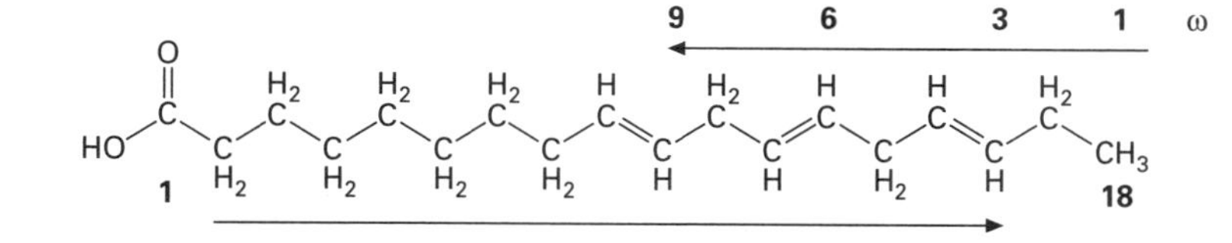

cis-9, 12, 15-Octadecatrienoic acid (α-linolenic acid): 18 : 3n–3

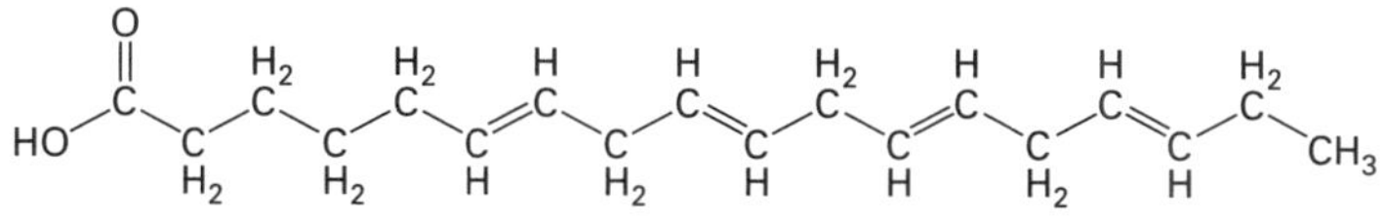

cis-6, 9, 12, 15-Octadecatetraenoic acid (stearidonic acid): 18 : 4n-3

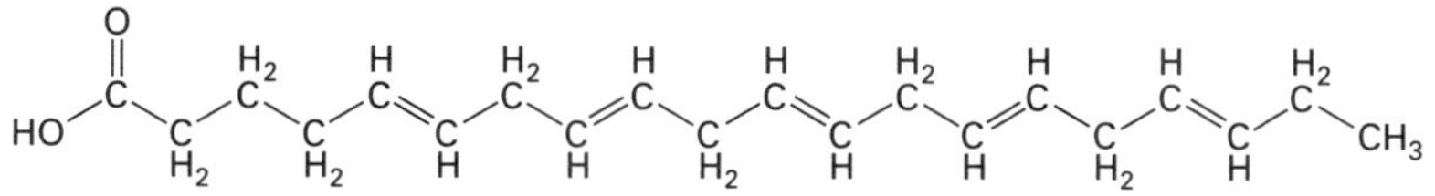

cis-5, 8, 11, 14, 17-Eicosapentaenoic acid (EPA) : 20 : 5n-3

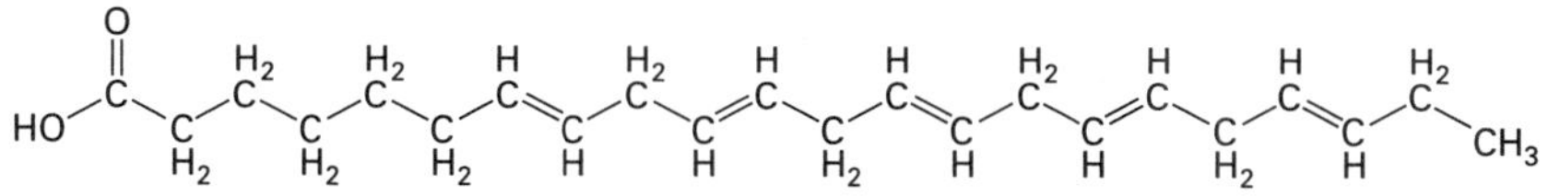

cis-7, 10, 13, 16, 19-docosapentaenoic acid (DPA): 22 : 5n-3

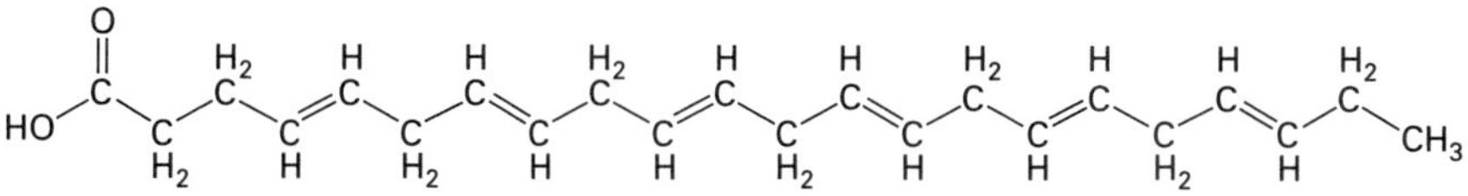

cis-4, 7, 10, 13, 16, 19-docosahexaenoic acid (DHA): 22 : 6n-3

Fig. 15.1 Structures of the omega-3 fatty acids. The bold numbering above the first fatty acid represents the omega (ω) labeling format, whereas the numbering under the first fatty acid represents the IUPAC (International Union of Pure and Applied Chemistry) number method.

first highlight the types and sources of omega-3 lipids, the primary health benefits of these lipids and consumer attitudes toward omega-3 lipids. Section 15.4 will focus on the development and manufacturing of omega-3 fortified breads. In this section, issues such as oxidative stability and the influence of omega-3 sources on dough rheology and on bread characteristics will be discussed. The chapter will conclude with future trends and additional sources of information. In writing this chapter, the authors found that limited peer-review articles exist in the literature on the topic of omega-3 fortification in bread. Thus, a significant focus of the chapter deals with work conducted in the authors' laboratory on flaxseed in bread.

15.2 Sources and health benefits of omega-3 lipids

15.2.1 Sources of omega-3 lipids

Omega-3 lipids are broadly grouped into two categories, namely fish and plant. The two groups can better be defined by the type of omega-3 present in the tissue. In fish, including microalgae, long-chain omega-3 lipids having 20–22 carbons and 5–6 double bonds predominate (Fig. 15.1). The short-chain omega-3 lipids having 18 carbons (i.e. ALA) predominate in plant oils (Fig. 15.1). Some genetically modified plants can produce long-chain omega-3 lipids, which will be discussed further under the future trends section of this chapter.

Aside from fish consumption, the major dietary source of fish oil has been in the form of capsules as a dietary supplement. The composition of fish omega-3 lipids varies significantly ranging from 300 to 2500 mg/100 g for cod and mackerel, respectively (Holub, 2002; USDA, 2006). Thus, the oil derived from fish will also provide varying omega-3 contents. The US Department of Agriculture (USDA, 2006) nutrient database shows that oils such as herring or menhaden contain approximately 141 mg of omega-3 lipids per gram of oil whereas salmon oil contains 381 mg omega-3 lipids per gram of oil (Table 15.1). Fatty acid variation in other fish has been reported (Bimbo, 1998).

An alternative source of long-chain omega-3 lipids comes from microalgae. Similar to fish, large variations in the oil content and omega-3 lipid composition, in both commercial and research oils, can be observed (Metting, 1996; Mansour *et al.*, 2005; Ward and Singh, 2005; Spolaore *et al.*, 2006). Metting (1996) reported that the lipid content of microalgae ranged from 1% to 70% on a dry weight basis. Mansour *et al.* (2005) evaluated nine species, representing four algae classes, grown under similar conditions for omega-3 production. These authors (Mansour *et al.*, 2005) found that the EPA and DHA contents ranged from 3.3% to 29.6% and 0.4% to 24.7% of the algae oil, respectively. From these data, it is evident that variation in fatty acid composition in microalgae is common; thus, the application of different sources of microalgae oil may not provide the same composition of omega-3 fatty acids.

Table 15.1 Concentrations of the omega-3 fatty acids in fish oil[a]

	Fatty acid content (mg/g)[b]					
Fish oil	ALA (18:3)	STA (18:4)	EPA (20:5)	DPA (22:5)	DHA (22:6)	Total
Herring	7.63	23.05	62.73	6.19	42.06	141.66
Cod liver	9.35	9.35	68.98	9.35	109.68	206.71
Sardine	13.27	30.25	101.37	19.73	106.56	271.18
Menhaden	14.9	27.39	131.68	49.15	85.62	308.74
Salmon	10.61	27.98	130.23	29.91	182.32	381.05

[a] Data based on the USDA nutrient database. Data were converted to a mg/g oil format from g/100 g.
[b] Eicosatetraenic acid (20:4n-3) is not included because this fatty acid is not differentiated from arachidonic acid (20:4n-6) in the calculation of the 20:4. See text for other abbreviations.

Yan (2003) patented an encapsulation method that could be used to enhance the stability of omega-3 lipids. In this method, a multi-core approach was taken by first encapsulating the omega-3 lipids and then applying a second shell over the agglomerated encapsulated lipids. This approach provides a product that has multi-layers of protection. Currently, Ocean Nutrition Canada uses this technology under the Powder-loc trademark to produce stable omega-3 encapsulated powders. The powder that results from this process can deliver 500–800 mg oil per gram of dry powder. In contrast, 200–350 mg oil per gram of dried powder is typically delivered for conventional spray-dried encapsulated powders (Barrow, 2005). The Wright Group (Crowley, LA) uses a proprietary technology called Super Micro Atomization Retention Technology (SMART) to produce SuperCoat® omega-3 products. The SuperCoat® encapsulation technology uses multiple component encapsulants to provide omega-3 stability. Companies providing powdered omega-3 products can be found in Table 15.2. Unlike fish and microalgae oils, plant oils containing ALA are not typically encapsulated due to the larger availability of the oilseeds and plant oils, which make the oils inexpensive compared with fish or microalgae oils. Thus, the economics of the encapsulation process are not offset by the initial high value of the oil as in the case of the fish oil. However, the whole or ground seed of the plant can be added as part of a bread formula and thus stability is of lesser concern than with the addition of oil directly.

There are over 20 different oilseeds and nuts that contain ALA. However, only about ten are commercially viable sources of ALA. Flaxseed or linseed (*Linum usitatissimum*) is an oilseed that contains roughly 38–45% oil and is one of the leading sources of ALA (Hauman, 1998). ALA accounts for 50–57% of the fatty acids in the oil (Daun *et al.*, 2003). Other common food oils, canola and soybean oils contain approximately 6.5–14.1% and 5.5–9.5%

Table 15.2 Examples of companies selling encapsulated omega-3 oils

Company	Brand name or product
Arista Industries, Inc., Wilton, CT, USA	Powdered tuna oil using maltodextrins
Bioriginal, Saskatoon, SK, Canada	Powdered DHA concentrate
Europharma, Inc., Green Bay, WI , USA	Vectomega – dry powder
Martek Biosciences, Columbia, MD, USA	35% DHA in oil or encapsulated form (microalgae)
National Starch, Bridgewater, NJ, USA	NOVOMEGA™, encapsulated menhaden oil
Nu-Mega Ingredients Pty Ltd, Queensland, Australia	Driphorm® HiDHA® microencapsulated tuna oil
Ocean Nutrition Canada, Dartmouth, NS, Canada	MEG-3™ brand omega-3 EPA/DHA fish oil
Roche, Basel, Switzerland	ROPUFA®
The Wright Group, Crowley, LA, USA	2:1 EPA to DHA, and 2:1 DHA to EPA powders

ALA, respectively (Budin *et al.*, 1995; Hauman, 1998). Several plant species have high levels of ALA but are not widely available. *Camelina sativa* (camelina) is an oilseed plant of the family Cruciferae. This plant is sometimes referred to as false flax but has plant characteristics similar to other Cruciferae plants such as mustard and canola. However, camelina contains more than twice the amount of ALA compared with canola. Camelina contains approximately 35% oil of which ALA accounts for 31–39% of the oil (Budin *et al.*, 1995; Fröhlich and Rice, 2005). Another oilseed with excellent ALA composition is the chia (*Salvia hispanica* L.) seed. This plant is a member of the mint family and is commercially produced in Argentina. Seeds contain between 28% and 33% oil with ALA contents in the range of 54–64% of the oil (Coates and Ayerza, 1998). *Perilla frutescens* (L.) Britton is an oilseed that has also attracted attention for its high ALA content (Brenner, 1993). Perilla seed yields approximately 38–48% oil, of which ALA accounts for 61–64% of the oil (Shin and Kim, 1994). Hemp (*Cannabis sativa* L.) seed oil contains approximately 15–25% ALA (Deferne and Pate, 1996; Oomah *et al.*, 2002). Although this plant is heavily regulated, Canada has allowed the cultivation of low δ-9-tetrahydrocannabinol (THC) cultivars. The regulatory restrictions placed on this plant will probably limit the use of hempseed oil in breads.

Other sources of ALA include blackcurrant (*Ribes nigrum* L.) seed oil (Castillo *et al.*, 2002, 2004), alfalfa seed oil and fenugreek seed oil (Hauman, 1998), which contain approximately 10–19%, 11–32% and 14–22% ALA, respectively. Basil seed oil is also high (43–65%) in ALA (Angers *et al.*, 1996). Although these oils contain relatively high levels of ALA, they are not currently exploited commercially due to the abundance of other oils. In addition to oilseed, nuts or nut oils can provide ALA. Walnut oil contains approximately 11–16% ALA (Amaral *et al.*, 2004) which is the highest content among nuts. Butternuts (8.7%) and black walnuts (3.8%) are the only other nuts with significant ALA contents (Hauman, 1998).

Both animal and plant sources of omega-3 can vary, as demonstrated by the range of ALA in plants and long-chain omega-3 in fish and microalgae previously discussed. Many of the oils are not standardized; thus, a detailed breakdown of the fatty acids beyond the label polyunsaturated should be included in product specifications. The various omega-3 lipids function differently in disease prevention, thus knowledge of the fatty acid profile is essential when adding omega-3 products to bread for the purpose of disease prevention.

15.2.2 Consumption and health benefits of omega-3 lipids

A number of researchers have reported summaries of dietary omega-3 intakes (Sinclair *et al.*, 1994; Kris-Etherton *et al.*, 2000; Sanders, 2000; Gebauer *et al.*, 2006). In general, there is a lack of omega-3 consumption by individuals consuming a Western diet. The consumption rates of long-chain omega-3 are approximately 100–190, 100–500 and 100–200 mg/day in Australia (Sinclair *et al.*, 1994), the United Kingdom (Sanders, 2000), and the United States and the Netherlands (Gibney, 1997; Kris-Etherton *et al.*, 2000; Gebauer *et al.*, 2006), respectively. The omega-3 intake in the European Union varies widely. The majority of the population have intakes in the range of 300–820 mg/day whereas in Spain and Portugal the daily intakes are higher at 1440 and 1840 mg/day, respectively (Gibney, 1997). In contrast to the low consumption of long-chain omega-3 lipids, ALA consumption is closer to the recommended intake of 1000–2000 mg/day (Gibney, 1997; Kris-Etherton *et al.*, 2000). Sanders (2000) reported that, despite repeated messages of the benefits of omega-3, the intake of EPA and DHA by men did not increase significantly between 1980 and 1992. In contrast, an increase in ALA intake was observed, which was most probably due to the increased consumption of canola oil. Using baseline consumption data, various organizations have proposed minimum recommended omega-3 intakes.

The recommended intake of omega-3 lipid has been widely debated. Since 1989 a number of organizations have proposed omega-3 intakes of between 1100 and 3800 mg/day (Gebauer *et al.*, 2006). The US Institute of Medicine (2002) and Health Canada (Morris, 2003a) recommend ALA intakes of 1100 and 1600 mg/day (1.1 and 1.6 g/day) for women and men, respectively. Further recommendations of 1.3 and 1.4 g/day have been extended to lactating and pregnant women, respectively. Healthy individuals who have no signs of ALA deficiency are the basis for ALA recommendations, which are less variable than recommendations for EPA and DHA.

Within the United States, recommendations for EPA and DHA combined vary significantly. Intakes of 140–1000 mg of long-chain omega-3 lipids per day have been proposed by the American Heart Association (Kris-Etherton *et al.*, 2002) and the US Institute of Medicine (2002), respectively. Differences in these values are based on the viewpoints of disease prevention by the AMA and the lack of deficiency symptoms by the US Institute of Medicine (2002). The European Society of Cardiology (Van de Werf *et al.*, 2003) and World Health Organization (2003) support higher consumptions (1000 mg/day) of EPA and DHA. The International Society for the Study of Fatty Acids and Lipids (2004) supports a consumption of greater than 400 mg of long-chain fatty acids. The general consensus is that at least 400 mg of long-chain omega-3 lipids are needed for disease prevention. Fortification of foods with omega-3 is the approach most likely to supply the necessary dietary omega-3 lipids; this is also the delivery method that is most convenient for consumers.

Lovegrove *et al.* (1997) reported that fortified foods such as bread and biscuits could be a means of delivering dietary omega-3 lipids. They found that the consumption of long-chain omega-3 fortified foods increased plasma and phospholipid EPA and DHA levels. In this particular report multiple foods were evaluated and the increase in long-chain omega-3 lipids could not be attributed to bread alone. Metcalf *et al.* (2003) reported that omega-

3 could be enhanced significantly through dietary intervention that included foods such as flaxseed, canola oil and fish oil fortified products, for example, milk and margarine. The enhanced omega-3 intake proved beneficial as higher concentrations of EPA and DHA were incorporated into phospholipid pools, mononuclear cells and platelets. Higgins *et al.* (1999) concluded that microencapsulated fish oil was biologically available based on data that showed a significant increase in plasma EPA and DHA levels after a microencapsulated fish oil was consumed. Wallace *et al.* (2000) found that platelet arachidonic acid (AA), EPA and DHA did not differ significantly between the subjects fed fish oil capsules or food enriched with microencapsulated fish oil. These data further support the bioavailability of omega-3 lipids from foods containing microencapsulated fish oil. Fortification of foods with microencapsulated fish oil, therefore, is an effective way of increasing n-3 polyunsaturated fatty acid (PUFA) intakes so that they are in line with current dietary recommendations. Microencapsulated products are the most likely delivery mechanism of fish oil in breads and bakery products. However, studies involving the effect of omega-3 fortification in breads on various biomarkers are limited.

Increased long-chain omega-3 lipids were also observed on plasma phospholipids in subjects fed bread containing 1 g of stabilized fish oil ESKIMO-3® (Cardinova, Uppsala, Sweden) for 2 and 4 weeks (Saldeen *et al.*, 1998). At 4 weeks, these researchers (Saldeen *et al.*, 1998) observed a 17% reduction in serum triacylglycerols in patients fed the bread containing fish oil. Feeding of bread containing 1.3 g of stabilized fish oils was found to enhance plasma high density lipoprotein (HDL) content in subjects in 4 weeks (Liu *et al.*, 2001). These researchers (Liu *et al.*, 2001) also found that the subjects fed the bread containing the 1.3 g of fish oil had lower plasma triacylglycerol contents compared with subject fed the control breads. Yep *et al.* (2002) reported a significant increase in the proportion of DHA in the plasma triacylglycerols 2 and 4 hours after consumption of bread containing 80 mg of omega-3 lipids from encapsulated tuna oil. In contrast, no significant increase in the proportion of EPA in the plasma triacylglycerols was observed. The chronic (over 4 weeks) intake of a low level (60 mg) of long-chain omega-3 lipids via bread showed significant increases in both EPA and DHA in the plasma triacylglycerols and phospholipids (Yep *et al.*, 2002). These studies demonstrate the bioavailability of omega-3 from products containing stabilized or encapsulated fish oil. Unlike fish oils, plant-based omega-3 fortification using flaxseed or nuts can be achieved using ground or chopped products without protection using encapsulation.

Cunnane *et al.* (1993) found that a daily consumption of ground flaxseed (50 g) or flaxseed oil (20 g) enhanced plasma and erythrocyte ALA contents in test subjects compared with the control diets. No differences in plasma ALA contents were found in subjects fed 50 g flaxseed or 20 g flaxseed oil. Thus, when evaluating plasma ALA, the form in which ALA is consumed may not be as important as the level because 20 g flaxseed oil and 50 g flaxseed have equivalent ALA amounts (12 g). However, a decreased postprandial glucose response was found in subjects fed the 50 g flaxseed diet suggesting that soluble fiber has a positive impact on health and supports the consumption of flaxseed as a source of n-3 fatty acids.

Similar to flaxseed, walnut consumption had a positive effect on plasma fatty acids (Almario *et al.*, 2001). They reported a significant increase in plasma ALA content. Positive alteration in lipoprotein composition was also observed (Almario *et al.* 2001). Kris-Etherton *et al.* (1999) and Simopoulos (2004) provided overviews of the role of nuts in disease prevention. In these reviews, the reduction in total and low-density lipoprotein (LDL) cholesterol and alteration in lipoprotein types were the primary benefits of nut consumption.

Collectively, the data support the bioavailability of omega-3 lipids from both plant and marine sources. The role of these lipids in disease prevention has been reviewed extensively (Nestel, 1990; Sanders, 2000; de Lorgeril *et al.*, 2001; Hu *et al.*, 2001; Kris-Etherton *et al.*, 2002; Lee and Lip, 2003; Holub and Holub, 2004; Gebauer *et al.*, 2006; Minihane, 2006). The main focus of these reviews was cardiovascular disease prevention by EPA and DHA, as the body of knowledge regarding the benefits is extensive. ALA was, in most cases, not considered a major contributor to cardiovascular disease prevention. However, ALA must be considered as an important dietary constituent based upon the potential mechanisms by which omega-3 lipids can prevent cardiovascular disease. Some of these mechanisms include suppression of inflammatory cytokines, reducing apolipoprotein B formation, blocking platelet activation factor, improving blood vessel flexibility and prevention of atherosclerosis (Connor, 2000; Holub, 2002). Additional sources of information on the health benefits of flaxseed can be found in reviews by Hall et al. (2006) and Morris (2003 a, b).

Atherosclerosis is a disease that is characterized by deposition and accumulation of cholesterol and other blood lipids in blood vessel walls. Factors affecting the development of atherosclerosis include high cholesterol, inflammatory compounds such as eicosanoids and cytokines, and reactive oxygen species (Ross, 1999). Caughey *et al.* (1996) reported that flaxseed oil inclusion into the diet of healthy volunteers resulted in 30% reduction in the production of the cytokines tumor necrosis factor α (TNF α) and interleukin 1-β (IL-1β). These cytokines were inversely related to EPA levels in mononuclear cells. An intake of 1.8 g EPA and DHA/day and 9.0 g ALA/day over 4 weeks resulted in a 20%, 26% and 36% reduction in IL-1 β, prostaglandin E_2 (PGE_2) and thromboxane B_2 (TXB_2), respectively (Mantzioris *et al.*, 2000). Zhao *et al.* (2005) also reported that ALA could reduce inflammatory markers in human monocytic THP-1 cells.

Zhao *et al.* (2004) reported that a high-ALA diet was significantly better than an average American diet in controlling inflammatory markers in 23 subjects. Serum C-reactive protein was also improved in subjects on the high-ALA diets compared with the average American diet (i.e. high-LA diet). Vascular cell adhesion molecule-1 and E-selectin were lower in subjects on the ALA diet when compared with the LA diet. These parameters were inversely related to serum EPA and to a lesser extent DPA. The conclusion of the study was that the reductions of cardiovascular disease by ALA occurs through several mechanisms based on the findings that the high ALA affected both lipids/lipoproteins and C-reactive protein/cell adhesion molecular parameters (Zhao *et al.*, 2004). Paschos *et al.* (2005) found similar reductions in inflammatory markers in dyslipidemic males fed 15 ml of flaxseed oil per day. These researchers (Paschos *et al.*, 2005) also observed that the degree of the anti-inflammatory response depended on the subject's apolipoprotein E genotype.

Wilkenson *et al.* (2005) found that dietary fatty acid intake could affect biological lipid status. ALA and EPA levels in the plasma erythrocytes increased by 225% and 150%, respectively, in subjects fed a high-ALA diet (i.e. 0.5 to 1 LA:ALA ratio). Hussein *et al.* (2005) also reported similar findings in 57 male subjects fed high-ALA diets (17 g/day). A 12.4% reduction in total plasma cholesterol and a 10% reduction in HDL cholesterol were observed in patients on the flaxseed oil diets. Clandinin *et al.* (1997) found that fish oil significantly lowered plasma triacylglycerols compared with flaxseed and olive oils. However, total cholesterol and LDL levels in humans were slightly lower, but not significantly, in flaxseed and olive oil diets compared with fish oil diets. In a placebo-controlled, parallel study involving 150 hyperlipidemic subjects, a significant reduction was observed in fasting plasma triacylglycerols after 2 months when subjects were given 1.7 g EPA + DHA/day (Finnegan and Minihane, 2003). However, after 6 months on this diet the significance in the reduction diminished. The 9.5 g ALA/day significantly increased the EPA concentrations in plasma phospholipids without significantly altering plasma triacylglycerols (Finnegan and Minihane, 2003).

The knowledge regarding the role of ALA in disease prevention is growing (Morris, 2003a,b; Gebauer *et al.*, 2006). Recent epidemiological reports (Baylin *et al.*, 2003; Albert *et al.*, 2005; Djousse *et al.*, 2005) further support a role of ALA in disease prevention. Mozaffarian *et al.* (2005) completed a 14-year follow-up involving 45 722 men in which they concluded that both plant and seafood omega-3 fatty acids could reduce the risk of coronary heart disease. In the light of the recent findings, ALA should be considered as part of the solution, along with EPA and DHA, regarding cardiovascular disease prevention (de Lorgeril and Salen, 2004).

15.3 Consumer awareness, potential market, products and claims

15.3.1 Consumer awareness

The awareness of omega-3 lipids has been growing recently due in part to the introduction of omega-3 fortified dairy products and infant formulas.

Menrad (2003) reported that between 25% and approximately 40% of consumers in Northern Europe were aware of the health benefits of omega-3 lipids. Gerdes (2005) reported that 58% of consumers participating in a Gallup poll were aware of omega-3 fatty acids. Recently, the International Food Information Council Foundation (IFIC, 2006) reported that 63% and 76% of Americans were aware of omega-3 lipids and fish oil, respectively. However, fish oil has a less favorable connotation than omega-3 (Molyneaux, 2006). Differentiation of the specific fatty acids by consumers was far less successful (IFIC, 2006). Although consumer differentiation of the specific fatty acids is currently low, one would expect that to change over time. For example, the IFIC (2006) reported that 91% of consumers were aware of saturated fats. Considering that the message to reduce saturated fats has been promoted for nearly 20 years, this suggests that consumers over time will become more knowledgeable about the specific fatty acids.

15.3.2 Potential market

Dornblaser and Jago (2006) reported that new introductions of omega-3 fortified products were highest in Europe (36% of the new omega-3 products), followed by North America (28%) and Asia-Pacific (21%). McGuigan (2005) reported that Frost & Sullivan estimated that the US omega-3 and omega-6 ingredients market is valued at $204.5 million. By 2010 the market is expected to reach $524 million with 65% of the market being marine oils. The European marine oil market in 2004 was valued at just over $264 million (McGuigan, 2005). As in the United States, the entire omega-3 market is expected to increase 8% annually until 2010. Although this growth is expected for all sectors, the bakery industry could be an important user of omega-3 products.

Benkouider (2004) reported that Euromonitor International estimated that functional breads will account for 2% of functional bakery and snack products by 2008. In Germany, functional breads are expected to grow by 15% by 2008 whereas functional breads in the United Kingdom are expected to remain flat. In the United States, omega-3 fortified breads will most probably remain a niche market (Gelski, 2006). However, Gelski (2006) reported that ACNielsen estimated the sales of omega-3 breads at $10.6 million in 2005. Although dollar values decreased from the previous year, volume actually increased by more than 350 000 units. The awareness of the value of omega-3 consumption may drive unit sales in the future.

15.3.3 Products and claims

The market for omega-3 fortified breads has matured in the sense that a number of bread products have been introduced worldwide. Kolanowski and Laufenberg (2006) presented a shortlist of 12 products from European countries. Many omega-3 enhanced products can be found using a simple internet search. A number of bakeries are now offering multi-ingredient breads with omega-3 as a primary focus (Gelski, 2006).

In 2004, the Food and Drug Administration (FDA, 2004) approved a qualified health claim addressing omega-3 fatty acids and coronary heart disease. The FDA approved claim reads as follows: 'Supportive but not conclusive research shows that consumption of EPA and DHA omega-3 fatty acids may reduce the risk of coronary heart disease. One serving of [name of food] provides [x] grams of EPA and DHA omega-3 fatty acids. [See nutrition information for total fat, saturated fat and cholesterol content.]' To be eligible for this claim, conventional foods and dietary supplements must contain EPA and DHA omega-3 fatty acids. To use the claim, other requirements have been identified related to the saturated fat, cholesterol, sodium and vitamin contents (FDA, 2004). The United Kingdom has also supported the use of a health claim in foods containing EPA and DHA. The Joint Health Claims Initiative (JHCI, 2005) supported the statement 'Eating 3 g weekly, or 0.45 g daily, long-chain omega-3 polyunsaturated fatty acids, as part of a healthy lifestyle, helps maintain heart health.' Similar to the FDA qualified claim, specific criteria must be met before the claim can be used (JHCI, 2005).

The use of nutrient claims is also permitted by the FDA for ALA. The claim can state the level such as 'high in', 'excellent source of' – but must specify the type of omega-3 present. Furthermore, at least 130 mg of EPA, or DHA, or 260 mg ALA per reference amount customarily consumed must be present in the product to meet the terms of the claim. Specific requirements are also stated for saturated fat and cholesterol.

15.4 Development and manufacturing technology

15.4.1 Methods for manufacturing breads – ingredient technology

Omega-3 supplements and ingredients are accepted as optional ingredients in bread making; however, these ingredients are most associated with nutrient enhancement. Gerdes (2005) discussed the use and addition of omega-3 fatty acids in functional bread products. Whole flaxseed, milled flaxseed, flaxseed oil, powdered microencapsulated omega-3 fatty acids, tuna oil and menhaden oil were given as potential ingredients. These ingredients can alter the physical and chemical properties of the baking system, and these effects should be determined prior to large-scale production.

Several concerns associated with adding omega-3 ingredients to bread formulae include the effect of the omega-3 on the physical properties of the dough and potential oxidation of the lipid during and after baking. The physical properties of the omega-3 oils are quite different compared with solid fats (i.e. shortenings), which are the preferred lipids in baking. The benefits of shortenings in breads have been well documented. The solid nature of the shortening has been identified as one reason for the improved functionality over liquid oils. Increasing the solid fats (i.e. hydrogenated soybean oils) content in an emulsion system used in a bread formula led to higher loaf volumes and lower rates of staling (Smith and Johansson, 2004). Inoue *et al.* (1995) observed similar findings in frozen dough systems. They stated that high maximum resistance and low extensibility in non-frozen dough made with a formula that contained canola oil was likely due to the inability of canola oil to lubricate the gluten and dough. Sluimer (2005) indicated that saturated fats with melting temperatures near 30 °C were optimal. Flaxseed oil, with an ALA content of 50–55% (Hall, 2002), has a melting point of approximately 0.1–2 °C (Firestone, 1999). Menhaden oil has a similar melting point of approximately 0.6–1.6 °C (Firestone, 1999). Thus, both oils would exhibit poor lubricating properties in dough systems resulting in behaviors observed by previous researchers (Inoue *et al.*, 1995; Smith and Johansson, 2004).

In contrast to the lubricating effects of shortenings, oil (i.e. liquid at room temperature) that exists in the triacylglycerol form can compete with protein during foam formation resulting in a reduced loaf volume. MacRitchie and Gras (1973) and MacRitchie (1977) observed that the increasing levels of neutral lipids (i.e. triacylglycerols) caused a reduction in loaf volume. The liquid oil probably competes with the protein for the interface between the gas and dough matrix (Junge *et al.*, 1981), thus reducing the loaf volume. Furthermore, the liquid oils are more likely to disperse more rapidly than the solid fats causing greater interactions with the hydrophobic regions of the proteins; thus resulting in a greater disruption of the foam interface, and a reduction in the lubricating effect that occurs with shortenings during gluten formation.

The fact that microencapsulation of flaxseed oil has limited market interest and that oil could interfere with dough development suggests that different ALA sources, other than the oil, are needed. Milled flaxseed would be a significant ingredient for ALA fortification in baking applications. Other oilseed flours could also be the source of omega-3 lipids. Milled flaxseed has a moisture content of 7–8% (Tulbek *et al.*, 2004) and a particle size range of 450–700 μm (Schorno *et al.*, 2004); this might generate blending issues with wheat flour, which has an average particle size of 149 μm. Milled flaxseed is generally classified as 20 mesh (600 μm) and 30 mesh (850 μm) mesh for baking applications. Milled flaxseed and wheat flour should be blended sufficiently before mixing with water and other ingredients. Garden (1993) reported on the bread making properties of wheat flour blended with ground flaxseed. She observed detrimental effects on bread properties of hard red spring wheat in the presence of flaxseed. Gerdes (2005) suggested that finer ground flaxseed should be used in bread formulations.

In addition to the physical properties, ground flaxseed significantly reduced crust and crumb color scores. This observation might be related to the Maillard browning reaction, which could possibly be caused by the reaction between flaxseed proteins and carbonyl compounds such as oxidized phenols (Garden, 1993). These results were consistent with experiments conducted in our laboratories (Hall *et al.*, unpublished data, 2005). We observed increased crust browning in bread as the flaxseed concentrations increased up to 20%. We speculated that the gum fraction (i.e. carbohydrate) was responsible for the increased browning via interaction with the wheat proteins. We also observed that milled flaxseed levels affect bread quality in terms of fermentation schedule where shorter fermentation times indicated better bread quality. Experiments conducted at Northern Crops Institute (NCI) laboratories indicated that 13% flaxseed addition significantly decreased proof time for bagel dough and produced bagels that were softer in texture than non-flaxseed bagels. In contrast, 10.2% flaxseed addition to oatmeal bread dough increased fermentation time by 15 minutes compared with dough with no flaxseed. The increased fermentation time may have been related to the removal of shortening from the flaxseed fortified formula. In some formulae containing flaxseed, additional yeast and vital wheat gluten may be required. Shorter mixing times could be implemented with flaxseed. NCI baking tests indicated higher levels of water absorption, which would have an impact on mixing and dough handling properties.

Bimbo (1989) reported the use of menhaden oil in biscuit dough, bread fats, emulsions and bread dough systems. The oil, compared with shortening, would be inferior for bread making applications (Sluimer, 2005) in terms of dispersibility, which might affect the oxidative properties, since the oil would be exposed to air during mixing. Considering that long-chain omega-3 lipids are about three times more reactive at 37 °C to the oxidation process than ALA (Cosgrove *et al.*, 1987), the rate of oxidation would be more rapid for fish oil application even during the mixing operation. Thus, oil should be added to the dough system as late as possible. Gerdes (2005) reported that menhaden oil was stable at 350 °F for 20 minutes. Furthermore, the oil might be incorporated as a solid fat, which could improve its heat stability at 375 °F for 40 minutes. However, hydrogenating the fish oil into a solid fat would reduce the level of omega-3, thus losing the health benefits of the omega-3. The stability of fish oil fortified bread has not been investigated extensively and additional research is needed to support stability claims.

Microencapsulated omega-3 oils containing EPA and DHA are currently being manufactured (Table 15.2). Microencapsulation of the oil would protect the oil against shear, temperature and oxygen during bread making. Thus, this would minimize the changes in flavor and odor, and eliminate the dispersibility issues of adding the fish oil directly to the formula. Powdered omega-3 fatty acids have a fine particle size range (100–149 μm) similar to flour, which would assist blending and mixing processes. Morita and Shirai (1997) reported that DHA powder can be successfully used in bread systems without any decomposition. In the last decade there has been significant research in the area of encapsulation, thus encapsulated products are becoming the primary source of EPA and DHA ingredients for the baking industry.

15.4.2 Physical dough parameters of bread formulae fortified with omega-3 lipids

The physical properties of dough are related to bread making quality, which influences loaf volume, specific volume, oven spring, break and shred, grain and bread firmness properties. Process optimization and target omega-3 weight per serving are imperative factors for breads fortified with omega-3 lipids. Garden (1993) reported that the dough strength of hard red spring wheat flour dough was stable with up to 5% milled flaxseed in the formula. After that, vital wheat gluten should be added. Farinograph absorption significantly ($P \leq 0.05$) increased, and dough mixing time (farinograph peak time) increased with milled flaxseed addition. As the flaxseed addition increased, so did the impact on these measured parameters. The observed dough characteristics also extended baking time. In contrast, a weakening in the farinograph curve was observed due to decrease in dough stability when over 5% milled flaxseed was added. Flaxseed flour and oat bran were utilized as a fat replacer in cake without any detrimental effects (Lee *et al.*, 2004). Flaxseed flour addition at 20% decreased viscosity; however, an increase in cake volume was observed. Flaxseed flour addition significantly ($P \leq 0.05$) darkened crumb color and increased yellowness values. Results indicated that flaxseed flour was an acceptable additive when used with Nutrim oat bran.

Ghosh *et al.* (2004) reported the effects of ground flaxseed on wheat flour tortilla quality. Flaxseed addition significantly increased water absorption and decreased dough strength, as observed by Garden (1993). Brightness values significantly decrease with flaxseed addition due to the presence of specks. Tensile properties of fresh tortillas containing 15% or more ground flaxseed were significantly higher than those of control and lower-level flaxseed treatments (5% and 10%). In contrast, tensile property tests did not show any significant differences after 10 days' storage.

In a study conducted in the authors' laboratories, the physical dough properties of flaxseed in hard red spring wheat flour dough systems were analyzed by Mixolab. Mixolab is a new piece of dough testing equipment recently developed by Chopin (Villeneuve la Garenne, France) and presented to the bread making industry in 2005. Mixolab gives the flexibility to analyze dough systems by mixing and heating the dough at the same time, which allows for the researcher to understand mixing and cooked stability properties. We evaluated the effects of 5–20% finely milled flaxseed (<550 μm) in wheat flour dough with a standard heating cycle after 8 minutes (Fig. 15.2–15.6). Flaxseed addition significantly increased torque values of the Mixolab, which is the direct indicator of water absorption. Control (Fig. 15.2), 5% flaxseed (Fig. 15.3), 10% flaxseed (Fig. 15.4), 15% flaxseed (Fig. 15.5) and 20% flaxseed (Fig. 15.6) gave torque values of 1.1, 1.17, 1.18, 1.23 and 1.22 N m, respectively. The stability of the dough significantly decreased with increasing flaxseed addition, whereas farinograph peak time which is related to optimum mixing time or dough development time, increased.

Mixolab tests allow the determination of the overall (or final) torque

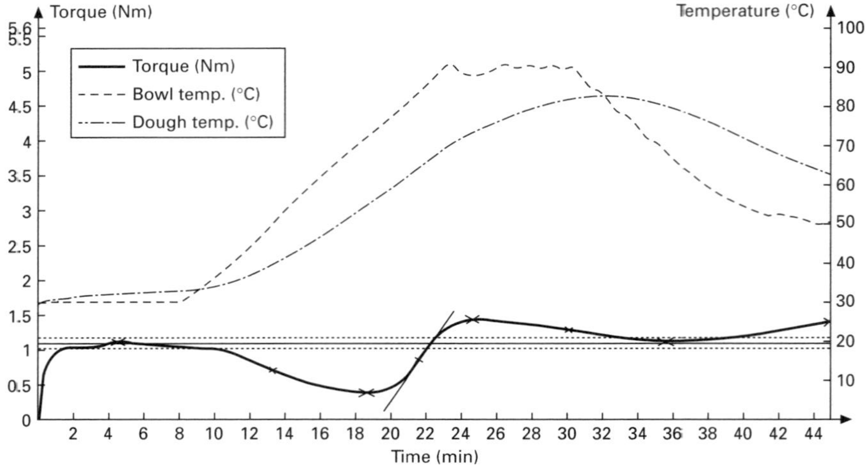

Fig. 15.2 Mixolab of control hard red spring wheat flour.

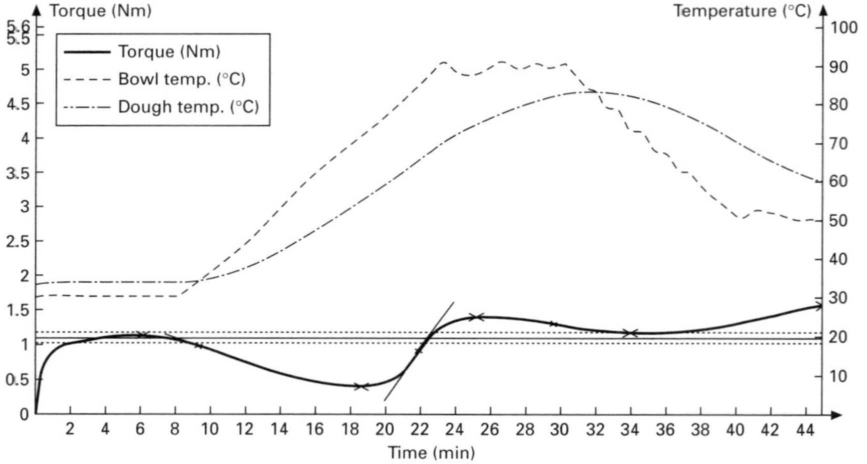

Fig. 15.3 Mixolab of hard red spring wheat flour fortified with 5% fine milled flaxseed.

value of the dough during heating. We observed pasting properties of highly abused dough after 18 minutes, where the initial torque value might relate to the cold paste viscosity. The final torque values of the control dough and the dough containing 5%, 10%, 15%, and 20% flaxseed were 1.45, 1.48, 1.53, 1.52 and 1.45 N m, respectively. In light of these experiments we speculate that finely milled flaxseed addition enhanced the dough matrix and interacted mainly with gluten and starch fractions, thus accounting for the increased torque values during mixing. We expect milled flaxseed to have hydrophobic properties due to its high level of oil (42%). However, high amounts of water might instead be absorbed by non-starch polysaccharides (gum) and flaxseed

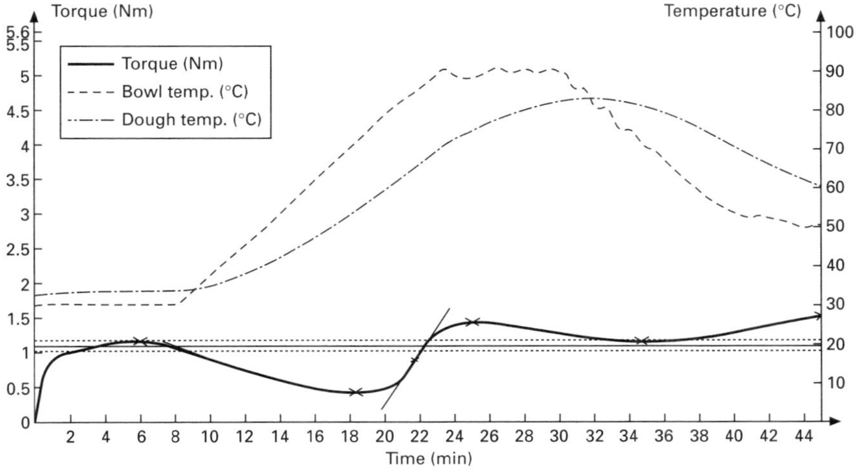

Fig. 15.4 Mixolab of hard red spring wheat flour fortified with 10% fine milled flaxseed.

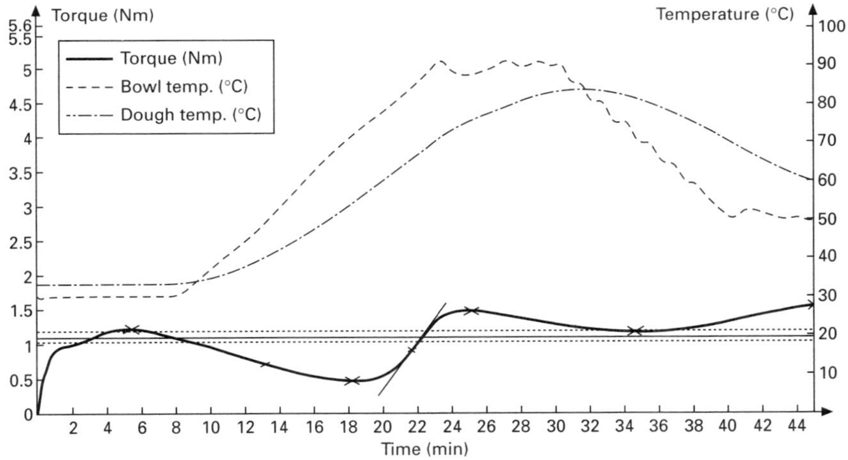

Fig. 15.5 Mixolab of hard red spring wheat flour fortified with 15% fine milled flaxseed.

proteins. This hypothesis might explain the observed decrease in final torque values. Further research is required to understand the effects of flaxseed fractions in dough systems.

In addition to Mixolab analysis, bread making experiments with flaxseed were conducted. Loaf volume and bread firmness values are given in Table 15.3. The control bread had the highest loaf volume, whereas loaf volume of breads containing 5%, 10% and 15% flaxseed were not significantly ($P \leq 0.05$) different. Bread made with 20% flaxseed had the lowest loaf volume and the darkest color. In contrast, firmness values of the control, 5%, 10%

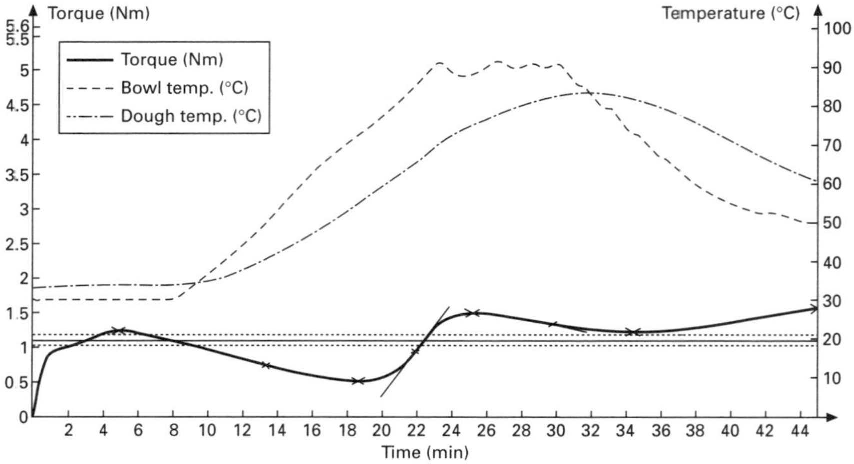

Fig. 15.6 Mixolab of hard red spring wheat flour fortified with 20% fine milled flaxseed.

Table 15.3 Loaf volume and bread firmness data for flaxseed bread

Sample	Loaf volume (cc)	Bread firmness (g cm)
Control	817^{a}	142^{a}
5% flaxseed	725^{b}	171^{ab}
10% flaxseed	688^{bc}	197^{ab}
15% flaxseed	640^{cd}	200^{ab}
20% flaxseed	620^{d}	236^{b}

Statistically significant results ($P \leq 0.05$) are indicated by different superscript letters.

and 15% flaxseed breads were statistically ($P \leq 0.05$) less firm than the 20% flaxseed bread. The grain structure of bread samples showed similar trends as loaf volume. The control and 5% flaxseed breads exhibited the best grain scores. Gas cells, inferior wall structures, and break and shred were observed with high levels of flaxseed addition. Our observations were consistent with Garden (1993) and Ghosh *et al.* (2004) in terms of dough and bread quality. In terms of ALA fortification with flaxseed, we speculate that 5% ground flaxseed addition would be the highest limit for breads without vital wheat gluten addition if loaf volume is to be maintained. Koca and Anil (2007) also found that flaxseed significantly affected dough quality beyond 5%, but concluded that breads up to 20% flaxseed were acceptable to sensory panelists. Thus, in 100 g of bread containing 5% and 20% flaxseed, the ALA levels would be 1 and 4 g/100 g, respectivety, which would meet the daily requirement for ALA.

Morita and Shirai (1997) tested DHA powder in wheat flour dough and bread systems. DHA powder levels of 0.2–0.3% showed significant increases in loaf volume. Values for elasticity, modulus, viscosity coefficient and relaxation time were significantly higher in dough with DHA powder. Farinograph results indicated no change in arrival and development times, whereas stability increased. In addition, no changes were reported for gelatinization temperature and enthalpy of starch in the dough. No enhancing agents such as vital wheat gluten were required for bread made with DHA powders (Morita and Shirai, 1997). The size of gas cells in the bread crumb, grain structure, overall appearance and firmness of the breads were not affected by the negligible fortification levels. Due to the proprietary processing and the limited research in this area, sufficient information regarding the use of encapsulated powders in the baking industry is limited. However, we speculate that encapsulated powders would be the primary source of long-chain omega-3 fortification in the future.

15.4.3 Difficulties in manufacturing omega-3-enriched breads – oxidation focus

It is difficult to utilize fish, algae and plant oils in bread making due to their higher propensity for lipid oxidation. Although fish and microalgae oils can provide a substantial level of long-chain omega-3, adding these oils directly to a bakery formulation is not practical. The sensitivity of fish oil to oxidation (Frankel *et al.*, 2002) precludes the addition of the oil directly to the formula. Research by Frankel *et al.* (2002) also showed that iron further enhanced the oxidation of the fish and algae oils. This point is of particular importance because bread contains iron, whether naturally present as in whole wheat or through enrichment or fortification. The ability of iron to promote oxidation in breads has not been investigated but must be considered in the formulation of bread containing omega-3 lipids. Jacobsen *et al.* (2000) also demonstrated fishy and rancid off-flavors in fish oil fortified mayonnaise. These authors (Jacobsen *et al.*, 2000) noted that the application of antioxidants created varying degrees of protection against off-flavor development. Thus, antioxidants may not be sufficient to protect the shelf stability of breads containing fish or algae oils. Heat can also stimulate degradation of oils. In bread systems, the application of heat is essential and thus the success of oil addition directly to a bread formula is minimal. In contrast, encapsulated fish or algae oils are better protected than the oil itself and could be a means to deliver omega-3 lipids through breads. Morita and Shirai (1997) reported that DHA powder suspension did not significantly change on heating at 120 °C for 30 minutes, during fermentation or in baking processes.

Fine and granular fractions of milled flaxseed should be utilized in baking applications. Chen *et al.* (1994) reported thermal and oxidative stability of whole and milled flaxseed and extracted flaxseed oil at 178 °C. Milled flaxseed exhibited the highest level of oxygen consumption. In contrast, whole seeds showed little change over a period of 90 minutes at 178 °C. ALA content decreased by 3.8% and 3.3% in milled flaxseed and flaxseed 0 : 1, respectively, during heat exposure. The oxidative susceptibility of milled flaxseed was related to particle size. Chen *et al.* (1994) found that coarse flaxseed fractions (>950 μm) showed the highest oxygen consumption, followed by fine fractions(<500 μm) and granular (intermediate) size fractions (500–710 and 710–850 μm). These researchers concluded that, under typical baking conditions, there is minimal loss of ALA.

In a baking study with fine milled flaxseed (<550 μm) conducted in our laboratory, we observed the formation of secondary aldehydes in relation to lipid oxidation. Propanal levels were monitored in bread samples. Flaxseed bread with 20% flaxseed addition had the highest propanal content (3.9 ppm), whereas control (2.92 ppm) and 5% flaxseed (2.96 ppm) breads gave the lowest values. We observed a linear increase in propanal content with flaxseed addition. Nevertheless, no statistically significant differences were determined between the samples. Hexanal levels followed a similar pattern. The control had the lowest hexanal content (18.9 ppm) whereas flaxseed bread samples had hexanal contents between 19.2 and 21.0 ppm. 2-Hexenal values were similar to hexanal contents. No statistical difference between the samples was detected in terms of hexanal and 2-hexenal contents. Flaxseed bread was monitored at 1, 4, 5, 9 and 14 days. Propanal levels in flaxseed breads on day 14 (4.6 ppm) were statistically higher than those of days 1 (3.4 ppm), 4 (3.2 ppm), 5 (3.1 ppm) and 9 (2.3 ppm). In contrast, flaxseed bread stored for 1 day had the highest level of hexanal (30.9 ppm), whereas flaxseed bread at day 9 (14.5 ppm) and day 14 (11.7 ppm) had hexanal levels that were statistically lower than the day 1 values. Ethanol and 2-hexenal levels were highest in flaxseed bread stored for 1 day, whereas the bread stored for 14 days gave the lowest concentration. We hypothesize that ALA degradation is minimal during baking and that the degradation is not detrimental to bread quality. Thus, ALA delivered via milled flaxseed is stable during baking.

15.4.4 Sensory characteristics of omega-3 breads

The sensory attributes of omega-3 breads are crucial for market acceptability. Carter (1993) noted that flaxseed has a nutty flavor. However, sensory characteristics have not been fully determined and consumer preference of flaxseed is largely dependent on seed quality. Appearance, color and flavor attributes depend on cultivars and growing conditions. Gerdes (2005) noted that milled flaxseed has a similar texture to wheat bran, which had no detrimental impact on flavor properties. Experiments conducted at the NCI showed that flaxseed addition (5.9–13%) provided unique flavor characteristics to bagels, bran muffins and cracked wheat hoagies. Flaxseed addition (6.4–10.2%) gave excellent flavor properties in oatmeal bread without masking the whole wheat aroma. Furthermore, roasted flaxseed can be considered as an ingredient for bread making, due to superior sensory properties compared with non-roasted flaxseed. Consumer preference testing indicated that roasted flaxseed had better flavor, texture and overall acceptability than the non-roasted control sample (Tulbek *et al.*, 2004). Roasting significantly improved the aroma and flavor properties of flaxseed. We believe that the presence of pyrazines in the roasted flaxseed affected consumer preference. Koslowska (1989) reported that the pleasant taste of roasted flaxseed was related to the presence of pyrazines. In a study conducted in our laboratory, bread samples containing flaxseed had similar propanal contents to the control and did not have a fishy flavor, which is related to ketones from ALA degradation.

Saldeen *et al.* (1998) noted that only 2 panelists out of 195 perceived a fishy flavor or aroma in fish oil fortified breads. Only 10% of the panelists rated bread containing fish oil as bad. However, the bread was made available daily to the subjects and thus it is likely that only limited oxidation would occur in such a short time. To extend the shelf-life of bread containing fish oil beyond a day, additional protection is probably needed. Kolanowksi *et al.* (2004) noted that a reduced-fat spread may be enriched with EPA and DHA by up to about 10.0 g/kg; however, fishy taste was related to fish oil concentration and could be improved by encapsulation of the omega-3 for bread making. Gerdes (2005) noted that the encapsulation process will allow the bakery industry to incorporate fish oil into products without affecting the taste or texture. Microencapsulated powders are dispersible in cold water and stable. These properties, in addition to neutral taste, make encapsulated powders ideal for enriching foods such as breads and reduced-fat products.

15.5 Future trends

The future trends in omega-3 fortified breads are primarily related to ingredient applications. There has been a significant progress in developing oilseeds that produce long-chain EPA and DHA. Fortifying breads with these potentially new oilseeds would provide an alternative to current oilseed and marine sources. Abbadi *et al.* (2004) genetically modified flaxseed to produce both EPA and DHA at levels slightly less than 1% of the oil weight. However, consuming 1 tablespoon of this oil would provide 300–400 mg of long-chain omega-3 fatty acids. Wu *et al.* (2005) genetically modified *Brassica juncea* seeds to produce oil with up to 15% EPA. Using genetic engineering, the fatty acid profile of canola oil was altered to produce stearidonic acid (Ursin, 2003). These examples demonstrate that one day genetically modified oilseed may be the source of omega-3 fatty acids (Truksa *et al.*, 2006).

The application of specific fatty acids to bread may also be a new approach to incorporate more omega-3 into the diet. Dornblaser and Jago (2006) reported that DHA introduction in products promoted cognitive development. In Denmark, Kohberg's introduced bread fortified with DHA alone. Part of the marketing includes the phrase 'the smart omega-3'. More introductions of products with a specific fatty acid or combination of fatty acids are likely to occur. In the United States, OmegaPure introduced an encapsulated ingredient with the benefits of EPA and DHA, whereas Kerry Ingredients developed canola oil powders for the functional food market as a source of ALA.

Although a number of omega-3 fortified breads are available on the market, little data exist in the literature to fully support their application. This is particularly true for rheological behavior and oxidation stability. Additional research is needed to evaluate shelf stability and oxidation beyond the baking process.

15.6 Sources of further information and advice

For information regarding flaxseed, the reader is directed to the reviews of Hall *et al* (2006) and Morris (2003a). The book *Fish, Omega 3 and Human Health* by Lands (2005) provides information on the health aspects of long-chain omega-3 fatty acids. The book *Lipids in Cereal Technology* by Barnes (1983) provides an excellent overview of the effects of lipids in dough rheology and bread baking. Finally, the application of Mixolab in rheological anaylsis can be found at www.chopin.fr.

15.7 References

Abbadi A, Domergue F, Bauer J, Napier J, Welti R, Zähringer U, Cirpus P and Heinz E (2004), 'Biosynthesis of very-long-chain polyunsaturated fatty acids in transgenic oilseeds: constraints on their accumulation', *Plant Cell*, **16**, 2734–48.

Albert C, Oh K, Whang W, Manson J, Chae C, Stampfer M, Willett W and Hu F (2005), 'Dietary α-linolenic acid intake and risk of sudden cardiac death and coronary heart disease', *Circulation*, **112**, 3232–8.

Almario R, Vonghavaravat V, Wong R and Kasim-Karakas S E (2001), 'Effects of walnut consumption on plasma fatty acids and lipoproteins in combined hyperlipidemia', *Am J Clin Nutr*, **74**, 72–9.

Amaral J, Cunha S, Alves M, Pereira J, Seabra R and Oliveira-Beatriz P (2004), 'Triacylglycerol composition of walnut (*Juglans regia L.*) cultivars: characterization by HPLC-ELSD and chemometrics', *J Agric Food Chem*, **52**, 7964–9.

Angers P, Morales M and Simon J (1996), 'Fatty acid variation in seed oil among *Ocimum* species', *J Am Oil Chem Soc*, **73**, 393–5.

Barnes P (1983), *Lipids in Cereal Technology*, Academic Press, New York.

Barrow C (2005), 'Microencapsulated fish oil', *Worldnutra Newsletter*, Worldnutra.com.

Baylin A, Kabagambe E, Ascherio A, Spiegelman D and Campos H (2003), 'Adipose tissue α-linolenic acid and nonfatal acute myocardial infarction in Costa Rica', *Circulation*, **107**, 1586–91.

Benkouider C (2004), 'Functional bread in Europe – a niche market with potential', Euromonitor International Inc., http://www.euromonitor.com/articles.aspx?folder=Functional_bread_in_Europe_-_a_niche_market_with_potential&industryfolder=Articles (accessed June 15, 2006).

Bimbo A (1989), 'Fish oils: past and present food uses', *J Am Oil Chem Soc*, **66**, 1717–26.

Bimbo A (1998), 'Guidelines for characterizing food-grade fish oil' *Inform*, **9**, 473–83.

Brenner D (1993), 'Perilla: Botany, uses and genetic resources', in Janick J and Simon J (Eds), *New Crops*, Wiley, New York, pp. 322–8.

Budin J, Breene W and Putman D (1995), 'Some compositional properties of camelina (*Camelina sativa* L. Crantz) seeds and oils', *J Am Oil Chem Soc*, **72**, 309–15.
Carter J (1993), 'Potential of flaxseed and flaxseed oil in baked goods and other products in human nutrition', *Cereal Foods World,* **38**, 753–9.
Castillo M, Dobson G, Brennan R and Gordon S (2002), 'Genotypic variation in fatty acid content of Blackcurrant seeds', *J Agric Food Chem*, **50**, 332–5.
Castillo M, Dobson G, Brennan R and Gordon S (2004), 'Fatty acid content and juice characteristics in black currant (*Ribes nigrum* L.) genotypes', *J Agric Food Chem*, **52**, 948–52.
Caughey G, Mantzioris E, Gibson R, Cleland L and James M (1996), 'The effect on human tumor necrosis factor α and interleukin 1 β production of diets enriched with n-3 fatty acids from vegetable oils or fish oil', *Am J Clin Nutr*, **63**, 116–22.
Chen Z Y, Ratnayake W M N and Cunnane S C (1994), 'Oxidative stability of flaxseed lipids during baking', *J Am Oil Chem Soc*, **71**, 629–632.
Clandinin M, Foxwell A and Goh Y (1997), 'Omega–3 fatty acid intake results in a relationship between the fatty acid composition of LDL cholesterol ester and LDL cholesterol content in humans', *Biochim Biophys Acta*, **1346**, 247–52.
Coates W and Ayerza R (1998), 'Commercial production of Chia in northwestern Argentina', *J Am Oil Chem Soc*, **75**, 1417–20.
Connor W (2000), 'Importance of n-3 fatty acids in health and disease', *Am J Clin Nutr*, **71**, 171S–175S.
Cosgrove J, Church D and Pryor W (1987), 'The kinetics of the autoxidation of polyunsaturated fatty acids', *Lipids*, **22**, 299–304.
Cunnane S, Ganguli S and Menard C (1993), 'High a-linolenic acid flaxseed (*Linum usitatissimum*): Some nutritional properties in humans', *Br J Nutr*, **69**, 443–53.
Daun J, Barthet J, Chornick T and Duguid S (2003), 'Structure, composition, and variety development of flaxseed', in Cunnane S and Thompson L (Eds), *Flaxseed in Human Nutrition*, 2nd edn, AOCS Press, Champaign, 1L, pp. 1–40.
de Lorgeril M and Salen P (2004), 'Use and misuse of dietary fatty acids for the prevention and treatment of coronary heart disease', *Rep Nutr Dev*, **44**, 283–8.
de Lorgeril M, Salen P, Laporte F and de Leiris J (2001), 'Alpha-linolenic acid in the prevention and treatment of coronary heart disease', *Eur Heart J Supp*, **3**, D26–D32.
Deferne J and Pate D (1996), 'Hemp seed oil: a source of valuable essential fatty acids', *J Int Hemp Assoc*, **3**, 4–7.
Djousse L, Rautaharju P, Hopkins P, Whitsel Eric, Arnett D, Eckfeldt J, Province M and Ellison R C (2005), 'Dietary linolenic acid and adjusted QT and JT intervals in the National Heart, Lung, and Blood Institute Family Heart Study', *J Am Col Cardiol*, **45**, 1716–22.
Dornblaser L and Jago D (2006), 'Global health and wellness trends: latest news in fortification', Mintel Trend and Tasting Presentations, Presented at the Food Expo during the Institute of Food Technologists' Annual Meeting, June 27, 2006, Orlando, FL.
Finnegan Y and Minihane A (2003), 'Plant- and marine-derived *n*-3 polyunsaturated fatty acids have differential effects on fasting and postprandial blood lipid concentrations and on the susceptibility of LDL to oxidative modification in moderately hyperlipidemic subjects', *Am J Clin Nutr*, **77**, 783–95.
Firestone D (1999), *Physical and Chemical Characteristics of Oils, Fats, and Waxes*, AOCS Press, Champaign, IL.
FDA (Food and Drug Administration) (2004), 'Omega–3 Fatty Acids and Coronary Heart Disease', Docket No. 2003Q-0401, http://www.cfsan.fda.gov/~dms/qhc-sum.html#omega3 (accessed June 15, 2006).
Frankel E, Satue-Gracia T, Meyer A and German B (2002), 'Oxidative stability of fish and algae oils containing long-chain polyunsaturated fatty acids in bulk and in oil-in-water emulsions', *J Agric Food Chem*, **50**, 2094–9.

Fröhlich A and Rice B (2005), 'Evaluation of *Camelina sativa* oil as a feedstock for biodiesel production', *Ind Crops Prod*, **21**, 25–31.
Garden J (1993), '*Flaxseed gum: Extraction, characterization and functionality*,' PhD Dissertation Department of Cereal Science, North Dakota State University, Fargo, ND.
Gebauer S, Psota T, Harris W and Kris-Etherton P (2006), 'n-3 fatty acid dietary recommendations and food sources to achieve essentiality and cardiovascular benefits', *Am J Clin Nutr*, **83**, 1526S–1535S.
Gelski J (2006), 'Blending bonanza', *Milling Baking News*, Feb 28, 1, 27–34.
Gerdes S (2005), 'Omega-3 fatty acids: The next big ingredient? The buzz surrounding omega-3 fatty acids grows as bakers throughout the country begin to formulate products with these healthful ingredients', http://www.bakingconferences.com/article/8081 (accessed June 15, 2006).
Ghosh P, Chakraborty M, Sorenson B, Berzonsky W A and Hall C (2004), 'Effects of ground flaxseed on the processing and quality parameters of wheat flour tortilla', AACC Annual Meeting, San Diego, CA, September 19–21, 2004.
Gibney M (1997), 'Incorporation of *n*-3 polyunsaturated fatty acids into processed foods', *Br J Nutr*, **78**, 193–5.
Hall III C (2002), 'Flaxseed as a functional food', in Proceedings of the 59th Flax Institute of the United States, Fargo, ND. Flax Institute of the United States, Fargo, ND, pp. 1–6.
Hall III C, Tulbek M and Xu Y (2006), 'Flaxseed', in Taylor S (Ed.), *Advances in Food and Nutrition Research*, Vol. 51, Elsevier, New York, pp. 1–99.
Hauman B (1998), 'Alternative sources of *n*-3 fatty acids', *Inform*, **9**, 1108–19.
Higgins S, Carroll Y, O'Brien N and Morrissey P (1999), 'Use of microencapsulated fish oil as a means of increasing *n*-3 polyunsaturated fatty acid intake', *J Hum Nutr Diet*, **12**, 265–71.
Holub B (2002), 'Clinical nutrition: 4. Omega-3 fatty acids in cardiovascular care', *Can Med Assoc J*, **166**(5), 608–15.
Holub D and Holub B (2004), 'Omega–3 fatty acids from fish oils and cardiovascular disease', *Mol Cell Biochem*, **263**, 217–25.
Hu F, Manson J and Willett W (2001), 'Types of dietary fat and risk of coronary heart disease: a critical review', *J Am Col Nutr*, **20**, 5–19.
Hussein N, Ah-Sing E, Wilkinson P, Leach C, Griffin B and Millward D (2005), 'Long-chain conversion of [^{13}C] linoleic acid and α-linolenic acid in response to marked changes in their dietary intake in men', *J Lipid Res*, **46**, 269–80.
IFIC (International Food Information Council Foundation) (2006), 'Food and health survey: consumer attitudes toward food, nutrition and health', http://www.ific.org/publications/factsheets/omega3fs.cfm (accessed June 15, 2006).
Inoue Y, Sapirstein H and Bushuk W (1995), 'Studies on frozen doughs. IV. Effect of shortening systems on baking and rheological properties', *Cereal Chem*, **72**, 221–6.
Jacobsen C, Hartvigsen K, Lund P, Adler-Nissen J, Holmer G and Meyer A (2000), 'Oxidation in fish-oil-enriched mayonnaise. 2. Assessment of the efficacy of different tocopherol antioxidant systems by discriminant partial least squares regression analysis', *Eur Food Res Tech*, **210**, 242–57.
JHCI (Joint Health Claims Initiative) Expert Committee (2005), 'Omega-3 PUFA and heart health - final report', http://www.jhci.org.uk/approv/omega.htm (accessed June 15, 2006).
Junge R, Hoseney R and Varriano-Marston E (1981), 'Effect of surfactants on air incorporation in dough and the crumb grain of bread', *Cereal Chem*, **58**, 338–42.
Koca A F and Anil M (2007), 'Effect of flaxseed and wheat flour blends on dough rheology and bread quality', *J Sci Food Agric*, **87**, 1172–5.
Kolanowski W and Laufenberg G (2006), 'Enrichment of food products with polyunsaturated fatty acids by fish oil addition', *Eur Food Res Technol*, **222**, 472–7.
Kolanowski W, Swiderski F, Jaworska D and Berger S (2004), 'Stability, sensory quality,

texture properties and nutritional value of fish oil-enriched spreadable fat', *J Sci Food Agric*, **84**, 2135–41.

Koslowska J (1989), 'The use of flax seed for food purposes', in *Flax in Europe Production and Processing*, Proceedings of the European Regional Workshop on Flax', Poznan, Poland June 19–21, 1989.

Kris-Etherton P, Yu-Poth S, Sabate J, Ratcliffe H, Zhao G and Etherton T (1999), 'Nuts and their bioactive constituents: effects on serum lipids and other factors that affect disease risk', *Am J Clin Nutr*, **70**, 504S–511S.

Kris-Etherton P, Taylor Denise, Yu-Poth S, Huth P, Moriarty K, Fishell V, Hargrove R, Zhao G and Etherton T (2000), 'Polyunsaturated fatty acids in the food chain in the United States', *Am J Clin Nutr*, **71**, 179S–188S.

Kris-Etherton P, Harris W and Appel L (2002), 'Fish consumption, fish oil, omega-3 fatty acids, and cardiovascular disease', *Circulation*, **106**, 2747–57.

Lands W (2005), *Fish, Omega 3 and Human Health,* 2nd edition, AOCS Press, Champaign, IL.

Lee K W and Lip G Y H (2003), 'The role of omega-3 fatty acids in the secondary prevention of cardiovascular disease', *Q J Med*, **96**, 465–80.

Lee S, Inglett G E and Carriere C J (2004), 'Effect of nutrim oat bran and flaxseed on rheological properties of cakes', *Cereal Chem*, **81,** 637–42.

Liu M, Wallin R and Saldeen T (2001), 'Effect of bread containing stable fish oil on plasma phospholipid fatty acids, triglycerides, HDL-cholesterol, and malondialdehyde in subjects with hyperlipidemia', *Nutr Res*, **21**, 1403–10.

Lovegrove J, Brooks C, Murphy M, Gould B and Williams C (1997), 'Use of manufactured foods enriched with fish oils as a means of increasing long-chain *n*-3 polyunsaturated fatty acid intake', *Br J Nutr*, **78**, 223–36.

MacRitchie F (1977), 'Flour lipids and their effects in baking', *J Sci Food Agric*, **28**, 53–8.

MacRitchie F and Gras P (1973), 'The role of flour lipids in baking', *Cereal Chem*, **50**, 292–302.

Mansour M, Frampton D, Nichols P, Volkman J and Blackburn S (2005), 'Lipid and fatty acid yield of nine stationary-phase microalgae: Applications and unusual C24–C28 polyunsaturated fatty acids', *J Appl Phycol*, **17**, 287–300.

Mantzioris E, Cleland L, Gibson R, Neumann M, Demasi M and James M (2000), 'Biochemical effects of a diet containing foods enriched with *n*-3 fatty acids', *Am J Clin Nutr*, **72**, 42–8.

McGuigan P (2005), 'Taking the 'fishy' out of fish oil', *Functional Foods & Nutraceuticals, September*, http://www.ffnmag.com/NH/ASP/strArticleID/778/strSite/FFNSite/articleDisplay.asp (accessed June 15, 2006).

Menrad K (2003), 'Market and marketing of functional food in Europe', *J Food Eng*, **56**,181–8.

Metcalf R, James M, Mantzioris E and Cleland L (2003), 'A practical approach to increasing intakes of *n*-3 polyunsaturated fatty acids: use of novel foods enriched with *n*-3 fats', *Eur J Clin Nutr*, **57**, 1605–12.

Metting F (1996), 'Biodiversity and application of microalgae', *J Ind Microbiol*, **17**, 477–89.

Minihane A (2006), 'Dietary fat composition and cardiovascular disease', *Food Sci Tech Bull: Funct Foods*, **3**(2), 13–22.

Molyneaux, M (2006), 'Trending toward tomorrow: mega market experts look ahead', Presented at the Institute of Food Technologists' Annual Meeting, Orlando, FL, June 27, 2006.

Morita N and Shirai J (1997), 'Effects of microencapsulated docosahexaenioc acid preparation on properties of dough and bread', *J Nut Sci Vit*, **43**, 591–600.

Morris D (2003a), '*Flax, a Health and Nutrition Primer'*, Flax Council of Canada, Winnipeg, Manitoba, pp. 9–19.

Morris D (2003b), 'Methodologic challenges in designing clinical studies to measure differences in the bioequivalence of *n*-3 fatty acids', *Mol Cell Biochem*, **246**, 83–90.

Mozaffarian D, Ascherio A, Hu F, Stampfer M, Willett W, Siscovick D and Rimm E (2005), 'Interplay between different polyunsaturated fatty acids and risk of coronary heart disease in men', *Circulation*, **111**, 157–64.

Nestel P (1990), 'Effects of n-3 fatty acids on lipid metabolism', *Ann Rev Nutr*, **10**, 149–67.

Oomah B, Bussonb M, Godfreya D and John C (2002), 'Drovera characteristics of hemp (*Cannabis sativa* L.) seed oil', *Food Chem*, **76**, 33–43.

Paschos G, Yiannakouris N, Rallidis L S, Davies I, Griffin B, Panagiotakos D B, Skopouli F N, Votteas V and Zampelas A (2005), 'Apolipoprotein E genotype in dyslipidemic patients and response of blood lipids and inflammatory markers to alpha-linolenic acid', *Angiology*, **56**, 49–60.

Ross R (1999), 'Atherosclerosis – an inflammatory disease', *N Engl J Med*, **340**, 115–26.

Saldeen T, Wallin R and Marklinder I (1998), 'Effects of a small dose of stable fish oil substituted for margarine in bread on plasma phospholipid fatty acids and serum triglycerides', *Nutr Res*, **18**, 1483–92.

Sanders T (2000), 'Polyunsaturated fatty acids in the food chain in Europe', *Am J Clin Nutr*, **71**, 176S–178S.

Schorno A, Manthey F, Wiesenborn D, Hall C and Hammond J (2004), 'Flaxseed milling', in Proceedings of the 60th Flax Institute of the United States, Fargo, ND. Flax Institute of the United States, Fargo, ND, pp. 15–23.

Shin H S and Kim S (1994), 'Lipid composition of perilla seed', *J Am Oil Chem Soc*, **71**, 619–22.

Simopoulos A (2004), 'Health effects of eating walnuts', *Food Rev Intern*, **20**, 91–8.

Sinclair A, O'Dea K and Johnson L (1994), 'Estimation of the *n*-3 PUFA status in a group of urban Australians by analysis of plasma phospholipid fatty acids', *Aust J Nutr Diet*, **51**, 53–7.

Sluimer P (2005), *Principles of Breadmaking Functionality of Raw Materials and Process Steps*, American Association of Cereal Chemists, St Paul, MN.

Smith P and Johansson J (2004), 'Influences of the proportion of solid fat in a shortening on loaf volume and staling of bread', *Food Proc Pres*, **28**, 359–67.

Spolaore P, Joannis-Cassan C, Duran E and Isambert A (2006), 'Commercial applications of microalgae', *J Biosci Bioeng*, **101**, 87–96.

The International Society for the Study of Fatty Acids and Lipids (2004), 'Recommendations for intake of polyunsaturated fatty acids in healthy adults', http://www.issfal.org.uk/Welcome/PolicyStatement3.asp (accessed 22 June 2006).

Truksa M, Wu G, Vrinten P and Qiu X (2006), 'Metabolic engineering of plants to produce very long-chain polyunsaturated fatty acids', *Transgenic Res*, **15**, 131–7.

Tulbek M C, Schorno A, Meyers S, Hall C and Manthey, F (2004), 'Chemical and sensory analysis and shelf life stability of roasted flaxseed', in Proceedings of the 60th Flax Institute of the United States, Fargo, ND. Flax Institute of the United States, Fargo, ND, pp. 31–36.

USDA (United States Department of Agriculture) (2006), 'USDA National Nutrient Database for Standard Reference, Release 18', Nutrient Data Laboratory United States Department of Agriculture, available at http://www.ars.usda.gov/ba/bhnrc/ndl (accessed June 15, 2006).

US Institute of Medicine, Food and Nutrition Board (2002), *Dietary Reference Intakes for Energy, Carbohydrate, Fiber, Fat, Fatty Acids, Cholesterol, Protein, and Amino Acids*', National Academies Press, Washington, DC, pp. 8–97.

Ursin V (2003), 'Modification of plant lipids for human health: development of functional land-based omega-3 fatty acids', *J Nutr*, **133**, 4271–4.

Van de Werf F, Ardissino D, Betriu A, Cokkinos D, Falk E, Fox K, Julian D, Lengyel M, Neumann F J, Ruzyllo W, Thygesen C, Underwood S R, Vahanian A, Verheugt F and

Wijns W (2003), 'Management of acute myocardial infarction in patients presenting with ST-segment elevation. The Task Force on the Management of Acute Myocardial Infarction of the European Society of Cardiology', *Eur Heart J*, **24**, 28–66.

Wallace J, McCabe A, Robson P, Keogh M, Murray C, Kelly P, Márquez-Ruiz G, McGlynn H, Gilmore W and Strain J (2000), 'Bioavailability of *n*-3 polyunsaturated fatty acids (PUFA) in foods enriched with microencapsulated fish oil', *Annals Nutr Met*, **44**, 157–62.

Ward O and Singh A (2005) 'Omega-3/6 fatty acids: Alternative sources of production', *Process Biochem*, **40**, 3627–52.

Wilkenson P, Leach C, Ah-Sing E, Hussain N, Miller G, Millward D and Griffin B (2005), 'Influence of α-linolenic acid and fish-oil on markers of cardiovascular risk in subjects with an atherogenic lipoprotein phenotype', *Atherosclerosis*, **181**, 115–24.

World Health Organization (2003), 'Diet, nutrition, and the prevention of chronic diseases', Joint WHO/FAO Expert Consultation, *World Health Organ Tech Rep Ser*, **916**, 89–90.

Wu G, Truksa M, Datla N, Vrinten P, Bauer J, Zank T, Cirpus P, Heinz E and Qiu X (2005), 'Stepwise engineering to produce high yields of very long-chain polyunsaturated fatty acids in plants', *Nat Biotech*, **23**, 1013–17.

Yan N (2003), 'Encapsulated agglomeration of microcapsules and their preparation', US Patent 6,974, 592 B2.

Yep Y, Li D, Mann N, Bode O and Sinclair A (2002), 'Bread enriched with microencapsulated tuna oil increases plasma docosahexaenoic acid and total omega-3 fatty acids in humans', *Asia Pac J Clin Nutr*, **11**, 285–91.

Zhao G, Etherton T and Martin K (2004), 'Dietary alpha-linolenic acid reduces inflammatory and lipid cardiovascular risk factors in hypercholesterolemic men and women', *J Nutr*, **134**, 2991–7.

Zhao G, Etherton T, Martin K, Vanden Heuvel J, Gillies P, West S and Kris-Etherton P (2005), 'Anti-inflammatory effects of polyunsaturated fatty acids in THP-1 cells', *Biochem Biophys Res Commun*, **336**, 909–17.

16

Soy-enriched bread

Y. Vodovotz, The Ohio State University, USA

16.1 Introduction

Epidemiological and experimental evidence suggests that consumption of soybean products may have a significant impact upon health (Messina *et al.*, 1994; Setchell, 1998; Birt *et al.*, 2001; Messina and Loprinzi, 2001; Scheiber *et al.*, 2001; Brouns, 2002; Hasler, 2002). The biological activity in soy has been associated, in part, with the presence of proteins and isoflavones (Coward *et al.*, 1993; Barnes, 1998; Fournier *et al.*, 1998; Setchell and Cassidy, 1999; Henkel, 2000). In 1999, the Food and Drug Administration declared that 6.25 g of soy protein per serving included in a diet low in saturated fat and cholesterol may reduce the risk of coronary heart disease by lowering blood cholesterol levels (FDA, 1999) and allowed product labeling to that effect (Henkel, 2000). Increased recognition of the link between soy and health has resulted in the development of a series of soy-containing functional foods (Klein *et al.*, 1995; Hendrich and Murphy, 2001; Zhou *et al.*, 2002). However, consumer acceptability of such foods remains low due to off-flavors and undesirable textures of many of these products as well as the need to incorporate a non-traditional food (bars, shakes, Asian foods) into the diet. Bread made partially with soy may represent a viable delivery system for isoflavones (as well as other phytochemicals) and soy proteins, since bread is a traditional and well-accepted food of widespread consumption. This chapter will outline the various physico-chemical changes occurring in bread upon addition of significant (>20% w/w) amounts of soy, the changes in phytochemicals both during manufacturing and storage of soy bread, consumer attitudes towards the bread and some possible future trends for the bakery industry regarding soy.

16.2 Physico-chemical properties of soy-enriched bread

The favorable elastic and extensible properties of bread depend on the quality and quantity of gluten in the bread flour. Gluten is composed of a mixture of: gliadin (40–45%), partly hydrophilic and partly hydrophobic in nature; glutenin (20–25%), which is more hydrophilic than gliadin; and residue protein (35–40%) called insoluble glutenin (Kasarda *et al.*, 1976). Gluten contains high concentrations of glutamine, proline and leucine. When wheat flour is hydrated and subjected to mechanical shear, the gluten proteins absorb water and form a continuous three-dimensional network that can trap gas bubbles (Ablett *et al.*, 1986; Pyler, 1988; Hoseney, 1994). Soybean flour has a chemical composition very different from wheat flour. Defatted soybean flour, for example, contains about 50% proteins (principally glycinin) and about 30% carbohydrates on an 'as is' mass basis. The carbohydrate fraction is composed of soluble sugars (sucrose 17%, stachyose 13%, raffinose 3%) and insoluble 'fiber' (67%). Starch comprises a very small portion of the carbohydrates. While wheat flour contains essentially starch and gluten that are modified during bread making in the presence of moisture, soy flour undergoes protein denaturation mainly during heating (Riganakos and Kontominas, 1997; Chen and Rasper, 1982). Additionally, soy protein is globular in nature and does not form a stable network in wheat bread, as opposed to gluten which does so through disulfide and hydrogen bonds.

In the USA, soy flour is used in bakery products not for nutritional reasons but rather for its functional characteristics. Levels of 1.5–2% (w/w) soy flour (sometimes soy protein concentrate) are used in bakery products, particularly in white bread, as a replacement for non-fat milk solids. Enzyme-active defatted soy flour is used as a bleaching and dough-improving agent. Defatted soy flours with 50–75% protein dispersibility are extensively used in bakery products to increase the water absorption capacity of flours in bread dough and cake batters. In cakes, they improve film forming and even distribution of air cells. As a result, even cake texture and tender crumb structure are achieved. In all these products, soy flour is used at a level of 2–5%.

At higher replacement levels, however, negative effects were observed in baked goods. These effects include lower loaf volume (Erdman *et al.*, 1977; Fleming and Sosulski, 1977; Buck *et al.*, 1987; Brewer *et al.*, 1992) and less desirable texture (Mizrahi, 1967). These deleterious sensory qualities were found to be proportional to soy addition to bread and were correlated to a change in the water absorption of bread ingredients (Doxastakis *et al.*, 2002) and/or dilution of the gluten fraction (Knorr and Betschart, 1978). Addition of soy to bread requires added water for adequate dough formation and therefore increases the moisture content of the bread (44%) as compared with a typical wheat bread (35%). As a result, Vittadini and Vodovotz (2003) observed a higher initial stiffness of fresh bread containing soy and an increase in the density of the bread in proportion to the amount of soy flour substitution (20–40%). In order to satisfy the FDA heart-healthy claim (FDA, 1999), at least 6.25 g of soy protein must be delivered per serving (50 g) of bread. Therefore, the challenge becomes to produce highly acceptable baked goods with at least 30% soy.

A soy bread formulated in our laboratory (Vodovotz and Ballard, 2002), meeting the FDA requirements for the heart-healthy claim, was found highly acceptable. Forty subjects were asked to rate the acceptability of the fresh soy and fresh wheat breads and both received high acceptability ratings with at least 70% of all subjects rating the samples moderately acceptable or higher (Smith, 2003). There was no statistically significant difference in acceptability between the soy and wheat breads made from fresh dough. These results deviate from past research by Dhingra and Jood (2001) and Brewer *et al.* (1992) who found that with increasing levels of soy substitution there was an increased beany flavor and an intense aftertaste. The lack of off-flavor in the soy bread tested in this study can be attributed to the different product formulation and processing method used (Vodovotz *et al.*, 2004).

Figure 16.1(a) shows the air cell distribution of a fresh slice of soy-enriched (at the 30% level) bread as compared with a wheat bread. The soy bread has a denser structure (loaf volume was less than half of the wheat bread) as compared with the wheat bread yet the air cell distribution was more homogeneous. Wheat bread had a maximum compression load (40% compression) of 0.63 ± 0.04 N (Newton) as compared with soy bread which had a maximum compression load of 3.62 ± 0.96 N. Fresh soy bread was concluded to be firmer (stiffer) than fresh wheat bread. This may be attributed to the lack of gluten network formation and smaller air cell structure of bread after soy addition (Buck *et al.*, 1987; Brewer *et al.*, 1992; Vittadini and Vodovotz, 2003) as confirmed by the higher density of the soy bread.

To understand how the addition of soy at these higher levels influences the macroscopic, microscopic and molecular properties of the final bread, various analytical techniques can be used (Roulet *et al.*, 1988; Biliaderis, 1990). It is imperative to consider the various events taking place concurrently within the bread system since each event occurs in a distinct timeframe and scale (e.g. macroscopic to molecular). Thermal analysis techniques (i.e. differential scanning calorimetry (DSC), dynamical mechanical analysis (DMA) and thermogravimetric analysis (TGA)) have been used to understand bread material and its components (i.e. water, protein, carbohydrates) that change chemically or physically as a function of temperature (Hamann *et al.*, 1990; Hatakeyama and Quinn, 1994; Haines, 1995; Hallberg, 1996; Vodovotz *et al.*, 1996; Fessas and Schiraldi, 2000).

Water content and distribution in baked goods play a critical role in dictating quality and stability of the products (Chinachoti, 1998). The moisture contents of fresh wheat and soy breads (obtained from the TGA weight loss curves, Fig. 16.2a) were 39.9 ± 0.2% and 44.7 ± 0.5%, respectively. The higher moisture content in fresh soy bread was attributed to the additional water added in soy bread formulation due to the high water-holding capacity of soy ingredients (Brewer *et al.*, 1992; Porter and Skarra, 1999; Vittadini and

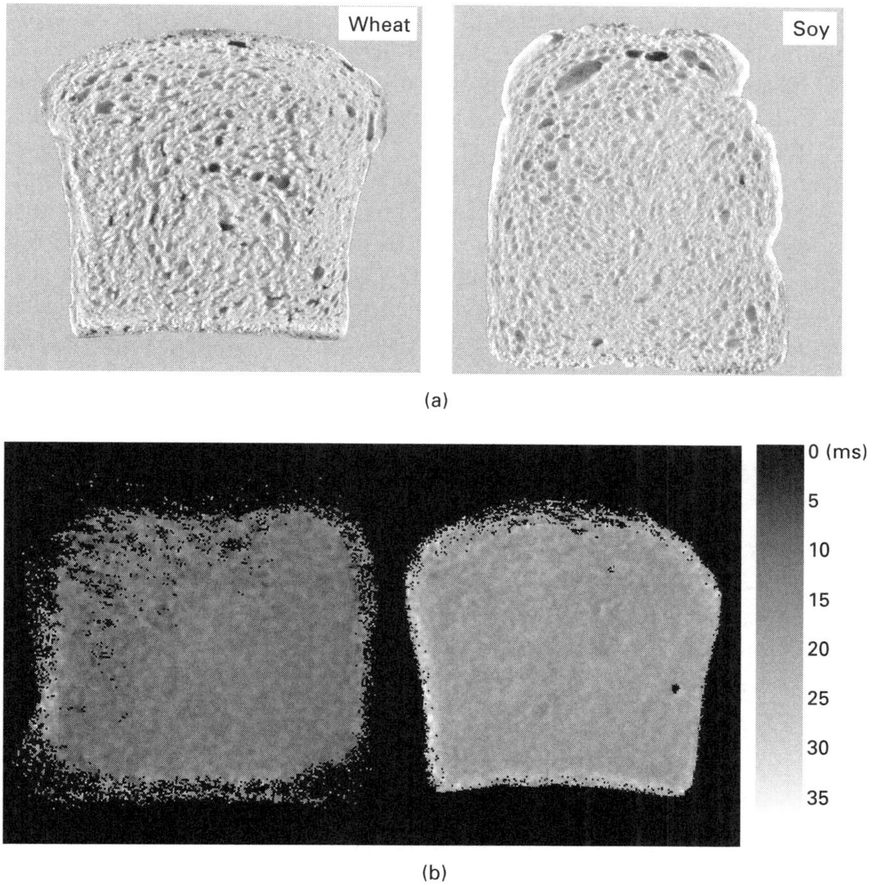

Fig. 16.1 (a) C-Cell Bread Imaging System (Calibre Control International Ltd, Warrington, UK) images of wheat and soy breads. (b) T_2 magnetic resonance imaging images of fresh (day 0 of storage) wheat (left) and soy (right) bread samples (Lodi *et al.* 2007).

Vodovotz, 2003). TGA derivative weight loss curves of water released during heating had a peak temperature (greatest evaporation) of ~100 °C for wheat bread and ~60 °C for soy bread (Fig. 16.2(a)) indicating a weaker association of water with the soy bread matrix as compared with wheat bread (Renkema, 2001; León *et al.*, 2003). Soy bread contains additional proteins from soy ingredients thereby increasing the water holding capacity of the bread components (Vittadini and Vodovotz, 2003). Smith (2003) compared the TGA derivative weight loss curves of several soy bread components – including wheat gluten, wheat starch, soy flour and soy protein isolates – and found the requirement of a higher temperature (approximately 140 °C) to release the water in wheat gluten than those in wheat starch, soy flour and soy protein isolates (approximately 60 °C), suggesting a stronger interaction of wheat gluten with water as compared with starch and soy proteins. A similar result was observed in the work of Fessas and Schiraldi (2000).

DSC analysis indicated the presence of a major endothermic transition at ~ 0 °C in both breads (Fig. 16.2(b)) that was attributed mainly to ice melting (or 'freezable' water) (Vodovotz *et al.*, 1996; Baik and Chinachoti, 2001; Vittadini and Vodovotz, 2003). Fresh wheat bread crumb was found to have 16.2 ±0.9% 'freezable' water (calculated as described previously; Vittadini and Vodovotz, 2003) as compared with fresh soy bread crumb with 21.6 ± 0.1%. The greater amount of 'freezable' water in soy bread crumb was probably from the additional water added in the soy bread formulation as compared with wheat bread since the percentage 'unfreezable' water content (calculated by subtracting the 'freezable' water content from the total moisture

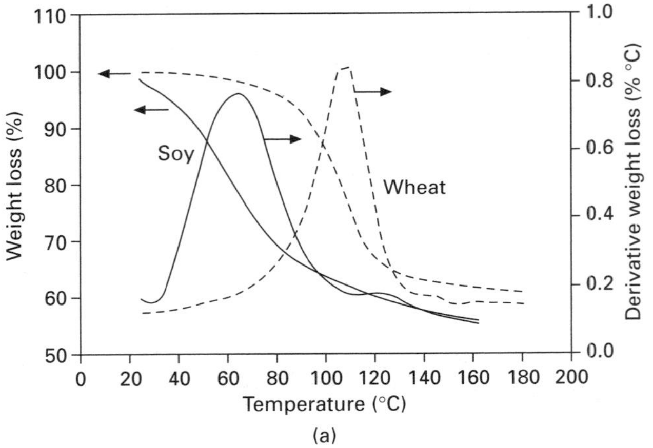

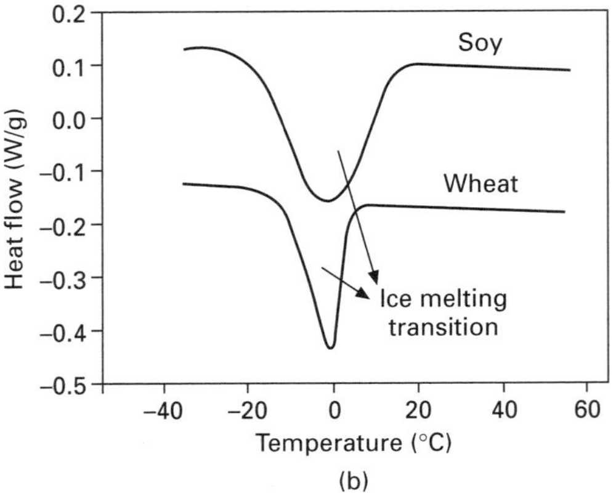

Fig. 16.2 (a) Typical thermogravimetric analysis thermograms for fresh wheat bread crumb (dashed lines) and soy bread crumb (solid lines) showing the percentage weight loss as well as the derivative weight loss curves. (b) Typical differential scanning calorimetry thermograms for fresh wheat and soy bread crumb showing a major endothermic transition at ~0 °C.

content of the bread) was the same (~23%) for both breads. In addition, these results are consistent with the TGA findings that the water in soy bread is more easily removed.

Proton intensity images acquired by magnetic resonance imaging (MRI) of wheat and soy breads showed a much more homogeneous distribution of the proton signal (most probably representing water) for the soy as compared with the wheat bread (Lodi and Vodovotz, 2007). Similar homogeneous appearance for the soy bread products was also observed for T_2 images (Fig. 16.1(b)). Water proton T_2 values for the fresh soy breads were ~18 ms as compared with the lower T_2 value (~10 ms) of the wheat bread. Additionally, these T_2 values in the wheat bread concentrated in the central part of the loaf while the outermost part of the wheat bread loaf, corresponding to the crust and a layer underneath it, generated a very low signal (too solid-like). A more homogeneous water distribution with greater 'freezable' water content or easily removed water, can significantly impact the soy bread's storage stability since water plays a key role in such events (Chinachoti, 1998).

16.3 Effect of soy addition on staling (firming) of bread

Bread represents a heat-set firm foam system and is subjected to physico-chemical instability during storage (Taub and Singh, 1998). Staling refers to the undesirable changes that occur (except for microbial spoilage) between the time bread is removed from the oven and the time it is consumed (D'Applonia and Morad, 1981). From a sensory standpoint, staling reduces the springback and softness of the crumb and increases the dry mouth-feel. Staling also reduces the crispness of the crust and increases its toughness. Recognizing the factors that affect the staling of wheat bread is necessary in order to understand the impact of soy addition to this process. These factors include: retrogradation of amylopectin (Maga, 1975; Kulp and Ponte, 1981; Zeleznak and Hoseney, 1987); moisture redistribution (Leung *et al.*, 1983; Zeleznak and Hoseney, 1986; Baik and Chinachoti, 2000, 2001); changes in gluten functionality (Maga, 1975; Kulp and Ponte, 1981); and the state of the amorphous phase (Hallberg and Chinachoti, 1992; Laine *et al.*, 1994).

Retrogradation is an exothermic process in which starch physically changes from a swollen, gel-like state to a more crystalline state (Krog *et al.*, 1989). The initial firmness of newly baked wheat bread is dominated by retrograded amylose (Slade and Levine, 1991); however, progressive increase in the crystallization of amylopectin (due to the development of more extensively ordered regions; Zobel and Kulp, 1996) is detected upon further storage (Manzacco *et al.*, 2002). Hallberg and Chinachoti (1992) have shown that amylopectin recrystallization is not necessarily reflected in an increase in crumb firmness when ingredients acting as gluten plasticizers are included in the bread formulation. Bread firmness is influenced by changes in crystallization as well as other factors.

Water, in particular, has been recognized as playing an important role in modulating the changes occurring in the bread matrix that result in staling. Water can diffuse and redistribute among the protein–starch and crumb–crust fractions of the bread and it can also evaporate (Willhoft, 1971; He and Hoseney, 1990; Piazza and Masi, 1995). When the newly baked bread is removed from the oven, the crust is practically moisture-free but its moisture content increases rapidly during cooling because of moisture migration from the crumb. Migration of water from the crumb to the crust continues during storage and it has been reported to affect the staling kinetics (Zobel and Kulp, 1996; Chen *et al.*, 1997; Baik and Chinachoti, 2000). Mechanical mixing during dough-making treatment affects water binding capacity and water partition among various dough constituents (e.g. gluten, starch, pentosans, etc.). During baking, some water is involved in starch gelatinization but a significant fraction remains with the other polymers and solutes, which compete for the available moisture. The water in bread has a high thermodynamic activity (above 95%), suggesting that it should be somewhat free to move along any gradient of the relevant chemical potential throughout the crumb structure (Fessas and Schiraldi, 2000). When polymers are thermodynamically incompatible, they can use the water available within their own phase to sustain conformational and phase transitions. Willhoft (1973) suggested that molecular changes in the gluten or starch fraction may lead to the formation of cross-links, which may be associated with the production of free water, i.e. the release of water originally bound to the polymer chain (Schiraldi *et al.*, 1996). This can facilitate the formation of new cross-links keeping the relevant binding sites closer (Schiraldi *et al.*, 1996; Zobel, 1973). As a result of the transformation of polymer components (such as those occurring during storage), water molecules remain entrapped within the crystal structure, or are more easily released toward phases containing other polymers, possibly in the state of viscous starch gel. It has therefore been suggested that free and entrapped water could play different roles in determining the staling rate.

Rasmussen and Hansen (2001) found that the development of bread firmness during storage was correlated to the changes in the 'freezable' water fraction of the bread in non-linear manner. On the other hand, various studies on bread staling found only a small change (0.02) in the water activity (a_w) of bread crumb during storage (Czuchajowska and Pomeranz, 1989; Rasmussen and Hansen, 2001), suggesting that water activity is not an appropriate indicator of bread staling. Water loss causes crumb firming but bread staling also occurs at constant moisture (He and Hoseney,1990; Martin *et al.*, 1991), providing evidence that firmness is not directly related to the loss of water. Chinachoti (1998) found that water in the bread becomes less mobile as a network of amorphous and crystalline solids is formed during storage. A study using the nuclear magnetic resonance (NMR) deuteron relaxation method (Leung *et al.*, 1983) suggested that there is an increase in water binding and a decrease in water mobility in staling bread. Therefore, the state and mobility of water may be better indicators of bread staling than moisture content and water activity.

There is less evidence regarding the role of proteins in bread staling. A model proposed by Hoseney (1994) related bread firming to the formation of H-bond cross-links between the continuous three-dimensional protein structure and the discontinuous remnants of starch granules (Peppas and Brannon-Peppas, 1994). Studies by Levine and Slade (1988) suggested that moisture migratess from gluten to starch, firming the former and softening the latter, during storage. However, Menon and Mujumdar (1987) found evidence that moisture also migrates from starch to gluten. Gerrard and coworkers (2001) found that breads made with alternative proteins (soy and milk) increased in firmness during storage at the same rate as traditional breads containing gluten proteins, suggesting that gluten was not essential to the firming process and that starch alone caused the bread to firm over time. The firmness of both wheat bread and the soy bread developed in our laboratory increased during 10-day storage (Fig. 16.3). For soy bread, there was a significant increase ($p < 0.05$) during the first day after baking and a gradual increase through day 1 to day 10; however, the difference in firmness between days was not found to be significant ($p > 0.05$). After 10 days, the firmness of soy bread increased by about 62%. For wheat bread, there was a significant difference ($p < 0.05$) in firmness through day 1 to day 10 and after 10 days of storage the wheat bread increased in firmness by 470%. Although the fresh soy bread was found to be firmer than the fresh wheat bread, the firmness of the soy bread increased at a much lower rate than the wheat bread, probably due to the higher density of the soy bread.

Soy addition was also found to affect the dynamics (i.e. mobility and redistribution) of the water in the bread and hence the mobility of polymer chains during the firming process, since soy proteins increase water absorption capacity in bread (Selvaraj and Shurpalekar, 1982; Vittadini and Vodovotz, 2003; Dhingra and Jood, 2004). In the soy bread formulated in our laboratory, the moisture content decreased by ~2% during 7 days of storage as compared

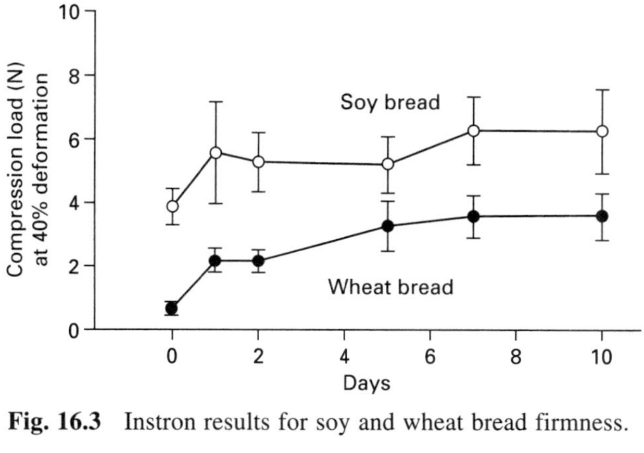

Fig. 16.3 Instron results for soy and wheat bread firmness.

with ~4% reduction in wheat bread (Fig. 16.4). The smaller decrease in moisture content for soy bread suggests that soy bread crumb is able to retain water better than wheat bread crumb and this may, at least partially, result in a slower rate of firmness increase upon storing at room temperature. DSC analysis showed an increase in 'freezable' water content from 21.6% to 27.7% (Fig. 16.4). This increase is unusual for a bread system (Baik and Chinachoti, 2000; Vittadini and Vodovotz, 2003) and is probably caused by the high affinity of soy for water (Brewer *et al.*, 1992; Renkema, 2001). The soy may act as a sponge that binds or entraps water initially, but with time releases some of the less tightly held water to the surroundings. Additionally, there was a redistribution of the water in the soy bread where the 'unfreezable' water decreased by 7.4% as the 'freezable' water increased (Fig. 16.4). In contrast, the percentages of 'freezable' and 'unfreezable' water in wheat decreased slightly during storage (Fig. 16.4). The water dynamics present in the soy bread most probably affects the storage stability and quality of the final product.

Vittadini and Vodovotz (2003) found that the greater the amount of soy addition, the less amylopectin recrystallization was observed in bread during storage, indicating that soy may play a role in modulating the staling process. The amylopectin recrystallization in our soy bread increased at a much lower rate than in wheat bread stored at room temperature (Fig. 16.5). Factors that may contribute to this difference include the diluting of amylopectin in soy bread by the addition of soy ingredients, a low moisture gradient between soy bread crumb and crust that may affect staling kinetics and the difference in water distribution among bread components (Vittadini and Vodovotz, 2003).

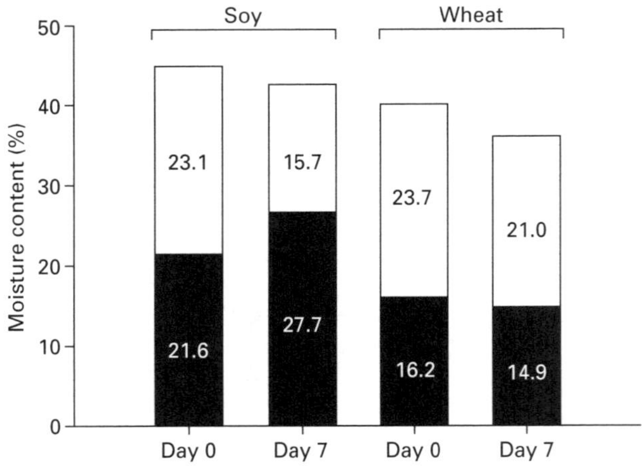

Fig. 16.4 Differential scanning calorimetry and thermogravimetric analysis results indicating the changes in moisture content (total bars), 'unfreezable' water (white bars) and 'freezable' water (black bars) of wheat and soy breads during 7 days of storage.

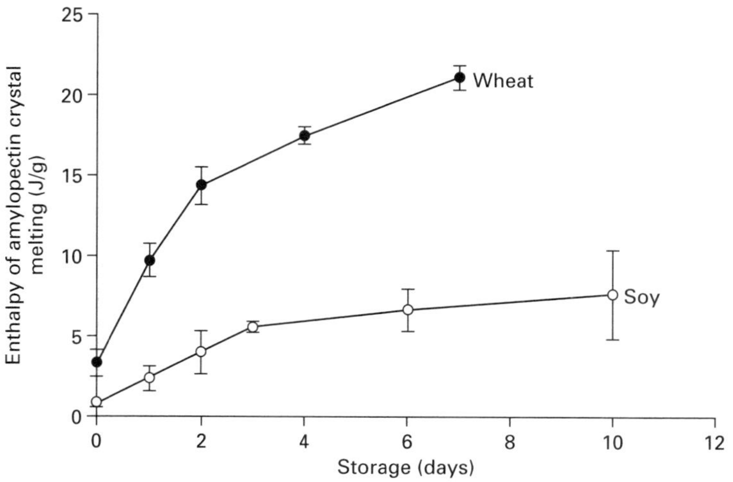

Fig. 16.5 Heat of melting of amylopectin crystals (normalized per unit weight of wheat flour) in soy bread and wheat bread obtained from the differential scanning calorimetry endothermic peak at 60 °C during 10 days of storage.

The greater water-holding capacity of soy that leads to a differential water distribution in the fresh and stored bread, as well as altered amylopectin recrystallization and a smaller change in stiffness during storage, results in an altered staling process and possibly an increased shelf-life of the soy bread product.

16.4 Health benefits of soy-enriched bread

There is mounting evidence for the potential protective effect of soy against a number of chronic diseases. Much of the focus has been placed on isoflavonoids, as described in detail in various reviews (Munro *et al.*, 2003; Chen and Rogan, 2004; McCue and Shetty, 2004; Uzzan and Labuza, 2004; Leung *et al.*, 2005; Ricketts *et al.*, 2005; Cassidy *et al.*, 2006), since soybean products are the most prominent sources of these phytoestrogens in the diet, ranging in concentration from 0.1 to 3.0 mg/g, dry weight (Coward *et al.*, 1993). Genistein, daidzein and glycitein are the three most prevalent isoflavones in soybeans, existing in four possible chemical forms: the aglycone, the β-glucoside, the malonylglucoside and the acetylglucoside. The amount and form of the isoflavones present in a food may be critical to the absorption and bioavailability of these compounds, and therefore to their eventual impact upon health.

After ingestion, the conjugated β-glucosides have been suggested to be hydrolyzed by β-glucosidase present in the human gut mucosa (Zubik and Meydani, 2003), and by bacterial β-glucosidase in the large intestine (Hutchins, *et al.*, 1995), yielding the three major aglycones: daidzein, genistein and glycitein. Throughout the digestive and absorption processes, mass balance experiments show that less than 5% of soy isoflavones are excreted in the aglycone isoflavone form (Setchell *et al.*, 2002), and are thus metabolized mostly to glucuronides and sulfate conjugates of the precursor isoflavones and microbial metabolites. Despite the fact that aglycones are readily absorbed in the upper intestinal tract, there is an ongoing discussion about the bioavailability of isoflavone aglycones compared with isoflavone glucosides (Turner *et al.*, 2003; Zhang *et al.*, 2003, 2004). Studies have suggested higher (Turner *et al.*, 2003), lower (Zhang *et al.*, 2003) or no difference in absorption (Zhang *et al.*, 2004) of the glucosides compared with the aglycones. This disagreement might be explained by the different forms in which isoflavones have been administered: pure (Coldham *et al.*, 2002) or within a food matrix (Zhang *et al.*, 2003). In addition, potential metabolites such as equol, *O*-desmethylangolensin and dihydrodaidzein have often not been included in the analyses and thus overall isoflavone absorption would be underestimated.

Conjugation patterns of isoflavones in soy foods are significantly affected by processing, and the presence of other ingredients (Wang and Murphy, 1996; Wang *et al.*, 1998; Grun *et al.*, 2001). In particular, dry heat results in the formation of acetylglucosides, and fermentation causes a loss of the glucosides to produce the aglycones (Coward *et al.*, 1998). It is therefore imperative that changes in isoflavone content and distribution are examined throughout food processing and storage. During development of our soy-containing bread we monitored the changes in isoflavone at the various processing steps. The primary source of isoflavones in the soy bread is soy flour. Malonylglucosides (47%) and β-glucosides (40%) were the prevalent isoflavones in soy flour with the aglycones representing only 3% of the total isoflavone content (Table 16.1). The isoflavone composition of soy milk powder differed in that β-glucosides (52%) and aglycones (31%) were most abundant.

Changing the isoflavone profile in raw soy ingredients requires selecting for soybean cultivars with high isoflavone contents as well as the desired isoflavone distribution. Raw soybeans contain mostly glucoside forms of isoflavone and a low percentage of aglycone forms. Isoflavone content in soybean is influenced by many factors, including genotypes, crop years, crop locations, storage period, and genotype × environment interactions (Wang and Murphy, 1994a; Hoeck *et al.*, 2000; Lee *et al.* 2003a; Riedl *et al.*, 2007). For example, Riedl and coworkers quantified the isoflavone content and distribution in major Ohio soybean varieties and then characterized specific varieties by location (Riedl *et al.*, 2007) as did other authors for various locations (Wang and Murphy, 1994b; Hoeck *et al.*, 2000 Vyn *et al.* 2002; Lee *et al.*, 2003a,b). Riedl *et al.*, 2007 found that isoflavone biosynthesis shifted towards daidzeins (mostly malonyl daidzin) from genistein and glycitein in soybeans with high total isoflavones. Such findings may help identify

Table 16.1 Isoflavone content (nmol/g, dry basis) of soy ingredients (soy flour and soy milk powder) and soy bread crumb

Isoflavone	Soy flour	Soy milk	Soy bread crumb
Daidzin	551 ± 33	866 ± 35	432 ± 20
Genistin	1222 ± 37	1926 ± 12	935 ± 8
Glycitin	120 ± 4	87 ± 8	139 ± 5
β-Glucosides	**1893**	**2879**	**1506**
Malonyldaidzin	827 ± 8	315 ± 25	404 ± 1
Malonylgenistin	1259 ± 13	593 ± 2	925 ± 8
Malonylglycitin	141 ± 4	37 ± 1	145 ± 0
Malonylglucosides	**2227**	**945**	**1474**
Acetyldaidzin	236 ± 24	85 ± 1	131 ± 1
Acetylgenistin	111 ± 0	73 ± 5	189 ± 3
Acetylglycitin	106 ± 4	70 ± 6	80 ± 4
Acetylglucosides	**453**	**228**	**400**
Daidzein	39 ± 1	551 ± 11	259 ± 3
Genistein	93 ± 1	1148 ± 46	345 ± 9
Glycitein	7 ± 0	5 ± 0	7 ± 1
Aglycones	**139**	**1704**	**611**
Total	**4712**	**5756**	**3991**

environmental variables associated with isoflavone production that lead to improved soybean crops with targeted isoflavone profiles to be used as soy ingredients.

After the different soy ingredients had been added to the soybread dough formulation, little change in the isoflavone profile or content was observed during mixing and kneading. However, proofing caused significant changes ($p < 0.05$) in isoflavone composition (Fig. 16.6) including an increase in aglycones (+153%) and a decrease in β-glucosides (–23%) as compared with soy dough before proofing. Both amounts of acetyl-β-glucosides and malonyl-β-glucosides of isoflavones changed slightly during proofing (+9% and –6%, respectively). After baking (160 °C, 50 min), there were increases in aglycones (+9%), β-glucosides (+7%) and acetylglucosides (+13%), and a decrease in malonylglucosides (–28%) when compared with soy bread dough before baking (Fig. 16.6). β-Glucosidase is the key enzyme activity that cleaves the glucose moiety from the isoflavone glucosides resulting in the isoflavone aglycones. In the soy bread formulation β-glucosidase activity has been shown to be present in the various ingredients including wheat flour, soy ingredients (soy flour, soy milk powder) and active dry yeast (*Saccharomyces cerevisiae*), but the soy-based ingredients exhibited the highest activities (Coward *et al.*, 1998).

Riedl *et al.* (2005) found that β-glucosidase activity in soy bread dough was dependent on the proofing temperature and duration. Both isoflavone β-glucosides and malonylglucosides were used as substrates for β-glucosidase to produce isoflavone aglycones. However, isoflavone β-glucosides were

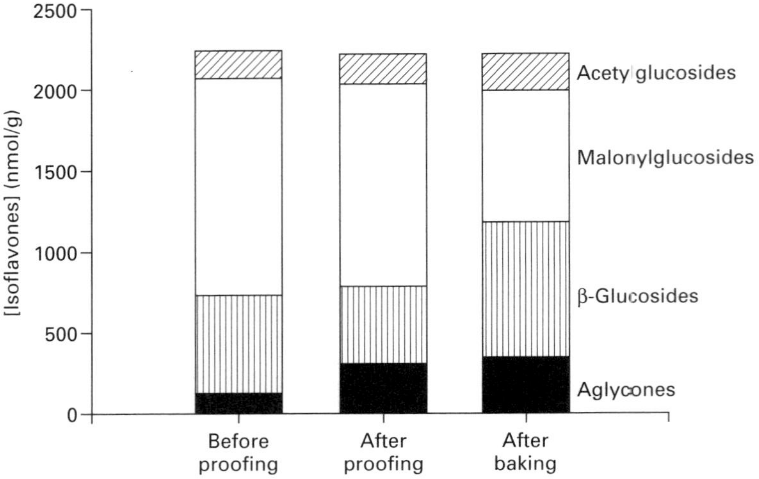

Fig. 16.6 Isoflavone concentration and profile in soy bread dough before and after proofing (48 °C, 1 hour) and soy bread after baking (160 °C, 50 min).

preferred to isoflavone malonylglucosides in this enzyme–substrate reaction. Therefore, the production of isoflavone aglycones could be manipulated to increase the aglycone content by controlling the duration and temperature of proofing. The proofing temperature of 48 °C for 2 hours was found to be optimal for aglycone production in soy bread among the proofing conditions studied (1–4 hours at 22, 32 and 48 °C).

An alternative to depending on the endogenous β-glucosidase content and activity to aid in cleavage of the glucose unit, is to add exogenous sources such as almond (a rich natural source of the enzyme). Almond addition was found to significantly increase β-glucosidase activity after proofing of soy bread dough with 2.5–10% almond (Fig. 16.7). After 1 hour proofing at 48 °C, the aglycone levels of soy bread containing 2.5, 5.0, 7.5 and 10% almond were 488, 615, 648 and 664.9 nmol/g (total weight), respectively, while it was 316 nmol/g in soy bread without almond but proofed under the same temperature and time conditions (Fig. 16.7). After baking at 160 °C for 50 min, β-glucosidase activity was not detected. Since no significant increase in aglycone content was detected at almond addition levels greater than 5%, this level was found to be an effective and economic way to enhance isoflavone aglycones in the soy bread formula.

Characterizing the isoflavone profile in the food product is a critical step in understanding their biological activity and digestive stability. Zhang *et al.*, (2003) examined the effect of soy bread extracts on the proliferation of an invasive androgen independent metastatic human prostate cancer cell line (PC-3). A decrease in PC-3 cell proliferation was only observed upon treatment with soy bread crust extract that contained a higher level of β-glucosides (43.6%) and 6-*O*″-acetylglucosides (23.2%) and a much lower level of 6-*O*″-malonylglucosides (8.6%) of isoflavones compared with soy

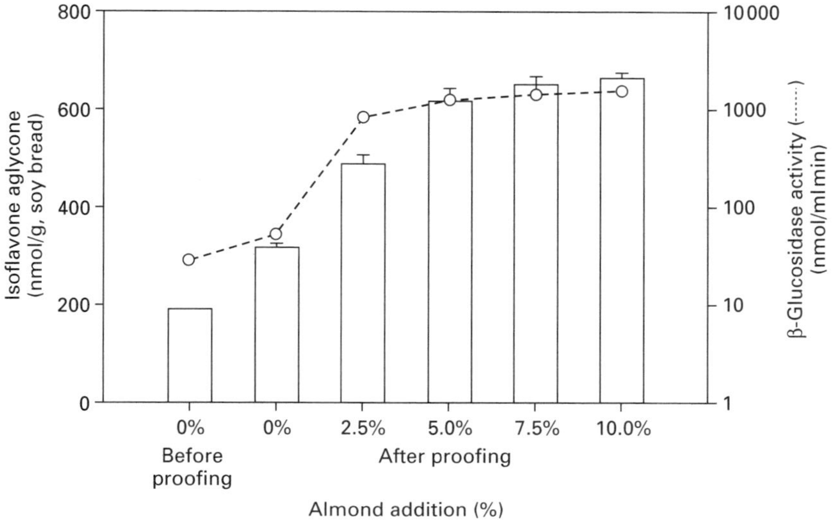

Fig. 16.7 Isoflavone aglycone (nmol/g, total weight) (bar graph) and β-glucosidase activity (dashed line) in soy bread containing almond after proofing at 48 °C for 1 hour, as compared with soy bread without almond before and after proofing.

bread crumb (32.7, 7.0 and 36.0%, respectively). In addition, the bread crust contained less moisture (16%) than the bread crumb (44%), and was also the major site for Maillard reactions, a possible reason for the differences in biological activity observed.

In order to better understand the stability of isoflavones in the digestion process, as well as the role of micelles in the bioaccessibility of these compounds, Walsh *et al.* (2003) used a model system to simulate oral, gastric and small intestinal phases. Daidzein, genistein, their corresponding β-glucosides, and glycitein in the soy bread were stable during all phases of the simulated digestion. The authors also suggested that micellarization during the small intestinal phase of digestion enhanced the bioaccessibility of isoflavonoid aglycones (especially genistein). Human clinical trials are now required to validate many of these findings.

16.5 Future trends

In the 2006 United Soybean Board (USB) annual consumer acceptability survey, they found that 82% of consumers rate soy products as healthy and 35% say they are aware of the FDA heart-healthy claim (USB, 2006). However, only 31% of those surveyed intentionally seek out soy products (USB, 2006). The lag may be due to consumers associating soy-containing products with off-flavor and unacceptable texture. In breads, the sensory acceptance of products containing more than 10–12% soy has been very low (Raidl and

Klein, 1983; Dhingra and Jood, 2004; Olaoye *et al.*, 2006) therefore, regardless of the nutrition-enhancing benefits of soy, palatability remains a major issue. However, these deleterious effects can be overcome. For example, Shogren *et al.* (2003) found that manipulating the bread ingredients produced an acceptable product when 30-40% soy inclusion was achieved. Similarly, the bread developed in our laboratory was found to be very acceptable as described previously. Therefore consumer acceptance of such products will be highly dependent on overcoming previous negative experiences with soy-containing foods.

Once sensory acceptability is obtained, various other strategies may be used to possibly enhance the nutritional qualities of the soy bread. Understanding the role and stability of the various other phytochemicals (such as saponins) in soy may indicate ways in which these compounds may be retained and increased in the product. Saponins are a structurally and biologically diverse class of glycosides of steroids and triterpenoids that are widely distributed in many plant food products. Their ability to bind bile acids suggests that saponins have the potential to serve as cholesterol-lowering agents and anti-carcinogens. Indeed, frequent consumption of soy products rich in saponins was reported as causing reduction of breast, colon and prostate cancers (Kerwin, 2004; Kim *et al.*, 2004). We have found that our soy bread contains about 0.1% soyasaponins. Their stability during the bread making process has not been investigated, nor have ways to increase the content of these phytochemicals.

A separate approach may entail combining different foods with known phytochemical contents to possibly enhance the nutritional benefit compared with consuming each food alone. For example, green tea contains polyphenols such as catechins that are antioxidants enhanced by the presence of metals (Hai, 2006). Addition of green tea to wheat bread yielded acceptable results (Jung, 2005) yet the combination of soy bread with green tea has not been studied and may yield significant improvements in health. Many other combinations are also possible as long as the product remains acceptable and the content of the phytochemicals is not degraded during processing.

Finally, changes in agricultural practices or crop manipulation may enhance the amount of phytochemicals in the soybean that would translate into a more functional soy ingredient. Results presented earlier showed that the isoflavone content and distribution may be manipulated. The same may be true for the other phytochemicals found in soy, as has been done successfully to increase the protein content in the bean.

16.6 References

Ablett, S., Attenburow, G.E. and Lillford, P.J. 1986. The significance of water in the baking process. In: *Chemistry and Physics of Baking*. Blanshard, J.M.V., Frazier, P.J., and Galliard, T. Eds. The Royal Society of Chemistry, London, pp. 30–41.

Baik, M. and Chinachoti, P. 2000. Moisture redistribution and phase transitions during bread staling. *Cereal Chemistry*, **77**(4): 484–488.
Baik, M. and Chinachoti, P. 2001. Effects of glycerol and moisture gradients on thermomechanical properties of white bread. *Journal of Agricultural and Food Chemistry*, **49**: 4031–4038.
Barnes, S. 1998. Evolution of the health benefits of soy isoflavones. *Proceedings of the Society for Experimental Biology and Medicine*, **217**: 386–392.
Biliaderis, C.G. 1990. Thermal analysis of food carbohydrates. In: *Thermal Analysis of Food*. Harwalkar, R. and Ma, C.Y. Eds. Elsevier Applied Science, London, pp. 168–220.
Birt, D.F., Hendrich, S. and Wang, W. 2001. Dietary agents in cancer prevention: flavonoids and isoflavonoids. *Pharmacology and Therapeutics*, **90**: 157–177.
Brewer, M.S., Potter, S.M., Sprouls, G. and Reinhard, M. 1992. Effect of soy protein isolate and soy fiber on color, physical and sensory characteristics of baked products. *Journal of Food Quality*, **15**: 245–262.
Brouns, F. 2002. Soya isoflavones: a new and promising ingredient for the health foods sector. *Food Research International*, **35**: 187–193.
Buck, J.S., Walker, C.E. and Watson, K.S. 1987. Incorporation of corn gluten meal and soy into various cereal based foods and resulting product functionality. Sensory and protein quality. *Cereal Chemistry*, **64**(4): 264–269.
Cassidy, A., Albertazzi, P., Nielsen, I.L., Hall, W., Williamson, G., Tetens, I., Atkins, S., Cross, H., Manios, Y., Wolk, A., Steiner, C. and Branca, F. 2006. Critical review of the health effects of soyabean phyto-oestrogens in post menopausal women. *Proceedings of the Nutrition Society*, **65**, 76–92.
Chen, A. and Rogan, W.J. 2004. Isoflavones in soy infant formula: a review of evidence of endocrine and other activity in infants. *Annual Review of Nutrition*, **24**: 33–54.
Chen, P.L., Long, Z., Ruan, R. and Labuza, T.P. 1997. Nuclear magnetic resonance studies of water mobility in bread during storage. *Lebensmittel-Wissenschaft und Technologie*, **30**: 178–183.
Chen, S.S. and Rasper, V.F. 1982. Functionality of soy proteins in wheat flour/soy isolate doughs. III. Protein and lipid binding during dough mixing. *Canadian Institute of Food Science and Technology*, **15**(4): 302–306.
Chinachoti, P. 1998. Water migration and food storage stability. In: *Food Storage Stability*. Taub, I.A and Singh, R. P. Eds. CRC Press, Boca Raton, FL.
Coldham, N.G., Darby, C., Hows, M., King, L.J., Zhang, A.Q. and Sauer, M.J. 2002. Comparative metabolism of genistin by human and rat gut microflora: detection and identification of the end-products of metabolism. *Xenobiotica*, **32**: 45–62.
Coward, L., Barnes, N.C., Setchell, K.D. R. and Barnes, S. 1993. Genistein, daidzein, and their β-glucoside conjugates: antitumor isoflavones in soybean foods from American and Asian diets. *Journal of Agricultural and Food Chemistry*, **41**: 1961–1967.
Coward, L., Smith, M., Kirk, M. and Barnes, S. 1998. Chemical modification of isoflavones in soyfoods during cooking and processing. *American Journal of Clinical Nutrition*, 68(Suppl.): 1486S–1491S.
Czuchajowska, A. and Pomeranz, Y. 1989. Differential scanning calorimetry, water activity, and moisture contents in crumb center and near-crust zones of bread during storage. *Cereal Chemistry*, **66**(4): 305–309.
D'Applonia, B.L. and Morad, M.M. 1981. Bread staling. *Cereal Chemistry*, **66**(4): 305–309.
Dhingra, S. and Jood, S. 2001. Organoleptic and nutritional evaluation of wheat breads supplemented with soybean and barley flour. *Food Chemistry*, **77**: 479–488.
Dhingra, S. and Jood, S. 2004. Effect of flour blending on functional, baking and organoleptic characteristics of bread. *International Journal of Food Science and Technology*, **39**: 213–222.
Doxastakis, G., Zafiriadis, I., Irakli, M., Marlani, H. and Tananaki, C. 2002. Lupin, soya

and triticale addition to wheat flour doughs and their effect on rheological properties. *Food Chemistry*, **77**: 219–227.
Erdman, J.M., O'Connor, M.P., Solomon, L.W. and Nelson, A.I. 1977. Production, nutritional value and baking quality of soy-egg flours. *Journal of Food Science*, **42**: 964–968.
FDA (Food and Drug Administration) (1999) Food labeling: health claims; soy protein and coronary heart disease, *Federal Register*, **64**, 57699–57733.
Fessas, D. and Schiraldi, A. 2000. Water properties in wheat flour dough I: Classical thermogravimetry approach. *Food Chemistry*, **72**: 237–244.
Fleming, S.E. and Sosulski, F.W. 1977. Breadmaking properties of flour concentrated plant proteins. *Cereal Chemistry*, **54**(5): 1124–1140.
Fournier, D.B., Erdman, J.W. and Gordon, G.B. 1998. Soy, its components, and cancer prevention: a review of the in vitro, animal and human data. *Cancer Epidemiological Biomarkers Prevention*, **7**(11): 1055–1065.
Gerrard, J.A., Abbot, R.C., Newberry, M. P., Gilpin, M. J., Ross, M. and Fayle, S.E. 2001. The effect of non-gluten proteins on the staling of bread. *Starch*, **53**: 278–280.
Grun, I.U., Adhikari, K., Li, C., Li, Y., Lin, B., Zhang, L. and Fernando, L.N. 2001. Changes in the profile of genistein, daidzein, and their conjugates during thermal processing of tofu. *Journal of Agricultural and Food Chemistry*, **49**: 2839–2843.
Hai, N.Y., Jun, J.Y. and Sheng, R.S. 2006. Effects of epi-gallocatechin on PC-3 cell cytoplasmic membrane in the presence of Cu^{2+}. *Food Chemistry*, **95**: 108–118.
Haines, P.J. Ed. 1995. Thermogravimetry. In: *Thermal Methods of Analysis: Principles, Applications and Problems*. Blackie Academic and Professional, London, New York, pp. 22–60.
Hallberg, L.M. 1996. The staling of long shelf-life bread as determined by thermal analysis. PhD Dissertation, University of Massachusetts, Amherst, MA.
Hallberg, L.M. and Chinachoti, P. 1992. Dynamic mechanical analysis for glass transitions in long shelf life bread. *Journal of Food Science*, **57**, 1–5.
Hamann, D.D., Purkayastha, S. and Lanier, T.C. 1990. Applications of thermal scanning rheology to the study of food gels. In: *Thermal Analysis of Foods*. Harwalkar, V.R., Ma, C.Y. Eds. Elsevier Science Publishers Ltd, New York, pp. 306–332.
Hasler, C.M. 2002. The cardiovascular effects of soy products. *Journal of Cardiovascular Nursing*, **16**: 50–63.
Hatakeyama, T. and Quinn, F.X. 1994. Differential thermal analysis and differential scanning calorimetry. In: *Thermal Analysis: Fundamentals and Applications to Polymer Science*. John Wiley, Chichester, UK, New York, pp. 5–18.
He, H. and Hoseney, R.C. 1990. Changes in bread firmness and moisture during long-term storage. *Cereal Chemistry*, **67**(6): 603–605.
Hendrich, S. and Murphy, P.A. 2001. Isoflavones: source and metabolism. In: *Handbook of Nutraceuticals and Functional Foods*. Wildman, R. E. C. Ed. CRC Press, Boca Raton, FL, pp. 55–75.
Henkel, J. 2000. Health claims for soy protein, questions about other components, *FDA Consumer*, **34**: 13–15, 18–20.
Hoeck, J.A., Fehr, W.R., Murphy, P.A. and Welke, G.A. 2000. Influence of genotype and environment on isoflavone contents of soybean. *Crop Science*, **40**: 48–51.
Hoseney, R.C. 1994. *Principles of Cereal Science and Technology*, 2nd edn. American Association of Cereal Chemists, St. Paul, MN. Chapters 2 and 10.
Hutchins, A.M., Slavin, J.L. and Lampe, J.W. 1995. Urinary isoflavonoid and lignan excretion after consumption of fermented and unfermented soy products. *Journal of the American Dietetic Association*, **95**: 545–551.
Kasarda, D.D., Bernardin, J.E. and Nimmo, C.C. 1976. Wheat protein. In: *Advances in Cereal Science and Technology*. Pomeranz, Y. Ed. American Association of Cereal Chemists, St. Paul, MN.
Kerwin, S.M. 2004. Soy saponins and the anticancer effects of soybeans and soy-based foods. *Current Medicinal Chemistry – Anti-Cancer Agents*, **4**: 263–72.

Kim, H.Y., Yu, R., Kim, J.S., Kim, Y.K., and Sung, M.K. 2004. Antiproliferative crude soy saponin extract modulates the expression of IκBα, protein kinase C, and cyclooxygenase-2 in human colon cancer cells. *Cancer Letters*, **210**: 1–6.

Kim, J.R., Choi, O.J. and Shim K.H. 2005. Quality properties of loaf bread added with fermented tea powder. *Journal of the Korean Society of Food Science and Nutrition*, **34**: 869–874.

Klein B.P., Perry, A.K. and Adair, N. 1995. Incorporating soy proteins into baked products for use in clinical studies. *Journal of Nutrition*, **125**(35): 666S–674S.

Knorr, D. and Betschart, A.A. 1978. The relative effect of an inert substance and protein concentrations upon leaf volume of breads. *Lebensmittel-Wissenschaft und Technologie*, **11**: 198–204.

Krog, N. Olesen, S.K., Ternaes, H. and Joensson, T. 1989. Retrogradation of the starch fraction in wheat bread. *Cereal Foods World*, **34**: 281–285.

Kulp, K. and Ponte, J.G. Jr. 1981. Staling of white pan bread: fundamental causes. *CRC Critical Reviews in Food Science and Nutrition*, **15**: 1–47.

Laine, R., Laine, M.J.K. and Roos, Y. 1994. Water plasticization and recrystallization of starch in relation to glass translation. In: *ISOPOW Practicum II*, June 19–24, Puebla, Mexico.

Lee, S.J., Ahn, J.K., Kim, S.H., Kim, J.T., Han, S.J., Jung, M.Y. and Chung, I.M. 2003a. Variation in isoflavone of soybean cultivars with location and storage duration. *Journal of Agricultural and Food Chemistry*, **51**: 3382–3389.

Lee, S.J., Yan, W.K., Ahn, J.K. and Chung, I.M. 2003b. Effects of year, site, genotype and their interactions on various soybean isoflavones. *Field Crop Research*, **81**: 181–92.

León, A., Rosell, C.M. and de Barber, C.B. 2003. A differential scanning calorimetry study of wheat proteins. *European Food Research and Technology*, **217**(1): 13–16.

Leung, H.K., Magnuson J.A. and Bruinsma, B.L. 1983. Water binding of wheat flour doughs and breads as studied by deuteron relaxation. *Journal of Food Science*, 48: 95.

Leung, L.K., Yuen, Y.M., Leung, H.Y. and Wang, Y. 2005. Dietary soya isoflavones and breast carcinogenesis: a perspective from a cell- culture model. *Nutrition Research Reviews*, **18**: 202–211.

Levine, H. and Slade, L. 1988. Influences of the glassy and rubbery states on the thermal, mechanical and structural properties of doughs and baked products. In: *Dough Rheology and Baked Product Texture*. Faridi, H. and Faubion, J.M. Eds. Van Nostrand Reinold, New York, pp. 157–330.

Lodi, A., Abduljalil, A. and Vodovotz, Y. 2007. Characterization of water distribution in bread using magnetic resonance imaging. *Magnetic Resonance Imaging* (in press).

Maga, J.A. 1975. Bread staling. *CRC Critical Review in Food Technology*, **5**: 443–492.

Manzacco, L., Nicoli, C.N. and Labuza, T. 2002. Study of bread staling by X-ray diffraction analysis.

Martin, M.L., Zeleznak, K.J. and Hoseney, R.C. 1991. A mechanism of bread firming. I. Role of starch swelling. *Cereal Chemistry*, **68**: 498–503.

McCue, P. and Shetty, K. 2004. Health benefits of soy isoflavonoids and strategies for enhancement: a review. *Critical Reviews in Food Science and Nutrition*, **44**: 361–367.

Menon, A.S. and Mujumdar, A.S. 1987. Drying of solids: principles, classification, and selection of dryers. In: *Handbook of Industrial Drying*. Mujumdar, A.S. Ed. Marcel Dekker, New York.

Messina, M.J. and Loprinzi, C.L. 2001. Soy for breast cancer survivors: a critical review of the literature. *Journal of Nutrition*, **131**: 3095S–3108S.

Messina, M.J., Persky, V., Setchell, K.D.R. and Barnes, S. 1994. Soy intake and cancer risk: a review of the in vitro and in vivo data. [Review]. *Nutrition and Cancer*, **21**(2): 113–131.

Mizrahi, S. 1967. The use of isolated soybean proteins in bread. *Cereal Chem*istry, **44**: 193.

Munro, I.C., Harwood, M., Hlywka, J.J., Stephen, A.M., Doull, J., Flamm, W.G. and Aldercreutz, H. 2003. Soy isoflavones: a safety review. *Nutrition Reviews*, Vol. **61**: 1–33.

Olaoye, O.A., Onilude, A.A. and Idowa, O.A. 2006. Quality characteristics of bread produced from composite flours of wheat, plantain, and soybeans. *African Journal of Biotechnology*, **5**(11): 1102–1106.

Peppas, N.A. and Brannon-Peppas, L. 1994. Water diffusion and sorption in amorphous macromolecular systems and foods. In: *Water in Foods*. Fito P., Mulet A., and McKenna. B, Eds. Elsevier Applied Science, New York, p. 189.

Piazza, L. and Masi. P. 1995. Moisture redistribution throughout the bread loaf during staling and its effect on mechanical properties. *Cereal Chemistry*, **72**(3): 320–325.

Porter, M.A. and Skarra, L.L. 1999. Reducing costs through the inclusion of soy flour in breads. *Cereal Foods World*, **44**: 632–637.

Pyler, E.J. 1988. *Baking Science and Technology*, 3rd edn, Vol. 1. Sosland Publishing Co., Merriam, KS, Chapters 24 and 25.

Raidl, M.A. and Klein, B.P. 1983. Effects of soy or field pea flour substitution on physical and sensory characteristics of chemically leavened quick breads. *Cereal Chemistry*, **60**(5): 367–370.

Rasmussen, P.H. and Hansen Å. 2001. Staling of wheat bread stored in modified atmosphere. *Lebensmittel-Wissenschaft und Technologie*, **34**: 487–491.

Renkema, J.M.S. 2001. Formation, structure and rheological properties of soy protein gels. PhD thesis, Wageningen University, The Netherlands.

Ricketts, M., Moore, D.D., Banz, W.J., Mezei, O. and Shay, N.F. 2005. Molecular mechanisms of the action of the soy isoflavones includes activation of promiscuous nuclear receptors. A review. *Journal of Nutritional Biochemistry*, **16**, 321–330.

Riedl, K.M., Zhang, Y.C., Schwartz, S.J., and Vodovotz, Y. 2005. Optimizing dough proofing conditions to enhance isoflavone aglycones in soy bread. *Journal of Agricultureal and Food Chemistry*, **53**(21): 8253–8258.

Riedl, K.M., Lee, J.H., Renita, M., St. Martin, S.K., Schwartz, S.J. and Vodovotz, Y. 2007. Isoflavone profiles, phenolic content and antioxidant activity of soybean seeds as influenced by cultivar and growing location in Ohio. *Journal of the Science of Food and Agriculture*, **87**, 1197–1206.

Riganakos, K.A. and Kontominas, M.G. 1997. Study of water sorption of flours (wheat and soy) using a hygrometric method: effect of relative humidity during heat treatment. *Zeitschrift für Lebensmittel-Untersuchung und Forschung A*, **204**: 369–373.

Roulet, P.H., MacInnes, W.M., WYrsch, P., Sanchez, R.M. and Raemy, A.A. 1988. Comparative study of the retrogradation kinetics of gelatinized wheat starch in gel and powder form using X-rays, differential scanning calorimetry and dynamic mechanical analysis. *Food Hydrocolloids*, **2**: 381–396.

Scheiber, M.D., Liu, J.H., Subbiah, M., Rebar, R.W. and Setchell, K.D.R. 2001. Dietary inclusion of whole soy foods results in significant reductions in clinical risk factors for osteoporosis and cardiovascular disease in normal postmenopausal women. *Menopause*, **8**: 384–392.

Schiraldi, A., Pizza, L. and Riva, M. 1996. Bread staling: a calorimetric approach. *Cereal Chemistry*, **73**: 32–39.

Selvaraj, A. and Shurpalekar, S.R. 1982. On improving the quality of soy fortified bread. *Journal of Food Science and Technology*, **19**: 242–246.

Setchell, K.D. 1998. Phytoestrogens: the biochemistry, physiology, and implications for human health of soy isoflavones. *American Journal of Clinical Nutrition*, 68(Suppl.): 1333S–1346S.

Setchell, K.D.R. and Cassidy, A. 1999. Dietary isoflavones: biological effects and relevance to human health. *Journal of Nutrition*, **129**(3): 758S–767S.

Setchell, K.D., Brown, N.M., Zimmer-Nechemias, L., Brashear, W.T., Wolfe, B.E., Kirschner, A.S. and Heubi, J.E. 2002. Evidence for lack of absorption of soy isoflavone glycosides in humans, supporting the crucial role of intestinal metabolism for bioavailability. *American Journal of Clinical Nutrition*, **76**: 447–453.

Shogren, R.I., Mohamed, A.A. and Carriere, C.J. 2003. Sensory analysis of whole wheat/soy flour breads. *Journal of Food Science*, **68**(6): 2141–2145.

Slade, L. and Levine, H. 1991. Beyond water activity: recent advances based on an alternative approach to the assessment of food quality and safety. *Critical Reviews in Food Science and Nutrition*, **30**: 115–350.

Smith, K.B. 2003. Effect of freezing on the physico-chemical properties of soy dough. MSc thesis, The Ohio State University, OH.

Taub, I.A. and Singh, R. P. 1998. Effect of storage on texture, and water migration and food storage stability. In: *Food Storage Stability*. CRC Press, Boca Raton, FL, pp. 193–267.

Turner, N.J., Thomson, B.M. and Shaw, I.C. 2003. Bioactive isoflavones in functional foods: the importance of gut microflora on bioavailability. *Nutrition Reviews*, **61**: 204–213.

USB (United Soybean Board). 2006. Consumer attitudes about nutrition, insights into nutrition, health and soyfoods. United Soybean Board 13th Annual National Report.

Uzzan, M. and Labuza, T.P. 2004. Critical issues in R&D of soy isoflavone enriched foods and dietary supplements. *Journal of Food Science*, **69**(2): 77–86.

Vittadini, E. and Vodovotz, Y. 2003. Changes in the physicochemical properties of wheat-and-soy-containing breads during storage as studied by thermal analyses. *Journal of Food Science*, **68**(6): 2022–2027.

Vodovotz, Y. and Ballard, C. 2002. Formula and process for making soy-based bakery products. A patent application, Docket No. 10–267845.

Vodovotz, Y., Hallberg, L. and Chinachoti, P. 1996. Effect of aging and drying on thermomechanical properties of white bread as characterized by dynamic mechanical analysis (DMA) and differential scanning calorimetry (DSC). *Cereal Chemistry*, **73**: 264–270.

Vodovotz, Y., Zhang, Y. C., and Schwartz, S. J. 2004. Methods for enhancing soy-containing foods and foods made thereby. A patent application, Docket No. 22727.

Vyn, T.J., Yin, X.H., Bruulsema, T.W., Jackson, C.J.C., Rajcan, I. and Brouder, S.M. 2002. Potassium fertilization effects on isoflavone concentrations in soybean [*Glycine max* (L.) Merr.]. *Journal of Agricultural and Food Chemistry*, **50**: 3501–3506.

Walsh, K.R., Zhang, Y.C., Vodovotz, Y., Schwartz, S.J. and Failla, M. 2003. Stability and bioaccessibility of isoflavones from soy bread during in vitro digestion. *Journal of Agricultural and Food Chemistry*, **51**(16): 4603–4609.

Wang, C., Ma, Q., Pagadala, S., Sherrard, M.S. and Krishnan, P.G. 1998. Changes of isoflavones during processing of soy protein isolates. *Journal of the American Oil Chemists Society*, **75**(3): 337–341.

Wang, H. and Murphy, P.A. 1994a. Isoflavone composition of American and Japanese soybeans in Iowa: effects of variety, crop year, and location. *Journal of Agricultural and Food Chemusty*, **42**: 1674–1677.

Wang, H.J. and Murphy, P.A. 1994b. Isoflavone content in commercial soybean foods. *Journal of Agricultural and Food Chemistry*, **42**: 1666–1673.

Wang, H.-J. and Murphy, P.A. 1996. Mass balance study of isoflavones during soybean processing. *Journal of Agricultural and Food Chemistry*, **44**, 2377–2383.

Willhoft, E.M.A. 1971. Bread staling. III. Measurement of the redistribution of moisture in bread by gravimetry. *Journal of Science and Food Agriculture*, **22**: 647–649.

Willhoft, E.M.A. 1973. Mechnism and theory of staling of bread and baked goods, and associated changes in textural properties. *Journal of Texture Studies*, **4**: 292–322.

Zeleznak, K.L. and Hoseney, R.C. 1986. The role of water in the retrogradation of wheat starch gels and bread crumb. *Cereal Chemistry*, **63**: 407–411.

Zeleznak, K.J. and Hoseney, R.C. 1987. Characterization of starch from bread aged at different temperature. *Starch*, **39**: 231–233.

Zhang, Y.C., Albrecht, D., Bomser, J., Schwartz, S.J., and Vodovotz, Y. 2003. Isoflavone profile and biological activity of soy bread. *Journal of Agricultural and Food Chemistry*, **51**: 7611–7616.

Zhang, Lee, J.H., Vodovotz, Y., and Shwartz, S.J. 2004. Changes in distribution of isoflavones and β-glucosidase activity during soy bread proofing and baking. *Cereal Chemistry*, **81**: 741–745
Zhou, J.R., Yu, L.Y., Zhong, Y., Gaston, S.M. and Blackburn, G.L. 2002. Soy and tea as functional foods in prevention of prostate cancer progression, *American Journal of Clinical Nutrition*, **75**: 230.
Zobel, H.F. 1973. A review of bread staling. *Baker's Digest*, **47**: 52.
Zobel, H.F. and Kulp, K. 1996. The staling mechanism. In: *Baked Good Freshness: Technology, Evaluation and Inhibition of Staling*. Hebeda, R.E. and Zobel H. Eds. Marcel Dekker, New York, Chapter 1.
Zubik, L. and Meydani M. 2003. Bioavailability of soybean isoflavones from aglycone and glucoside forms in American women. *American Journal of Clinical Nutrition*, **77**: 1459–1465.

17

Inulin in bread and other cereal-based products

D. Meyer, Sensus, The Netherlands and J. de Wolf, Cosun Food Technology Centre, The Netherlands

17.1 Introduction: trends in food consumption and product development

We cannot deny the facts. Not a single day passes without news about the obesity epidemic. According to the World Health Organization, approximately 300 million people worldwide are believed to be obese and 750 million overweight. These figures are growing explosively. A recent American study showed that obesity is more dangerous to one's health than poverty, smoking or consuming large quantities of alcohol. European studies show similar facts. Evidence suggests that obesity significantly increases the risk of heart disease, diabetes and other life-threatening conditions. These alarming figures and their long-term consequences have stimulated nutritionists to take a closer look at solutions through nutrition and diets. There are several types of product that are designed to help people lose or maintain weight:

- light products,
- low-calorie products,
- low-fat products,
- sugar-free products.

It will be clear that all kinds of fibres are very attractive ingredients for the development of such products: they are low in calories and some offer technological features that allow them to be used as fat or sugar replacers.

Although still controversial, it is thought that the high consumption of easily digestible carbohydrates is one of the factors related to the obesity epidemic (e.g. Brand Miller *et al.*, 2002), and it is partly on this basis that products with an increased content of non-digestible carbohydrates are offered on the food market. These products may contribute to weight management. Polyols and fibres are often used in these products to reduce the content of digestible carbohydrates (sugars and starch). These low-carb or low glycaemic index (GI) products offer excellent opportunities for the application of inulin and fructo-oligosaccharides (FOS) as sugar replacers or as part of a starch replacement system.

Non-digestible carbohydrates are part of the dietary fibre complex and according to all general dietary guidelines and recommendations, adequate fibre consumption is an indispensable part of a healthy diet. General food surveys show that the recommendations for fibre consumption are rarely met. For the average European consumer there is a gap of approximately 10–15 g between the recommended (30 g) and actual daily fibre intake. However, in other parts of the world there is also a considerable gap between actual and recommended fibre intake (see Tungland & Meyer, 2002). For instance, in the USA fibre consumption reaches about 15 g/d maximum (Marlett & Slavin, 1997), whereas recent recommendations are 38 g/d for adult men and 26 g/d for adult women (US Institute of Medicine, 2005). These new recommendations have now also been taken up in other countries (Health Council of The Netherlands, 2006), thereby further increasing the gap between actual and recommended fibre consumption.

This gap can be narrowed by the consumption of fibre-enriched food products. However, increasing fibre content in a food product is a difficult challenge for product development, since many fibre sources cannot easily be incorporated in foods due to adverse effects on texture and taste. It is of crucial importance that high-fibre foods have at least similar organoleptic features to standard foods. In this chapter we will not only show how inulin and FOS can be used successfully for the development of cereal-based products to lower the glycaemic response, but also how these carbohydrates can be successfully incorporated in food products to increase their fibre content.

17.2 Inulin: structure, occurrence and general properties

Inulin is a mixture of linear chains of fructose units mostly containing one terminal glucose unit, and characterised by β-(2-1) bonds between the fructose units (Fig. 17.1). The term inulin covers all β-(2-1)-linked linear fructans, with variable degrees of polymerisation (DP). Inulin is present as a reserve carbohydrate in many plants and vegetables (Vijn & Smeekens, 1999). It occurs, for instance, in wheat, onions, garlic and chicory. The inulin content ranges from less than 1% in bananas, 1–4% in wheat, 1–7% in onions and leeks to 15–20% in chicory (van Loo *et al.*, 1995). DP may extend from 3 to 250, mainly depending on the plant species.

Inulin has been available as an ingredient for foods for about 25 years. This type of inulin is extracted with hot water from chicory roots and the resulting juice is subsequently purified and spray dried (Boeckner *et al.*,

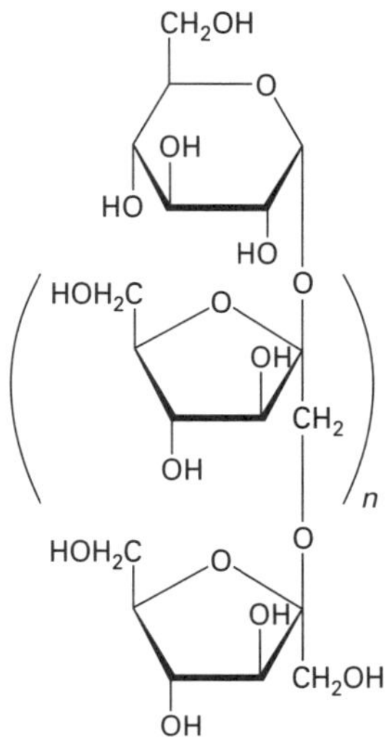

Fig. 17.1 Structural formula of inulin GF_n with terminal glucose residue (degree of polymerisation (DP) = $n + 2$).

2000). It shows a DP ranging from 2 to 60. The partial enzymatic hydrolysis product with a DP of 2–8 is called oligofructose or FOS. FOS with a DP of 2–5 can also be produced from sucrose with fructosyltransferases (Hirayama & Hidaka, 1993). Long-chain inulin (with an average DP of about 23) can be produced by applying specific separation technologies.

Due to the β-(2-1) bonds, inulin is not digestible (or is hardly digestible) by digestive enzymes in the small intestine; it reaches the colon intact where it is fermented to short-chain fatty acids (SCFA) and gases by colonic bacteria. The absorption of the SCFA and subsequent metabolism by the host, salvages some of the original energy in inulin. However, only about 1.5 kcal/g is delivered from inulin (compared with 3.9 kcal/g from fructose) which explains its low caloric value (e.g. Roberfroid, 1999).

Being indigestible carbohydrates, inulin and FOS fit well within the current concept of dietary fibre. By increasing faecal biomass and water content of stools, they may improve bowel habits (e.g. den Hond *et al.*, 2000; Kleessen *et al.*, 1997). In addition, they are classified as prebiotics; the prebiotic effect has been defined as the capacity to selectively stimulate the growth and activity of specific species of bacteria in the gut, usually bifidobacteria and lactobacilli, with benefits to health (Gibson *et al.*, 1995; Roberfroid *et al.*, 1998).

17.2.1 Technological and nutritional properties

Both inulin and oligofructose are increasingly used in new food products in all segments of the food market. Examples can be found in the dairy and bakery market, in cereals and cereal bars, in beverages and in table spreads, but also in infant formulae and in medical foods. Inulin and oligofructose are used for several nutritional reasons: as a soluble dietary fibre; because of their prebiotic and health-related benefits; because of their low caloric value (see above); and as a sugar or fat replacer. Furthermore, because of their low glycaemic response, inulin and FOS are suitable for diabetics. The potential health benefits include:

- increased mineral absorption, especially of calcium (e.g. Kim *et al.*, 2004) and magnesium (Tahiri *et al.*, 2001), and increase of bone mineral density (Abrams *et al.*, 2005);
- lowering of serum lipids, such as cholesterol or triglycerides, which may benefit heart health (e.g. Causey *et al.*, 2000; Delzenne and Williams, 2002);
- positive effect on the immune system in mouse models with a systemic infection of *Listeria monocytogenes* (Buddington *et al.*, 2002) and in patients with inflammatory bowel disease (Welters *et al.*, 2002) or children at risk for eczema (Moro *et al.*, 2006);
- lowered risk for colon cancer, especially in combination with certain probiotic bacteria (e.g. Femia *et al.*, 2002);
- a reduced production of potential liver toxins like ammonia, or potential carcinogens (van Nuenen *et al.*, 2003);
- positive effect on satiety, which may contribute to weight management (Archer *et al.*, 2004; Cani *et al.*, 2006).

Oligofructose/FOS are highly soluble and possesses technological properties that are closely related to those of sugar and glucose syrups; hence it is often used as a sugar replacer, especially in combination with high-intensity sweeteners, such as aspartame. Inulin has a lower solubility compared with FOS; it shows fat-like characteristics when used in the form of a gel in water. Thus, apart from their nutritional advantages, the recognised technological properties of inulin and oligofructose are: their capacity as a texturiser especially in low-fat food systems (Kip *et al.*, 2006); as a sugar or fat replacer; for taste improvement; and their ability to be used as fillers/binders for tablets.

17.3 Trends in bread consumption

The increasing interest in healthy foods has caused a change in the buying behaviour of the European bread consumer. Watching the baker's shop window, more and more brown and whole-wheat bread can be seen. However, there are also people who prefer white bread. For instance, for children there is nothing like a slice of white bread covered with a thick layer of peanut butter. Also, elderly people might give their preference to white bread above brown because of the easier palatability of the former type of bread.

For those who prefer white bread, but who would also like to follow a healthy lifestyle, inulins offer the opportunity to develop a fibre-enriched white bread with the same dietary fibre content as whole-wheat bread. The fibre contents of various types of bread range from about 1% for white bread to 4% in brown bread and 5% for wholemeal bread. Inulin can be added to bread (and to many other bakery products) in amounts up to 8%. Even at these relatively high concentrations, products are obtained with good organoleptic properties. This contrasts with the use of, for instance, wheat fibres, the application of which is limited due to their less suitable technological and sensory properties.

17.4 Development of inulin-enriched bread

It is well known that the inclusion of inulin in bread affects dough and bread properties (e.g. Meyer, 2003a). Inulin influences the processing of the dough, as well as dough development and bread structure, and measures have to be taken to obtain inulin-enriched dough and bread with the correct properties. Equally important is the final inulin content of the bread: the effect of baker's yeast on inulin has also been investigated, as it is well known that the invertase from this micro-organism is able to split short-chain inulins (Mitterdorfer *et al.*, 2001). These issues will be described below.

17.4.1 Effect on dough properties

In order to obtain further insight into dough properties, we carried out farinograph measurements with inulin of different average chain length and flour with different protein contents. The results (Table 17.1) show that the stability of the dough increases with increasing inulin chain length (DP) and that water absorption also increased with the addition of inulin: less water is required for a specified dough consistency following the addition of inulin. The chain length of the carbohydrate did not affect water absorption, nor

Table 17.1 Farinograph data

	Flour type Edelweiss (12.5 % protein)		Flour type Primula (14 % protein)	
Sample	Water absorption (%)	Stability (min)	Water absorption (%)	Stability (min)
Reference	61.4	10.4	62.0	10.3
5 % DP8 inulin	53.0	11.0	54.0	13.2
5 % DP10 inulin	52.0	13.7	53.0	15.3
5 % DP23 inulin	5.0	13.8	55.4	14.8
5 % DP23 inulin coarse grain	53.2	14.9	54.4	14.9

Different types of inulin were mixed with the flour and water was added. All inulins were from Sensus (Frutafit® CLR, DP8; Frutafit® HD, DP10; Frutafit® TEX, DP23) and the flour types were obtained from Meneba (The Netherlands). The reference does not contain inulin.

either did the protein content of the flour. Dough consistency during kneading was influenced by the chain length of the inulin and by the protein content of the flour. With high-gluten flour, the effects of inulin addition on dough consistency were less pronounced (Fig. 17.2a and b).

Using less water is not a proper solution as this will have a negative impact on further dough development and bread structure. On the other hand, with more water added to the dough, it may become too weak, which will hamper its processing. An additional ingredient was therefore added to support adequate water distribution during dough development and further processing. Carboxymethylcellulose (CMC) was chosen as this additional ingredient, as it has also been used by other researchers (de Man and Weegels, 2005). CMC absorbs excess water during kneading of the dough and releases it during baking, thus leading to suitable dough and bread properties.

17.4.2 Bread development

Starting from a standard recipe based on high-quality flour with high protein content, the effect of inulins with various chain lengths on bread properties was investigated. Contrary to what was expected from the farinograph experiments, the addition of long-chain inulin did not require adaptations, but with inulin of shorter chain length increasing amounts of CMC were needed. Although the water management in the dough was affected by DP23 inulin, the resulting bread properties proved to be acceptable, and hence no additional water was used.

To obtain a proper dough workability, CMC addition was required when inulin with an average DP of 10 or lower was used. Dough firmness was affected differently by the various inulins, with the highest firmness obtained when long-chain inulin was added (Fig. 17.3). The figure also shows how dough firmness and bread firmness after 1 day are related, and that different inulins have different effects.

Not surprisingly, bread volume is also influenced by inulin inclusion (Fig. 17.4): all types of inulin give a reduction in the volume, with DP10 inulin showing the largest decrease and DP23 the least change.

17.4.3 Hydrolysis by yeast

Inulin hydrolysis was measured using a simple model system consisting of a suspension of *Saccharomyces cerevisiae* and different types of inulin in water. It has been described that invertase of *S. cerevisiae* cleaves low-DP inulin (Hensing *et al.*, 1993), and it was shown in our studies that hydrolysis increased with decreasing chain length. After 3 h of incubation, almost 90% of the short-chain inulin was lost, whereas for long-chain inulin hydrolysis was only 5–10% under these *in vitro* conditions (see Fig. 17.5). The data reported by Mitterdorfer *et al.* (2001), who investigated yeast growth on various fructans, corroborate these findings as *S. cerevisiae* showed better

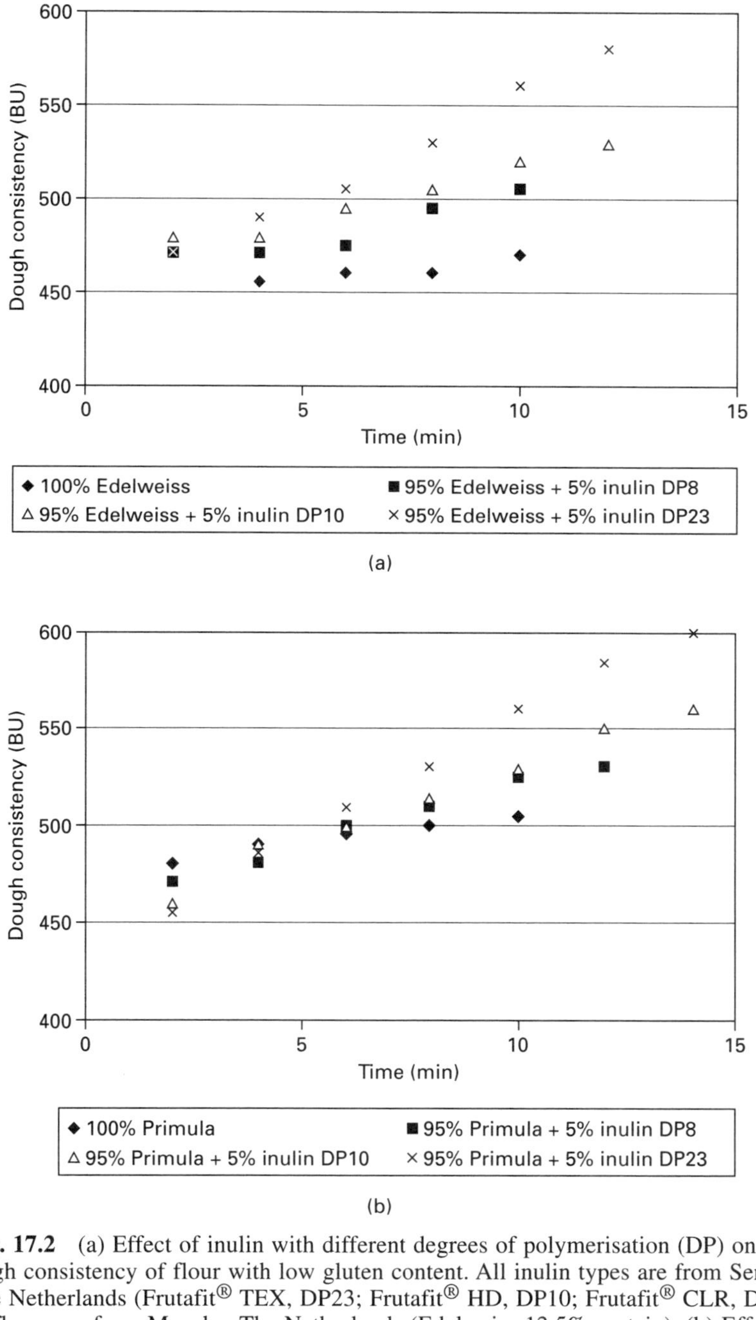

Fig. 17.2 (a) Effect of inulin with different degrees of polymerisation (DP) on the dough consistency of flour with low gluten content. All inulin types are from Sensus, The Netherlands (Frutafit® TEX, DP23; Frutafit® HD, DP10; Frutafit® CLR, DP8) and flour was from Meneba, The Netherlands (Edelweiss 12.5% protein). (b) Effect of different types of inulin on dough consistency of high gluten content flour. Inulin types are as for panel (a); flour was from Meneba (Primula 14% protein). BU, Brabender units.

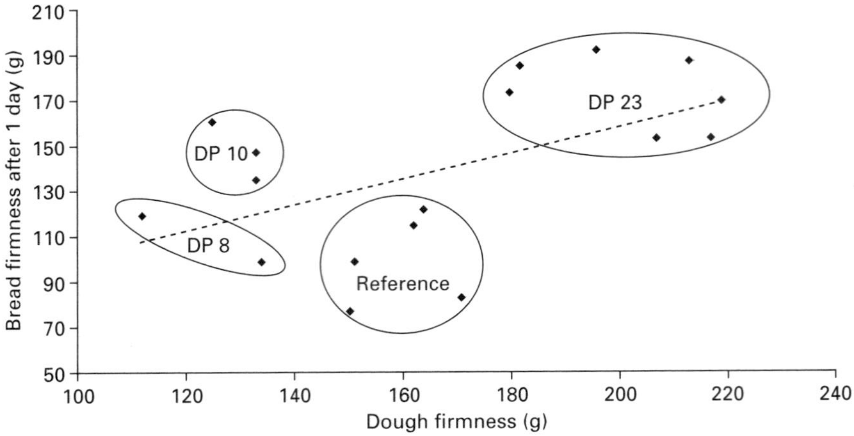

Fig. 17.3 Effect of addition of different types of inulin (DP8, DP10 and DP23) on dough firmness and bread firmness after 1 day. Reference does not contain inulin.

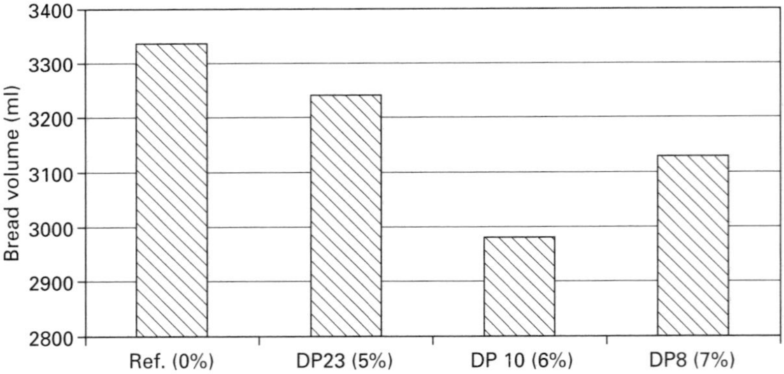

Fig. 17.4 Loaf volume with addition of inulin with different DP values.

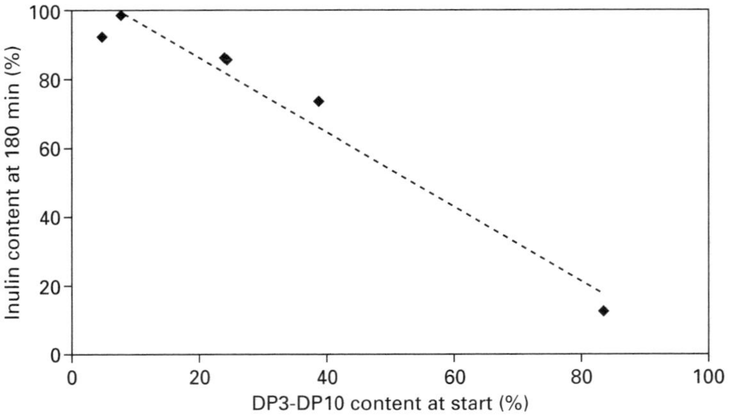

Fig. 17.5 Relationship between inulin chain length and hydrolysis by yeast.

growth on oligofructose than on native inulin. Praznik *et al.* (2002) compared the loss of inulin from Jerusalem artichoke and of inulin added as an ingredient during bread making. They also found that this loss was greater with decreasing inulin chain length.

The problem of inulin hydrolysis can be compensated directly by increasing inulin content. As an alternative, one can use yeast strains with less or no invertase activity, an approach chosen by Hara *et al.* (1986); however, the use of special types of yeast makes this route less attractive.

17.5 Inulin in product development of other cereal-based products

17.5.1 Development of a fibre-enriched cereal bar

Granola-type cereal bars typically consist of cereals, fruits, rice crisps and a sugar-based binding syrup. To increase fibre content, an aim has been to substitute sugars like sucrose and glucose syrup in the binding syrup with inulin and FOS (see also Izzo and Niness, 2001). This replacement should not lead to a decrease in product quality; the bar still has to maintain product characteristics comparable with its low-fibre counterpart and it should be able to be produced on a standard production line. The latter feature determines the technological requirements of the binding syrup.

The following parameters need to be carefully controlled to obtain a well-balanced bar:

- viscosity,
- stickiness,
- water activity,
- sweetness.

In our experiments, different amounts of sucrose and/or glucose syrup were first substituted: glucose syrup was replaced by FOS syrup and sucrose by inulin powder. The influence of different ratios of FOS and inulin on viscosity was established at different temperatures, with 60 °C representing typical processing conditions and 20 °C final product conditions. Increasing levels of FOS in the glucose syrup decreased viscosity at both temperatures (Fig. 17.6). Higher viscosity at 20 °C resulted in a harder cereal bar in these experiments. Viscosity is also one of the factors contributing to stickiness. FOS has a lower viscosity than inulin in solution, hence changing the FOS/inulin ratio is a useful tool in setting the final product characteristics; by choosing this ratio correctly, the viscosity can be set at the required value (Fig. 17.6). Some glycerol can be added to reach the required viscosity and to improve softness and flexibility. An acceptable recipe is shown in Table 17.2.

With this bar, acceptable crunchiness is guaranteed by its low water activity, the humectancy delivered by the glycerol, and the water-binding capacity of

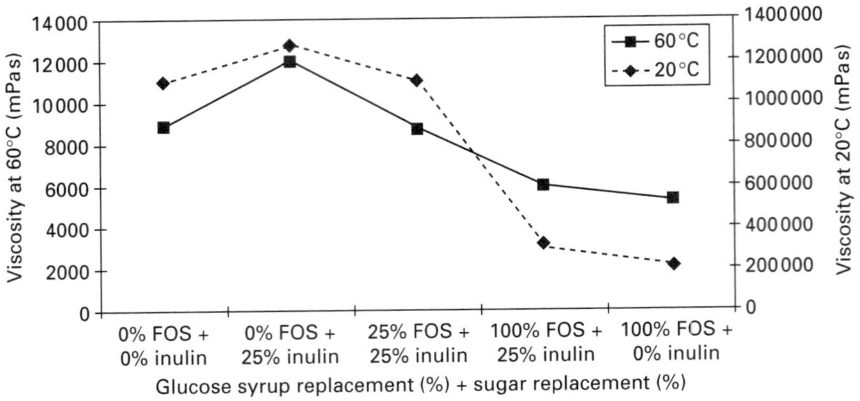

Fig. 17.6 Viscosity of cereal bar binding syrup at 60 °C and 20 °C with different levels of sugar replacement (by inulin) and glucose syrup replacement (by FOS).

Table 17.2 Recipes for cereal bars

Ingredient	Standard cereal bar (% w/w)	High-fibre bar (% w/w)	Reduced GR bar (% w/w)
Binding syrup			
Glucose syrup 60 DE	16.6	11.7	–
FOS syrup[1]	–	3.9	23.6
DP8 inulin[1]	–	4.8	6.7
Granulated sugar	15.2	11.6	–
Glycerol	–	–	1.5
Water[2]	6.0	1.8	5.9
Emulsifier[3]	0.02	0.02	0.02
Fat[4]	2.2	2.2	2.2
Flavour[5]	q.s.	q.s.	q.s.
Cereals			
Crisp granola mix[6]	60.0	60.0	60.0
Caloric value (kcal/100 g)	349	329	278

GR, glycaemic response; q.s., *quantum satis*.
[1]FOS syrup (Sensus Frutalose® L85), DP8 inulin (Sensus Frutafit® CLR).
[2]Water content is the total of added water and water from the syrups. used, after cooking of the binding syrup.
[3]Sucrose ester Sisterna SP70.
[4]Akolena® (Karlshamns).
[5]Caramel (Unifine Food & Bake).
[6]Oat flakes 40% (Meneba), rice crisps 16% (Tefco), raisins 4%.

the longer inulin chains. The water activity was very close to that of the reference bar (0.54 versus 0.56) and these values did not change over an 8-month period. Thus shelf-life of the bars is guaranteed over a long period of time.

17.5.2 Development of a low-glycaemic load, high-fibre cereal bar

A cereal bar with a low glycaemic load, developed using the low glycaemic properties of inulin and FOS, is described below. In this product concept the sugar and glucose syrup were completely replaced by inulin and FOS. The guidelines used in developing the high-fibre bar described above proved equally applicable, and the resulting recipe is given in Table 17.2. From this table it becomes apparent that the energy content is further reduced and that the fibre content is further increased compared with the high-fibre bar.

17.5.3 Measurement of glycaemic response

The glycaemic response (GR) of the bars was measured by *in vivo* trials. Serving size was such that the volunteers consumed 25 g available carbohydrate from the reference bar and comparison was made with consumption of 25 g glucose as a reference. The blood glucose response following consumption of reduced-GR bars containing the same weight (25 g) of FOS + inulin is shown in Fig. 17.7. The addition of inulin and FOS reduced the GR significantly: whereas the GR of the reference bar was 105, that of the reduced-GR bar was 65; i.e. a 40% reduction was achieved with this bar. Further lowering of the GR may be achieved by altering the cereal mix as well, but this was beyond the scope of these investigations.

17.5.4 Cakes

When incorporating inulins in cakes, fat reduction is the first target. To replace or reduce fat with any degree of success, the substitute ingredient must not only result in a lowered caloric value but it also must provide good

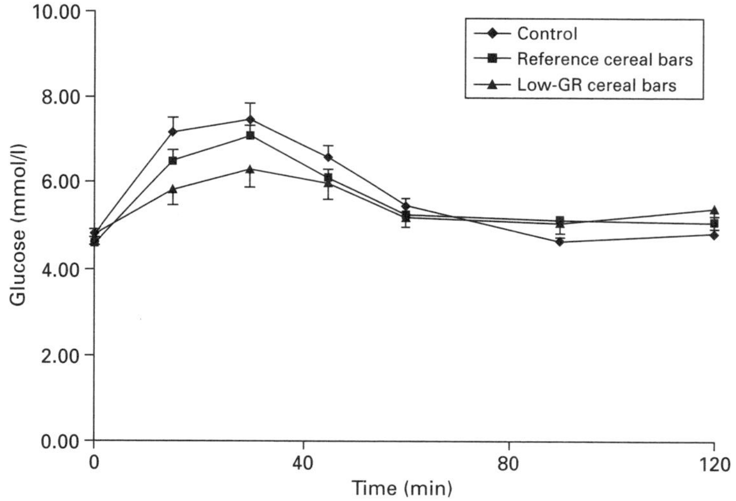

Fig. 17.7 Blood glucose response curves of the control (glucose) and cereal bars. Control was 25 g of glucose, and with the reference bar 25 g of digestible carbohydrate was consumed. For the low-GR bar the same amount was consumed as for the reference bar. All data from Sensus.

functional properties such as heat stability, emulsification, aeration, lubricity, spreadability, and, most importantly, texture and mouthfeel (Silva, 1996). Native and long-chain inulin can form a particle gel with a fat-like texture and this water-texturising characteristic is also used in cakes. Therefore, by replacing fat with inulin and water, inulin can be part of a fat replacement system in cakes.

Sugar replacement with inulin is another option for creating cakes with a healthier image. As sugar is a disaccharide, it is desirable to select inulins with a shorter chain length for replacing sugar at rates of as much as 25% (Kim, 1998). However, at higher substitution levels, the reducing power of inulin results in too high a level of Maillard browning and the development of off-flavors. Furthermore, overall sweetness is reduced, because shorter chain inulins have a relative sweetness of 30–65 in comparison with sucrose. This loss of sweetness cannot be compensated for by the use of intensive sweeteners in fine bakery products, as this is not allowed according to European Union food law.

The addition of inulin to a cake formulation for fibre enrichment purposes is the other option for introducing a health benefit in cakes. Inclusion levels should be based on fibre claim possibilities as laid down in food law. To make such claims fibre levels are needed lower than those mentioned above for fat or sugar replacement, so the technological impact will be limited.

Overall it can be concluded that, with inulin and oligofructose, fat and sugar reduction in cakes is easily possible, and that fibre enrichment is also an interesting option for creating cakes with a healthier image.

17.5.5 Biscuits

Both fat and sugar reduction in biscuits (cookies) can be targeted with inulins or FOS. Fat replacement is possible, but to a lesser extent in comparison with cake. As the moisture content of biscuits is low, the benefit of using the water-texturising effect of inulins is therefore limited or negligible. Combination with sugar replacement to realise healthier biscuits is a better option. Sugar has an important role in determining the texture, shape and size of the biscuits. It limits gluten development making dough processing easier. During baking, it dissolves and melts resulting in the final shape and size of the biscuit. As in cakes, the shorter chain inulins can replace sugar to levels of approximately 25% (Gallagher *et al.*, 2003). Chain length of the inulin is important for the shape of the biscuits: with longer chain inulin a more irregular shape will be obtained (see Fig. 17.8). From this figure it is also evident that colour formation is affected. The taste profile and texture are also influenced by the use of inulin; with DP8 inulin the organoleptic properties are very close to the reference biscuit, whereas with native inulin DP10 the taste is acceptable but crunchiness is much less. This may be related to the higher FOS content and less wide chain length distribution in DP8 inulin.

In addition to sucrose, biscuit formulations often use invert sugar or glucose

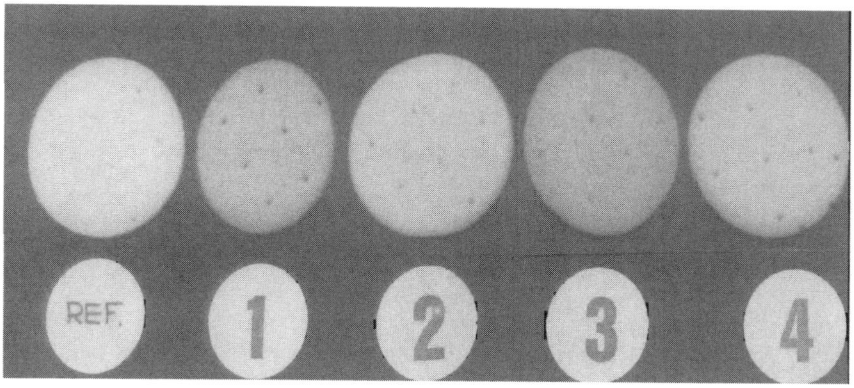

Fig. 17.8 Effect of different types of inulin on shape and colour of biscuits. REF., reference without inulin (sucrose based); 1, inulin DP10; 2, inulin DP8; 3, FOS 85% + 15% mono/disaccharides; 4, FOS 60% + 40% mono/disaccharides.

syrup for taste and colour reasons. Maillard browning is enhanced to a limited level resulting in a specific baked flavour. FOS can be used to replace both glucose syrup and invert sugar that are used for this purpose. In wafers, inulin can be used to enhance crispiness with dosage levels of approximately 1%.

17.5.6 Low glycaemic index cakes and biscuits

As mentioned earlier 'low-GI' foods attract a lot of attention. Product development for such products is aimed at reducing the content of digestible carbohydrates, in other words reducing the sugar and digestible starch content. Flour replacement for reduction of the starch content is possible to a large extent using inulin and resistant starch or dextrin. Native or long-chain inulins can be used well for their texturising effects.

Complete sugar replacement is possible using a combination of maltitol or isomalt with inulin, with a recommended ratio of 60:40 to 70:30. Inulin dilutes the possible laxative effect of the polyol. In addition, the necessity of providing warnings on the label (as legally prescribed) can be avoided if polyol content is reduced. A guideline cake batter formulation is given in Table 17.3. This recipe results in a cake with only 7 g of digestible carbohydrates per serving size (two slices), as compared with 28 g for a serving of the standard cake. Not surprisingly, the GR is considerably lowered as well (not shown).

Reduction of GR in biscuits can be attained through adaptations of the processing conditions and/or changes in the formulation (Danone, 2003). Producers who do not want to change processing conditions should consider inulin as a sugar or flour (starch) replacer with a low GI. As the first target is to limit starch gelatinisation, reduction of the water content is important. Inulin's ability to retain moisture may contribute to this effect.

Table 17.3 Guideline formulation for a low glycaemic response cake batter

	Concentration (% w/w)	
Ingredient	Reference	Low GR
Whole egg	24.8	24.8
Sugar	24.8	–
Maltitol	–	17.3
Margarine	24.7	19.7
Sunflower oil	–	5.0
Resistant starch	–	13.3
Gluten	–	2.6
DP8 inulin	–	7.4
Flour	24.8	6.8
Dried egg white	–	2.0
Salt	0.25	0.3
SAPP 15	0.3	0.4
Sodium bicarbonate	0.25	0.3
Xanthan gum	0.1	0.3
Flavour	q.s.	q.s.
Caloric value (portion)	132 kcal	108 kcal
Glycaemic response* (glucose = 100)	50 ± 5	25 ± 6

DP8 inulin (Frutafit® CLR). GR, glycaemic response; SAPP 15, sodium acid pyrophosphate (baking salt); g.s., *quantum satis*.
*GR was determined based on 25 g of digestible carbohydrate with glucose set at 100%. For the low-GR cake, a portion size identical to the reference cake was used to determine the glycaemic response.

17.5.7 Fillings

Plain fine bakery products are less common nowadays; filled and decorated bakery products are produced fulfilling the consumer's need for variation. These fillings offer another possibility to increase the healthy image of bakery products. Different types of fillings can be found in fine bakeries. The water activity of the bakery product in or on which it will be applied is the main factor determining the recipe. The most commonly applied fillings are fat-based. These have a low water activity and are applied on top of or in-between layers of biscuit. Fat is the continuous phase and keeps the other powdered ingredients together. Fat reduction is limited to approximately 25% on total product. Inulin may serve as a healthy bulking agent in fat fillings because of its fibre properties and low caloric value. Sugar reduction is another option to reduce the caloric content of a filling.

Cake has a high water activity of approximately 0.8. Therefore, water has to be used in fillings for cakes to increase the water activity to this level. These types of fillings are classified as emulsified fillings. In most cases fat and emulsified fillings are also aerated for volume and sensory reasons.

A guideline recipe for a low-fat filling is given in Table 17.4. Inulin has taken over the structuring function of the (crystallised) fat in this formulation.

Table 17.4 Guideline formulation for a low-fat aerated cake filling

Ingredient	Concentration (% w/w)
Fructose syrup	42.1
Inulin	13.5
Water	13.0
Skimmed milk powder	13.0
Sugar	11.0
Vegetable fat	6.0
Emulsifier	1.0
Carrageenan	0.2
Flavour	q.s.
Colour	q.s.

q.s., *quantum satis*.

Depending on the required texture, native or long-chain inulin can be applied. The fructose syrup in this recipe is derived from chicory. When a reduction in sweetness is desired, fructose can be replaced by 25–50% FOS.

Finally, inulin or FOS can be applied in fruit fillings for bakery applications. Inulin or FOS can replace part of the sugars and serve as a fibre source. These fillings also contain relatively high amounts of dry matter and therefore higher solubility is required and FOS is the most suitable ingredient. The enhancement of fruit flavour by inulin/FOS is an additional technological benefit in this application.

17.6 Nutrition and health claims for inulin-containing cereal products

17.6.1 Nutrition claims

The fibre properties of inulin and FOS allow the development of products with claims such as 'source of fibre' or 'rich in fibre'. National regulations dictate what the total fibre content must be to make these type of claims. For instance, in the latest European Union proposal for health and nutrition claims (European Commission, 2006), a product must contain at least 3% total fibre in order to carry the claim 'source of fibre', and 6% fibre for a 'high in fibre' claim. Similar regulations exist in other countries.

Applying inulins as sugar or fat replacers lowers the sugar or fat content of the food product as well as the caloric content. These features may be used for nutrition-type claims such as 'low in sugar', 'reduced fat content' or 'low in energy'. As with fibre claims, in many countries regulations exist that set the rules for these type of claims (e.g. European Commission, 2006; US FDA, 1994).

For the low-GR products mentioned above, it would be interesting to apply low-GI or GR claims. However, such claims are only allowed in very few countries and then only under fairly stringent conditions for testing the GI of the products (Australia, Sweden, South Africa). Reduced carbohydrate claims are described in the recently published European Union regulation (European Commission, 2006) but they refer to the total carbohydrate content and not to the digestible carbohydrate content, specifically.

17.6.2 Inulin-containing bread with a health claim

In prepared fibre-enriched bread with inulin, one can make use of the prebiotic properties of this fibre ingredient. Indeed, it has proved possible to develop a product with a prebiotic claim. In close collaboration between a large Dutch retailer, an industrial baker and an inulin supplier, inulin-containing bread was developed with an approved gut health effect (Meyer, 2003b): 'Consumption of three slices of Vitaalbrood® flora per day supports a well-balanced gut flora composition and colonic function by selectively stimulating the growth of *Bifidobacterium*. Vitaalbrood® flora contains at least 5 g Frutafit®-inulin per 100 gram'.

When making a claim based on this effect, the wording should be carefully chosen. One has to find a proper balance between what is scientifically proven (in this case the bifidogenic effect) and what the consumer understands. Unpublished data (Sensus, 2002) show that, in The Netherlands, only about 10% of the population has some knowledge about 'Bifidus', whereas about 40% have heard about 'gut flora', and 60% about 'fibres'. (It is also interesting to note that the term 'functional food' was hardly known to the general public.) Not surprisingly, therefore, the actual wording was more focussed on gut flora than on the bifidogenic effect. The latter was formulated on the label as 'supports a well-balanced gut flora'.

An inulin supplier (Sensus) was responsible for setting up the dossier for approval of the health effect according to the Dutch Code of Practice for health claims on food products (Voedingscentrum, 1998), and also supported product development as described above as well as the information for the consumer. This tripartite collaboration of retailer, industrial baker and ingredient supplier resulted not only in the first approved gut health claim (Voedingscentrum, 2003), but also showed that, with the combination of an officially approved health claim and a communication campaign, the functional food market and consumer knowledge can be further increased. It also led to increased sales volumes and distribution penetration of inulin-enriched bread.

17.7 References

Abrams, S.A., Griffin, I.J., Hawthorne, K.M., Liang, L., Gunn, S.K., Darlington, G. and Ellis, K.J. (2005) A combination of prebiotic short- and long-chain inulin-type fructans enhances calcium absorption and bone mineralization in young adolescents. *Am. J. Clin. Nutr.* **82**: 471–476.

Archer, B.J., Johnson, S.K., Devereux, H.M. and Baxter, A.L. (2004) Effect of fat replacement by inulin or lupin-kernel fibre on sausage patty acceptability, post-meal perceptions of satiety and food intake in men. *Br. J. Nutr.* **91**: 591–599.

Boeckner, L.S., Schnepf, M.I. and Tungland, B.C. (2001). Inulin: A review of nutritional and health implications. *Adv. Food Nutr. Res.* **43**: 1–63.

Brand-Miller, J.C., Holt, S.H.A., Pawlak, D.B. and McMillan, J. (2002) Glycemic index and obesity. *Am. J. Clin. Nutr.* **76**: S281–S285.

Buddington, K., Donahoo, J.B. and Buddington, R.K. (2002) Dietary oligofructose and inulin protect mice from enteric and systemic pathogens and tumor inducers. *J. Nutr.* **132**: 472–477.

Cani, P.D., Joly, E., Horsmans, Y. and Delzenne, N.M. (2006) Oligofructose promotes satiety in healthy humans: a pilot study. *Eur. J. Clin. Nutr.* **60**: 567–572.

Causey, J.L., Feirtag, J.M., Gallaher, D.D., Tungland, B.C. and Slavin, J.L. (2000) Effects of dietary inulin on serum lipids, blood glucose and the gastrointestinal environment in hypercholesterolemic men. *Nutr. Res.* **20**: 191–201.

Danone (2003) Slowly digestible carbohydrates: benefits of a diet with a low glycemic index. *Danone Nutritopics*, **28**.

de Man, M. and Weegels, P.L (2005) High-fibre bread comprises a combination of carboxymethylcellulose and at least one other type of fibre material to improve softness of the crumb and provide prolonged softness in time. Patent Application WO 05/ 023007.

Delzenne, N.M. and Williams, C.M. (2002) Prebiotics and lipid metabolism. *Curr. Opin. Lipidol.* **13**: 61–67.

den Hond, E., Geypens, B. and Ghoos, Y. (2000) Effect of high performance chicory inulin on constipation. *Nutr. Res.* **20**: 731–736.

European Commission (2006) Regulation No. 1924/2006 of the European Parliament and of the Council of 20 December 2006 on Nutrition and Health Claims Made on Foods, Brussels, Belgium, http://eur-lex.europa.eu/LexUriServ/site/en/oj/2006/l_404/ l_40420061230en00090025.pdf.

Femia, A.P., Luceri, C., Dolara, P., Giannini, A., Biggeri, A., Salvadori, M., Clune, Y., Collins, K.J., Paglierani, M. and Caderni, G. (2002) Antitumorigenic activity of the prebiotic inulin enriched with oligofructose in combination with the probiotics *Lactobacillus rhamnosus* and *Bifidobacterium lactis* on azoxymethane-induced colon carcinogenesis in rats. *Carcinogenesis* **23**: 1953–1960.

Gallagher, E., O'Brien, C.M., Scannell, A.G.M. and Arendt, E.K. (2003) Evaluation of sugar replacers in short dough biscuit production. *J. Food Engineer* **23**: 261–263.

Gibson, G.R., Beatty E.R., Wang, X. and Cummings J.H. (1995) Selective stimulation of bifidobacteria in the human colon by oligofructose and inulin. *Gastroenterology* **108**: 975–982.

Hara, K., Kono, T., Nakatomi, Y., Niimoto, H. and Umeda, F. (1986) Baker's yeast which does not ferment sucrose – suitable for making bread containing fructooligosaccharide(s). European Patent EP 197497.

Health Council of The Netherlands (2006) *Guidelines for Dietary Fiber Intake*. Publication No. 2006/03.

Hensing, M.C.M., Rouwenhorst, R.J., Scheffers, W.A. and Dijken, J.P. van (1993) Production and localization of inulinase in *Kluyveromyces* yeast. In: Fuchs, A. (ed.), *Inulin and Inulin-Containing Crops*. Elsevier Science Publishers, Amsterdam, pp. 241–250.

Hirayama, F. and Hikada, H. (1993) Production and utilization of microbial fructans. In: Suzuki M. and Chatterton, N.J. (eds.), *Science and Technology of Fructans*. CRC Press, Boca Raton, Florida, pp. 274–302.

Izzo, M. and Niness, K. (2001) Formulating nutrition bars with inulin and oligofructose. *Cereal Foods World*, March: 102–106.

Kim, Y.-A. (1998) Effects of fructo-oligosaccharide and isomalto-oligosaccharide on quality and staling of cake. *Han'guk Sikp'um Yongyang Kwahak Hoechi* **25**: 875–888 (English summary).

Kim, Y.-Y., Jang, K.-H., Lee, E.-Y., Cho,Y., Kang, S.A., Ha, W.-K. and Choue, R. (2004) The effect of chicory fructan fiber on calcium absorption and bone metabolism in Korean postmenopausal women. *Nutr. Sci.* **7**: 151–157.

Kip, P., Meyer, D. and Jellema, R.H. (2006) Inulins improve sensoric and textural properties of low-fat yoghurts. *Int. Dairy J.* **16**: 1098–1103.

Kleessen, B., Sykura, B., Zunft, H.J. and Blaut, M. (1997) Effects of inulin and lactose on fecal microflora, microbial activity, and bowel habit in elderly constipated persons. *Am. J. Clin. Nutr.* **65**: 1397–1402.

Marlett, J.A. and Slavin, J.L. (1997) Position of the American Dietetic Association: health implications of dietary fiber. *J. Am. Diet. Assoc.* **97**: 1157–1159.

Meyer, D. (2003a) Frutafit®inulin applications in bread. *Innov. Food Technol.* **18**: 38–40.

Meyer, D. (2003b) A health claim for inulin-containing bread: experience with the Dutch code of practice for health claims. Abstracts of the 9th European Nutrition Conference (Rome, 1–4 October 2003). *Ann. Nutr. Metab.* **47**: 616.

Mitterdorfer, G., Kneifel, W. and Viernstein, H. (2001) Utilization of prebiotic carbohydrates by yeasts of therapeutic relevance. *Lett. Appl. Microbiol.* **33**: 251–255.

Moro, G., Arslanoglu, S., Stahl, B., Jelinek, J., Wahn, U. and Boehm, G. (2006) A mixture of prebiotic oligosaccharides reduces the incidence of atopic dermatitis during the first six months of age. *Arch. Dis. Child.* **91**: 814–819.

Praznik, W., Cieslik, E. and Filipiak–Florkiewicz, A. (2002) Soluble dietary fibres in Jerusalem artichoke powders: composition and application in bread. *Nahrung* **46**: 151–157.

Roberfroid, M.B. (1999) Caloric value of inulin and oligofructose. *J. Nutr.* **129**: S1436–S1437.

Roberfroid, M.B., van Loo. J.A. and Gibson, G.R. (1998) The bifidogenic nature of chicory inulin and its hydrolysis products. *J. Nutr.* **128**: 11–19.

Silva, R.F. (1996). Use of inulin as a natural texture modifier. *Cereal Foods World* **41**: 792–794.

Tahiri, M., Tressol, J.C., Arnaud, J., Bornet, F., Bouteloup-Demange, C., Feillet-Coudray, C., Ducros, V., Pepin, D., Brouns, F., Rayssiguier, A.M. and Coudray, C. (2001) Five-week intake of short-chain fructo-oligosaccharides increases intestinal absorption and status of magnesium in postmenopausal women. *J. Bone Min. Res.* **16**: 2152–2160.

Tungland, B.C. and Meyer, D. (2002) Nondigestible oligo- and polysaccharides (dietary fiber): their physiology and role in human health and food. *Compr. Rev. Food Sci. Food Safety* **3**: 73–92, http://members.ift.org/NR/rdonlyres/0FBCEF73–7765–41A5-89B9-2FE55410D5F7/0/crfsfsv01n3p073092ms20001598.pdf.

US Institute of Medicine (2005) Dietary, functional, and total fiber. In: *Dietary Reference Intakes for Energy, Carbohydrate, Fiber, Fat, Fatty Acids, Cholesterol, Protein, and Amino Acids (Macronutrients).* National Academies Press, Washington, DC, Chapter 7, pp. 339–421, http://www.nap.edu/catalog/10490.html.

US FDA (United States Food and Drug Administration) (1994) *A Food Labeling Guide*, http://www.cfsan.fda.gov/~dms/flg-toc.html.

van Loo. J., Coussement, P., de Leenheer. L., Hoebregs, H. and Smits, G. (1995) On the presence of inulin and oligofructose as natural ingredients in the western diet. *Crit. Rev. Food Sci. Nutr.* **35**: 525–552.

van Nuenen, M., Meyer, P.D. and Venema, K. (2003) The effect of various inulins and *Clostridium difficile* on the metabolic activity of the human colonic microbiota in vitro. *Microb. Ecol. Health Dis.* **15**: 137–144.

Vijn, I. and Smeekens, S. (1999) Fructan: more than a reserve carbohydrate? *Plant Physiol.* **120**: 351–359.

Voedingscentrum (1998) Code of Practice for Assessing the Scientific Evidence for Health Benefits in Health Claims on Food and Drink Products.

Voedingscentrum (2003) Assessment Report Vitaalbrood® flora, http://www.voedingscentrum.nl/NR/rdonlyres/85938396-FFD6-44B0-8143-769E3B25A14C/0/beoordelingsrapport_vitaalbrood_florapdf.pdf.

Welters, C.F.M., Heineman, E., Thunnissen, F.B.J.M., van den Bogaard, A.E.J.M., Soeters, P.B. and Baeten, C.G.M.I. (2002) Effect of dietary inulin supplementation on inflammation of pouch mucosa in patients with an ileal pouch–anal anastomosis. *Dis. Colon Rectum* **45**: 621–627.

18

High-fibre pasta products

C. S. Brennan, Massey University, New Zealand

18.1 Introduction

The convenience of pasta in terms of consumer preparation, and the relatively uncomplicated ingredient formulation and processing conditions required for the manufacture of pasta, mean that pasta is an excellent model food system and vehicle for the inclusion of dietary fibres. Although this inclusion is associated with problems in relation to cooking characteristics and the texture of pasta, it provides a tangible opportunity to develop nutritious, fibre-rich, carbohydrate-rich but glycaemic index (GI)-low food products, suitable for the health-conscious consumer. This chapter explores some of the work conducted on fibre addition to pasta and its implications in terms of food structure and function.

18.2 Pasta origins and its importance as a food commodity

Traditional pasta is composed of two basic ingredients: wheat (durum) semolina and water. These ingredients are mixed into a crumble-dough before being formed either by manual manipulation or extrusion, and cut into appropriate shapes. The product is then either sold as a fresh (wet) commodity or can be dried down to be used as a shelf-stable, low-moisture, food product. The ease of pasta manufacture and the simplicity of raw material formulation make pasta a relatively cheap food product to manufacture. Indeed, the limited number of raw ingredients makes pasta an attractive model food system in terms of studying potential nutritional and technological effects of the incorporation of dietary fibre into the food system.

There are a range of processing parameters that also need to be considered when determining pasta quality (such as the control of the extrusion temperature, die conformation and pressure, mixing time and screw speed, drying conditions and the presentation of the pasta in either a dried or fresh state to the consumer). Variations in die design enable a plethora of shapes and sizes of pasta products (spaghetti, linguine, vermicelli, penne and gnocchi to mention a few). Each of these factors would have an effect on how the raw materials combine and form the product structure, with a resulting effect on product texture and digestibility. In addition to these process-related factors, there is the potential to add other ingredients such as egg, flavours and colours, and non-durum-wheat material (for example, in the manufacture of gluten-free pasta) to manipulate the colour, flavour and texture of pasta.

The variation in pasta shape and size, the ease of manufacture and simplicity of cooking make pasta a global food commodity. Pasta is part of the staple diet in many countries and has found global consumer acceptance, as is illustrated by the import and export values of the commodity as presented in Table 18.1. Not surprisingly, Italy dominates the global export quantity and also the value of pasta products, with the United States of America (USA) and Canada making the top three exporters of pasta. The USA, Germany and France are the top three importers of pasta globally, illustrating the economic importance of pasta and pasta-related products.

18.3 Typical composition of pasta and noodles

The average chemical composition of a range of pasta (and pasta-related products) is given in Table 18.2. As can be seen from the table, the contribution of pasta to the diet is mainly in terms of carbohydrates (not surprising considering the composition of the raw materials used). From an energy intake point of view, pasta and spaghetti have a relatively low energy value when compared with other carbohydrate counterparts such as white or wholemeal bread (931 and 922 kJ, respectively). Pasta also contains an appreciable level of protein, and in certain types (such as wholemeal spaghetti) can be a contributor of dietary fibre to our diet. One of the food industry's interests recently has been the design of high-fibre or fibre-enriched food items using commonly consumed foods. With this in mind, pasta presents itself as an ideal product in which to incorporate fibre ingredients.

The importance of fibre in terms of improving the nutritional content of foods has been extensively studied (Trowell, 1972; DeVries and Faubion, 1999; DeVries and Prosky, 1999; Guillon and Champ, 2000; Brennan, 2005; Casiraghi *et al.*, 2006). Considerable attention has been focused on the use of fibre as a calorie dilutant in food products, with fibre components generally being regarded as non-calorie-containing ingredients. Thus diets rich in dietary fibre have been linked to lowering the calorie value of products. Related to this, fibre-rich diets have also been linked to the reduction of starch degradation

Table 18.1 Export and import values of pasta products

	Export value ($US)	Export quantity (tonnes)	Import value ($US)	Import quantity (tonnes)
Italy	1196682	1429765	24958	26930
USA	89170	88894	333184	271513
Canada	85219	49483	89588	85787
Belgium	66747	57965	64075	46181
Switzerland	50384	21560	61115	34683
Spain	39110	47656	43210	23167
France	33291	30273	197461	242590
Greece	31585	54311	8972	9714
Germany	29215	19146	248039	301068
Japan	22571	9017	187643	139469
Netherlands	18493	12766	38620	35274
Australia	17217	13296	39761	35343
Austria	16999	17134	32346	24961
China	15385	18892	1829	1965
Denmark	9132	2996	19886	25963
Chile	7516	14816	7690	14809
UK	6175	2364	135484	120952
Hungary	5597	3157	4736	5677
Ireland	4422	1789	7512	4166
Brazil	4045	5077	8836	11663
Portugal	3594	5420	11297	10690
Malaysia	3398	2299	2763	3375
Sweden	2480	1247	41721	44518
Poland	2453	2608	12808	14702
Pakistan	1997	4520	76	104
India	1442	1388	2920	2755
Finland	966	1302	10444	9396
Luxembourg	555	196	6377	2816
New Zealand	143	64	5283	3993
Norway	20	18	7882	7536

Source: FAO statistics, faostat.fao.org.

Table 18.2 Chemical composition of cooked pasta (per 100 g)

Pasta type	Water (g)	Protein (g)	Fat (g)	Carbohydrate (g)	Starch (g)	Dietary fibre (g)	Energy value (kJ)
Macaroni	78.1	3.0	0.5	18.5	18.2	0.9	1483
Noodles (egg)	84.3	2.2	0.5	13.0	12.8	0.6	1656
Pasta (fresh)	61.5	6.6	1.5	31.8	30.7	1.9	677
Spaghetti (white)	73.8	3.6	0.7	22.2	21.7	1.2	442
Spaghetti (wholemeal)	69.1	4.7	0.9	23.2	21.9	3.5	485

Source: Food Standards Agency (2002).

during digestion, the subsequent control of glucose and insulin responses and reduction of cholesterol levels. In addition, fibre-rich diets contain large amounts of material that resists digestion and can act as sources of fermentation for intestinal bacterial. This, together with the ability of fibre ingredients to alter the viscosity profile of digesta, has other beneficial nutritional effects such as acting as a adjunct to reduced cancer rates (Brennan, 2005). Pasta has been just one cereal commodity where there have been focused efforts on incorporating added dietary fibre to improve the level of fibre.

18.4 Pasta and glycaemic index

The ease of cooking and preparation of pasta partly explain its popularity as a convenience food. It has long been regarded as a healthy food attracting a wide consumer base. This viewpoint that pasta is healthy partly arises from the consumer's perception of pasta being part of a 'Mediterranean' diet, and is partly due to the understanding that this carbohydrate-rich food has a relatively low GI or glycaemic load. In recent years, there has been a movement by consumers to regard most carbohydrate-rich foods as calorie-dense food items, labelled (in the consumer's eye at least) as high-GI foods. Related to this there has been a public perception that high-GI foods are linked to the rising levels of obesity that are being observed globally. However, the GI of pasta products is relatively low (Table 18.3), a factor that has been used in the marketing of some pasta products.

Evidence showing that pasta affects blood glucose levels differently to most carbohydrate-rich foods came from the early work by Jenkins *et al.* (1983), who confirmed that pasta exerts a lower glycaemic effect when ingested than one would assume based on carbohydrate loading and that it has a potential benefit in terms of weight control. Tappy *et al.* (1996) demonstrated that the GI of foods could be altered by the inclusion of

Table 18.3 Comparative glycaemic index and glycaemic loading of pasta. Adapted from Foster-Powell, *et al.* (2002)

Pasta type	Glycaemic index (glucose = 100)	Glycaemic loading (per serving)	Serving size (g)
Maize pasta (gluten-free)	78	32	180
Fettucine (egg)	40	18	180
Noodles (instant)	47	19	180
Macaroni	47	23	180
Spaghetti (white)	32	15	180
Spaghetti (wholemeal)	32	14	180
Bread (white fibre enriched)	68	9	30
Bread (resistant starch rich)	77	11	30
Bread (wholemeal)	69	8	30

dietary fibre. In their work they incorporated oat beta-glucan into breakfast cereals, illustrating that the presence of fibre could reduce the post-prandial glycaemic response by up to 50%. This modulation of glucose response has been further estimated by Jenkins *et al.* (2000), who indicated that 1 g of fibre (beta-glucan) per 50 g of ingested carbohydrates could reduce the GI of food by four units.

The specific relationship between fibre ingredients and the reduction in GI has to be established on a fibre by fibre basis as generalisations can be made that may not necessarily hold true when further verification is attempted. However, the incorporation of fibre into pasta can enhance its appeal as a food product rich in slowly digestible carbohydrates. It is possible, therefore, to use fibre enrichment of pasta to manipulate energy intake and weight gain and possibly to have an effect on the perception of satiety (Barkeling *et al.*, 1995; Delargy *et al.*, 1997).

18.5 The influence of pasta processing on pasta structure

The low GI value of a pasta appears not only to be related to the composition of the raw ingredients but also to the processing conditions used during pasta manufacture. Several researchers have suggested that compaction of food structure is an explanation as to why pasta has low GI values. The implication is that pasta has a tightly bound protein and starch matrix that holds the material together in a compact fashion (Pagani *et al.*, 1986; Wolever *et al.*, 1986; Fardet *et al.*, 1998, 1999a,b; Tudorica *et al.*, 2002). The idea behind this suggestion is that the compact nature of the pasta inhibits water movement into the pasta, reducing the amount of starch hydration during cooking, the degree of starch gelatinisation, swelling, dispersal, and hence the potential degradation of starch. The ease of starch digestion being related to the degree of starch gelatinisation.

In relation to the possible compaction of structure observed in pasta, scanning electron microscopy images of raw pasta products illustrate that, although the structure of pasta is relatively compact, there is a porosity in the pasta structure that is not present in other carbohydrate foods such as bread (Fig. 18.1). This would imply that the extrusion process has a greater effect on the structure of pasta than just that of shape formation. In particular, the preponderance of relatively swollen starch granules in pasta compared with the image of gelatinised starch granules in bread (Fig. 18.1) helps to explain potential differences in the nutritional quality (i.e. the GI) of pasta and bread.

However, it would appear that starch granules in pasta products behave differently depending on whether they are at the outer surface of the pasta product or in the internal sections. Cunin *et al.* (1995) demonstrated that gelatinisation of starch granules tended to occur more readily at the outer regions of pasta strands, whereas incomplete gelatinisation was evident in the central regions (potentially due to the limited amount of water absorption

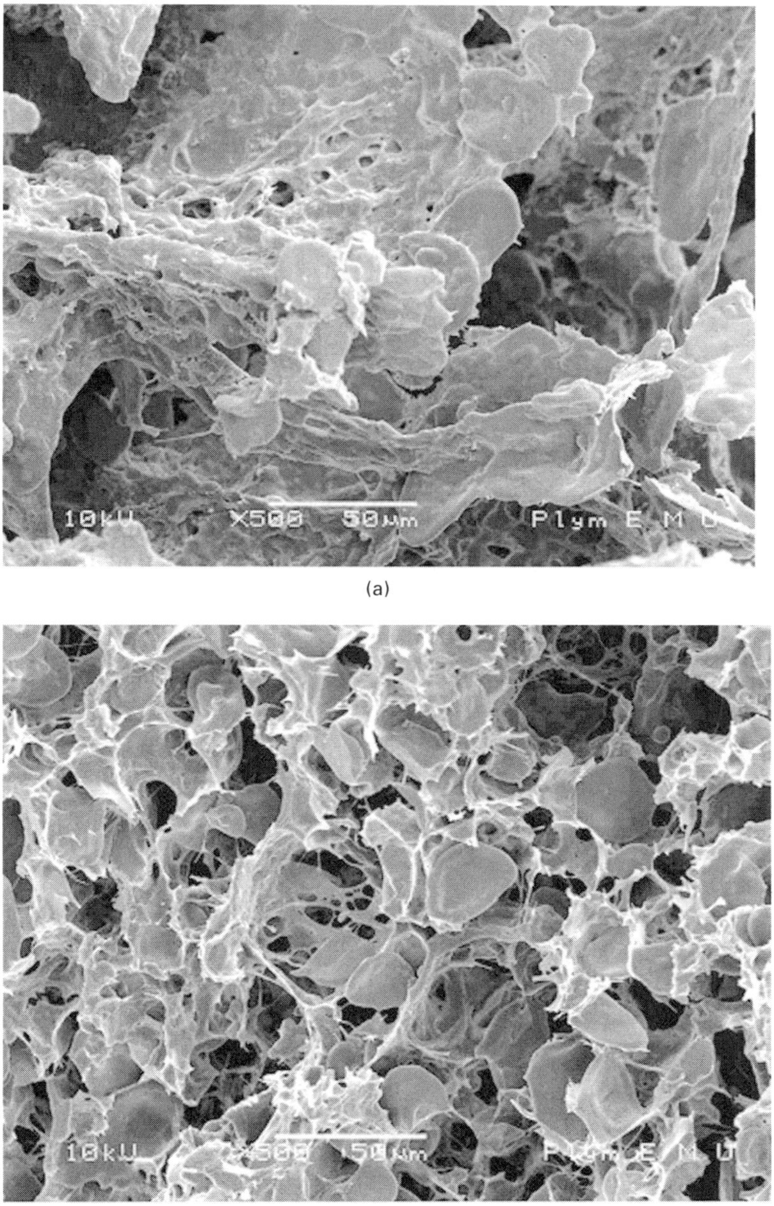

Fig. 18.1 Scanning electron micrographs of (a) cooked bread and (b) pasta.

in this area). Research from our laboratories also suggests that this is a possible mechanism by which starch degradation is manipulated, and that the incorporation of fibre into pasta products can further enhance this effect by altering pasta structure (Brennan *et al.*, 2004).

Most extrusion processes subject the raw material (in this case starch and other carbohydrates) to relatively high pressures, temperature fluctuations and high shear (Wang *et al.*, 1993; Jin *et al.*, 1994). All of these have a significant effect on starch and carbohydrate degradation. It is well known that extrusion technologies can be used to pre-digest starch molecules so that reductions in the molecular weight of the amylose and amylopectin content of starch can be seen after extrusion (Politz *et al.*, 1994). This degradation of the starch molecule and reduction in molecular weight to shorter and simpler sugar units could explain the significance of extrusion in the GI of extruded breakfast cereal products. Although the extrusion process used for the production of pasta is less harsh (in terms of having a lower processing temperature, and potentially less mechanical energy input) than that of breakfast cereal production, some modification of the starch molecule may occur during production. This modification would be related to the particle size of the semolina used in the raw materials, the amount of water present in the system and the control of barrel temperatures during processing, all of which would affect overall pasta quality (eating and nutritional quality) (Fig. 18.2). In particular, the role that extrusion plays in 'filming' the outer surface of pasta products still needs to be explored adequately. This film formation on

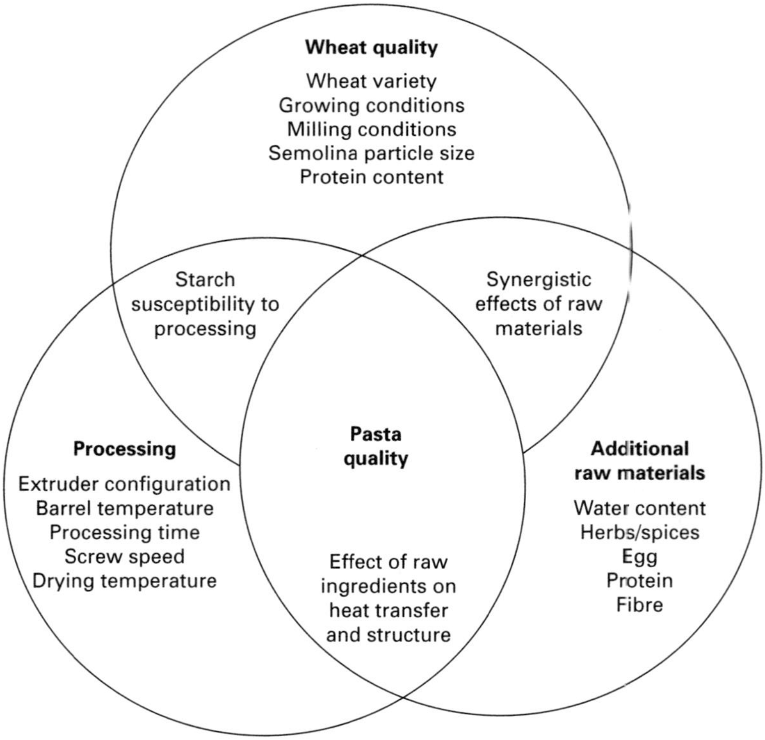

Fig. 18.2 Factors affecting pasta quality.

the outer surfaces of the pasta during extrusion would significantly affect the ease of water penetration during cooking.

As Fig. 18.2 suggests, raw material quality – such as wheat flour quality, semolina particle size, level of water addition and protein content – will all affect pasta quality. Much research has been conducted on the importance of starch type, starch damage, protein content, semolina quality and starch gelatinisation on cooking time and quality of pasta (Dexter and Matsuo, 1977; Cubadda, 1988; Chung *et al.*, 2003; El-Khayat *et al.*, 2003, 2006; Samaan *et al.*, 2006). Regarding processing, both starch content and composition have roles to play in terms of overall pasta quality. Part of this is related to dough hydration (water absorption from both the starch and protein components of the dough) and how hydration dynamics is associated with the final structure of the pasta, i.e. the degree of starch swelling and gelatinisation – and hence the accessibility of the pasta structure to degrading enzymes. Additionally, the low extrusion temperatures that are normally monitored and adhered to during processing (from 30–45 °C) would limit starch damage, swelling and gelatinisation during processing; hence the preponderance of relatively unswollen starch granules in the structure of raw pasta (Fig. 18.1). The addition of raw ingredients that have the ability to minimise starch gelatinisation during processing would have an impact on the glycaemic impact of pasta (Table 18.3). Dietary fibre is just one such ingredient that has the potential to alter the water-holding capacity of a product and compete for water with protein and starch during dough development. Thus, much attention has been focused recently on the incorporation of dietary fibre ingredients into pasta.

18.6 Dietary fibre enrichment of pasta

Dietary fibre is regarded as the fraction of non-starch polysaccharides resistant to digestion in the small intestine and fermentable in the large intestine (celluloses, hemicelluloses, pectins, modified celluloses, oligosaccharides and polyfructans such as inulin, gums and mucilages). It also includes oligosaccharides with various degrees of polymerisation (DP) and material bound to the plant cell wall (lignin, waxes, cutin and suberin). Materials with analogous characteristics to dietary fibre are included in the new definition under the term 'analogous carbohydrates' (Fig. 18.3). Many of these fibre components are already commercially exploited within the food industry as gums or stabilisers, and their use in cereal, dairy and water-based food systems is well documented. However, their use as stabilisers tends to be at relatively low levels of inclusion (below 1%) and higher levels would be needed in order to achieve the nutritional benefits attributed to dietary fibres.

Dietary fibre

Non-starch polysaccharides and resistant oligosaccharides

- Cellulose
- Hemicellulose
- Arabinoxylans
- Arabinogalactans
- Polyfructose
- Inulin
- Oligofructans
- Galacto-oligosaccharides
- Gums
- Mucilages
- Pectins

Analogous carbohydrates

- Indigestible dextrins
- Resistant maltodextrins
- Resistant potato dextrins
- Synthesised carbohydrate compounds
- Polydextrose
- Methyl cellulose
- Hydroxypropylmethyl cellulose
- Indigestible (resistant) starches

Lignin

Substances associated with the non-starch polysaccharides and lignin complex in plants

- Waxes
- Phytates
- Cutin
- Saponins
- Suberin
- Tannins

Fig. 18.3 Constituents of dietary fibre. Adapted from Anon. (2001).

18.6.1 Effects on pasta cooking performance

The ability of dietary fibre to hydrate and behave either as gel-like entangled networks, or sponge-like water stores is well documented and the hydration properties of just some of the most commonly used dietary fibres are given in Table 18.4. The water retention and water absorption capacity of dietary fibres differ significantly, depending upon the botanical source of the fibre (related to the composition of the polysaccharides), the method of extraction (native, de-pectinated or de-lignified) and also the particle size of the fibre used. Thus, when incorporating dietary fibre components into pasta, the origin and characteristics of the fibre have to be taken into account when considering the ability of the fibre to retain water and hence the potential contribution such fibre inclusion will have on the overall formation of pasta structure. Alteration of pasta structure will, in turn, affect pasta texture and overall product quality. For instance, incorporation of fibre ingredients into

Table 18.4 Hydration characteristics of some dietary fibres. Adapted from Guillon and Champ (2000), Robertson *et al.* (2000), and Grigelmo-Miguel and Martin-Belloso (1999)

Source of fibre	Treatment	Particle size (μm)	Swelling (mg/l)	Water retention (g water/g dry pellet)	Water absorption (ml water/g dry fibre)
Sugar beet fibre	–	385	21.4	22.6	8.8
	–	205	15.9	19.2	7.3
Citrus fibre	–	540	15.7	10.4	7.0
	–	139	10.4	10.7	4.6
Apple fibre	–	540	9.6	6.9	3.8
	–	80	5.6	7.1	2.7
Pea hull	–	67	7.5	3.94	
	Native	–	6.2	4.2	
Wheat bran	–	900	11.9	6.8	1.0
	–	320	5.9	3.0	0.9
	Native	–	7.0	7.0	
	Delignified	–	11.0	10.4	
Oat bran	–	–	5.5	3.5	
Resistant starch					
Novelose	–	40	5.6	2.9	3.0
Eridania	–	84	7.4	3.1	3.9

pasta will have varying effects on the dry matter of the cooked pasta, and its swelling index and cooking losses (Table 18.5). The intensity of this effect is dose related, so that as more fibre is utilised then the greater will be the effect on food structure.

Any alteration of pasta structure will also have an effect on pasta texture. Thus, addition of dietary fibre to pasta will have a dramatic effect on the firmness (Fig. 18.4) of the cooked pasta. With regards to pasta firmness, the soluble, high molecular weight, gel-forming dietary fibres (such as guar, locust bean and xanthan gum) have the potential to increase firmness (the breaking point of a pasta during stretching), whereas the largely insoluble (pea fibre and cellulose) and the low molecular weight soluble fibres (inulin) decrease the overall firmness of pasta. The latter observation is probably linked to the way that incorporation of insoluble fibre ingredients fragments the association of protein and starch during structure formation, hence causing a weakened protein–starch dough matrix within the pasta. The research of Edwards *et al.* (1995) showed similar findings with regards to an increase in pasta firmness when xanthan gum was added at levels of 1% or 2%. More recently, Fardet *et al.* (1999b) made pasta from freeze-dried flour fractions in which soluble fibres accounted for 7%, and indicated that the firmness of the pasta increased (and cooking loss decreased) with fibre inclusion. Izdorczyk *et al.* (2005) also illustrated that incorporation of fibre-rich fractions from roller milling of waxy and high-amylose starch barley reduced the cooking time of noodles and improved their overall cooking characteristics.

Table 18.5 Effect of fibre addition on pasta cooking properties

Sample	Dry matter (%)	Dry matter of the cooking water (%)	Swelling index (g water/g DM)	Cooking losses (%)
Effect of the type of NSP				
Control	32.5±1.03	0.93±0.06	2.18±0.24	8.8±1.2
Pea fibre	31.8^{a}	0.96^{a,b}	2.15^{b}	10.2^{a,}
Inulin	32.4^{a}	0.90^{a,b}	2.13^{b}	9.7^{a}
Guar gum	29.6^{a,b}	0.88^{a,b}	2.39^{a,b}	9.3^{a}
Bamboo fibre	31.5^{a}	0.93^{a,b}	2.21^{b}	9.9^{a}
β-Glucan (as Hi Sol)	31.4^{a,b}	1.08^{a}	2.19^{b}	11.5^{a}
Xanthan gum	28.4^{b}	0.75^{b}	2.72^{a}	7.6^{b}
Locust bean gum	32.2^{a}	0.84^{a,b}	2.11^{b}	9.3^{a}
Compound effect of the level of NSP addition				
2.5%	31.8	0.88	2.18^{B}	9.6
5.0%	31.6	0.87	2.18^{B}	9.7
7.5%	30.8	0.92	2.30^{A,B}	10.6
10.0%	30.1	0.96	2.44^{A}	10.7

NSP, non-starch polysaccharide. Within the same column, values with the same letter are not significantly different ($P > 0.05$).

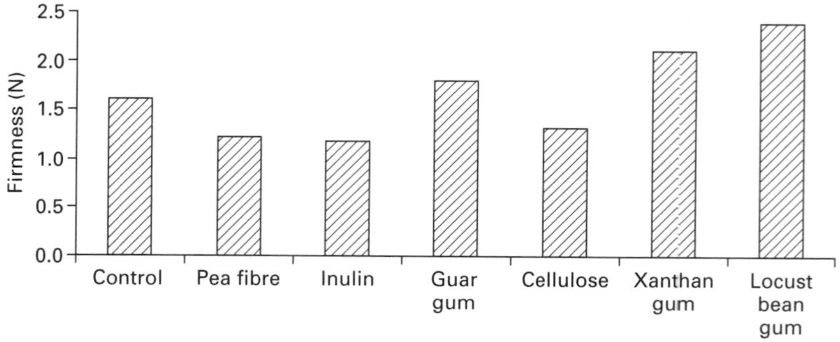

Fig. 18.4 Effect of fibre addition on pasta firmness.

Such effects are likely to be attributed to the ability of fibres to hydrate and retain water whilst forming entangled polysaccharide–protein networks around the starch granules, leading to a stronger cohesiveness between starch and protein within the pasta structure. Thus, from a technological basis, dietary fibre could be added into pasta without a loss in pasta firmness, and with the possibility of maintaining or improving cooking properties (cooking losses). However, the selection of fibre type, and also the level of fibre inclusion would be of importance when considering such additions.

18.6.2 Effect of fibre addition on the glycaemic index of pasta

As highlighted earlier in this chapter, both pasta and dietary fibre-enriched products have the potential to be regarded as low-GI food items and to regulate starch degradation during digestion. It is likely that this is achieved through a number of mechanisms, which act concurrently. Certainly, the hydration of fibre, and the trapping of water by fibre components, have the potential to reduce the amount of 'free' water available, which will have a significant effect in reducing the susceptibility of starch to gelatinisation, and which will in turn affect the swelling potential of the starch granules. This would have an obvious role in altering the digestibility of starch within the pasta matrix. Numerous research groups have investigated the possibility of using fibre in the process of reducing starch digestion. A number of these research groups have used pasta as a model carbohydrate system.

Many studies in the 1980s and 1990s illustrated the potential use of soluble dietary fibres (such as guar gum) in cereal foods for the reduction of starch degradation and the control of blood glucose levels (Jenkins *et al.*, 1983; Burley *et al.*, 1984; Gatti *et al.*, 1984; Peterson *et al.*, 1987; Ellis *et al.*, 1988; Wood, 1994; Wood *et al.*, 1994); they also illustrated that other soluble dietary fibres (beta-glucan material extracted from oats) had a similar effect to guar gum in reducing the glycaemic response following the ingestion of carbohydrate beta-glucan-enriched foods. Similar results have been observed by Bourdon *et al.* (1999). These researchers indicated that, when men were fed with barley-enriched pasta (three times as much fibre as a low-fibre control meal), a slower rate of carbohydrate absorption was observed.

Research within our laboratory (Cleary and Brennan, 2006) has also illustrated that beta-glucan-enriched pastas have the potential to reduce both the rate and extent of starch degradation, and hence the amount of reducing sugars released during *in vitro* digestion (Table 18.6). This reduction in starch degradation appears to be related not only to the structure of the pasta but also to the amount of starch gelatinisation that has occurred during processing and cooking (Table 18.7).

Differential scanning calorimetry (DSC) data shown in Table 18.7 illustrate that the incorporation of beta-glucan into pasta has the effect of shifting the onset, end point and peak of gelatinisation downwards compared with the control pasta, resulting in a reduction in the total enthalpy of gelatinisation in the fibre-enriched samples. A similar observation was noted by Tudorica *et al.* (2002) in pasta enriched with pea, inulin and guar components: the degree of enthalpy of fibre-enriched raw pasta decreased with the level of fibre added to the dough system.

The reductions seen in starch swelling and starch gelatinisation during cooking of pasta, and the resulting decrease of starch degradation during ingestion, have benefits in terms of lowering the overall GI of pasta products (Fig. 18.5). Incorporation of soluble dietary fibre fractions has the ability to reduce the rate and extent of sugars released from pasta during starch digestion and hence lower the potential GI of pasta by 20–40% (compared with a control pasta). This has the result of making an already low-GI product, lower in terms of potential glycaemic loading.

Other potential mechanisms by which some dietary fibres reduce starch

Table 18.6 Reducing sugars released (RSR) (expressed as maltose equivalents, as a percentage of total available carbohydrates) from durum wheat pastas substituted with 2.5%, 5%, 7.5% and 10% barley beta-glucan (BBG) fibre fraction

Time (min)	Control	2.5% BBG	5% BBG	7.5% BBG	10% BBG
30	0.00 ± 0.00^{a}	0.00 ± 0.00^{a}	0.00 ± 0.00^{a}	0.00 ± 0.00^{a}	0.00 ± 0.00^{a}
60	1.61 ± 0.52^{a}	0.94 ± 0.19^{b}	0.50 ± 0.21^{b}	0.12 ± 0.13^{b}	0.78 ± 0.33^{b}
90	2.43 ± 0.18^{a}	3.06 ± 0.53^{a}	2.05 ± 0.36^{a}	2.18 ± 0.74^{a}	2.37 ± 0.43^{a}
120	4.16 ± 0.05^{a}	$3.67 \pm 0.12^{a,b}$	3.44 ± 0.15^{b}	2.78 ± 0.19^{c}	4.51 ± 0.65^{a}
150	7.02 ± 0.32^{a}	7.42 ± 0.50^{a}	6.42 ± 0.24^{a}	6.17 ± 0.55^{a}	7.00 ± 0.63^{a}
180	10.80 ± 0.42^{a}	11.60 ± 0.15^{a}	8.73 ± 0.70^{b}	8.37 ± 0.50^{b}	9.32 ± 0.38^{b}
210	13.42 ± 0.48^{a}	14.01 ± 0.74^{a}	11.90 ± 0.76^{b}	10.70 ± 0.45^{b}	11.80 ± 0.64^{b}
240	17.42 ± 0.56^{a}	17.49 ± 0.49^{a}	14.38 ± 0.35^{b}	13.46 ± 0.18^{b}	14.47 ± 0.28^{b}
270	19.91 ± 0.09^{a}	20.25 ± 0.31^{a}	17.21 ± 0.51^{b}	16.73 ± 0.06^{b}	16.58 ± 0.49^{b}
300	24.17 ± 0.60^{a}	23.14 ± 0.12^{a}	20.10 ± 0.51^{b}	17.71 ± 0.10^{c}	18.62 ± 0.13^{c}

All measurements are mean values (±SD) of triplicate determinations. Mean values within the same row not followed by the same letter are significantly different ($P < 0.05$). Source: Cleary and Brennan (2006).

Table 18.7 Starch gelatinisation characteristics of cooked pasta substituted with barley beta-glucan fibre fraction

	Control	2.5% BBG	5% BBG	7.5% BBG	10% BBG
T_{onset} (°C)	$54.00 \pm 0^{a,b}$	53.50 ± 0.40^{a}	$53.80 \pm 0.26^{a,b}$	54.27 ± 0.21^{b}	55 ± 0
$T_{endpoint}$ (°C)	70.90 ± 0.96^{a}	$69.73 \pm 0.40^{a,b}$	69.17 ± 0.29^{b}	68.67 ± 0.29^{b}	68.63 ± 0.23^{b}
T_{peak} (°C)	$61.23 \pm 0.21^{a,b,c}$	$61.07 \pm 0.06^{a,c}$	60.83 ± 0.23^{a}	61.73 ± 0.15^{b}	$61.57 \pm 0.38^{b,c}$
J/g (°C)	4.37 ± 0.29^{a}	4.50 ± 0.15^{a}	3.68 ± 0.15^{b}	3.48 ± 0.14^{b}	3.41 ± 0.23^{b}

Values are means ± SD. Mean values in the same row followed by the same letter are not significantly different ($P > 0.05$).

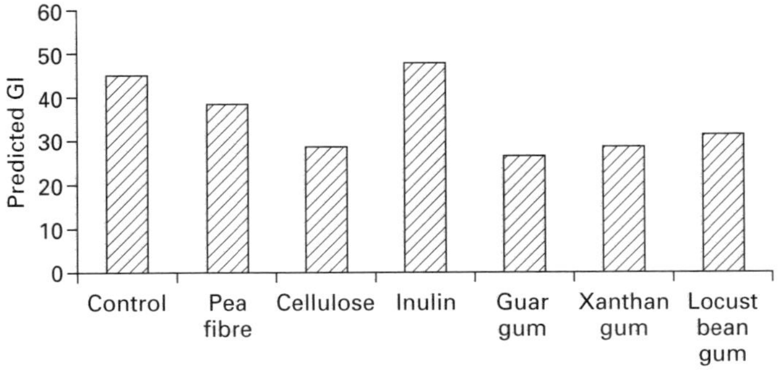

Fig. 18.5 Effect of dietary fibre addition on the reduction of predicted glycaemic index.

digestibility include that dietary fibre ingredients act with a degree of thermodynamic incompatibility to starch and/or protein, which results in the formation of a protective coat around starch granules and/or the starch–protein matrix, and in turn limits enzyme accessibility to starch during degradation. Additionally, the viscosity-altering nature of dietary fibres when in solution cannot be ignored as a possible way in which they impede pasta hydration during cooking and resultant ingestion, mixing and destruction of pasta structure during digestion and reduction of pasta particle size whilst eating, and hence the ability of starch degrading enzymes to reach starch granules. Further research is required to elucidate the major factors involved in the reduction of starch digestibility.

18.7 Future trends

The diversity of dietary fibre ingredients available to the food industry offers an interesting dilemma for food technologists as to which fibre to select for a particular pasta formulation. To help solve this quandary, careful evaluation of the use of fibre ingredients is still required to determine the specific interactions of these products in pasta products. As illustrated earlier in this chapter, the selection of fibre ingredients may have a role in altering the structure, texture and nutritional quality of the food product, thus use of novel ingredients (such as resistant starches and partially hydrolysed fibres) needs further investigation. Much of the present research has focused on the application of a single form of fibre to a pasta formulation and the examination of the effects of that fibre on pasta quality. However, it is possible that combinations of fibres may be used more effectively than single fibre sources, to yield a complexity of fibre–starch interactions. This, in turn, may have significant effects on product structure, mastication, satiety and gut health as well as enabling a more controlled regulation of the GI of foods.

Another area that has not featured extensively in the literature so far is the fortification of pasta with other minerals and vitamins to improve the overall nutritional quality of pasta. Fortification of pasta with folate, calcium and other minerals and phytochemicals is a possibility in order to deliver these nutritional advantages to a wider global consumer market in a relatively inexpensive convenience food item (Cho *et al.*, 2002; Normen *et al.*, 2002). If pasta is to be used as a vehicle for mineral fortification, research also needs to be conducted on the bioavailability of these compounds in pasta products, and the possible effects that dietary fibre components may have on the bioavailability of these nutrients.

18.8 Summary

Pasta is a cheap, convenient food product which presents itself as an ideal vehicle for improving the nutritional quality of diets. Even though it is already regarded as a low-GI food product, incorporation of dietary fibre into pasta products has the potential to reduce the GI of pasta further and, in turn, for pasta to be used effectively in weight control, regulation of appetite and also in the modulation of glucose responses to meals. The mechanisms by which fibre-enriched foods can modulate the glycaemic response of individuals are largely related to the inhibition or reduction of starch gelatinisation. Fibre compounds compete with the starch and protein in the pasta for water during cooking and hydration, and it is likely that it is this competition that limits the extent of gelatinisation and the disruption of starch structure. This, in turn, reduces the rate of starch degradation and hence sugar release. As such, pasta is an ideal carbohydrate food product for the current health-conscious consumer.

18.9 References

Anon. (2001) The definition of dietary fibre. *Cereal Foods World* **46** 112–129.

Barkeling, B., Granfelt, Y., Bjorck, I., & Rossner, S. (1995) Effects of carbohydrates in the form of pasta and bread on food intake and satiety in man. *Nutritional Research* **15** 467–476.

Bourdon, I., Yokoyama, W., Davis, P., Hudson, C., Backus, R., Richter, D., Knuckles, B., & Schneeman, B.O. (1999) Postprandial lipid, glucose, insulin, and cholecystokinin responses in men fed barley pasta enriched with beta-glucan. *American Journal of Clinical Nutrition* **69** 55–63.

Brennan, C.S., Kuri, V., & Tudorica, C.M. (2004) The importance of food structure on the glycaemic responses of carbohydrate rich foods. In *Dietary Fibre Bio-active Carbohydrates for Food and Feed.* Editors J.W. van der Kamp, N.-G. Asp, J. Miller Jones, & G. Schaafsma. Wageningen Academic Publishers, Wageningen, The Netherlands, pp. 113–126.

Brennan, C.S. (2005) Dietary fibre, glycaemic response, and diabetes. *Molecular Nutrition and Food Research* **49** 560–570.

Burley, V., Leeds, A.R., Ellis, P.R., & Peterson, D.B. (1984) Wholemeal guar bread – acceptability and efficacy combined – a study of blood-glucose, plasma-insulin and palatability in normal subjects. *Proceedings of the Nutrition Society* **43** a48.

Casiraghi, M.C., Garsetti, M., Testolin, G., & Brighenti, F. (2006) Post-prandial responses to cereal products enriched with barley beta-glucan. *Journal of the American College of Nutrition* **25** 313–320.

Cho, S., Johnson, G., & Song, W.O. (2002) Folate content of foods: Comparison between databases compiled before and after new FDA fortification requirements. *Journal of Food Composition and Analysis* **15** 293–307.

Chung, O.K., Ohm, J.B., Lookhart, G.L., & Bruns, R.F. (2003) Quality characteristics of hard winter and spring wheats grown under an over-wintering condition. *Journal of Cereal Science* **37** 91–99.

Cleary, L., & Brennan, C.S. (2006) The influence of a (1-3) (1-4)-β-D-glucan rich fraction from barley on the physico-chemical properties and in vitro reducing sugars release of durum wheat pasta. *International Journal of Food Science and Technology* **41** 910–918.

Cubadda, R. (1988) Evaluation of durum wheat, semolina and pasta in Europe. In *Durum Wheat: Chemistry and Technology*. (Editors G. Fabriani & C. Lintus. AACC Publishers, St Paul, Minnesota, pp. 217–228.

Cunin, C., Handschin, F., Walther, P., & Escher, F. (1995) Structural changes of starch during cooking of durum wheat pasta. *Food Science and Technology – Lebensmittel-Wissenschaft und Technologie* **28** 323–328.

Delargy, H.J., Osullivan, K.R., Fletcher, R.J., & Blundell, J.E. (1997) Effects of amount and type of dietary fibre (soluble and insoluble) on short-term control of appetite. *International Journal of Food Sciences and Nutrition* **48** 67–77.

DeVries, J.W., & Faubion, J.M. (1999) Defining dietary fibre: a report on the AACC/ILSI NA consensus workshop. *Cereal Foods World* **44** 506–507.

DeVries, J.W., & Prosky, L. (1999) A historical perspective on defining dietary fibre. *Cereal Foods World* **44** 367–369.

Dexter, J.E., & Matsuo, R.R. (1977) Influence of protein content on some durum wheat quality parameters. *Canadian Journal of Plant Science* **57** 717–727.

Edwards, N.M., Biliaderis, C.G., & Dexter, J.E. (1995) Textural characteristics of wholewheat pasta and pasta containing non-starch polysaccharides. *Journal of Food Science* **60** 1321–1324.

El-Khayat, G.H., Samaan, J., & Brennan, C.S. (2003) Evaluation of vitreous and starchy syrian durum wheat grains: the effect of amylose content on starch characteristics and flour pasting properties. *Starch*. **55** 358–365.

El-Khayat, G.H., Samaan, J., Manthey, F.A., Fuller, M.P., & Brennan, C.S. (2006) Durum wheat quality I: Correlations between physical and chemical characteristics of Syrian durum wheat cultivars. *International Journal of Food Science and Technology* **41** (S2) 22–29.

Ellis, P.R., Kamalanathan, T., Dawoud, F.M., Strange, R.N., & Coultate, T.P. (1988) Evaluation of guar biscuits for use in the management of diabetes – tests of physiological-effects and palatability in non-diabetic volunteers. *European Journal of Clinical Nutrition* **42** 425–435.

Fardet, A., Hoebler, C., Baldwin, P.M., Bouchet, B., Gallant, D.J., & Barry, J.L. (1998) Involvement of the protein network in the in vitro degradation of starch from spaghetti and lasagne: a microscopic and enzymic study. *Journal of Cereal Science* **27** 133–145.

Fardet, A., Abecassis, J., Hoebler, C., Baldwin, P.M., Buleon, A., Berot, S., & Barry, J.L. (1999a) Influence of technological modifications of the protein network from pasta on in vitro starch degradation. *Journal of Cereal Science* **30** 133–145.

Fardet, A., Hoebler, C., Armand, M., Lairon, D., & Barry, J.L. (1999b) In vitro starch degradation from wheat-based products in the presence of lipid complex emulsions. *Nutrition Research* **19** 881–892.

Food Standards Agency (2002) *McCance and Widdowson's The Composition of Foods*. The Royal Society of Chemistry, Cambridge, UK.

Foster-Powell, K., Holt, H.A., & Brand-Miller, J.C. (2002) International table of glycaemic index and glycaemic load values: 2002. *American Journal of Clinical Nutrition* **76** 5–56.

Gatti, E., Catenazzo, G., Camisasca, E., Torri, A., Denegri, E., & Sirtori, C.R. (1984) Effects of guar-enriched pasta in the treatment of diabetes and hyperlipidemia. *Annals of Nutrition and Metabolism* **28** 1–10.

Grigelmo-Miguel, N., & Martin-Belloso, O. (1999) Comparison of dietary fibre from by-products of processing fruits and greens and from cereals. *Food Science and Technology – Lebensmittel-Wissenschaft und Technologie* **32** 503–508.

Guillon, F., & Champ, M. (2000) Structural and physical properties of dietary fibres, and consequences of processing on human physiology. *Food Research International* **33** 233–245.

Izydorczyk, M.S., Lagasse, S.L., Hatcher, D.W., Dexter, J.E., & Rossnagel, B.G. (2005) The enrichment of Asian noodles with fiber-rich fractions derived from roller milling of hull-less barley. *Journal of Science Food and Agriculture* **85** 2094–2104.

Jenkins, D.J.A., Wolever, T.M.S., Jenkins A.L., Lee R., Wong G.S., & Josse R., (1983) Glycaemic response to wheat products: reduced response to pasta but no effect of fiber. *Diabetes Care* **11** 149–159.

Jenkins, D.J.A., Axelsen, M., Kendall, C.W.C., Augustin, L.S.A., Vuksan, V., & Smith, U. (2000) Dietary fibre, lente carbohydrates and the insulin-resistant diseases. *British Journal of Nutrition* **83**, S157–S163.

Jin, Z., Hsieh, F., & Huff H.E. (1994) Extrusion cooking of corn meal with soy fiber, salt and sugar. *Cereal Chemistry* **71** 227–234.

Normen, L., Bryngelsson, S., Johnsson, M., Evheden, P., Ellegard, L., Brants, H., Andersson, H., & Dutta, P. (2002) The phytosterol content of some cereal foods commonly consumed in Sweden and in the Netherlands. *Journal of Food Composition and Analysis* **15** 693–704.

Pagani, M.A., Gallant, D..J., Bouchet, B., & Resimin, P. (1986) Ultrastructure of cooked spaghetti. *Food Microstructure* **5** 111–129.

Peterson, D.B., Ellis, P.R., Baylis, J.M., Fielden, P., Ajodhia, J., Leeds, A.R., & Jepson, E.M. (1987) Low-dose guar in a novel food product – improved metabolic control in non-insulin-dependent diabetes. *Diabetic Medicine* **4** 111–115.

Politz, M.L., Timpa, J.D., White, A.R., & Wasserman, B.P. (1994) Non-aqueous gel permeation chromatography of wheat starch in dimethylacetamide (DMAC) and LiCl: extrusion- induced fragmentation. *Carbohydrate Polymers* **24** 91–99.

Robertson, J.A., de Monredon, F.D., Dysseler, P., Guillon, F., Amado, R., & Thibault, J.F. (2000) Hydration properties of dietary fibre and resistant starch: A European collaborative study. *Food Science and Technology – Lebensmittel-Wissenschaft und Technologie* **33** 72–79.

Samaan, J., El-Khayat, G.H., Manthey, F.A., Fuller, M.P., & Brennan, C.S. (2006). Durum wheat quality II: The relationship of kernel physicochemical composition to semolina quality and end product utilisation. *International Journal of Food Science and Technology* **41** (S2) 47–55.

Tappy, L., Gugolz, E., & Wursch, P. (1996) Effects of breakfast cereals containing various amounts of beta-glucan fibers on plasma glucose and insulin responses in NIDDM subjects. *Diabetes Care* **19**, 831–834.

Trowell, H. (1972) Crude fibre, dietary fibre and atheroslerosis. *Atherosclerosis* **16** 138–140.

Tudorica, C.M., Kuri, V., & Brennan, C.S. (2002) Nutritional and physicochemical characteristics of dietary fiber enriched pasta. *Journal of Agricultural and Food Chemistry* **50** 347–356.

Wang, S., Casulli, J., & Bouvier, J.M. (1993) Effect of dough ingredients on apparent viscosity and properties of extrudates in a twin-screw extrusion cooking. *International Journal of Food Science and Technology* **28** 465–479.

Wolever, T.M., Jenkins, D.J., Kalmusky, J., Giordano, C., Giudici, S., Jenkins, A.L., Thompson, L.U., Wong, G.S., & Josse, R.G. (1986) Glycemic response to pasta: effect of surface area, degree of cooking, and protein enrichment. *Diabetes Care* **9** 401–404.

Wood, P.J. (1994) Evaluation of oat bran as a soluble fiber source – characterization of oat beta-glucan and its effects on glycemic response. *Carbohydrate Polymers* **25** 331–336.

Wood, P.J., Braaten, J.T., Scott, F.W., Riedel, K.D., Wolynetz, M.S., & Collins, M.W. (1994) Effect of dose and modification of viscous properties of oat gum on plasma-glucose and insulin following an oral glucose-load. *British Journal of Nutrition* **72** 731–743.

19

Functional cereal products for those with gluten intolerance

E. K. Arendt and F. Dal Bello, University College Cork, Ireland

19.1 Introduction

Gluten is a protein component in wheat, one of the staple foods for most populations in the world, and other cereals (mainly rye and barley). Unfortunately, in genetically susceptible individuals, the ingestion of gluten and related proteins triggers an immune-mediated enteropathy known as coeliac disease (CD).

19.1.1 Coeliac disease

CD is defined by crypt hyperplasia, jejunal mucosa villous atrophy, and inflammatory infiltrate in the lamina propria associated with an increased number of intraepithelial lymphocytes (Gobbi, 2005). The ingestion of gluten induces an inflammatory response, resulting in the destruction of the villous structure of the small intestine (Shan *et al.*, 2002), which in turn leads to a flat jejunal mucosa. The gliadin and glutenin peptides of wheat gluten and similar alcohol-soluble proteins (prolamins) of barley and rye cause damage to the small intestinal mucosa (Murray, 1999; Fasano and Catassi, 2001; Farrell and Kelly, 2002; Vader *et al.*, 2002). Clinical data are now available suggesting that the great majority of coeliac patients tolerate oats, although some concern remains about the safety of this cereal for coeliacs (Janatuinen *et al.*, 2002) and oats remain currently on the Codex list of gluten-containing cereals.

The symptoms of CD include severe symptoms of malabsorption such as steatorrhoea (characterised by the passage of pale, bulky stools), abdominal discomfort, weight loss or gain (Fassano and Catassi, 2001; Fasano, 2005), tiredness, anaemia and severe diarrhoea. Table 19.1 (Feighery, 1999) outlines the most commonly encountered reactions to gluten by sufferers of CD. The accepted diagnostic criteria for CD, as defined by the European Society for Paediatric Gastroenterology, Hepatology and Nutrition are based on detection of flat mucosa on small bowel biopsy and disappearance of symptoms following a gluten-free diet (Walker-Smith *et al.*, 1990). According to Murray (1999), CD is the end result of three processes that culminate in intestinal mucosal damage: genetic predisposition, environmental factors and immunologically based inflammation. With prolonged exposure to gluten in the diet, the consequences of CD both internally and externally increase overtime. The internal effects are the continued presence of a flat intestinal mucosa quite often followed by a reduction in enzyme activity and hence lack of absorption of vitamins and minerals, which in turn may lead to different deficiencies. The external effects include dermatitis herpetiformis or the appearance of pale skin, dry hair roots or abdominal distention, especially in children younger than 2 years of age.

19.1.2 Epidemiology

CD was once thought to be a rare condition and confined to childhood (Green and Jabri, 2003), with a uniform clinical presentation of weight loss and diarrhoea. Recent data, using highly sensitive immunological screening methods, have shown that CD is more common than previously recognised in Europe (Grodzinsky, 1996; Martucci *et al.*, 2002), the United States (Not *et al.*, 1998) and South America (Gandolfi *et al.*, 2000). Table 19.2 (Fasano and Catassi, 2001) illustrates the prevalence of CD based on clinical diagnosis

Table 19.1 Symptoms (and related signs) of coeliac disease. Adapted from Feighery (1999)

Infancy (0–2 years)	Childhood	Adulthood
Diarrhoea (miserable, pale) Abdominal distension (enlarged abdomen)	Diarrhoea or constipation Anaemia	Diarrhoea or constipation Anaemia
Failure to thrive (low weight, lack of fat, hair thinning)	Loss of appetite (short stature, osteoporosis)	Aphthous ulcers, sore tongue and mouth (mouth ulcers, glossitis, stomatitis)
Anorexia, vomiting		Dyspepsia, abdominal pain, bloating (weight loss)
Psychomotor impairment (muscle wasting)		Fatigue, infertility, neuropsychiatric symptoms (anxiety, depression) Bone pain (osteoporosis) Weakness (myopathy, neuropathy)

Table 19.2 Prevalence of coeliac disease based on clinical diagnosis or screening data. From Fasano and Catassi (2001)

Geographic	Prevalence on clinical diagnosis	Prevalence on screening data
Denmark	1:10 000	1:500
Finland	1:1000	1:130
Germany	1:2300	1:500
Italy	1:1000	1:184
Netherlands	1:4500	1:198
Norway	1:675	1:250
Sweden	1:330	1:190
United Kingdom	1:300	1:112
United States	1:10 000	1:111
Worldwide average	1:3345	1:266

and screening data. However, a review by Farrell and Kelly (2002) suggests that the true prevalence of CD is difficult to ascertain, because many patients have atypical symptoms or none at all. Currently, CD prevalence has been estimated to be 1 in *c*. 100 people worldwide (Mustalahti *et al*., 2002; Hamer, 2005; Sollid and Khosla, 2005; Catassi and Fasano, 2007). Such a rate establishes CD as one of the most common forms of food intolerance.

19.1.3 The iceberg model

The iceberg is a common model (Fig. 19.1) used to explain the epidemiology of CD (Visakorpi *et al*., 1996). The relatively few cases that have been properly clinically diagnosed with the classical symptoms such as chronic diarrhoea, unexplained iron deficiency, failure to thrive and the presence of a flat mucosa, form the small tip of the iceberg (Fasano and Catassi, 2001). Patients who have been recently diagnosed and who are now living under gluten-free diets and show a normal mucosa form the lower part of the tip. The ratio of diagnosed to undiagnosed cases gives the so-called 'waterline'. A relatively large group of 'silent' cases are situated below the waterline. These cases have not yet been identified and show a small bowel mucosa lesion but may remain undiagnosed because the clinical symptoms are lacking (Feighery, 1999). The lower portion of the iceberg consists of a small group of people with latent CD. These patients show a normal mucosa while ingesting gluten. However, they have the potential to develop the disease at any stage in their life.

19.1.4 Treatment and dietary requirements

The only way that CD can be treated is the total lifelong avoidance of gluten ingestion (Mowat, 2003), even though the Codex Alimentarius tolerates 200 ppm of gluten per food. The Codex Standard for gluten-free foods was

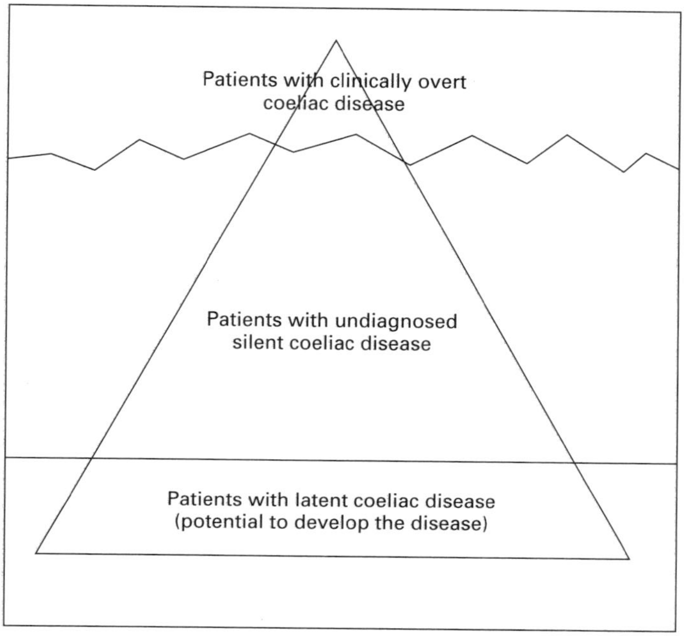

Fig. 19.1 Iceberg model depicting prevalence of coeliac disease. Adapted from Feighery (1999).

adopted by the Codex Alimentarius Commission of the World Health Organization (WHO) and by the Food and Agriculture Organization (FAO) in 1976. In 1981 and 2000, draft revised standards stated that so-called gluten-free foods are described as:

> (a) consisting of, or made only from ingredients that do not contain any prolamins from wheat or all *Triticum* species such as spelt, kamut or durum wheat, rye, barley, oats or their crossbred varieties, with a gluten level not exceeding 20 ppm; or (b) consisting of ingredients from wheat, rye, barley, oats, spelt or their crossbred varieties, which have been rendered gluten-free, with a gluten level not exceeding 200 ppm; or (c) any mixture of two ingredients as mentioned in (a) and (b) with a gluten level not exceeding 200 ppm.

In this context, the WHO/FAO standard gluten was defined as a protein fraction from wheat, rye, barley, oats or their crossbred varieties (e.g. Triticale) and derivatives thereof, to which some persons are intolerant and that is insoluble in water and 0.5 M NaCl. Prolamins are defined as the fraction from gluten that can be extracted by 40–70% aqueous ethanol. The prolamin from wheat is gliadin, from rye is secalin, from barley is hordein and from oats is avenin. The prolamin content of gluten is generally taken as 50%. In the United States and Canada, the gluten-free diet is devoid of any gluten, and is based on naturally gluten-free ingredients such as rice.

Since the burden of illness related to CD is doubtless higher than previously thought, efforts to reduce the human intolerance to cereals are of medical, nutritional and economic interest. Patients with CD are unable to consume some of the most common products on the market today, namely breads, baked goods and other food products made with wheat flour (Lovis, 2003). Hidden ingredients, for example by-products or processed foods that contain wheat and gluten-derivates as thickeners and fillers, must also be avoided. These include hamburgers, salad dressings, cream sauces, dried soup mixes or canned soups and processed cheese. Medications may include wheat proteins as binders and therefore must also be excluded. Since CD can, in fact, result in lactose intolerance due to the lack of lactase production (Murray, 1999), individuals with lactose intolerance due to CD must avoid cow's milk. Since November 2005, food labelling rules require prepacked food sold in the United Kingdom and the rest of the European Union to show clearly on the label if they (or one of the ingredients) contain any cereal containing gluten – this applies even if the cereal has been specially treated to remove gluten (http://www.eatwell.gov.uk/).

19.1.5 Gluten and coeliac disease

Even if the demand for gluten-free foods is constantly rising, the majority of gluten-free products available on the market are of very poor quality. Gluten is the main structure-forming protein in wheat flour and is often termed the 'structural' protein for breadmaking. It is responsible for the elastic characteristics of dough, and contributes to the appearance and crumb structure of many baked products. The gluten proteins are complex storage proteins that share some common repetitive sequences. Peptides rich in glutamines and prolines are potent activators of the immune response in CD (Murray, 1999). The gluten proteins present in the wheat flour are embedded in the flour particles along with the other flour components, mainly starch granules (Fig. 19.2). The structure of gluten is complex, and is stabilised by intermolecular disulphide, hydrogen and hydrophobic bonds. The properties of gluten become apparent when flour is hydrated, giving an extensible dough, with good gas-holding properties, as well as a good texture and crumb structure in baked bread (Faubion and Hoseney, 1990; Stear, 1990). The protein fractions in gluten are glutenin and gliadin. The former is a rough, rubbery mass when fully hydrated, while gliadin produces a viscous, fluid mass on hydration. Gluten, therefore, exhibits cohesive, elastic and viscous properties that combine the extremes of the two components. The gluten matrix is a major determinant of the important properties of dough (extensibility, resistance to stretch, mixing tolerance, gas-holding ability), which encloses the starch granules and fibre fragments. Gluten removal results in major problems for bakers and, currently, many gluten-free products available on the market are of low quality, exhibiting poor mouthfeel and flavour (Arendt *et al.*, 2002). This presents a major

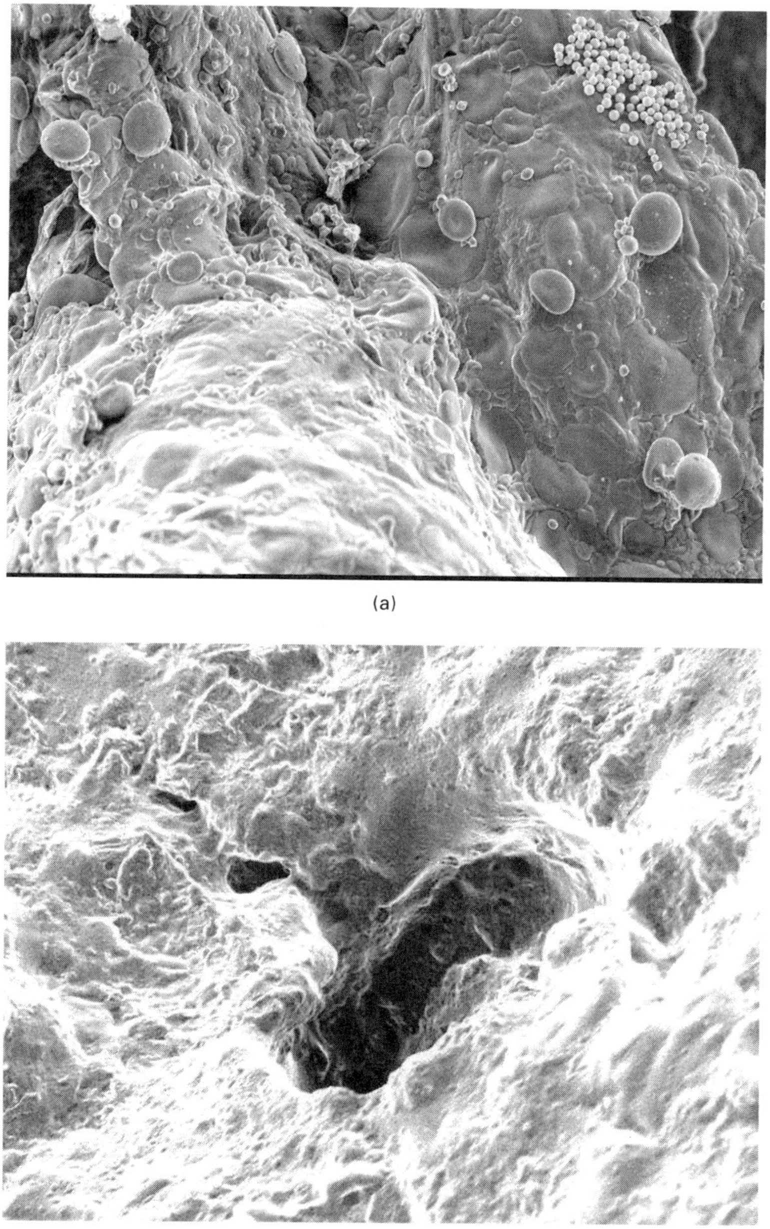

(a)

(b)

Fig. 19.2 Scanning electron micrographs of wheat bread (a) and gluten-free bread (b) Magnification, ×430.

challenge to the cereal technologist and baker alike, and has led to the search for alternatives to gluten in the manufacture of gluten-free bakery products.

19.2 Difficulties in producing gluten-free breads

In view of the fact that gluten is responsible for the viscoelastic properties of bread, its replacement is one of the biggest challenges when developing gluten-free cereal products. The absence of gluten often results in a liquid batter rather than a dough pre-baking, and can result in baked bread with a crumbling texture, poor colour and other quality defects post-baking. In recent years there has been significantly more research and development on gluten-free products, involving a diverse approach that has included the use of starches, dairy products, gums and hydrocolloids, other non-gluten proteins, prebiotics and combinations thereof, as alternatives to gluten, to improve the structure, mouthfeel, acceptability and shelf-life of gluten-free bakery products. Even so, a marketing review by Arendt *et al.* (2002) reported that most of the gluten-free products are of inferior quality and very often showed off-flavours with a crumbly and very dry structure.

Gluten-free breads are usually characterised by deficient quality characteristics as compared with wheat breads (Gujral *et al.*, 2003a). Problems related to volume and crumb texture are associated with gluten-free breads, even when rice flour is used, which seems to be the best raw material for this type of bread (Gujral *et al.*, 2003b). Several studies have been conducted (Gallagher *et al.*, 2002, 2003, 2004; Schober *et al.*, 2003, 2005; Moore *et al.*, 2004a, b; McCarthy *et al.*, 2005) whereby novel ingredients – such as dairy powders, sorghum, rice, starches, pseudocereals, etc. – in combination with hydrocolloids, replaced gluten. These studies show how gluten-free breads require a different technology. Due to the lack of a gluten network, the properties of the dough are more fluid than wheat doughs and closer in viscosity to cake batters (Cauvain, 1998; Moore *et al.*, 2004a). These batter-type doughs have similar handling properties to cake batters rather than typical wheat doughs. Furthermore, gas holding is more difficult and the use of gums, stabilisers and pre-gelatinised starch have been suggested to provide gas occlusion and stabilising mechanisms (Cauvain, 1998; Satin, 1998).

19.3 Ingredients suitable for gluten-free bread production

Currently, different gluten-free flours and ingredients are under investigation for their suitability to produce gluten-free bread of good quality. In this section the principal ingredients and their technological and nutritional relevance are discussed.

19.3.1 Pseudocereals

There are two major subclasses of flowering plants, i.e. monocots (one seed leaf) and dicots (two seed leaves). Wheat, rye and barley are monocots, whereas buckwheat, amaranth and quinoa are dicots and very distantly related to grains in the monocot subclass (Kasarda, 2001). Because of their unique chemical structures, buckwheat, amaranth and quinoa are classified as pseudocereals. Many gluten-free cereal foods (e.g. bread, pasta, cold cereal) are made from refined flour and/or starch and most are not enriched with iron and B vitamins (Thompson, 1999). As a result, a gluten-free diet may contain inadequate amounts of fibre, iron, thiamine, riboflavin, niacin and folate (Thompson, 1999, 2000). Buckwheat, amaranth and quinoa are all good sources of fibre and iron. In addition, the riboflavin content of quinoa and the niacin content of buckwheat flour compare favourably with those of enriched wheat flour (US Department of Agriculture, 2000). Therefore, the addition of buckwheat, amaranth and quinoa would add nutritional value to the diet of an individual with CD.

Buckwheat

Buckwheat has been suggested by Carroll and Hamilton (1975) for the treatment of CD, as it does not contain gluten-like proteins (Francischi *et al.*, 1994; Kreft and Kreft, 2000) and therefore can be used in the production of gluten-free products. Additionally, the unique protein structure and amino acid composition imply that buckwheat might be a very valuable functional food resource and will be of benefit in treating some chronic diseases such as diabetes, hypertension and other cardiovascular diseases (Li and Zhang, 2001; Wijngaard and Arendt, 2006). Buckwheat was recently used in combination with xanthan gum to produce high-quality gluten-free bread (Fig. 19.3C) (Moore *et al.*, 2004a). The addition of buckwheat was reported to increase the water absorption for the bread formulation; however, this was at the expense of volume. Even though the authors concluded that buckwheat-containing bread was firmer in texture, the staling rate was lower in comparison to the starch-based commercial gluten-free bread (Fig. 19.3), thus indicating that buckwheat is suitable for the production of high-quality gluten-free bread. Recently, its application for the production of gluten-free beers has also been investigated (Maccagnan, 2004; Phiarais *et al.*, 2005; Wijngaard *et al.*, 2005) and these studies have showed that buckwheat has also great potential as a brewing ingredient.

Amaranth

The potential amaranth for the production of gluten-free products has been investigated (Tosi *et al.*, 1996; Gambus *et al.*, 2002). By replacing 10% of the cornstarch with amaranth flour, Gambus *et al.* (2002) found an increase in protein and fibre levels by 32 and 152% respectively, while sensory quality was unaffected. The results indicate that amaranth flour can be used to enhance the protein and fibre contents of gluten-free breads.

Quinoa

Taylor and Parker (2002) discussed the application of quinoa as a novel ingredient in the production of enriched gluten-free bakery goods. Caperuto

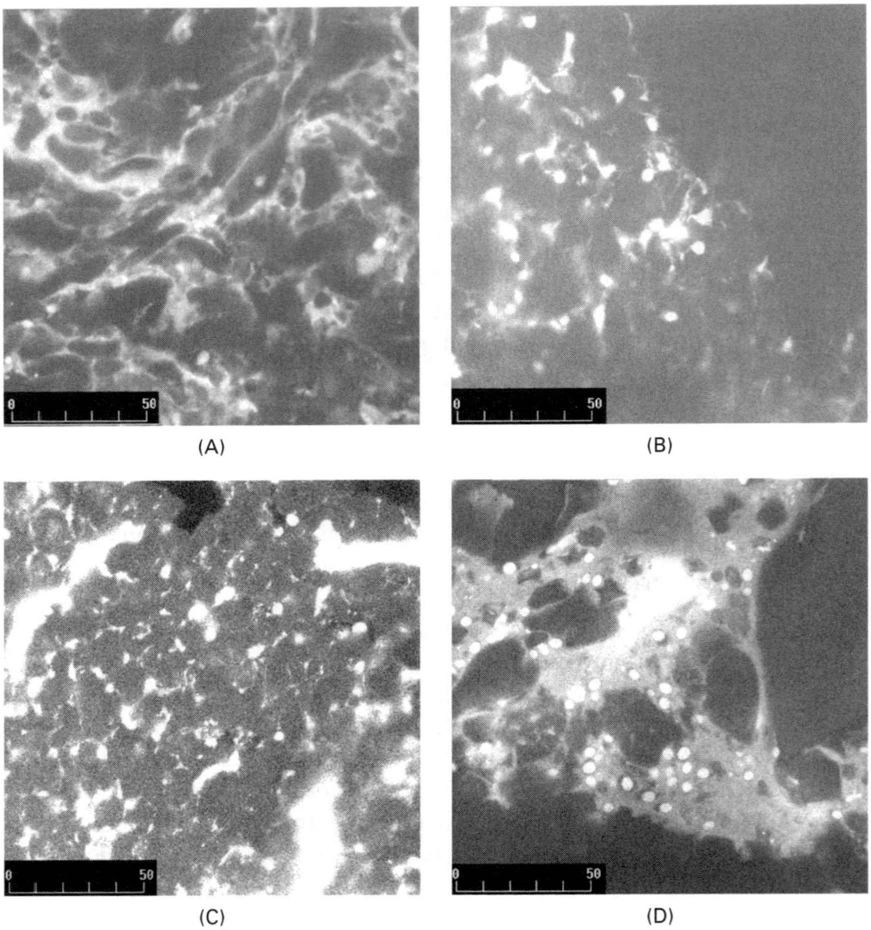

Fig. 19.3 Wheat bread control (A) and gluten-free breads (bread from commercial gluten-free flour (B), non-dairy recipe (C), dairy recipe (D)). Panels show microscopical structure as detected by confocal laser-scanning microscopy (magnification bar corresponds to 50 μm). Adapted from Moore *et al.* (2004a).

et al. (2000) produced gluten-free spaghetti with quinoa. The spaghetti was a composite of quinoa flour (10%) and corn flour, with either egg albumen or curdlan β-glucan as binder. The spaghetti had reasonable physical properties in comparison with soft wheat spaghetti and was well accepted by a consumer sensory panel. Even at the low level of quinoa addition, the lysine and methionine contents were significantly improved in comparison with corn protein. Quinoa, thus, is an excellent source of energy and nutrient density. However, the use of quinoa in the production of gluten-free bread has not been extensively investigated.

19.3.2 Sorghum

Like other cereal grains, the primary component of sorghum is carbohydrate in the form of starch (Rooney and Waniska, 2000). Recent research on sorghum has shown that it has tremendous potential as a 'functional' food (Bean *et al.*, 2006). Specific types of sorghum have a very high content of tannins, which are polyphenolic compounds with potential health benefits for humans (Awika and Rooney, 2004). Other phytochemicals found in sorghum include phenolic acids, anthocyanins, phytosterols and policosanols. These compounds are reported to have a potential impact against cancer and obesity as well as promoting cardiac health (Hargrove *et al.*, 2004; Hwang *et al.*, 2004).

Sorghum is often recommended as a safe food for coeliac patients, because it is more closely related to corn than to wheat, rye and barley (Kasarda, 2001). However, the majority of studies dealing with leavened loaf breads containing sorghum have focused on composite breads from wheat and sorghum, in which a maximum of only 30% low-tannin sorghum is regarded as acceptable (Munck, 1995). While such breads have been found acceptable by consumers (Carson *et al.*, 2000), they are inappropriate for coeliac patients. Only a limited number of studies have addressed wheat-free sorghum breads and most have used relatively complex recipes incorporating xanthan gum (Satin, 1988), carboxymethylcellulose (CMC) and skimmed milk powder (Cauvain, 1998), egg (Keregero and Mtebe, 1994; Cauvain, 1998) or rye pentosans (Casier *et al.*, 1997). Schober *et al.* (2005) investigated the impact of different sorghum varieties on the quality of gluten-free bread. Significant differences were observed between the various sorghum varieties, and the addition of skim milk powder and xanthan gum led to an improvement of the gluten-free bread quality.

19.3.3 Oats

There is much debate on whether or not oats may be used in the production of gluten-free products. A review by Thompson (2003) reports that most adults with CD can consume moderate amounts of uncontaminated oats, without experiencing damage to the intestinal mucosa. Studies have shown that oats are neither toxic nor immunogenic in CD (Janatuinen *et al.*, 1995; Srinivasan *et al.*, 1996; Hardman *et al.*, 1997). According to Hardman *et al.* (1997), oats belong to a different tribe, Aveneae, than that of wheat, rye and barley. Oat prolamin (avenin) has a lower proline content than the prolamins present in wheat, rye and barley (gliadin, secalin and hordein, respectively). The sequence Gln–Gln–Gln–Pro–Phe–Pro is found in prolamins of wheat, rye and barley, but so far has not been found in oats. The ability of oat prolamins to induce an immune response in intestinal biopsy specimens *in vitro* as well as the crossreactivity of monoclonal antibodies to enterotoxic wheat peptides with oat peptides are some of the reasons cited for continued caution (Parnell *et al.*, 1998). One of the major concerns when dealing with oats is that the risk of contamination of oats with wheat, rye, or barley before reaching the consumer is considered high. Even though the Gluten Intolerance Group, Celiac Disease Foundation (Celiac Disease Foundation/Gluten Intolerance Group, 2000), Celiac Sprue Association (Celiac Sprue Association, 2001) and the American Dietetic Association (Inman-Felton, 1999; American Dietetic Association and Dieticians of Canada, 2000) all advise against the use of oats, the Finnish Coeliac Society considers them safe for consumption by adults (Finnish Coeliac Society, 2001), as does the Coeliac Society in the United Kingdom (in moderate amounts) (Coeliac UK, 2001).

Oats have many beneficial properties. According to Oomah and Mazza (1999), oat β-glucan has beneficial effects in the treatment of diabetes and lowers serum cholesterol. Oats also contain a multitude of other compounds – such as phytates, phenolic compounds, vitamins and minerals – conferring further physiological benefits. These include powerful antioxidants that are utilised by the pharmaceutical and cosmetics industries. Furthermore, high consumption of oat meal, oat bran or oat flour can reduce the risk of coronary heart disease.

19.3.4 Rice

Rice flour has been found to be one of the most suitable cereal grain flours for preparing foods for coeliac patients. The suitability of rice flour is attributed to its low level of prolamins. For baking applications, starch granule properties assume a major role in dictating the suitability of rice flour, especially if it is present in a large enough quantity (usually over 10% substitution) to influence the texture of a product (Bean and Nishita, 1985). Optimal rice bread formulations have been developed using CMC and hydroxypropylmethylcellulose (HPMC), which meet wheat bread reference standards for specific volume, crumb and crust colour, firmness and moisture (Collar *et al.*, 1999; Gallagher *et al.*, 2002).

Rice flours can be used in baking applications in many different forms. They can be either white rice flours, where the bran is removed or whole-grain brown rice flour where the bran is intact. The type of rice flour used for the production of gluten-free bread is vital, as the properties of the rice have an impact on the quality of gluten-free breads. A study by Moore *et al.* (2004a) suggested that the production of high-quality gluten-free bread can be achieved from a blend of gluten-free rice flours (brown rice flour) and starches (potato, corn) in combination with the correct hydrocolloid.

19.3.5 Corn

Corn is the most important cereal grain, followed by wheat, in terms of yield per hectare (FAOSTAT, 2005). Corn flour is composed of the endosperm, which generally contains between 75 and 87% starch and 6–8% protein (Shukla and Cheryan, 2001). Zeins, the storage proteins of corn, represent 60% of the proteins and are located in protein bodies (Lending and Larkins,

1989). The endosperm proteins are characterised by high contents of glutamine, leucine and proline, and are practically devoid of lysine and tryptophan. Since corn is considered a gluten-free cereal, it can be used in the production of gluten-free bread. Ács *et al.* (1996a, b) used binding agents (xanthan, guar gum, locust bean gum and tragacanth) to substitute gluten in a gluten-free bread formulation based on cornstarch. The authors found that the binding agents resulted in an increase in loaf volume and loosening of the crumb structure. Ground corn is used in products such as corn meal, corn starch, masa farina and corn flour. Corn flour, in turn, is used in the manufacturing of food products such as tortillas, pancake mixes, cornbread mixes, corn chips and cereal, as well as in home cooking for stews, soups and casseroles.

19.3.6 Tef

The tef grain characteristics that influence food, nutritional and technological properties include colour, pericarp and seed coat, endosperm texture and hardness (Babatunde Obilana and Manyasa, 2002). The bulk of the grain consists of the bran and germ. This makes tef nutrient dense, as the bran and germ are the most nutritious parts of any grain. The starch granules are stored in the pericarp of the grain. The starch in the endosperm is stored in compound starch grains, which are made up of many polygonal starch granules 2–6 μm across (Babatunde Obilana and Manyasa, 2002). Tef contains 11% protein, 80% complex carbohydrate and 3% fat. It is an excellent source of essential amino acids, especially lysine – the amino acid that is most often deficient in grain foods. Tef is also an excellent source of fibre and iron, and has many times the amount of calcium, potassium and other essential minerals found in an equal amount of other grains. Due to the fact that tef does not contain gluten, tef can be used for the production of gluten-free bread. Its gluten-free flour is primarily processed into a pancake-like, fermented flat bread called injera and can be mixed with sorghum and pearl millet for another flat bread (kisra) in Sudan and Eritrea. Unleavened tef bread is used as an adjunct in making homemade beverages and beer (tella). Tef is further used to make waffles, chocolate cakes, muffins, biscuits, cookies, casseroles, soups, stews and puddings (Ketema, 1993).

19.3.7 Dietary fibre

Cereals are an important source of dietary fibre, contributing to about 50% of the fibre intake in Western countries (Nyman *et al.*, 1989). The role of dietary fibre in providing roughage and bulk, and in contributing to a healthy intestine has long been recognised. Diets that contain moderate quantities of cereal grains, fruits and vegetables are likely to provide sufficient fibre. Due to the fact that gluten-free products generally are not enriched/fortified, and are frequently made from refined flour or starch, they may not contain the same levels of nutrients as the gluten-containing counterparts they are intended to replace. Therefore, uncertainty still exists as to whether coeliac patients living on a gluten-free diet are ensured a nutritionally balanced diet, especially regarding the dietary fibre intake. Grehn *et al.* (2002) screened the intake of nutrients and foods of 49 adults diagnosed with CD and following a gluten-free diet. They were found to have a lower intake of fibre when compared with a control group of people on a normal diet. Similarly, Lohiniemi *et al.* (2000) found that the average fibre consumption amongst coeliacs in Sweden was lower than recommended. In their studies with coeliac adolescents, Mariani *et al.* (1998) and Thompson (2000) concluded that adherence to a strict gluten-free diet worsens the already nutritionally unbalanced diet of adolescents.

The enrichment of gluten-free baked products with dietary fibres has been a topic of research for food technologists. Studies have shown that the addition of high-fibre foods can give texture, gelling, thickening, emulsifying and stabilising properties to certain foods (Sharma, 1981; Dreher, 1987). Inulin is a non-digestible fructooligosaccharide that is classified as a dietary fibre. Inulin is used either as a macronutrient substitute or as a supplement added in foods mainly for its prebiotic properties (Gibson and Roberfroid, 1995; Zimeri and Kokini, 2003). Inulin also acts as a prebiotic by stimulating the growth of 'healthy' bacteria in the colon (Gibson and Roberfroid, 1995). When added to wheat bread, it improves loaf volume and sliceability, increases dough stability and produces a uniform and finely grained crumb texture (Anon., 1999). Gallagher *et al.* (2002) incorporated inulin (8% inclusion level) into a wheat starch-based gluten-free formulation. The dietary fibre content of the bread increased from 1.4 (control) to 7.5% (control + inulin). The overall quality of the gluten-free bread was also improved by the addition of inulin.

19.3.8 Starch sources in gluten-free bread

In cereal-based products, starch gelatinisation, as well as subsequent reorganisation, controls the texture and stability of the final product. Starches such as wheat starch (less than 200 ppm), rice, potato and cornstarch are most commonly used in the production of gluten-free products.

Wheat starch

Starch is the most abundant carbohydrate component in wheat and wheat flour as well as in many foods. The extent and manner in which starch granules swell and undergo phase transitions are influenced by formulation change, and particularly, by the amount of available water (Chiotelli and Le Meste, 2003). Wheat starch is in discrete granules within the cells of the endosperm. Modern milling practices are designed to remove the germ and bran, producing flour that essentially contains the endosperm and, thus, is rich in starch (Lineback and Rasper, 1988). Wheat starch has been utilised as a replacement for wheat flour in gluten-free products. The separation of wheat starch from wheat flour is a crucial process. Since wheat starch is still widely used in gluten-free products (with the exception of the United States), it is important that it contains less than 200 ppm gluten (Codex Alimentarius Commission, 2000).

Kulp *et al.* (1974) used a basic formula to produce bread from pure wheat starch with additives and found that the best quality loaves were produced with the addition of xanthan gum. However, many individuals sensitive to the gliadin fraction of gluten cannot tolerate even a very small amount of this protein remaining in isolated wheat starch. Reports have highlighted that the long-term effects of regular ingestion of small amounts of gliadin (e.g. from wheat starch) are harmful to patients with CD (Skerritt and Hill, 1992; Chartrand *et al.*, 1997; Lohiniemi *et al.*, 2000). However, the effects of small amounts of gliadin consumption vary among individuals and those who experience symptoms when consuming wheat starch-based gluten-free foods should use products that are naturally gluten-free (Ciclitira *et al.*, 1985).

Potato starch

Potato starch has desirable characteristics that differ significantly from those of starch from other plant sources (Madsen and Christensen, 1996), i.e. high swelling power, water-binding capacity and freeze–thaw stability. It is well known that small amounts of added potato solids help to retain the freshness of bread and also impart a distinctive, pleasing flavour and improve toasting qualities (Willard and Hix, 1987). Potato starch has a bland flavour and forms high-viscosity pastes that are susceptible to shear. The potato granules are large and are extremely susceptible to breakage. The high molecular weight amylose and phosphate groups esterified to amylopectin contribute to the high transparency, swelling power, water-binding capacity and freeze–thaw stability of potato starches (Craig *et al.*, 1989). As potato starch does not contain gluten, it is considered a safe component for gluten-free foods.

19.3.9 Protein sources in gluten-free breads

The replacement of gluten with other protein sources such as soya and dairy proteins is another approach used in the production of gluten-free products. Currently, dairy proteins, surimi, soybean and egg proteins are commonly included in gluten-free bread recipes.

Dairy proteins

Dairy proteins have functional properties similar to gluten. They are capable of forming networks and they have good swelling properties. Dairy proteins are highly functional ingredients and due to their versatility can be readily used in many food products (Gallagher *et al.*, 2004). They can be used in bakery products for both nutritional and functional benefits, including flavour and texture enhancement and storage improvement (Cocup and Sanderson,

1987; Mannie and Asp, 1999; Kenny *et al*., 2001; Gallagher *et al*., 2004). They contribute a number of critical characteristics to a food product, e.g. the emulsifying and stabilising ability of caseinates, the gelling properties of whey protein concentrates and isolates, the water-absorption capacity of high-heat, non-fat dry milk and the browning of lactose during heat processing (Chandan, 1997). Furthermore, they may be used in gluten-free bread formulae in combination with emulsifiers (Fig. 19.3D) to increase water absorption and therefore enhance the handling properties of the batter (Gallagher *et al*., 2003). Gallagher *et al*. (2003) found that dairy powders used in gluten-free bread formulations resulted in improved volume, appearance and sensory aspects. However, the supplementation of gluten-free breads with high-lactose-content powders is not suitable for persons with CD who have significant damage to their intestinal villi (Ortolani and Pastorello, 1997).

Surimi

Surimi is a concentrate of myofibrillar proteins obtained after mixing and water washing of fish flesh (Han-Ching and Leinot, 1993). Gormley *et al*. (2003) focused on the supplementation of a control gluten-free bread based on rice flour and potato starch with surimi (as a structure enhancer and protein replacer) at a 10% inclusion level (of starch weight) and found that the addition of surimi improved the overall quality of the bread. Furthermore, regarding sensory analysis, no significant negative effects were found with the addition of the surimi to the bread.

Soybean and egg proteins

Soybean and egg proteins are alternative protein sources used in baking applications. Soya has several functional properties. Soya isoflavones may be used in functional foods that target a reduction in the risks of osteoporosis. Furthermore, soya isoflavones are reported to reduce the risk of cardiovascular disease, reduce the oxidation of low-density lipoproteins (LDLs) and prevent breast cancer (Omoni and Aluko, 2005). Finally, soya isoflavones possess antioxidant properties. Soya flour is widely used in the production of gluten-free products. Ranhorta *et al*. (1975) discussed the application of soya protein to gluten-free bread. These authors formulated wheat starch-based gluten-free breads with 20, 30 and 40% soya protein isolate (containing 88% protein). The breads had more protein and fat than wheat bread and showed satisfactory baking results. Sanchez *et al*. (2002) reported that the inclusion of 0.5% soya to gluten-free bread enhances the crumb grain score, bread volume and overall bread score.

Research by Jonagh *et al*. (1968) has shown that proteins such as egg albumen at low concentration are able to link starch granules together. When used at high concentrations, egg albumin was found to improve gas retention properties in gluten-free bread. Kato *et al*. (1990) suggested that egg proteins form strong cohesive viscoelastic films, which are essential for stable foaming. According to Moore *et al*. (2004a), egg proteins, part of the soya protein or colloidal solutions such as caseins swell and form viscous solutions in a gluten-free bread system. Confocal laser-scanning microscopy has revealed film-like continuous protein structures similar to wheat gluten (Fig. 19.4). Investigating the interaction between egg powder and the enzyme transglutaminase (TGase) in a gluten-free bread system, Moore *et al.* (2004c) found that crosslinking occurred when 1 U TGase/g protein was added to the egg powder-based gluten-free breads. This crosslinking was due to covalent bonds being produced between lysine and glutamine residues in the egg proteins by the enzyme.

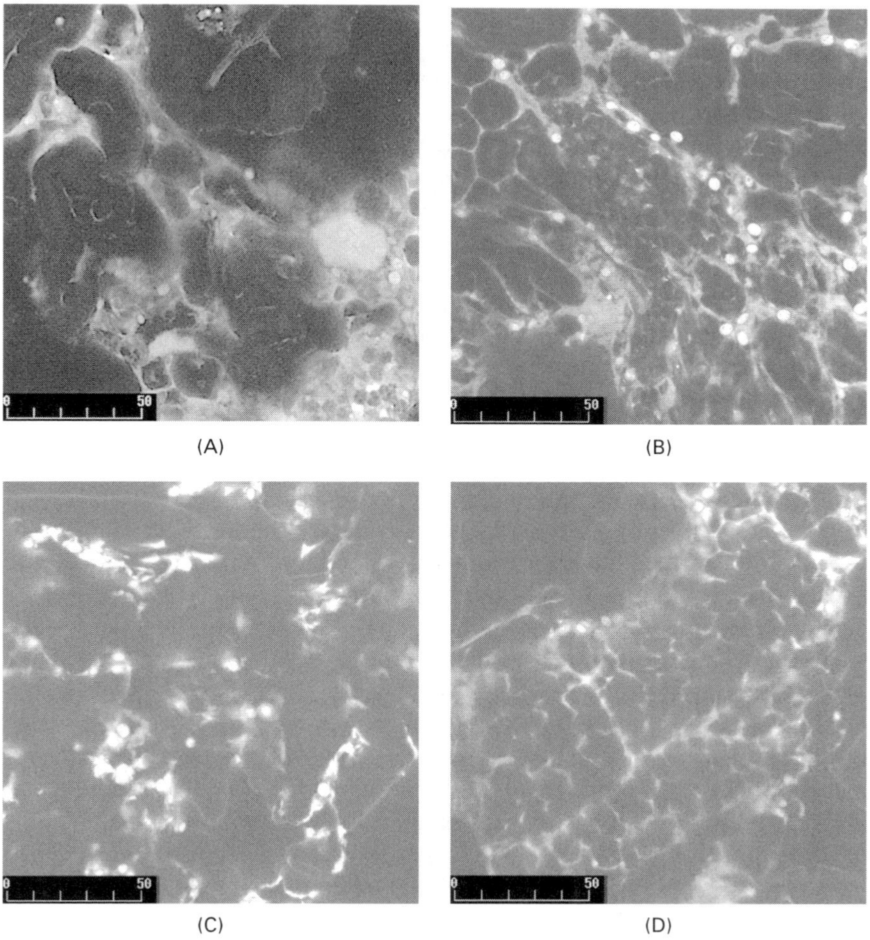

Fig. 19.4 Confocal laser-scanning micrographs of gluten-free breads: egg, 0 U transglutaminase (TGase)/g protein (control) (A); egg, 1 U TGase/g protein (B); skim milk powder (SMP), 0 U TGase/g protein (control) (C); SMP, 10 U TGase/g protein (D). The magnification bar corresponds to 50 μm for the × 63 objective. Adapted from Moore *et al.* (2004a).

19.4 Improving the quality of gluten-free bread

The main problems associated with gluten-free products are that they are crumbly, and have a rapid onset of staling (Arendt *et al.*, 2002). In order to overcome these problems, a number of options are available to the food scientist. This section deals with some of the strategies available to improve the quality of gluten-free breads.

19.4.1 Hydrocolloids

Hydrocolloids have been essential ingredients in the formulation of gluten-free cereal products. Hydrocolloids or gums are substances consisting of hydrophilic long-chain, high molecular weight molecules, usually with colloidal properties, that in water-based systems produce gels or thickened liquids, i.e. highly viscous suspensions or solutions with low dry-substance content (Hoefler, 2004). They are derived from seeds, fruits, plant extracts, seaweeds and microorganisms, being of polysaccharide or protein nature (Norton and Foster, 2002). Hydrocolloids or gums serve two basic functions in food systems: they stabilise the product and they improve the texture of the product. Other functional properties include: retarding starch retrogradation, increasing moisture retention and extending the overall quality of the product over time (Armero and Collar, 1996a,b; Davidou *et al.*, 1996).

The lack of a gluten network determines the properties of the gluten-free dough, which is more fluid than wheat doughs and closer in viscosity to cake batters (Cauvain, 1998; Moore *et al.*, 2004a) and thus has also to be handled in a similar way to cake batters rather than doughs. Most gluten-free breads are formulated using gluten-free starches and require the addition of hydrocolloids to provide structure and gas retention properties in the dough (Ylimaki *et al.*, 1998; Moore *et al.*, 2004a). Hydrocolloids have been found to affect dough rheological performance – since they mimic the viscoelastic properties of gluten in bread doughs (Smith, 1971; Toufeili *et al.*, 1994; Collar *et al.*, 1998) – and also swelling, gelatinisation, pasting properties and staling of starch. Polymeric substances such as xanthan gum or HPMC are required for the production of gluten-free breads (Toufeili *et al.*, 1994; Guarda *et al.*, 2004, McCarthy *et al.*, 2005). For gluten-free bread production, the most common hydrocolloids used are pectin, guar gum, xanthan gum and locust bean gum (Smith, 1971; Igoe, 1982; Okubo *et al.*, 1994, Moore *et al.*, 2004a; McCarthy *et al.*, 2005). Schwarzlaff *et al.* (1996) used combinations of guar gum and locust bean gum to partially replace flour in bread. They found that the introduction of guar gum resulted in a crumb structure with a more even cell size distribution, while locust bean gum inclusion increased the height of the bread loaves. Both gums retarded bread staling. Kang *et al.* (1997) investigated the effect of various hydrocolloids on a rice bread formulation. All investigated gum types gave successful formation of rice bread, with HPMC showing an optimum volume expansion. HPMC is a popular hydrocolloid for gluten replacement in gluten-free bread systems which, on exposure to water or biological fluid, becomes hydrated, swells and forms a gel barrier layer (Kavanagh and Corrigan, 2004).

In the preparation of gluten-free bread, xanthan gum has been found to confer a crumb structure similar to that of bread made from normal wheat flour (Christiansson *et al.*, 1974; Christiansson, 1976). Van Vliet *et al.* (1992) suggested that xanthan may act by improving the strain hardening properties of the 'dough'. Xanthan gum has been used as a thickener, or mixed with other gel-forming polysaccharides, e.g. konjac gum. Konjac gum is a polysaccharide that has the ability to form cohesive, tough films that remain stable in hot and cold systems (Tye, 1991). Konjac is therefore considered a useful hydrocolloid in the production of gluten-free bread. Recently, Moore *et al.* (2004a) produced good quality gluten-free bread using a mixture of both konjac and xanthan gum in a dairy-based gluten-free formulation.

19.4.2 Enzymes

Enzymes are naturally present in foods as long as they are not removed or inactivated. Amylases, proteases, hemicellulases, lipases and oxidases influence the whole baking process (Hozová *et al.*, 2002). The functions of enzymes are widespread throughout the baking industry. These include decolourising (bleaching) of dough (oxidoreductases), improvement of the volume and texture of dough, substitution of bromides and maintenance of shelf-life (Grossman and De Barber, 1997; Sahlström and Brathen, 1997; Delcros *et al.*, 1998; Géllinas *et al.*, 1998; Vemulappali and Hoseney, 1998; Corsetti *et al.*, 2000; Rossel *et al.*, 2001).

Starch-hydrolysing enzymes

To date, there are only few published reports on the utilisation of amylase or protease enzymes and their impact on gluten-free foods. Gujral *et al.* (2003a,b) tested two starch-hydrolysing enzymes from *Bacillus* spp. (α-amylase of intermediate thermostability and cyclodextrin glycosyl transferase (CGTase)). Their effectiveness in retarding rice bread staling was assessed. It was found that the presence of CGTase, in particular, decreased amylopectin retrogradation and showed a significant anti-staling effect. A favourable increase in loaf specific volume following addition of this enzyme was also noted. The results of these studies indicate that starch hydrolysing enzymes may be used to retard staling of gluten-free bread.

Transglutaminase

TGase (protein-glutamine γ-glutamyl-transferase) is a promising enzyme for the food industry. TGase catalyses acyl-transfer reactions introducing covalent crosslinks between proteins (Nonaka *et al.*, 1989), as well as peptides and various primary amines. Crosslinking occurs when the ε-amino groups of lysine residues in proteins act as an acyl-receptor. ε-(γ-Glu)Lys bonds (isopeptide bonds) are formed both intra- and inter-molecularly (Ando *et al.*, 1989). TGase has the ability to link proteins of different origin: casein and albumin from milk, animal proteins from eggs and meat, soya proteins and wheat proteins. TGase has been mainly used in the dairy and meat industry to improve the water-holding capacity. TGase is a highly functional enzyme and may be obtained from a range of different sources, e.g. animal tissue, fish, plant and microorganisms (Kuraishi *et al.*, 1996). The TGases used in baking applications are microbial aminotransferases. Recently, Moore *et al.* (2004c) reported that microbial TGase can be used to create a network-like structure in gluten-free bread (Fig. 19.4) and therefore has the potential to improve the quality of gluten-free bread. However, the use of TGase in gluten-free bread is very much dependent on the protein source used in the bread formulation.

19.4.3 Sourdough

Sourdough fermentation has a well-established role in improving the flavour and structure of bread. The use of sourdough results in the improvement of the aroma, taste, nutritional value and shelf-life of the bread. Moreover, sourdough improves the texture and palatability of whole-grain, fibre-rich or gluten-free products, stabilises or increases levels of various bioactive compounds, retards starch bioavailability (lower glycaemic index) and improves mineral bioavailability.

Currently, different approaches are under investigation to produce sourdough breads that can be tolerated by coeliacs. One approach involves the fermentation of non-toxic flours and the application of the resulting sourdough to improve the quality of gluten-free bread. The effects of addition of sourdough produced using a mixture of non-toxic brown rice flours, corn starch, buckwheat and soya flour to a gluten-free bread recipe have been investigated by Moore *et al.* (2005). The authors evaluated the influence of sourdoughs fermented by different strains of lactic acid bacteria on the textural quality of gluten-free bread. Changes in dough structure could not only be detected by small deformation viscoelastic measurements but also by confocal laser-scanning microscopy (Fig. 19.5). A significant difference in phase angle was observed between sourdough and non-fermented batter. This result showed that the fermentation causes an increase in elasticity over the initial 24 h of fermentation. Moreover, the protein fraction of the gluten-free sourdough was degraded over time (Fig. 19.5). This process was, however, far less obvious in a gluten-free system than with gluten isolated from wheat-based sourdough (Fig. 19.5). Interestingly, following the application of sourdough, the onset of staling was delayed, thus indicating that addition of sourdough to a gluten-free recipe improves the quality of the resulting bread.

19.4.4 Optimisation of gluten-free breads

Response surface methodology (RSM) is a statistical tool that is particularly appropriate for product development work. Successful application of RSM

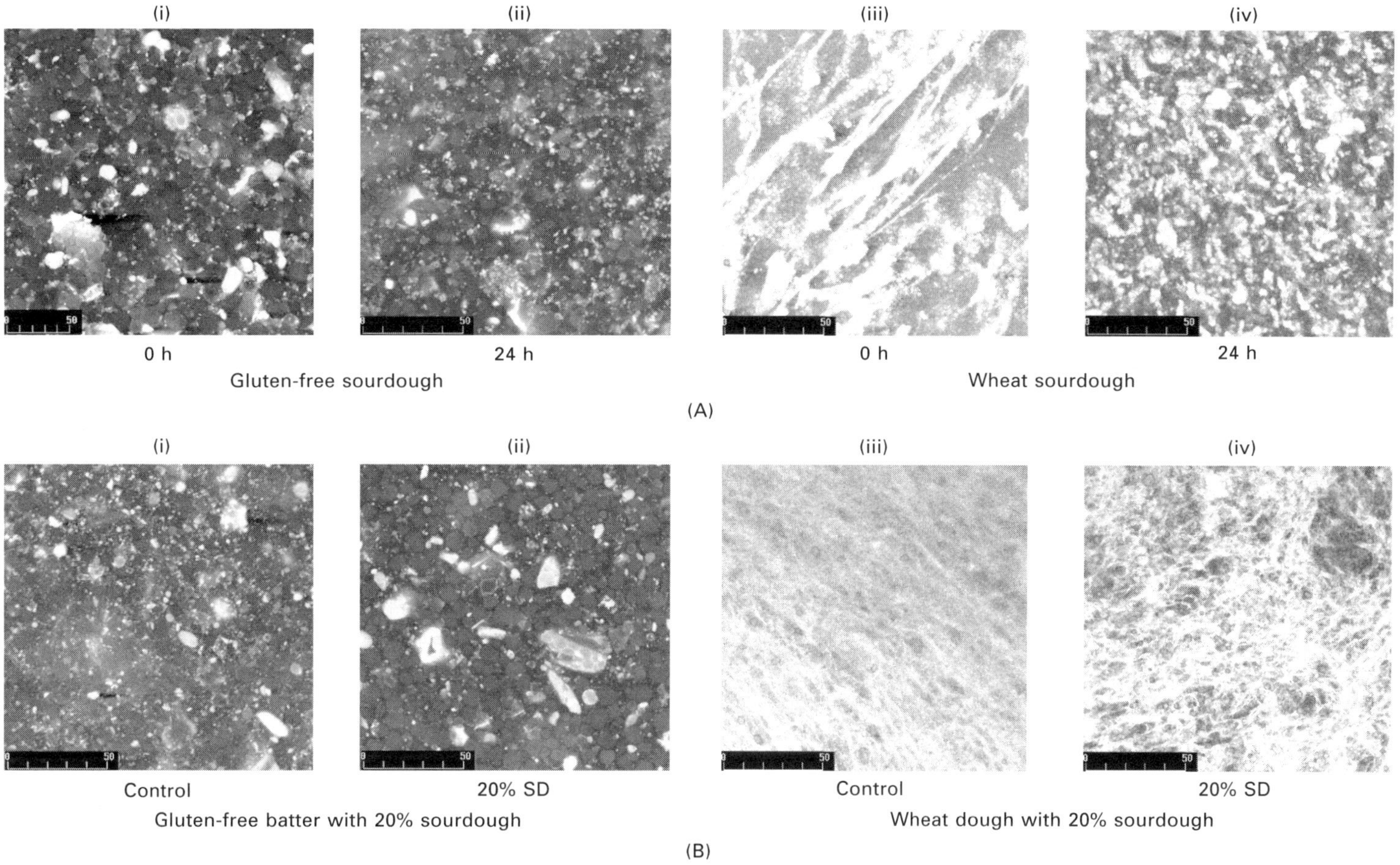

Fig. 19.5 Confocal laser-scanning micrographs of (A) gluten-free sourdough and wheat sourdough at 0 h and after 24 h (magnification bar corresponds to 50 μm) and (B) gluten-free batter with 20% sourdough (SD) and wheat dough with 20% sourdough. Adapted from Clarke *et al*. (2004).

in the production of different types of wheat bread has been reported (Lee and Hoseney 1982; Clarke *et al.*, 2002, 2004; Gallagher *et al.*, 2003, 2004). Ylimaki *et al.* (1991) used RSM to produce and objectively measure gluten-free breads based on three types of rice flour (varying in grain size and grinding method). Among their results, they found that optimal loaves were formulated with medium grain, finely ground rice flour, low levels of HPMC and low levels of CMC. These breads were the most similar to wheat flour breads, based on crust and crumb colour, Instron firmness and loaf moisture. The same authors used the same three rice flours in a second trial. Gluten-free yeast breads were produced based on the rice flours (80%) and potato starch (20%). Using sensory measurements from a trained panel, RSM was applied to find CMC, HPMC and water combinations for the different rice flours. It was found that gluten-free loaves made with medium-grain rice flours were of a higher standard with respect to moistness, cohesiveness, flavour, colour and cell structure than those made from long-grain rice flour (Ylimaki *et al.*, 1991).

RSM was recently used to optimise a gluten-free bread formulation primarily based on rice flour, potato starch and skim milk powder (McCarthy *et al.*, 2005). HPMC and water were the predictor variables. From the data obtained, optimal ingredient levels were determined. The optimised formulation contained 2.2% HPMC and 79% water flour/starch base and measured responses compared favourably with predicted values. Shelf-life analysis of the optimised formulation over 7 days revealed that, as crumb firmness increased, crust firmness and crumb moisture decreased. This study clearly indicates that RSM is a suitable tool for analysing the effect of various parameters for optimisation of gluten-free bread recipes.

19.5 Future trends

Currently, research and development are focused on improving the mouthfeel, flavour and rheology of gluten-free products. There are, however, other possibilities for producing gluten-free products of superior quality. One such alternative is the development of grains that have low or no content of immunotoxic epitopes, but with good baking quality. Such grains can potentially be developed by (a) selective breeding of ancient varieties (Molberg *et al.*, 2005); (b) transgenic technology involving mutation of sequences giving rise to immunostimulatory sequences (Vader *et al.*, 2003); (c) small interfering RNA technology; or (d) by incorporation of non-toxic gluten genes into harmless organisms such as rice (Sollid *et al.*, 2005).

Recently a new approach has been investigated based on sourdough fermentation to degrade toxic epitopes of wheat flour (Di Cagno *et al.*, 2004, 2005). This approach requires the virtually complete hydrolysis of gliadins. However, this extension of protein hydrolysis compromises the technological functionality of wheat flour, and thus for product formulation the fermented sourdough has to be applied in combination with unfermented non-toxic flours. Two types of bread containing *c.* 2.5% of wheat sourdough and unfermented oat, buckwheat and millet flours were manufactured with baker's yeast or sourdough lactobacilli and used for *in vivo* double-blind acute challenge on CD patients (Di Cagno *et al.*, 2004). Thirteen out of the 17 patients showed a marked alteration of the intestinal permeability after ingestion of baker's yeast bread, but after consuming sourdough bread the intestinal permeability of these patients did not differ from the baseline values. The same findings were obtained by using the commercial probiotic bacteria VSL#3® as starters for wheat sourdough fermentation (De Angelis *et al.*, 2006), showing that proteolysis due to selected lactic acid bacteria can, in fact, degrade the gliadin fraction of wheat flour, allowing fermented wheat flour to potentially be used as a base for future gluten-free recipes. Even if this approach may not be directly applicable for the industrial production of gluten-free bread, the results collected so far strongly indicate that select lactic acid bacteria can be used to degrade any potential contaminants present in gluten-free flours.

19.6 Conclusions

According to epidemiological studies, approximately 1 in 100 of the world population is suffering from CD. Gluten is the triggering factor and, to date, the only effective treatment is a strict adherence to a gluten-free diet throughout the patient's lifetime. However, gluten has unique properties in relation to production of good-quality cereal products, which makes its replacement particularly challenging for the food technologists. In recent years there has been extensive work in using several ingredients – such as starches, dairy products, gums and hydrocolloids, other non-gluten proteins and combinations thereof – as alternatives to gluten, to improve the structure, mouthfeel, acceptability and shelf-life of gluten-free bakery products. From the studies reported in this review, it becomes apparent that a mixture of different gluten-free flours, gums and proteins is necessary to replace gluten and produce good-quality gluten-free cereal products. Currently, research and development are focused on improving the mouthfeel, flavour and rheology of gluten-free products. The development of grains that have low or no content of immunotoxic epitopes, or the degradation of toxic epitopes by selected lactic acid bacteria are two opportunities currently under investigation to create gluten-free products of quality comparable with common products.

19.7 References

Ács, E., Kovacs, Z.S. and Matuz, J. 1996a. Bread from corn starch for dietetic purposes. I. Structure formation. *Cereal Res. Comm.* **24**: 441–449.

Ács, E., Kovacs, Z.S. and Matuz, J. 1996b. Bread from corn starch for dietetic purposes. II. Formation of the visual and technological properties. *Cereal Res. Comm.* **24**: 451–459.

American Dietetic Association and Dietitians of Canada. 2000. *Manual of Clinical Dietetics*, 5th edn. American Dietetic Association and Dietitians of Canada, Chicago, Illinois.

Ando, H., Adachi, M., Umeda, K., Matsuura, A., Nonaka, M., Uchio, R., Tanaka, A. and Motoki, M. 1989. Purification and characteristics of novel transglutaminase derived from microorganisms. *Agric. Biol. Chem.* **53**: 2613–2617.

Anon. 1999. Inulin: added value. *Eur. Baker*, **32**:40–44.

Arendt, E.K., O'Brien, C.M., Schober, T.J., Gallagher, E. and Gormley, T.R. 2002. Development of gluten-free cereal products. *Farm Food* **12**: 21–27.

Armero, E. and Collar, C. 1996a. Anti-staling additives, flour type and sourdough process effects on functionality of wheat doughs. *J. Food Sci.* **61**: 299–303.

Armero, E. and Collar, C. 1996b. Anti-staling additive effects on fresh wheat bread quality. *Food Sci. Technol. Int.* **2**: 323–333.

Awika, J.M. and Rooney, L.W. 2004. Sorghum phytochemicals and their potential impact on human health. *Phytochemistry*. **65**: 1199–1221.

Babatunde Obilana, A. and Manyasa, E. 2002. Millets. In: Belton, P. and Taylor, J. (eds). *Pseudocereals and Less Common Cereals: Grain Properties and Utilization Potential*. Springer-Verlag, Berlin, Heidelberg, New York.

Bean, M.M. and Nishita, D.K. 1985. Rice flours for baking. In: Juliano, B.O. (ed.) *Rice Chemistry and Technology*, 2nd edn. AACC International, St Paul, Minnesota, Ch. 14, pp. 539–556.

Bean, S.R., Schober, T.J., Arendt, E.K. and Fenster, C. 2006. Use of sorghum in bakery products. *Am. Inst. Baking Tech. Bull.* **28**: 1–6.

Caperuto, L.C., Amaya-Farfan, J. and Camargo, C.R.O. 2000. Performance of quinoa (*Chenopodium quinoa* Willd) flour in the manufacture of gluten-free spaghetti. *J. Sci. Food Agric.* **81**: 95–101.

Carroll, K.K. and Hamilton, R.M.G. 1975. *J. Food Sci.* Effects of dietary protein and carbohydrate on plasma cholesterol levels in relation to atherosclerosis **40**: 18–22.

Carson, L., Setser, C. and Sun, X.S. 2000. Sensory characteristics of sorghum composite bread. *Int. J. Food Sci. Technol.* **35**: 465–471.

Casier, J.P.J., de Paepe, G., Willems, H., Goffings, G. and Noppen, H. 1997. Bread from starchy tropical crops II. Bread production from pure millet and sorghum flours, using cereal endosperm-cellwall-pentosan as a universal baking factor. In: Dendy, D.A.V., (ed.). *Proceedings of a Symposium on Sorghum and Millets for Human Food*. Tropical Products Institute, London, pp. 127–131.

Catassi, C. and Fasano, A. 2007. Celiac disease. In: Arendt, E.K. and Dal Bello, F. (eds). *Gluten free cereal products and beverages*. Elsevier, The Netherlands (in press).

Cauvain, S.P. 1998. Other cereals in breadmaking. In: Cauvain, S.P. and Young, L.S. (eds). Technology of Breadmaking. Blackie Academic & Professional, London, pp. 330–346.

Celiac Disease Foundation/Gluten Intolerance Group. 2000. Quick Start Diet Guide for Celiac Disease: Available at: http://www.celiac.org/downloads/.

Celiac Sprue Association. Accessed February 16, 2001. Basics for the Gluten-Free Diet. Available at: http://www.csaceliacs.org.

Chandan, R. 1997. Properties of milk and its components. In: *Dairy-Based Ingredients. Egan Press Handbook*. AACC International, St Paul, Minnesota, pp. 1–10.

Chartrand, L.J., Russo, P.A., Dulhaime, A.G. and Seidman, E.G. 1997. Wheat starch intolerance in patients with coeliac disease. *J. Am. Diet. Ass.* **97**: 612–618.

Chiotelli, E. and Le Meste, M. 2003. Effect of triglycerides on gelatinisation and rheological properties of concentrated potato starch preparations. *Food Hydrocoll.* **17**: 629–639.

Christiansson, D.D. 1976. Baked foods fortified with vegetable protein. *Baker's Dig.* **50**: 34–36.

Christiansson, D.D., Gardner, H.W., Warner, K., Boundy, B.K. and Inglett, G.E. 1974. Xanthan gum in protein-fortified starch bread. *Food Technol.* **28**: 23–26, 28–29.

Ciclitira, P.J., Cerio, R., Ellis, H.J., Maxton, D., Nelufer, J.M. and Macarthy, J.M. 1985. Evaluation of a gliadin-containing gluten-free product in coeliac patients. *Hum. Nutr. Clin. Nutr.* **39**: 303–308.

Clarke, C.I., Schober, T.J. and Arendt, E.K. 2002. Effect of single strain and traditional mixed strain starter cultures on rheological properties of wheat dough and on bread quality. *Cereal Chem.* **79**: 640–647.

Clarke, C.I., Schober, T.J. Dockery, P. O'Sullivan, K. and Arendt, E.K. 2004. Wheat sourdough fermentation: Effects of time and acidification on fundamental rheological properties. *Cereal Chem.* **81**: 409–417.

Cocup, R.O. and Sanderson, W.B. 1987. Functionality of dairy ingredients in bakery products. *Food Technol.* **41**: 86–90.

Codex Alimentarius Commission 2000. Draft Revised Standard for Gluten-Free Foods, CX/NFSDU 00/4. FAO/WHO, Rome.

Coeliac UK. Accessed January 24, 2001. Guidelines on Coeliac disease and Oats. Available at: Coeliac UK homepage, http://www.coeliac.co.uk.

Collar, C., Armero, E. and Martinez, J. 1998. Lipid binding of formula bread doughs. *Z. Lebensm. Unters. Forsch.* **207**: 110–121.

Collar, C., Andreu, P., Martinez, J. and Armero, E. 1999. Optimization of hydrocolloid addition to improve wheat bread dough functionality: a response surface methodology study. *Food Hydrocoll.* **13**: 467–475.

Corsetti, A., Gobbetti, M., De Marco, B., Balestrieri, F., Paoletti, F., Russi, L. and Rossi, J. 2000. Combined effects of sourdough lactic acid bacteria and additives on bread firmness and staling. *J. Agri. Food Chem.* **48**: 3044–3051.

Craig, S.A.S., Maningat, C.C. Seib, A.P. and Hoseney, C.R. 1989. Starch paste clarity. *Cereal Chem.* **66**: 173–182.

Davidou, S., Le Meste, M. Debever, E. and Bekaert, D. 1996. A contribution to the study of staling of white bread: effect of water and hydrocolloid. *Food Hydrocoll.* **10**: 375–383.

De Angelis, M., Rizzello, C.G., Fasano, A., Clemente, M.G., De Simone, C., Silano, M., De Vincenzi, M., Losito, I. and Gobbetti, M. 2006. VSL#3 probiotic preparation has the capacity to hydrolyze gliadin polypeptides responsible for Celiac Sprue. *Biochim. Biophys. Acta.* **1762**: 80–93.

Delcros, J.F., Rakotozafy, L. and Boussard, A. 1998. Effect of mixing conditions on the behaviour of lipoxygenase, peroxidase, and catalase in wheat flour. *Cereal Chem.* **75**: 85–93.

Di Cagno, R., De Angelis, M., Auricchio, S., Greco, L., Clarke, C., De Vincenzi, M., Giovannini, C., D'Archivio, M., Landolfo, F., Parrilli, G., Minervini, F., Arendt, E.K. and Gobbetti, M. 2004. Sourdough bread made from wheat and nontoxic flours and started with selected lactobacilli is tolerated in celiac sprue patients. *Appl. Environ. Microbiol.* **70**: 1088–1096.

Di Cagno, R., de Angelis, M., Alfonsi, G., de Vincenzi, M., Silano, M., Vincentini, O. and Gobbetti, M. 2005. Pasta made from durum wheat semolina fermented with selected lactobacilli as a tool for a potential decrease of the gluten intolerance. *J. Agric. Food Chem.* **53**: 4393–4402.

Dreher, M.L. 1987. *Handbook of Dietary Fibre: an Applied Approach.* Marcel Dekker, New York.

FAOSTAT. 2005. © FAO Statistics Division 2005, http://faostat.fao.org/.

FAO (Food and Agriculture Organization). 1992. Maize in Human Nutrition: Report Series 25. Food and Agriculture Organization, Rome.

Farrell, J.R. and Kelly, P.C. 2002. Coeliac Sprue. *New Engl. J. Med.* **346**: 3: 180–188.

Fasano, A. 2005. Clinical presentation of coeliac disease in the pediatric population. *Gastroenterology* **128**: 68–73.

Fasano, A. and Catassi, C. 2001. Current approaches to diagnosis and treatment of celiac disease: An evolving spectrum. *Gastroenterology* **120**: 636–651.

Faubion, J.M. and Hoseney, R.C. 1990. The viscoelastic properties of wheat flour doughs. In: Faridi H. and Faubion J.M. (eds). *Dough Rheology and Baked Product Texture* Van Nostrand Reinhold, New York, pp. 211–239.

Feighery, C. 1999. Fortnightly review. Celiac disease. *Br. Med. J.* **319**: 236–239.

Finnish Coeliac Society. Accessed February 16, 2001. Coeliac Disease. Available at: http://www.sci.fi/keliakia/fincoel.htm.

Francischi, M.L.P., De Salgado, J.M. and Costa, C.P. 1994. Immunological analysis of serum for buckwheat fed coeliac patients. *Plant Foods Hum. Nutr.* **46**: 207–211.

Gallagher, E., Polenghi, O. and Gormley, T.R. 2002. Novel rice starches in gluten-free bread. In: *Proceedings of the International Association of Cereal Chemists Conference*, pp. 24–26.

Gallagher, E., Gormley, T.R. and Arendt, E.K. 2003. Crust and crumb characteristics of gluten-free breads. *J. Food Eng.* **56**: 153–161.

Gallagher, E., Gormley, T.R. and Arendt, E.K. 2004. Recent advances in the formulation of gluten-free cereal-based products. *Trends Food Sci. Technol.* **15** 143–152.

Gambus, H., Gambus, F. and Sabat, R. 2002. The research on quality improvement of gluten-free bread by amaranthus flour addition. *Zywnosc*, **9**: 99–112.

Gandolfi, L., Pratesi, R., Cordoba, J.C., Tauil, P.L., Gasparin, M. and Catassi, C. 2000. Prevalence of celiac disease among blood donors in Brazil. *Am. J. Gastroenterol.* **95**: 579–580.

Géllinas, P., Poitras, E., McKinnon, C.M. and Morin, A. 1998. Oxidoreductases and lipases as dough-bleaching agents. *Cereal Chem.* **75**: 810–814.

Gibson, G.R. and Roberfroid, M.B. 1995. Dietary modulation of the human colonic microbiota: introducing the concept of prebiotics. *J. Nutr.* **125**: 1401–1412.

Gobbi, 2005. Coeliac disease, epilepsy and cerebral calcifications. *Brain Dev.* **27**: 189–200.

Gormley, T.R., Elbel, C., Gallagher, E. and Arendt, E.K. 2003. Fish surimi as an ingredient in gluten-free breads. *Proceedings of the First Joint Trans Atlantic Fisheries Technology Conference*, Iceland, pp. 246–247.

Green, R.H.P. and Jabri, B. 2003. Coeliac Disease. *Lancet* **362**: 383–391.

Grehn, S., Fridell, K., Lilliecreutz, M. and Hallert, C. 2002. Dietary habits of Swedish adult celiac patients treated by a gluten-free diet for 10 years. *Scand. J. Nutr.* **45**: 178–182.

Grodzinsky, E. 1996. Screening for celiac disease in apparently healthy blood donors. *Acta Pediatr.* **412**: 36–38.

Grossman, M.V. and De Barber, B.C. 1997. Bread staling. Simultaneous effect of bacterial alpha-amylase and emulsifier on firmness and pasting properties of bread crumbs. *Archivos Latinoamericanos Nutric.* **47**: 229–233.

Guarda, A., Rosell, C.M., Benedito, C. and Galotto, M.J. 2004. Different hydrocolloids as bread improvers and antistaling agents. *Food Hydrocoll.* **18**: 241–247.

Gujral, H.S., Haros, M. and Rosell, C.M. 2003a. Starch hydrolysing enzymes for retarding the staling for rice bread. *Cereal Chem.* **80**: 750–754.

Gujral, H.S., Guardiola, I., Carbonell, J.V. and Rosell, C.A. 2003b. Effect of cyclodextrinase on dough rheology and bread quality from rice flour. *J. Agric. Food Chem.* **51**: 3814–3818.

Hamer, R.J. 2005. Coeliac disease: background and biochemical aspects. *Biotech. Adv.* **23**: 401–408.

Han-Ching, L. and Leinot, A. 1993. Surimi composition and technology: present status and nutritional condiderations. *Int. J. Food Sci. Nutr.* **44**: S55–S63.

Hardman, C.M., Garioch, J.J., Leonard, J.N., Thomas, H.J.W., Walker, M.M., Lortan, J.E., Lister, A. and Fry, L. 1997. Absence of toxicity of oats in patients with dermatitis herpetiformis. *New Engl. J. Med.* **377**: 1884–1887.

Hargrove, J.M., Greenspan, P. and Hartle, D.K. 2004. Nutritional significance and metabolism of very long chain fatty alcohols and acids from dietary waxes. *Exp. Biol. Med.* **229**: 215–226.

Hoefler, A.C. 2004. *Hydrocolloids. Egan Press Handbook*. AACC International, St Paul, Minnesota, Ch. 1–3, pp. 1–42.

Hozová, B., Janãoviãová, J., Dodok, L., Buchtová, V. and Staruch, L. 2002. Use of transglutaminase for improvement of quality of pastry produced by frozen-dough technology. *Czech. J. Food Sci.* **20**: 215–222.

Hwang, K.T., Kim, J.E., Soh, Y.J., Lee, H.O., Kim, S.B. and Weller, C.L. 2004. Policosanols in rice, rice bran, sorghum, and sorghum bran produced in Korea. *IFT Annual Meeting*, July 12–16, Las Vegas, Nevada.

Igoe, R.S. 1982. Hydrocolloids interactions useful in food systems. *Food Technol.* **36**: 72–74.

Inman-Felton, A.E. 1999. Overview of gluten-sensitive enteropathy (celiac sprue). *J. Am. Diet. Assoc.* **99**: 352–362.

Janatuinen, E.K., Pikkarainen, P.H., Kemppainen, T.A., Kosma, V.M., Jarvinen, R.M.K., Uusitupa, M.I.J. and Julkunen, R.J.K. 1995. A comparison of diets with and without oats in adults with celiac disease. *New Engl. J. Med.* **333**: 1033–1037.

Janatuinen, E.K., Kemppainen, T.A., Julkunen, R.J., Kosma, V.M., Maki, M., Heikkinen, M. and Uusitupa, M.I. 2002. No harm from five years ingestion of oats in celiac disease. *Gut* **50**: 332–335.

Jonagh, G., Slim, T. and Greve, H. 1968. Bread without gluten. *Baker's Dig.* **6**: 24–29.

Kang, M.Y., Choi, Y.H. and Choi, H.C. 1997. *Korea J. Sci. Technol.* **29**: 700–704.

Kasarda, D.D. 2001. Grains in relation to celiac disease. *Cereal Foods World.* **46**: 209–210.

Kato, A., Ibrahim, H., Watanabe, H., Honma, K. and Kobayashi, K. 1990. Enthalpy of denaturation and surface functional properties of heated egg white proteins in the dry state. *J. Food Sci.* **55**: 1280–1282.

Kavanagh, N. and Corrigan, O.I. 2004. Swelling and erosion properties of hydroxypropylmethylcellulose (Hypromellose) matrices – influence of agitation rate and dissolution medium composition. *Int. J. Pharm.* **279**: 141–152.

Kenny, S., Wehrle, K., Auty, M. and Arendt, E.K. 2001. Influence of sodium caseinate and whey protein on baking properties and rheology of frozen dough. *Cereal Chem.* **78**: 458–463.

Keregero, M.M. and Mtebe, K. 1994. Acceptability of wheat–sorghum composite flour products: An assessment. *Plant Food Hum. Nutr.* **46**: 305–312.

Ketema, S. 1993. Tef crop improvement, nutrition and utilization. In: Riley, K.W., Gupta, S.C., Seetharam, A. and Mushonga, J.N. (eds). *Advances in Small Millets*. Oxford and IBH Publishing, New Delhi, pp. 61–65.

Kreft, S. and Kreft, M. 2000. Localization and morphology of the buckwheat embryo. *Fagopyrum* **17**: 15–19.

Kulp, K., Frank, H.N. and Thomas, L.A. 1974. Preparation of bread without gluten. *Baker's Dig.* **48**: 34–38.

Kuraishi, C., Sakamoto, J. and Soeda, T. 1996. The usefulness of transglutaminase for food processing. In: Takeoka, G.R., Teranishi, R., Williams, P.J. and Kobayashi, A. (eds). *Biotechnology for Improved Foods and Flavours*, ACS Symposium Series 637. American Chemical Society, Washington, DC, pp. 29–38.

Lee, C.C. and Hoseney, R.C. 1982. Optimization of the fat-emulsifier system and the gum–egg white–water system for a laboratory scale single stage cake mix. *Cereal Chem.* **59**: 392–395.

Lending, C.R. and Larkins, B.A. 1989. Changes in the zein composition of protein bodies during maize endosperm development. *Plant Cell* **1**: 1011–1023.
Li, S. and Zhang, H.Q. 2001. Advances in the development of functional foods from buckwheat. *Crit. Rev. Food Sci. Nutr.* **41**: 451–464.
Lineback, D.R. and Rasper, V.F. 1988. Wheat carbohydrates. In: Pomeranz, Y. (ed). *Wheat Chemistry and Technology*. American Association of Cereal Chemists, St Paul, Minnesota, pp. 277–372.
Lohiniemi, S., Maki, M., Kaukinen, K., Laippala, P. and Collin, P. 2000. Gastrointestinal symptoms rating scale in celiac patients on wheat starch-based gluten-free diets. *Scand. J. Gastroenterol.* **35**: 947–949.
Lovis, L.J. 2003. Alternatives to wheat flour in baked goods. *Cereal Foods World* **48**: 61–63.
Maccagnan, G., Pat, A., Collavo, F., Ragg, G.L. and Bellini, M.P. 2004. Gluten-free beer containing rice malt and buckwheat. European Patent, EP 0949328B1.
Madsen, M.H. and Christensen, D.H. 1996. Changes in viscosity properties of potato starch during growth. *Starch* **48**: 245–249.
Mannie, E. and Asp, E.H. 1999. Dairy ingredients in baking. *Cereal Foods World* **44**: 143–146.
Mariani, P., Grazia, V.M., Montouri, M., Cipolletta, E., Calvani, L., Bonamico, M. and La Vecchia, A. 1998. The gluten-free diet: a nutritional risk factor for adolescents with celiac disease. *J. Ped. Gastroenterol. Nutr.* **27**: 519–523.
Martucci, S., Biagi, F., Di Sabatino, A. and Corazza, G.R. 2002. Coeliac disease. *Dig. Liver Dis.* **34**: S150–153.
McCarthy, D.F., Gallagher, E., Gormley, T.R., Schober, T.J. and Arendt, E.K. 2005. Application of response surface methodology in the development of gluten-free bread. *Cereal Chem.* **82**: 609–615.
Molberg, O., Uhlen, A.K., Jensen, T., Flaete, N.S., Fleckenstein, B., Arentz-Hansen, H., Raki, M., Lundin, K.E. and Solliol, L.M. *et al.* 2005. Mapping of gluten T-cell epitopes in the bread wheat ancestors; implications for coeliac disease. *Gastroenterology.* **128**: 393–401.
Moore, M.M., Schober, T.J., Dockery, P. and Arendt, E.K. 2004a. Textural comparison of gluten-free and wheat based doughs, batters and breads. *Cereal Chem.* **81**: 567–575.
Moore, M.M., Schober, T.J., Juga, B., Dockery, P., Ulmer, H.M., Hammes, W.P. and Arendt, E.K. 2004b. Influence of lactic acid bacteria on the quality of gluten-free batter and bread. Paper read at *12th ICC Cereal and Bread Congress*, Harrogate, UK.
Moore, M.M., Heinbockel, M., Dockery, P., Ulmer, H.M. and Arendt. E.K. 2004c. The influence of transglutaminase on gluten-free bread quality. Paper read at *34th Food Science and Technology Conference*, University College Cork, Ireland.
Moore, M.M., Heinbockel, M., Dockery, P., Ulmer, H.M. and Arendt, E.K. 2005. Network formation in gluten-free bread with the application of transglutaminase. *Cereal Chem.* **83**: 28–36.
Mowat, A.M. 2003. Celiac disease – a meeting point for genetics, immunology, and protein chemistry. *Lancet* **361**: 1290–1292.
Munck, L. 1995. New milling technologies and products: whole plant utilization by milling and separation of the botanical and chemical components. In: Dendy, D.A.V. (ed.). *Sorghum and Millets: Chemistry and Technology* AACC International, St Paul, Minnesota.
Murray, J.A. 1999. The widening spectrum of celiac disease. *Am. J. Clin. Nutr.* **69**: 354–365.
Mustalahti, K., Lohiniemi, S., Collin, P., Vuolteenaho, N., Laippala, P. and Markku, M. 2002. Gluten-free diet and quality of life in patients with screen-detected coeliac disease. *Eff. Clin. Pract.* **5**: 105–113.
Nonaka, M., Tanaka, H., Okiyama, A., Motoki, M., Ando, H., Umeda, K. and Matsuura, A. 1989. Polymerization of several proteins by Ca^{2+}-independent transglutaminase derived from microorganisms. *Agric. Biol. Chem.* **53**: 2619–2623.

Norton, I.T. and Foster, T.J. 2002. Hydrocolloids in real food systems. In: Williams, P.A. and Phillips, G.O. (eds). *Gums and Stabilizers in the Food Industry 11*. Royal Society of Chemistry, London, pp. 187–200.
Not, T., Horvath, K., Hill, I.D., Partanen, J., Hammed, A., Magazzu, G. and Fasano, A. 1998. Celiac disease risk in the USA: high prevalence of antiendomysium antibodies in healthy blood donors. *Scand. J. Gastroenterol.* **33**: 494–498.
Nyman, M.I., Björck, I., Siljeström, M. and Asp, N.G. 1989. Dietary fibre in cereals – composition, fermentation and effect of processing. In: *Proceedings from an International Symposium on Cereal Science and Technology*, Lund University, Sweden, pp. 40–54.
Okubo, T., Ishihara, N., Takanashi, H., Fujisawa, T., Yamamoto, T. and Mitsuoka, T. 1994. Effects of partially hydrolysed guar gum intake on human intestinal microflora and its metabolism. *Biosci. Biotechnol. Biochem.* **58**: 1364–1369.
Omoni, A.O. and R.E. Aluko. 2005. Soybean foods and their benefits: potential mechanisms of action. *Nutr. Rev.* **63**: 272–283.
Oomah, B.D. and Mazza, G. 1999. Health benefits of phytochemicals from selected Canadian crops. *Trends Food Sci. Technol.* **10**: 193–198.
Ortolani, C. and Pastorello, E.A. 1997. Symptoms of food allergy and food intolerance. In: *Study of Nutritional Factors in Food Allergies and Food Intolerance*. CEC, Luxembourg pp. 26–45.
Parnell, N., Ellis, H.J. and Ciclitira, P. 1998. Absence of toxicity of oats in patients with dermatitis herpetiformis. *New Engl. J. Med.* **338**: 1470–1471.
Phiarais, B.P.N., Wijngaard, H.H. and Arendt, E.K. 2005. The impact of kilning on enzymatic activity of buckwheat malt. *J. Inst. Brew.* **111**: 290–298.
Ranhorta, G.S., Loewe, R.J. and Puyat, L.V. 1975. Preparation and fortification of soy-fortified gluten-free bread. *J. Food Sci.* **40**: 62–64.
Rooney, L.W. and Waniska, R.D. 2000. Sorghum food and industrial utilization. In: Smith, C.W. and Frederiksen, R.A. (eds). *Sorghum: Origin, History, Technology, and Production*, John Wiley & Sons, Inc., New York, pp. 689–750.
Rossel, C.M., Haros, M., Escriva, C. and Benedito De Barber, C. 2001. Experimental approach to optimise the use of alpha-amylases in breadmaking. *J. Agr. Food Chem.* **49**: 2973–2977.
Sahlström, S. and Brathen, E. 1997. Effects of enzyme preparations for baking, mixing time and resting time on bread quality and bread staling. *Food Chem.* **58**: 75–80.
Sanchez, H.D., Osella, C.A. and de la Torre. 2002. Optimisation of gluten-free bread prepared from cornstarch, rice flour and cassaca starch. *J. Food Sci.* **67**: 416–419.
Satin, M. 1998. Bread without wheat. Novel ways of making bread from cassava and sorghum could reduce the Third World's dependence on imported wheat for white bread. *New Sci.* **28**: 56–59.
Schober, J.T., O'Brien, C.M., McCarthy, D., Darnedde, A. and Arendt, E.K. 2003. Influence of gluten-free flour mixes and fat powders on the quality of gluten-free biscuits. *Eur. Food Res. Technol.* **216**: 369–376.
Schober, J.T., Messerschmidt, M., Bean, S.R., Park, S.H. and Arendt, E.K. 2005. Gluten-free bread from sorghum: quality differences among hybrids. *Cereal Chem.* **82**: 394–404.
Schwarzlaff, S.S., Johnson, J.M., Barbeau, W.E. and Duncan, S. 1996. Guar and locust bean gums as partial replacers of all-purpose flour in bread: an objective and sensory evaluation. *J. Food Qual.* **19**: 217–229.
Shan, L., Molberg, O., Parrot, I., Hausch, F., Filiz, F., Gray, G.M., Sollid, L.M. and Khosla, C. 2002. Structural basis for gluten intolerance in celiac sprue. *Science* **297**: 2275–2279.
Sharma, S.C. 1981. Gums and hydrocolloids in oil–water emulsion. *Food Technol.* **35**: 59–67.
Shukla, R. and Cheryan, M. 2001. Zein: the industrial protein from corn. *Ind. Crops Prod.* **13**: 171–192.

Skerritt, J.H. and Hill, A.S. 1992. How 'free' is 'gluten-free'? Relationship between Kjeldahl nitrogen values and gluten protein content for wheat starches. *Cereal Chem.* **69**: 110–112.
Smith, E.B. 1971. Gluten-free breads for patients with uremia: *J. Am. Diet. Ass.* **59**: 572–574.
Sollid, L.M. and Khosla, C. 2005. Future therapeutic options for celiac disease. *Nat. Clin. Pract. Gastroenter. Hepatol.* **2**: 140–147.
Srinivasan, U., Leonard, N., Jones, E., Kasarda, D.D., Weir, D.G., O'Farrelly, C. and Feighery, C. 1996. Absence of oats toxicity in adult celiac disease. *Bio-Med. J.* **313**: 1300–1301.
Stear, C.A. 1990. Effect of dough additives: Measurement and control techniques; Wheat and rye sours and sourdough processing. In: *Handbook of Breadmaking Technology*. Elsevier Applied Science, London.
Taylor, J.R.N. and Parker, M.L. 2002. Quinoa. In: Belton, P. and Taylor J. (eds). *Pseudocereals and Less Common Cereals, Grain Properties and Utilization Potential.* Springer-Verlag, Berlin, pp. 93–122.
Thompson, T. 1999. Thiamin, riboflavin and niacin contents of the gluten-free diet: Is there cause for concern? *J. Am. Diet. Assoc.* **99**: 858–862.
Thompson, T. 2000. Folate, iron and dietary fibre contents of the gluten-free diet. *J. Am. Diet. Assoc.* **100**: 1389–1396.
Thompson, T. 2003. Oats and the gluten-free diet. *J. Am. Diet. Assoc.* **103**: 376–379.
Tosi, E.A., Ciappini, M.C. and Masciarelli, R. 1996. Utilisation of whole amaranthus (*Amaranths cruentus*) flour in the manufacture of biscuits for celiacs. *Alimentaria* **34**: 49–51.
Toufeili, I., Dagher, S., Shadarevian, S., Noureddine, A., Sarakbi, M. and Farran, T.M. 1994. Formulation of gluten-free pocket-type flat breads: Optimization of methylcellulose, gum arabic, and egg albumen levels by response surface methodology. *Cereal Chem.* **71**: 594–601.
Tye, J.R. 1991. Konjac flour: Properties and applications. *Food Technol.* **3**: 86–92.
US Department of Agriculture, Agricultural Research Service. 2000. USDA Nutrient Database for Standard Reference. Release 13. Available at: Nutrient Data home page, http://www.nal.usda.gov/fnic/foodcomp.
Vader, W., Kooy, Y., Van Veelen, P., De Ru, A., Harris, D., Benckhuijsen, W., Pena, S., Mearin, L., Drijfhout, J.W. and Koning, F. 2002. The gluten response in children with celiac disease is directed toward multiple gliadin and glutenin peptides. *Gastroenterology*, **122**: 1729–1737.
Vader, L.W., Stepniak, D.T., Bunnik, E.M., Kooy, Y.M., de Haan, W., Drijfhout, J.W. Van Veelan, P.A. and Koning, F. 2003. Characterization of cereal toxicity for celiac disease patients based on protein homology in grains. *Gastroenterology*. **125**: 1105–1113.
Van Vliet, T., Janssen, A.M., Bloksma, A.H. and Walstra, P. 1992. Strain hardening of dough as a requirement for gas retention. *J. Text. Studies* **23**: 439–460.
Vemulappali, V. and Hoseney, R.C. 1998. Glucose oxidase effects on gluten and water solubles. *Cereal Chem.* **75**: 859–862.
Visakorpi, J.K., Kuitunen, P. and Pelkonen, P. 1996. Intestinal malabsorption: a clinical study of 22 children over 2 years of age. *Acta Paediatr. Scand.* **59**: 273–280.
Walker-Smith, J.A., Guandalini, S., Schmitz, J., Shmerling, D.H. and Visakorpi, J.K. 1990. Revised criteria for diagnosis of celiac disease. Report of Working Group of European Society of Paediatric Gastroenterology and Nutrition. *Arch. Dis. Child.* **65**: 909–911.
Wijngaard, H.H. and Arendt, E.K. 2006. Review: Buckwheat. *Cereal Chem.* **83**: 391–401.
Wijngaard, H.H., Ulmer, H.M. and Arendt E.K. 2005. The effect of germination temperature on malt quality of buckwheat. *J. Am. Soc. Brew. Chem.* **63**: 31–36.
Willard, M.J. and Hix, V.M. 1987. Potato flour. In Talburt, W.F. and Smith, O. (eds) *Potato processing*, 4th ed. Van Nostrand Reinhold Co., New York, pp. 665–681.

Ylimaki, G., Hawrysh, Z.J., Hardin, R.T. and Thomson, A.B.R. 1991. Response surface methodology in the development of rice flour yeast breads: Sensory evaluation. *J. Food Sci.* **5**: 751–759.

Ylimaki, G., Hawrysh, Z.J., Hardin, R.T. and Thomson, A.B.R. 1998. *J. Food Sci.* **53**: 1800–1805.

Zimeri, J.E. and Kokini, J.L. 2003. Rheological properties of inulin–waxy maize starch systems. *Carbohydr. Polym.* **52**: 67–85.

20

Converting oats to high-fibre products for use in functional foods

G. E. Inglett, D. G. Stevenson, US Department of Agriculture, USA and S. Lee, Sejong University, Republic of Korea

20.1 Introduction: health benefits of functional high-fibre oat products

Functional foods are generally considered to be foods or dietary supplements that provide health benefits beyond basic nutrition when consumed on a regular basis at effective levels. The biologically active components of functional foods impart health benefits or desirable physiological effects. In industrialised nations, population migration trends have resulted in the majority of people being subjected to an urban lifestyle that is not conducive to good dietary choices or sufficient exercise. This has led to an alarming increase in the incidence of various health disorders. The American Heart Association reports that in 2003, in the United States alone, 71.3 million people had at least one form of cardiovascular disease (CVD) and CVD caused 911 614 deaths. This was 37.3% of the total death toll, making CVD the single leading cause of death. Diabetes in the United States is also increasing, with 73 249 diabetes-related deaths in 2002 (American Heart Association). Increases in CVD and diabetes are correlated to the rise in obesity. In 2002, 28.1% of men and 34.0% of women in the United States were obese (Hedley *et al.* 2004) with 80% of type II diabetes and 70% of CVD related to obesity. CVD

Names are necessary to report factually on available data; however, the USDA neither guarantees nor warrants the standard of the product, and the use of the name by USDA implies no approval of the product to the exclusion of others that may also be suitable. All programmes and services of the US Department of Agriculture are offered on a nondiscriminatory basis without regard to race, colour, national origin, religion, sex, age, marital status, or handicap.

and diabetes are also increasing in the two most populous countries, China and India (Popkin *et al.* 2001, Rajeshwari *et al.* 2005).

Commercial oat fractionation has been practised in dry milling plants since early times by separating the hull from the flour and bran. Only in recent history has the prospect of milling oats to obtain functional food ingredients emerged. Functional components have become attractive because of increased interest in dietary fibres for maintenance of good health. The β-glucans in oat bran have been shown to provide many human health benefits such as serum cholesterol lowering (de Groot *et al.* 1963, Anderson and Chen 1986, Mälkki *et al.* 1992, Ripsin *et al.* 1992, Uusitupa *et al.* 1992, Behall *et al.* 1997, Pomeroy *et al.* 2001), reduced coronary heart disease (Berg *et al.* 2003), reduced diabetic symptoms (Wood *et al.* 1994, Pick *et al.* 1996, Tappy *et al.* 1996, Reyna *et al.* 2003, Prosise *et al.* 2004), lowered blood pressure (Saltzman *et al.* 2001), cancer prevention (Adom and Liu 2002) and improved gastric emptying and nutrient absorption (Mälkki and Virtanen 2001). The Food and Drug Administration (FDA) recognised the importance of oat soluble fibre, generally called β-glucan, from oat flour and bran by allowing a health claim on food labels (Food and Drug Administration 1997). The cholesterol-lowering efficacy of oats is dependent on consumption of a sufficient quantity of oatmeal or oat bran over an extended period. Increased use of oat fibre in food applications would enable a sustainable consumption of oats to help combat CVD, diabetes, obesity and many other diet-related diseases.

20.2 Isolating and concentrating oat β-glucans using solvent extraction methods

An increase in β-glucan concentration can be achieved by removal of fat from oats with organic solvents (Wood *et al.* 1989, Knuckles *et al.* 1992, Wu and Stringfellow 1995), followed by dry milling. The concentration can be increased further by wet milling in an organic solvent (Wood *et al.* 1989, Collins and Paton 1992, Myllymäki *et al.* 1994, Mälkki and Myllymäki 1998), or by wet milling in cold water (Lehtomäki *et al.* 1992). Early procedures used mixtures of pentane, hexane, heptane and cyclohexane to remove oat oil (Oughton 1980) to enrich the β-glucan content. Presently, the most common solvent used is ethanol due to its low cost and suitability for production of food ingredients. Ethanol or 2-propanol precipitates β-glucan and produces a thread-like deposit that is easily separated by screening. Typical organic solvent concentrations reported are 80–95% ethanol or 80–90% 2-propanol (Myllymäki *et al.* 1994). Animal and clinical studies have shown that the cholesterol-reducing effect of oat products obtained after organic solvent extraction is weaker than that of products obtained by dry milling methods (Mälkki *et al.* 1992, Törrönen *et al.* 1992, Uusitupa *et al.* 1992). Some studies found that β-glucans can be suspended in alcohol without their viscosity deteriorating (Myllymäki *et al.* 1994), but other studies found that β-glucan molecular weight and viscosity are reduced after extraction in 75–100% ethanol, which can lead to the complete loss of cholesterol-lowering effects (Beer *et al.* 1995a).

20.3 Isolating and concentrating oat β-glucans using milling methods

The most simple procedure for enriching β-glucan content is dry milling of oats. Several methods have been reported for enriching β-glucan content, by dry milling followed by sieving or air classification; these methods have set the platform for present-day commercial production (Hohner and Hydon 1977, Shinnick *et al.* 1988, Wood *et al.* 1989, Shinnick *et al.* 1990, Wood *et al.* 1991, Knuckles *et al.* 1992, Paton and Lenz 1993, Wu and Stringfellow 1995, Doehlert and Moore 1997). Recently, further advancements in dry fractionation have been patented (Mälkki *et al.* 2004). In this procedure, oats are ground by roller or impact milling and classified by air or sieving to collect fine fractions <200 μm in diameter, which are discarded. The coarse particles still contain a considerable amount of starch and so they are processed by milling to concentrate the soluble fibre further, after the moisture content of the particles has been reduced. An oat moisture content of 12% is optimal for separation of the bran (Doehlert and Moore 1997). A further milling stage is required to disintegrate the material effectively and this is achieved using impact milling by increasing rotor tangential speed, changing the clearance between the rotor and the peripheral space, changing the feeding rate or changing the size of the exit openings. Roller mills with deeper corrugations can also be used and/or mills with a higher difference between the speed of the rollers. Removal of the fine material minimises dampening during additional impact forces. The fine starchy particles are then removed either through sieve openings of 100–180 μm or by air classification resulting in 75% of particles having a diameter of less than 125 μm. The final coarse fraction has a dry matter β-glucan content of 12–19%. Agglomeration of the unwanted fine material can occur during milling resulting in clogged sieves. These particles of fine material are then included with the coarse β-glucan-rich particles (Wood *et al.* 1989). The primary factors contributing to clogging and adhesion are air humidity, total and surface humidity of the oat grains, extent of starch damage and gelatinisation, and constructional details of the equipment. Generation and transfer of heat must be minimised to control the agglomeration phenomenon. In order to inactivate enzymes that reduce storage stability, the β-glucan concentrate is heated by fluidised bed heating or extrusion cooking. The latter heating method has the advantages of more effective inactivation and improved β-glucan solubility due to the mechanical treatment.

Extrusion cooking also provides an opportunity to prepare consumer products such as breakfast cereals by mixing other components with the concentrated soluble fibre.

A dry milling procedure followed by air classification to remove some starch was developed by Swedish Oat Fibre AB, Väröbacka, Sweden (Duss and Nyberg 2004, Ohr 2004). The fat stabilisation technique used in the process produces an oat bran ingredient resistant to lipid oxidation with a 14% β-glucan content, marketed as OatWell® 14% Oat Bran (Table 20.1). Further air classification is used to produce oat products enriched with 16% and 22% β-glucan, marketed as OatWell® 16% and OatWell® 22% Oat Bran (Table 20.1).

Table 20.1 Composition of commercial fractionated oat fibre products

Oat product	% β-Glucan	% Carbohydrate	% Protein	% Fat	% Dietary fibre
Cerogen™[1]	70–90	93	4	0	77–90
C-trim20	21	63	27	7	21
C-trim30	32	79	14	2	32
C-trim50	48	80	18	10	48
Glucagel®[2] (barley*)	75–80	94	2	0	91
Natureal®[3]	20	34	25	10	37
Nurture® 1500[4]	15	61	28	3	30
Nutrim-OB[5]	11	69	10	1	11
Oat β-glucan (Ceapro)[6]	94	94	1	0	94
Oat bran concentrate (Quaker)[5]	9	54	18	7	12
Oatrim[5]	5	73	10	4	10
OatsCreme™[7]	1	30	3	2	1
OatVantage™[4]	54	84	11	2	54
OatWell® 14%[8]	14	64	20	5	30
OatWell® 16%[8]	16	64	20	5	34
OatWell® 22%[8]	22	64	20	5	44
Viscofiber®[9]	50	86	4	5	82
Z-Trim™[10]	0	86	1	0	86

[1] Information received from Roxdale Foods Ltd, Auckland, New Zealand.
[2] Information received from Gracelinc Ltd, Gracefield, New Zealand.
[3] Information received from communication with Finn Cereal, Vantaa, Finland.
[4] Information received from communication with Nurture Inc., Missoula, MT.
[5] Carrière and Inglett, (2000).
[6] Information received from communication with Ceapro Inc., Edmonton, Canada.
[7] Information received from communication with American Oats Inc., Minneapolis, MN.
[8] Information received from communication with CreaNutrition, Zug, Switzerland and Oat Ingredients llc, Denver, CO.
[9] Information received from communication with Cevena Bioproducts Inc., Edmonton, Canada.
[10] Inglett and Carrière (2001).
* Currently a barley product but similar product achievable with oats.

20.4 Isolating and concentrating oat β-glucans using alkaline and acidic extractions

Soaking oat grains in an acidic aqueous medium with pH 3–7 at 40 °C–70 °C has been used to enrich β-glucan (Burrows *et al.* 1984, Paton and Lenz 1993). Oat grains are initially soaked in water until an amount at least equivalent to the weight of the grain is absorbed, allowing endogenous cell wall degrading enzymes to separate the husk from the endosperm portion, thereby allowing the endosperm contents to disperse in solution upon crushing of the grain. The slightly acidic conditioning step is used to inhibit enzymes involved in the degradation of starch, proteins and lipids. This is in complete contrast to later methods that primarily attempted to remove as much of the non-hydrocolloid component as possible. The imbibed grain is subjected to pressure to allow the separation of bran by sieving and centrifugation. Initial methods to enrich β-glucan from oat flour using basic conditions focused on sodium carbonate extractions (Hohner and Hyldon 1977, Wood *et al.* 1978, 1989, Welch *et al.* 1988). Other earlier alkaline extractions focused on methods of concentrating oat proteins that resulted in the separation of oat bran (Cluskey *et al.* 1976, Wu *et al.* 1977), and a later study extracted oat β-glucan at pH 9.2 with little protein and starch (Dawkins and Nnanna 1993). Simple and efficient methods for extracting high levels of β-glucan were later developed that employed strong bases such as sodium hydroxide (Bhatty 1995, 1996, Beer *et al.* 1997). Oat flour or bran is steeped in 0.25–1.0 N base solution for 2–25 h at 15 °C–35 °C. The extract is acidified to pH 4–7 enabling starch removal from the extract by addition of amylases and later adding polar alcohol to precipitate β-glucans. Sodium hydroxide extractions were found to lower oat β-glucan molecular weight further compared with sodium carbonate (Wood *et al.* 1989, Beer *et al.* 1997); thus the latter may be preferable.

Many methods of β-glucan enrichment proved impractical for commercial manufacturing processes because of high costs and/or low yields. Precipitation of β-glucan using alcohol resulted in high costs and losses during reclamation. To overcome this problem, a method was developed by Nurture Inc. that enriches β-glucan in an aqueous environment, utilising alkaline conditions (Potter *et al.* 2001, 2002, 2003). The procedure involves an alkaline aqueous extract of β-glucan from oats or barley, preferably with a pH of 10, that is neutralised with acid, heated between 60 °C and 100 °C, and subsequently cooled between 25 °C and 45 °C resulting in flocculation. Centrifugation is utilised to remove the flocculate and the β-glucan solution is further purified by ultrafiltration using 0.2 μm pore size or evaporated using drum or spray dryers. From this alkaline extraction method, Nurture Inc. has produced at Missoula, MT, the commercially available product, OatVantage™; this has a β-glucan content greater than 54% β-glucan content, is a beige powder with light oat odour and neutral taste. Applications for OatVantage™ include encapsulation, tableting, beverages, functional foods, and nutraceutical and dietary supplements.

Commercial cellulosic functional fibre ingredients have been prepared by chemical and physical alteration of oat and maize hull pericarp cell walls (Inglett 1997, Inglett and Carrière 2001, Carrière and Inglett 2003) by methods that differed from earlier ones (Dintzis *et al.* 1979). A two-stage operation – thermal alkaline degradation with intense shear force followed by alkaline peroxidation and shear – was essential for obtaining the desired functional gel. The dried gel readily disperses in water and exhibits high viscosity under high-shear procedures, such as colloid mills and homogenisation (Inglett 1998b). Z-Trim™, a commercial cellulosic fibre is predominantly produced from oats but is patented for cereals and agricultural byproducts. Cellulosic gels are important for adding increased viscoelastic properties to foods without increasing their caloric content. Z-Trim™ is currently manufactured and marketed by FibreGel Technologies Inc., a subsidiary of Circle Group Holding, Inc., at Mundelein, IL.

An enriched β-glucan extract has been modified to retard gel-forming properties by employing a biological buffer system consisting of a zwitterionic salt that varies in pH with temperature (Redmond 2001). The biological buffer used is typically HEPES or MOPS or any Good buffer (Good *et al.* 1966) and has a concentration ranging from 0.1 to 20 mM. A β-glucan solution is heated to 60°C–65°C to disperse the hydrocolloidal solution prior to adding the buffer. The temperature coefficient (pK_a/C) of buffers ensures the variation of pH with temperature that allows the destabilisation of gum gel formation from 20 °C to 96 °C due to reversal of hydrogen bond polarities and a stabilising gel effect as the solution nears critical freezing point (Redmond 2001). The oat β-glucan extract with retarded gel formation is marketed by Ceapro Inc., based in Edmonton, Canada. Ceapro Inc. has recently developed a highly concentrated, clear and viscous oat β-glucan extract (Table 20.1) in which a nanofiltration step (<1 μm screen) collects submicron β-glucan particles (Redmond and Fielder 2004, 2005a, 2005b).

20.5 Isolating and concentrating oat β-glucans using thermal and frozen extraction methods

A new generation soluble β-glucan product, called Nutrim-OB (OB, oat bran), was developed that qualified for a health claim (Inglett 1998a), since Oatrim (an enzyme hydrolysed oat product) did not qualify for the FDA health benefit claim until 2003 (Food and Drug Administration 2003). Nutrim-OB was prepared by heat-shearing oat flour or bran aqueous slurry in a series of treatments that solubilise oat β-glucans but maintain basic bran composition. The slurry viscosity is reduced by >90%, rendering it flowable through sieve pores (40–400 μm) or alternatively insoluble crude fibre particles are separated by centrifugation. These methods, coupled with subsequent drum-drying provide high yields (Inglett 1998b, 2000). Nutrim-OB is currently manufactured and marketed by Van Drunen Farms, Momence, IL and there is potential to coprocess with other cereals and oilseeds to provide materials with novel functional properties (Inglett 1999, Inglett *et al.* 2002, 2004). Solubilised Nutrim-OB has lower molecular weight β-glucans than uncooked material (Knuckles *et al.* 1998). Far from reducing biological activity, processing engenders β-glucans more biologically active (Yokoyama *et al.* 1998). Nutrim-OB maintains similar antioxidant properties to those of the starting oat material (Prior *et al.* 1998). Nutrim-OB has unusually high viscosity at 5–15% solids and ambient temperatures (Carrière and Inglett 1998b), with viscosity values that are many times higher than for Oatrim (Carrière and Inglett 1998c). Nutrim-OB produced from defatted oat fines, flour or bran exhibited shear-thinning throughout the shear rate range studied during a thixotropic loop experiment (Carrière and Inglett 2000). High-sugar-ratio cake batters formulated with Nutrim-OB had higher viscosity and unaltered shear-thinning compared with cake batters without hydrocolloids (Lee *et al.* 2004). The pasting properties of oat bran concentrate and Nutrim-OB without and with lipid extraction by supercritical carbon dioxide have been studied (Stevenson *et al.* 2006). Oat bran concentrate (OBC) showed a typical pasting curve exhibited by starchy materials whereas Nutrim-OB, in which starch is already gelatinised, showed a highly viscous peak during heating with little setback; this behaviour can be attributed to β-glucans. A scanning electron micrograph of Nutrim-OB revealed large porous particles (250–400 μm length) with ragged sides.

Another thermal method for β-glucan enrichment has been recently reported by Quaker Oats/Rhone Poulenc using a starting oat slurry with <8% β-glucan content, consisting of oat flour, flakes, bran or defatted grain (Cahill *et al.* 2003, 2004). Oat slurry is heated to 47 °C–60 °C, then the pH is adjusted to 8.1 and the temperature maintained for 50–120 min. Alkaline and heat conditions typically extract 50–80% of the total grain β-glucan content. Acidification of the slurry to pH 4.2–4.8 stops extraction and allows separation by centrifugation of a β-glucan-rich fraction with concentration on a dry weight basis of 18–30%. The β-glucan fraction is then cooled to 7°C and subsequently concentrated by spray-drying. Using a fluidised bed, the final β-glucan product is agglomerated into particles ranging from 75 to 840 μm which facilitates incorporation of the functional β-glucan ingredient into wet and dry foods.

Three new oat β-glucan hydrocolloids (designated as C-trim20, C-trim30 and C-trim50) obtained through thermal-shearing processing have been developed that have potential for use as functional food ingredients (Inglett 2004). C-trim products differ from Nutrim-OB as their production process has an additional centrifugation step prior to jet-cooking that eliminates a higher proportion of starch, and centrifugation and sieving steps after jet-cooking that determine the final β-glucan concentration. C-trim oat products are currently produced and marketed by Van Drunen Farms, Momence, IL. Using steady and dynamic shear measurements of cookie dough, C-trim20 exhibited shear-thinning and increased dynamic viscoelastic moduli with increased concentration (Lee and Inglett 2005, Lee *et al.* 2005c).

A β-glucan enrichment process that uses a freeze/thaw step and does not inactivate endogenous enzymes has been developed (Morgan 2002). Commercial focus has been on barley (Table 20.1) but the technology is patented for all β-glucan-containing cereals. Cereal flour is mixed with water at 50 °C for 1 h. Enzymes associated with cereal grain partially hydrolyse β-glucans, lowering average molecular weight. The enzymes responsible for hydrolysis are most likely cellulases (Wilhelmi and Morgan 2001). The solids, from extract, are removed by centrifugation and the supernatant is frozen. On thawing a precipitate is recovered. The product, Glucagel®, is undergoing commercial production by Gracelinc Ltd. Glucagel® has a β-glucan purity of 75–80% with typical molecular mass of 130 kDa. Glucagel® forms soft thermoreversible, translucent gels that melt and set at about 60 °C (Morgan and Ofman 1998). Applications of Glucagel® include bakery, dairy, dressings and edible films (Morgan 2000).

Recently, a warm-water extraction method has been developed by Roxdale Foods Ltd, Auckland, New Zealand, that can use both barley and oats for β-glucan enrichment. The product, marketed as Cerogen™ (Table 20.1), contains 70–90% β-glucan. Cerogen™ is a unique patented form of β-glucan, that is cold-water soluble and on standing will form soft translucent gels. In a recent study undertaken by the Human Nutrition Unit at Auckland University, Cerogen™ was shown to lower glycaemic response in healthy adult males.

20.6 Isolating and concentrating oat β-glucans using biotechnological enzymatic methods

Isolating and concentrating dietary fibre, such as soluble oat fibre, has been achieved by combining milling and enzymatic hydrolysis of grain components. Enzymes utilised include α-amylase, β-glucanase, proteases, xylanase, pectinase, phytase, β-amylase and glucoamylase. Thermal processing in conjunction with other technologies has been widely used to achieve highly concentrated commercial β-glucan oat products. Among the many enzymatic methods developed to produce functional oat β-glucan products, the earliest was Oatrim, developed (Inglett 1990, 1991, 1992) by the US Department of Agriculture (USDA) with licences granted. Oatrim is currently marketed as Beta Trim® by Danisco USA at Ardsley, NY and distributed by Skidmore Sales and Distributing Inc., West Chester, OH. To produce Oatrim, thermostable α-amylase is held at 95 °C for 10–60 min to convert gelatinised starch from oat flour or bran to maltooligosaccharides. After enzymatic hydrolysis, the α-amylase is inactivated by passing the mixture through a steam injection pressure cooker (jet cooker) at 140 °C. The soluble fibre (β-glucan) and maltooligosaccharides, recovered by centrifugation, are called hydrolysed oat flour, or Oatrim. Besides maltooligosaccharides and β-glucan, Oatrim contains lesser amounts of protein and lipids (Table 20.1) (Inglett 1993). Oatrim's blood cholesterol-lowering properties were found in chick (Inglett and Newman 1994) and hamster (Yokoyama *et al.* 1998) studies. Oatrim fibre lowers the glycaemic index of products in which it is used (Reyna *et al.* 2003). Clinical studies found that 5 g of oat β-glucan from Oatrim, consumed daily, gave a wide variety of health benefits including a hypocholesterolaemic effect of lowering total blood cholesterol, weight control and a decrease in urine oxidation products (Behall *et al.* 1993, 1997, Scholfield *et al.* 1993, Hallfrisch *et al.* 1995).

The rheological properties of Oatrim suspensions have been characterised in recent years (Carrière and Inglett 1998a, 1998c, 1999, Lee *et al.* 2005a). Depending upon β-glucan content, Oatrim suspensions can exhibit shear-thinning, or shear-thickening behaviours (Carrière and Inglett 1999). At high β-glucan levels (10%), Oatrim suspensions exhibit shear-thickening behaviour that is not observed at lower levels (5%). By itself, β-glucan strictly exhibits only shear-thinning behaviour, therefore the interaction between β-glucans and maltodextrins in Oatrim contributes to the shear-thickening properties.

Several enzymatic procedures have been developed to enrich oat fibre components since the production of Oatrim. A method of enrichment of β-glucan has been developed in which milled oat flour has β-glucananse and pentosanase enzymes inactivated (Mälkki and Myllymäki 1998). The inactivated oat flour is formulated with water and heated to 60 °C–85 °C to extract β-glucan. Polar organic solvents are used to precipitate the soluble fibre, which is dried and later hydrated with insufficient amount of water to allow starch gelatinisation to occur. Before heating, the proteolytic enzyme, trypsin, is added resulting in enriched β-glucan material due to degradation of protein. Proteolytic enzymes have been shown to lower both the molecular weight and viscosity of barley β-glucans, with thermolysin having a greater effect than papain, chymotrypsin and trypsin (Forrest and Wainwright 1977). Trypsin has been shown to decrease oat β-glucan viscosity (Mälkki *et al.* 1992) with the magnitude of the decrease varying with different oat cultivars (Autio *et al.* 1992). Protein and peptides tightly bound to β-glucans (Vårum and Smidsrød 1988) are responsible for the decreased viscosity due to the presence of glutamate and histidine, which possess charged groups. The lack of β-glucanase activity and economics are the main reasons why trypsin is chosen over other proteases (Mälkki and Myllymäki 1998).

A method for concentrating oat soluble dietary fibre developed by Inglett (1991) utilising α-amylase has been further advanced (Triantafyllou 2003). In this method, β-amylase is employed to reduce 50–65% by weight of the starch content to maltose in oat flour kept at 30 °C. Pullulanase, α-amylase and proteases can also be present in conjunction with β-amylase. Added enzymes are inactivated by briefly heating the oat suspension to 80 °C–95 °C, then the temperature is held at 52°–65 °C to degrade starch. The remaining insoluble material is removed by centrifugation and low molecular weight material, such as maltose, is removed by ultrafiltration, to obtain an extract with a β-glucan:glucose weight ratio of 15:1.

Recently, a concentrated oat or barley β-glucan product with high viscosity, called Viscofiber® (Table 20.1), was developed by Cevena Bioproducts Inc., Edmonton, Canada. Viscofiber® is produced by forming a flour slurry in 40–50% ethanol with a flour to alcohol ratio of 1:2 to 1:10. The slurry is filtered using a 40–75 μm screen and the fibre residue is subjected to a combination of sonication, protease and α-amylase treatment (Vasanthan and Temelli 2004, Vasanthan *et al.* 2004). Enzymes and ethanol washes extract starch degradation products and protein while leaving soluble β-glucans intact within cereal cell walls, preserving the high viscosity of native β-glucan which has been shown to be important in eliciting functional effects (Wood *et al.* 2000).

Recently, an oat-based functional syrup was developed by American Oats Inc. (Whalen *et al.* 2002, Whalen 2004). Oat flour is separated using a US #100 or finer mesh that removes sufficient large particles to reduce the bran component by 30–50% in terms of weight. The presence of bran causes dark colouration of the syrup, which lowers consumer appeal when it is incorporated into foods. After separation by sieving, oat flour is then formulated with water to form a slurry, and cooked for 2 h at a low temperature (65 °C–70 °C) to minimise Maillard browning reactions and flavour defects from bran, protein and fat. Glucoamylase is added in conjunction with α-amylase to the syrup converted from the slurry and cooked for a further 1–3 h. The enzyme combination saves processing time by improving the thinning and liquefying action, and rapidly forming sugars. Glucoamylase is added at a higher concentration than α-amylase to attain high glucose yields in a short time and to reduce or eliminate off-flavours that would be produced during longer processing. Once the desired sweetness level has been reached, the syrup is cooled to 10 °C, then diluted with water to 14% weight solids and clarified using a milk clarifier (cream separator) producing an almost white-coloured, oat-based functional syrup that enhances the ability of the functional syrup to be incorporated into a variety of foods.

Fractionation of oat β-glucans by a two-step enzymatic process has recently been reported (Kvist *et al.* 2005). The first step involves a combination of wet milling and the enzymes α-amylase and amyloglucosidase. The oat extract is then sequentially subjected to centrifugation and ultrafiltration, which physically separates the main bran fraction. The second step involves separating the soluble fraction from insoluble material by wet milling and a combination of xylanase, β-glucanase, pectinase and phytase, followed by further sequential centrifugation and ultrafiltration.

20.7 Application of high-value functional high-fibre oat products in functional foods

Selective polysaccharide and oligosaccharide products are used in a variety of food applications, such as processed meats, pasteurised cheeses, milks, frozen desserts, salad dressings, mayonnaise, sauces, gravies, soups, margarine, breads, waffles, granola bars, muffins, cookies, brownies, cakes and beverages. Some examples of these applications are discussed below. The primary benefits of incorporating OatWell® in foods are increased delivery of soluble and insoluble fibre, increased protein, fat replacement and shelf-life enhancement. OatWell® has good applications in cereal bars, snacks, breakfast cereals, breads, cookies, powdered drinks and pasta. The health benefits demonstrated by incorporating OatWell® into the diet are a lowering of cholesterol (Beer *et al.* 1995b, Pick *et al.* 1996, Amundsen *et al.* 2003) and diabetes control (Tappy *et al.* 1996, Battilana *et al.* 2001, Jenkins *et al.* 2002, Kabir *et al.* 2002). Another defatted and fractionated oat product with a 15% β-glucan content is manufactured by Nurture Inc. (Missoula, MT), marketed under the name Nurture® 1500 (Table 20.1), and has applications in fruit juices, skim milk, shakes, powdered beverages, breads, cakes and pasta. Dry-milling and air classification are used to produce Natureal® (Table 20.1), an oat bran product with a 20% β-glucan content that is manufactured and marketed by Finn Cereal, Vantaa, Finland (Lehtomäki *et al.* 1990, 1992). Natureal® forms highly viscous solutions that contain naturally occurring vitamin E and is well suited for application in health-promoting foods such as beverage powders, bakery, confectionery, meal replacement products and cereals.

Nutrim-OB can be formulated in functional foods or nutritional supplements to increase β-glucan levels to 0.75 g per serving. Nutrim-OB can substitute for fat and imparts moistness, softness and cohesiveness in baked foods, such as muffins and cookies (Warner and Inglett 1998, Lee *et al.* 2004), cheese (Konuklar *et al.* 2004a, 2004b) and Asian noodles (Inglett *et al.* 2005). A 5% Nutrim-OB dispersion can completely substitute for dairy or coconut cream (Maneepun *et al.* 1998a, 1998b). A limiting factor is the decrease in flavour when high levels of Nutrim-OB are used to replace butter or coconut cream. With 50% fat replacement, Nutrim-OB can be successfully incorporated without undesirable sensory properties.

Oatrim is an excellent fat replacer because the β-glucan and maltooligosaccharide combination produces a fat-like texture (Hallfrisch and Behall 1997). It has one calorie per gram compared with the nine calories per gram in fat (Inglett and Grisamore 1991), and mimics the appearance and taste of traditional higher-fat foods. Oatrim powder can be converted to a shortening-like gel by heating and cooling a 25% dispersion, and the gel is heat-stable for baking and pasteurising operations, substituting for shortening in food recipes on an equal weight basis. Oatrim powder or gel can be used in food applications such as processed meats, pasteurised cheeses, milks, frozen desserts, salad dressings, mayonnaise, sauces, gravies, soups, margarine, breads, waffles, granola bars, muffins, cookies, brownies, cakes and beverages (Inglett and Newman 1994, Inglett *et al.* 1994, 1996, 2000, Pszczola 1996, Swanson *et al.* 1999, Dawkins *et al.* 2001, Welty *et al.* 2001, Lee *et al.* 2005a). Studies on high-sugar-ratio cake batters formulated with Oatrim containing 5% β-glucan showed that higher levels of incorporation resulted in high shear-thinning at low shear rates (Lee *et al.* 2005a). Cake batters formulated with Oatrim also had fewer bubbles that were similar in diameter.

Viscofiber® has a light colour, bland flavour and high viscosity that make it suitable for food applications such as nutritional bars, beverages, soups, breakfast cereals, baked products, pasta, ice cream, yoghurt and dietary supplements – as well as cosmetic, medical and pharmaceutical applications. Viscofiber® can impart high viscosity to foods, a textural attribute with the potential to combat the global rise in obesity since viscous foods have been found to reduce caloric intake (Davidson and Swithers 2005). Viscofiber® β-glucans can be retained within grain cell walls, forming a honeycomb structure that is preserved using an encapsulant (Vasanthan and Temelli 2005). Viscofiber® is kosher and has GRAS (generally recognised as safe) status in the United States.

American Oats Inc., based in Minneapolis, MN, currently markets OatsCreme™, an oat-based functional syrup that is well suited for use in dairy-free frozen desserts. Another product, C-trim from FutureCeuticals, Momence, IL, makes a high-viscosity natural product for improving the health benefits of functional foods. Cookies containing C-trim20 spread less than controls due to increased springiness, and have higher water content, but up to 10% flour replacement with C-trim20 did not alter sensory properties (Lee *et al.* 2005c). C-trim20 incorporated into cake formulations increased the elasticity of cake batter but did not alter cake volume and texture (Lee *et al.* 2005b).

A comparison of the rheological and sensory characteristics of foods formulated with oat β-glucan-enriched extract produced by dry-milling or the use of α-amylase has recently been studied (Lyly *et al.* 2003, 2004). The average molecular weight of β-glucan extracted by dry-milling (2 000 000) was ten-fold higher than that of β-glucan extract made by α-amylase extraction (200 000). Soups and beverages made with β-glucan extracted by dry-milling were more viscous and had higher perceived thickness compared with soups made from enzyme-extracted β-glucan. Additionally, finished foods formulated with β-glucan extracts have been shown to have varying molecular weight distribution depending on cooking conditions (Åman *et al.* 2004).

20.8 Future trends

Despite an increase in fad diets, consumers consistently identify dietary fibre as an important component of a healthy diet. Coronary heart disease continues to be a major cause of mortality in industrialised nations and diabetes is increasing sharply in China and India. The incidence of obesity in adults and children is rising globally at alarming rates. To combat these and other health disorders, health advocacy groups will need to emphasise diets with increased fibre as part of a healthy lifestyle. The selective high-value, functional high-fibre oat products discussed in this chapter can be useful for those populations seeking better health and longevity. To satisfy the food industry's demand for low-cost, safe and versatile oat fibre ingredients, oat fibre production technologies in the future will need to focus on reducing processing costs, substitution of current chemicals with green chemistry techniques and development of oat fibres with broader functionalities. Greater understanding needs to be gained of the relationship between oat fibre molecular weight and health benefits since conflicting publications presently exist. Processing methods that minimise oat fibre destruction also need further research.

20.9 References

Adom, K. K. and Liu, R. H. 2002. Antioxidant activity of grains. *J. Agric. Food Chem.* **50**: 6182–6187.

Åman, P., Rimsten, L. and Andersson, R. 2004. Molecular weight distribution of β-glucan in oat-based foods. *Cereal Chem.* **81**: 356–360.

Amundsen, A. L., Haugum, B. and Andersson, H. 2003. Changes in serum cholesterol and sterol metabolites after intake of products enriched with an oat bran concentrate within a controlled diet. *Scand. J. Nutr.* **47**: 68–74.

Anderson, J. W. and Chen, W. L. 1986. Cholesterol-lowering properties of oat products. In: *Oats: Chemistry and Technology*. F. H. Webster, ed. AACC: St Paul, MN.

Autio, K., Myllymäki, O., Suorrti, T., Saastamoinen, M. and Poutanen, K. Physical properties of beta-glucan preparates isolated from different Finnish oat varieties 1992. *Food Hydrocoll.* **5**: 513–522.

Battilana, P., Ornstein, K., Minehira, K., Schwarz, J. M., Acheson, K., Schneiter, P., Burri, J., Jéquier, E. and Tappy, L. 2001. Mechanisms of action of β-glucan in postprandial glucose metabolism in healthy men. *Eur. J. Clin. Nutr.* **55**: 327–333.

Beer, M. U., Arrigoni, E. and Amadò, R. 1995a. Extraction of oat bran: effects of process on yield, molecular weight distribution, viscosity and (1→3)(1→4)-β-D-glucan content of the gum. *Cereal Chem.* **73**: 58–62.

Beer, M. U., Arrigoni, E. and Amadò, R. 1995b. Effects of oat gum on blood cholesterol levels in healthy young men. *Eur. J. Clin. Nutr.* **49**: 517–522.

Beer, M. U., Wood, P. J. and Weisz, J. 1997. Molecular weight distribution and (1→3)(1→4)-β-D-glucan content of consecutive extracts of various oat and barley cultivars. *Cereal Chem.* **74**: 476–480.

Behall, K. M., Scholfield, D. J. and Clark, W. M. 1993. Effect of beta-glucan level in 'Oatrim' added to a controlled basic diet on blood lipids and uric acid of male and female subjects. Abstract. *XV International Congress of Nutrition*. Adelaide Book 2. Smith-Gordon Press: Adelaide, Australia.

Behall, K. M., Scholfield, D. J. and Hallfrisch, J. G. 1997. Effect of beta-glucan level in oat fibre extracts on blood lipids in men and women. *J. Am. Coll. Nutr.* **6**: 46–51.

Berg, A., König, D., Deibert, P., Grathwohl, D., Berg, A., Baumstark, M. W. and Franz, I. W. 2003. Effect of an oat bran enriched diet on the atherogenic lipid profile in patients with an increased coronary heart disease risk. *Ann. Nutr. Metab.* **47**: 306–311.

Bhatty, R. S. 1995. Laboratory and pilot plant extraction and purification of β-glucans from hull-less barley and oat brans. *J. Cereal Sci.* **22**: 163–170.

Bhatty, R. S. 1996. Methods for extracting cereal β-glucans. US patent 5,518,710.

Burrows, V. D., Fulcher, R. G. and Paton, D. 1984. Processing aqueous treated cereals. US patent 4,435,429.

Cahill, A. P., Fenske, D. J., Freeland, M. and Hartwig, G. W. 2003. Beta-glucan process, additive and food product. US patent 6,531,178.

Cahill, A. P., Fenske, D. J., Freeland, M. and Hartwig, G. W. 2004. Beta-glucan process, additive and food product. US patent 6,749,885.

Carrière, C. J. and Inglett, G. E. 1998a. Solution viscoelastic properties of Oatrim-10 and cooked oat bran. *Cereal Chem.* **75**: 354–359.

Carrière, C. J. and Inglett, G. E. 1998b. Solution viscoelastic properties of an oat-based beta-glucan-rich hydrocolloidal extractive: Nutrim OB. In: *Proceedings of the 216th National Meeting of the American Chemical Society, Agricultural and Food Chemistry Division*, Boston, MA. Abstract, p. 36.

Carrière, C. J. and Inglett, G. E. 1998c. Nonlinear solution viscoelastic properties of amylolytic hydrolyzed oat-based beta-glucan-containing materials. In: *Proceedings of the 216th National Meeting of the American Chemical Society, Agricultural and Food Chemistry Division* Boston, MA. Abstract, p. 37.

Carrière, C. J. and Inglett, G. E. 1999. Nonlinear solution viscoelastic properties of oat-based beta-glucan-amylodextrin blends. *Carbohydr. Polym.* **40**: 9–16.

Carrière, C. J. and Inglett, G. E. 2000. Effect of processing conditions on the viscoelastic behavior of Nu-TrimX: an oat-based beta-glucan-rich hydrocolloidal extractive. *J. Texture Stud.* **31**: 123–140.

Carrière, C. J. and Inglett, G. E. 2003. Constitutive analysis of the nonlinear viscoelastic properties of cellulosic fibre gels produced from corn or oat hulls. *Food Hydrocoll.* **17**: 605–614.

Cluskey, J. E., Wu, Y. V., Inglett, G. E. and Wall, J. S. 1976. Oat protein concentrates for beverage fortification. *J. Food Sci.* **41**: 799–804.

Collins, F. W. and Paton, D. 1992. Method of producing stable bran and flour products from cereal grains. US patent 5,169,660.

Davidson, T. L. and Swithers, S. E. 2005. Food viscosity influences caloric intake compensation and body weight in rats. *Obesity Res.* **13**: 537–544.

Dawkins, N. L., Gager, J., Cornillon, J. P., Kim, Y., Howard, H. and Phelps, O. 2001. Comparative studies on the physicochemical properties and hydration behavior of oat gum and oatrim in meat-based patties. *J. Food Sci.* **66**: 1276–1282.

Dawkins, N. L. and Nnanna, I. A. 1993. Oat gum and beta-glucan extraction from oat bran and rolled oats: temperature and pH effects. *J. Food Sci.* **58**: 562–566.

de Groot, A. P., Luyken, R. and Pikaar, N. A. 1963. Cholesterol-lowering effect of rolled oats. *Lancet* **2**: 303–304.

Dintzis, F. R., Legg, L. M., Deatherage, W. L., Baker, F. L., Inglett, G. E., Jacob, R. A., Reck, S. J., Muñoz, J. M., Klevay, L. M., Sandstead, H. H. and Shuey, W. C. 1979. Human gastrointestinal action on wheat, corn and soy hull bran – preliminary findings. *Cereal Chem.* **56**: 123–127.

Doehlert, D. C. and Moore, W. R. 1997. Composition of oat bran and flour prepared by three different mechanisms of dry milling. *Cereal Chem.* **74**: 403–406.

Duss, R. and Nyberg, L. 2004. Oat soluble fibres (β-glucans) as a source for healthy snack and breakfast foods. *Cereal Foods World* **49**: 320–325.

Food and Drug Administration. 1997. *Fed Regist.* **62**: 3584–3601.

Food and Drug Administration 2003. *Fed Regist.* **68**: 44207–44209.

Forrest, I. S. and Wainwright, T. 1977. The mode of binding of β-glucans and pentosans in barley. *J. Inst. Brew.* **83**: 279–286.

Good, N. E., Winget, G. D., Winter, W., Connolly, T. N., Izawa, S. and Singh, R. M. M. 1966. Hydrogen ion buffers and biological research. *Biochemistry* **5**: 467–477.

Hallfrisch, J. and Behall, K. M. 1997. Evaluation of foods and physiological responses to menus in which fat content was lowered by replacement with Oatrim. *Cereal Foods World* **42**: 100–103.

Hallfrisch, J., Scholfield, D. J. and Behall, K. M. 1995. Diets containing soluble oat extracts improve glucose and insulin responses of moderately hypercholesterolemic men and women. *Am. J. Clin. Nutr.* **61**: 379–384.

Hedley, A. A., Ogden, C. L., Johnson, C. L., Carroll, M. D., Curtin, L. R. and Flegal, K.

M. 2004. Prevalence of overweight and obesity among US children, adolescents, and adults, 1999–2002. *JAMA* **291**: 2847–2850.
Hohner, G. A. and Hyldon, R. G. 1977. Oat groat fractionation process. US patent 4,028,468.
Inglett, G. E. 1990. Hypocholesterolemic beta-glucan–amylodextrins from oats as dietary fat-replacements. In: *Proceedings of Symposium on Beta-Glucans: Biotechnology and Nutrition, 199th National Meeting of the American Chemical Society*, Boston, MA. Abstract.
Inglett, G. E. 1991. Method for making a soluble dietary fibre composition from oats. US patent 4,996,063.
Inglett, G. E. 1992. Method of making soluble dietary fibre compositions from cereals. US patent 5,082,673.
Inglett, G. E. 1993. Amylodextrins containing beta-glucan from oat flours and bran. *J. Food Chem.* **47**: 133–136.
Inglett, G. E. 1997. Development of a dietary fibre gel for calorie-reduced foods. *Cereal Foods World* **42**: 382–385.
Inglett, G. E. 1998a. Nutrim OB, a new beta-glucan-rich hydrocolloid as a phytonutrient for increasing health benefit of functional foods. In: *Proceedings of the 216th National Meeting of the American Chemical Society, Agricultural and Food Chemistry Division*, Boston, MA. Abstract, p. 30.
Inglett, G. E. 1998b. Dietary fibre gels for preparing calorie reduced foods and method for preparing the same. US patent 5,766,662.
Inglett, G. E. 1999. Nutraceuticals: The key to healthier eating. *ChemTech*, **29**: 38–42.
Inglett, G. E. 2000. Soluble hydrocolloid food additives and method of making. US patent 6,060,519.
Inglett, G. E. 2004. Low-carbohydrate digestible hydrocolloidal fibre compositions. US patent application serial number 11/020,349.
Inglett, G. E. and Carrière, C. J. 2001. Cellulosic fibre gels prepared from the cell walls of maize hulls. *Cereal Chem.* **78**: 471–475.
Inglett, G. E., Carrière, C. J., Maneepun, S. and Boonpunt, T. 2002. Nutritional value and functional properties of a hydrocolloidal soybean and oat blend for use in Asian foods. *J. Sci. Food Agric.* **83**: 86–92.
Inglett, G. E., Carrière, C. J., Maneepun, S. and Tungtrakul, P. 2004. A soluble fibre produced from rice bran and barley flour as a fat replacer in Asian foods. *Int. J. Food Sci. Technol.* **39**: 1–10.
Inglett, G. E. and Grisamore, S. B. 1991. Maltodextrin fat substitute lowers cholesterol. *Food Technol.* **45**(6):104.
Inglett, G. E., Maneepun, S. and Vatanasuchart, N. 2000. Evaluation of hydrolyzed oat flour as a replacement for butter and coconut cream in bakery products. *Food Sci. Technol. Int.* **6**: 457–462.
Inglett, G. E. and Newman, R. K. 1994. Oat beta-glucan–amylodextrins: Preliminary preparations and biological properties. *Plant Foods Hum. Nutr.* **45**: 53–61.
Inglett, G. E., Peterson, S. C., Carrière, C. J. and Maneepun, S. (2005). Rheological, textural, and sensory properties of Asian noodles containing an oat cereal hydrocolloid. *Food Chem.* **90**: 1–8.
Inglett, G. E., Warner, K. and Newman, R. K. 1994. Sensory and nutritional evaluations of Oatrim. *Cereal Foods World* **39**: 755–759.
Inglett, G. E., Warner, K. and Newman, R. K. 1996. Soluble-fibre ingredient from oats: uses in foods and some health benefits. *Zywnosc Technologia Jakosc* **2**: 175–182.
Jenkins, A. L., Jenkins, D. J. A., Zdravkoviæ, U., Würsch, P. and Vuksan, V. 2002. Depression of the glycemic index by high levels of β-glucan fibre in two functional foods tested in type 2 diabetes. *Eur. J. Clin. Nutr.* **56**: 622–628.
Kabir, M., Oppert, J.-M., Vidal, H., Bruzzo, F., Fiquet, C., Würsch, P., Slama, G. and Rizkalla, S. 2002. Four-week low-glycemic index breakfast with a modes amount of soluble fibres in type 2 diabetic men. *Metabolism* **51**: 819–826.

Knuckles, B. E., Chiu, M. M. and Betschart, A. A. 1992. β-glucan-enriched fractions from laboratory-scale dry milling and sieving of barley and oats. *Cereal Chem.* **69**: 198–202.

Knuckles, B. E., Chiu, M. C. and Inglett, G. E. 1998. Physical characteristics of beta-glucans in oat and barley after treatment by the Oatrim and Nutrim OB processes. In: *Proceedings of the 216th National Meeting of the American Chemical Society, Agricultural and Food Chemistry Division*, Boston, MA. Abstract, p. 36.

Konuklar, G., Inglett, G. E., Carrière, C. J. and Felker, F. C. 2004a. Use of a β-glucan hydrocolloidal suspension in the manufacture of low-fat cheddar cheeses: manufacture, composition, yield and microstructure. *Int. J. Food Sci. Technol.* **39**: 109–119.

Konuklar, G., Inglett, G. E., Warner, K. A. and Carrière, C. J. 2004b. Use of a β-glucan hydrocolloidal suspension in the manufacture of low-fat cheddar cheeses: textural properties by instrumental methods and sensory panels. *Food Hydrocoll.* **18**: 535–545.

Kvist, S., Carlsson, T., Lawther, J. M. and DeCastro, F. B. 2005. Process for the fractionation of cereal brans. US patent application 20050089602.

Lee, S. and Inglett, G. E. 2005. Rheological and physical evaluation of jet-cooked oat bran in low calorie cookies. *Int. J. Food Sci. Technol.* **40**: 1–7.

Lee, S., Inglett, G. E. and Carrière, C. J. 2004. Effect of Nutrim oat bran and flaxseed on rheological properties of cakes. *Cereal Chem.* **81**: 637–642.

Lee, S., Kim, S. and Inglett, G. E. 2005a. Effect of shortening replacement with Oatrim on the physical and rheological properties of cakes. *Cereal Chem.* **82**: 120–124.

Lee, S., Kinney, M. P. and Inglett, G. E. 2005b. Rheological characterization of a new oat hydrocolloid and its application in cake baking. *Cereal Chem.* **82**: 717–720.

Lee, S., Warner, K. and Inglett, G. E. 2005c. Rheological properties and baking performance of new oat β-glucan-rich hydrocolloids. *J. Agric. Food Chem.* **53**: 9805–9809.

Lehtomäki, I., Karinen, P., Bergelin, R. and Myllymäki, O. 1990. A beta-glucane enriched alimentary fibre and process for preparing the same. European patent EP0377530.

Lehtomäki, I., Karinen, P., Bergelin, R. and Myllymäki, O. 1992. Beta-glucane enriched alimentary fibre and a process for preparing the same. US patent 5,106,640.

Lyly, M., Salmenkallio-Marttila, M., Suortti, T., Autio, K., Poutanen, K. and Lähteenmäki, L. 2003. Influence of oat β-glucan preparations on the perception of mouthfeel and on rheological properties in beverage prototypes. *Cereal Chem.* **80**: 536–541.

Lyly, M., Salmenkallio-Marttila, M., Suortti, T., Autio, K., Poutanen, K. and Lähteenmäki, L. 2004. The sensory characteristics and rheological properties of soups containing oat and barley β-glucan before and after freezing. *Lebensm-Wiss. Technol.* **37**: 749–761.

Mälkki, Y., Autio, K., Hänninen, O., Myllymäki, O., Pelkonen, K., Suortti, T. and Törrönen, R. 1992. Oat bran concentrates: physical properties of β-glucan and hypocholesterolemic effects in rats. *Cereal Chem.* **69**: 647–653.

Mälkki, Y. and Myllymäki, O. 1998. Method for enriching soluble dietary fibre. US patent 5,846,590.

Mälkki, Y., Myllymäki, O., Teinilä, K. and Koponen, S. 2004. Method for preparing an oat product and a foodstuff enriched in the content of beta-glucan. US patent 6,797,307.

Mälkki, Y. and Virtanen, E. 2001. Gastrointestinal effects of oat bran and oat gum. A review. *Lebensm.-Wiss. Technol.* **34**: 337–347.

Maneepun, S., Boonpunt, T. and Inglett, G. E. 1998a. Sensory and nutritional evaluation of co-processed Nutrim OB and soy flour in Thai dishes. In: *Proceedings of the 216th National Meeting of the American Chemical Society, Agricultural and Food Chemistry Division*, Boston, MA. Abstract, p. 31.

Maneepun, S., Vatanasuchart, N. and Inglett, G. E. 1998b. Nutrim OB, a substitute for coconut cream in Thai desserts: A sensory and nutritional evaluation. In: *Proceedings of the 216th National Meeting of the American Chemical Society, Agricultural and Food Chemistry Division*. Abstract, p. 31.

Morgan, K. R. 2000. Cereal β-glucans. In: *Handbook of Hydrocolloids*. G. O. Phillips and P. A. Williams eds. CRC Press. Boca Raton, FL, pp. 287–307.
Morgan, K. R. 2002. Beta-glucan products and extraction processes from cereals. US patent 6,426,201.
Morgan, K. R. and Ofman, D. J. 1998. Glucagel, a gelling β-glucan from barley. *Cereal Chem.* **75**: 879–881.
Myllymäki, O., Mälkki, Y. and Autio, K. 1994. Process for fractionating crop into industrial raw material. US patent 5,312,636.
Ohr, L. M. 2004. Nutraceuticals and functional foods. *Food Technol.* **58**(11): 73–76.
Oughton, R. W. 1980. Process for the treatment of comminuted oats. US patent 4,211,801.
Paton, D. and Lenz, M. K. 1993. Processing: current practice and novel processes. In: *Oat Bran*. P. J. Wood ed. AACC International: St Paul, MN, pp. 25–47.
Pick, M. E., Hawrysh, Z. J., Gee, M. I., Toth, E., Garg, M. L. and Hardin, R. T. 1996. Oat bran concentrate bread products improve long-term control of diabetes: a pilot study. *J. Am. Diet. Assoc.* **96**: 1254–1261.
Pomeroy, S., Tupper, R., Cehun-Aders, M. and Nestel, P. 2001. Oat beta glucan lowers total and LDL-cholesterol. *Aust. J. Nutr. Diet.* **58**: 51–55.
Popkin, B. M., Horton, S., Kim, S., Mahal, A. and Shuigao, J. 2001. Trends in diet, nutritional status, and diet-related noncommunicable diseases in China and India: the economic costs of the nutrition transition. *Nutr. Rev.* **59**: 379–390.
Potter, R. C., Fisher, P. A., Hash, K. R. and Neidt, J. D. 2001. Method for concentrating beta-glucan. US patent 6,323,338.
Potter, R. C., Fisher, P. A., Hash, K. R. and Neidt, J. D. 2002. Polysaccharide compositions and uses thereof. US patent 6,485,945.
Potter, R. C., Fisher, P. A., Hash, K. R. and Neidt, J. D. 2003. Method for concentrating beta-glucan film. US patent 6,624,300.
Prior, R. L., Cao, G. and Inglett, G. E. 1998. Antioxidant capacity of cereal grain ingredients and commercial breakfast cereals. In: *Proceedings of the 216th National Meeting of the American Chemical Society, Agricultural and Food Chemistry Division*, Boston, MA. Abstract, p. 37.
Prosise, R. L., Christmas, K. P., Elsen, J. J., Helmers, R. L., Kester, J. J. and Wehmejer, T. J. 2004. Compositions for reducing hypercholesterolemia and controlling of postprandial blood glucose and insulin levels. US patent application 20040086547.
Pszczoła, D. E. 1996. Oatrim finds application in fat-free, cholesterol-free milk. *Food Technol.* **50**(9): 80–81.
Rajeshwari, R., Nicklas, T. A., Pownall, H. J. and Berenson, G. S. 2005. Cardiovascular diseases – a major health risk in Asian Indians. *Nutr. Res.* **25**: 515–533.
Redmond, M. J. 2001. Cereal beta glucan compositions and methods of formulation. US patent 6,284,886.
Redmond, M. J. and Fielder, D. A. 2004. Improved extraction method for cereal beta (1-3) beta (1-4) glucan. Canadian patent pending PCT/CA2004/000666.
Redmond, M. J. and Fielder, D. A. 2005a. Cereal beta glucan compositions, methods of preparation and uses thereof. US patent application 20050031674.
Redmond, M. J. and Fielder, D. A. 2005b. Oat extracts: refining, compositions and methods of use. US patent application 20050042243.
Reyna, N. Y., Cano, C., Bermúdez, V. J., Medina, M. T., Souki, A. J., Ambard, M., Nuñez, M., Ferrer, M. A. and Inglett, G. E. 2003. Sweeteners and beta-glucans improve metabolic and anthropometrics variables in well controlled type 2 diabetic patients. *Am. J. Ther.*. **10**: 438–443.
Ripsin, C. M., Keenan, J. M. and Jacobs, D. R. 1992. Oat products and lipid lowering. *JAMA*. **24**: 3317–3325.
Saltzman, E., Das, S. K., Lichtenstein, A. H., Dallal, G. E., Corrales, A., Schaefer, E. J., Greenberg, A. S. and Roberts, S. B. 2001. An oat containing hypocaloric diet reduces systolic blood pressure and improves lipid profile beyond effects of weight loss in men and women. *J. Nutr.* **131**: 1465–1470.

Scholfield, D. J., Behall, K. M. and Armero, E. C. 1993. Long-term consumption of soluble beta-glucan by hyperlipidemic subjects: Evaluation of blood pressure, pulse, body weight, urea nitrogen, and creatinine. *FASEB J.* **7**: A738.

Shinnick, F. L., Ink, S. L. and Marlett, J. A. 1990. Dose response to a dietary oat bran fraction in cholesterol-fed rats. *J. Nutr.* **120**: 561–568.

Shinnick, F. L., Longacre, M. J., Ink, S. L. and Marlett, J. A. 1988. Oat fibre: composition versus physiological function in rats. *J. Nutr.* **118**: 144–151.

Stevenson, D. G., Eller, F. J., Radosavljević, M., Jane. J. and Inglett, G. E. 2006. Characterization of oat bran concentrate and Nutrim-OB without and with lipid extraction. *Int. J. Food Sci. Technol.* (in press).

Swanson, R. B., Carden, L. A. and Parks, S. S. 1999. Effect of a carbohydrate-based fat substitute and emulsifying agents on reduced-fat peanut butter cookies. *J. Food Qual.* **22**: 19–29.

Tappy, L., Gügolz, E. and Würsch, P. 1996. Effects of breakfast cereals containing various amounts of β-glucans fibres on plasma glucose and insulin responses in NIDDM subjects. *Diabetes Care* **19**: 831–834.

Törrönen, R., Kansanen, L., Uusitupa, M., Hänninen, O., Myllymäki, O., Härkönen, H., and Mälkki, Y. 1992. Effects of an oat bran concentrate on serum-lipids in free-living men with mild to moderate hypercholesterolemia. *Eur. J. Clin. Nutr.* **46**: 621–627.

Triantafyllou, A. O. 2003. Method for isolation of a beta-glucan composition from oats and products made therefrom. US patent 6,592,914.

Uusitupa, M. I. J., Ruuskanen, E., Mäkinen, E. A., Laitinen, J., Toskala, E., Kervinen, K. and Kesäniemi, Y. A. 1992. A controlled study on the effect of β-glucan-rich oat bran on serum lipids in hypercholesterolemic subjects: relation to apolipoprotein E phenotype. *J. Am. Coll. Nutr.* **11**: 651–660.

Vårum, K. M. and Smidsrød, O. 1988. Partial chemical and physical characterisation of (1→3),(1→4)- β-D-glucans from oat (*Avena sativa* L.) aleurone. *Carbohydr. Polym.* **9**: 103–117.

Vasanthan, T. and Temelli, F. 2004. Grain fractionation methods and producets (misspelled on application). US patent application 20040101935.

Vasanthan, T. and Temelli, F. 2005. Grain fibre compositions and methods of use. US patent application 20050208145.

Vasanthan, T., Temelli, F. and Burkus, Z. 2004. Preparation of high viscosity beta-glucan concentrates. US patent application 20040001907.

Warner, K. and Inglett, G. E. 1998. Flavor and texture characteristics of foods containing Nutrim OB oat bran hydrocolloid. In: *Proceedings of the 216th National Meeting of the American Chemical Society, Agricultural and Food Chemistry Division*, Boston, MA. Abstract, p. 31.

Welch, R. W., Peterson, D. M. and Schramka, B. 1988. Hypocholesterolemic and gastrointestinal effects of oat bran fractions in chicks. *Nutr. Rep. Int.* **38**: 551–561.

Welty, W. M., Marshall, R. T., Grun, I. U. and Ellersieck, M. R. 2001. Effect of milk fat, cocoa butter, or selected fat replacers on flavor volatiles of chocolate ice cream. *J. Dairy Sci.* **84**: 21–30.

Whalen, P. J. 2004. Process for preparing an oat-based functional syrup. US patent 6,685,974.

Whalen, P. J., Maxwell, D. L. and Wilbur, D. 2002. Process for forming an oat-based frozen confection. US patent 6,395,314.

Wilhelmi, C. and Morgan, K. 2001. The hydrolysis of barley β-glucan by the cellulase EC 3.2.1.4 under dilute conditions is identical to that of barley solubilase. *Carbohydr. Res.* **330**: 373–380.

Wood, P. J., Beer, M. U. and Butler, G. 2000. Evaluation of role of concentration and molecular weight of oat beta-glucan in determining effect of viscosity on plasma glucose and insulin following an oral glucose load. *Br. J. Nutr.* **84**: 19–23.

Wood, P. J., Braaten, J. T., Scott, F. W., Riedel, K. D., Wolynetz, M. S. and Collins, M.

W. 1994. Effect of dose and modification of viscous properties of oat gum on plasma glucose and insulin following an oral glucose load. *Br. J. Nutr.* **72**: 731–743.
Wood, P. J., Siddiqui, I. R. and Paton, D. 1978. Extraction of high-viscosity gums from oats. *Cereal Chem.* **55**: 1038–1049.
Wood, P. J., Weisz, J. and Fedec, P. 1991. Potential for β-glucan enrichment in brans derived from oat (*Avena sativa* L.) cultivars of different (1→3),(1→4)-β-D-glucan concentrations. *Cereal Chem.* **68**: 48–51.
Wood, P. J., Weisz, J., Fedec, P. and Burrows, V. D. 1989. Large-scale preparation of oat fractions enriched in (1→3)(1→4)-β-D-glucan. *Cereal Chem.* **66**: 97–103.
Wu, Y. V., Sexson, K. R., Cluskey, J. E. and Inglett, G. E. 1977. Protein isolate from high-protein oats: preparation, composition and properties. *J. Food Sci.* **42**: 1383–1386.
Wu, Y. V. and Stringfellow, A. C. 1995. Enriched protein- and β-glucan fractions from high-protein oats by air classification. *Cereal Chem.* **72**: 132–134.
Yokoyama, W. H., Knuckles, B. E., Chiu, M. C. and Inglett, G. E. 1998. Plasma cholesterol reduction in hamsters by dietary fibre fractions of oat and barley. In: *Proceedings of the 216th National Meeting of the American Chemical Society, Agricultural and Food Chemistry Division*. Abstract, p. 32.

21

Improving the use of dietary fiber and other functional ingredients to lower the glycemic index of cereal products

S. K. Patil, S. K. Patil and Associates, USA

21.1 Introduction

In human and pet nutrition, few issues have captured attention in the way that the health effects of dietary fibers have in recent years. As well as being discussed in the context of their effects on intestinal function and related health conditions such as bowel cancer, dietary fibers and other similar ingredients, they are often discussed today in the context of their effect on carbohydrate metabolism, glucose release and insulin response following consumption of carbohydrate-rich foods such as cereal products. These issues have an effect on major health conditions such as diabetes and obesity.

The glycemic index (GI), is a method of describing the blood glucose response to a given carbohydrate-containing food. The GI is a way of classifying carbohydrate-rich foods previously categorized as either simple or complex, or as sugars or starches. There has been increased utilization of the GI concept to measure glucose in the bloodstream and insulin response; however, there are some important limitations. Published GI values can vary; for example, due to individual differences, source/type of food, processing, preparation techniques and other variations, and this issue concerns some researchers and health care professionals. There are many research papers and publications that support GI benefits, and supporters of GI believe that low-GI foods help control weight and reduce the risk of diabetes; however, there is still conflicting research that points to a lack of evidence, particularly for weight loss.

Cereal grains are major contributors to carbohydrate intake, and altering the carbohydrate quality and fiber content of cereal products in the diet is likely to have the most tangible effect on dietary GI. Much choice and flexibility exists for food product developers in the selection of cereal products regarding type of grain, degree of refinement and type of processing (Prochaska *et al.* 2000). Fibers from various sources are also available, as are a range of functional cereal- and carbohydrate-derived functional ingredients. New wheat and barley whole grain flours, for example, that offer higher fiber content and some of the other nutritional benefits of beta-glucan and phytonutrients, have recently been introduced. In addition to fibers, polyols and some sugars such as fructose and tagatose are also considered as low-GI ingredients. Food-processing techniques employed in the manufacture of cereal products also tend to result in the disruption of the food matrix and the gelatinization of starch granules, thereby making them readily digestible and consequently cereal products generally have high GI values. The challenge for those aiming to produce low-GI cereal products is to select ingredients wisely and to identify techniques for cereal processing that result in starch that is slowly digested, thereby achieving low-GI products.

Currently, there is much interest in whole grain foods, and the strong growth of the low-GI food market suggests that this trend will be more enduring than the previous trend for low-carbohydrate foods. The goal of this chapter is to briefly review current knowledge on the physiological and nutritional aspects of GI and then to discuss efforts to enhance the use of fiber and other functional ingredients to lower the GI in cereal-based foods. The chapter will offer advice from a practical industrial perspective on the technological challenges involved in facing this very important and exciting area of nutritional food product development.

21.2 Glycemic index and glycemic load

21.2.1 Glycemic index

The GI ranks carbohydrates according to their effect on blood glucose levels after consumption. The concept was first developed and reported by Jenkins and his colleagues at the University of Toronto in 1981 (Jenkins *et al.* 1981). Values for the GI are based on postprandial (blood) glucose response over a 2 h period, related back to the curves produced after intake of a white bread meal, or consumption of a standard quantity of glucose. Thus, the GI of a food is a useful tool in determining the speed at which the carbohydrates in the food are digested and absorbed as glucose.

The Food and Agriculture Organization (FAO) of the United Nations defines GI as:

> the incremental area under the blood glucose response curve of a 50 grams carbohydrate portion of a test food expressed as a percent of the response to the same amount of carbohydrate from a standard food taken by the same subject. For most glycemic index data, the area under the curve has been calculated as the incremental area under the blood glucose response curve (IAUC).

Glucose has a GI value of 100 and is generally considered the standard. When tests use white bread as the standard, they give GI values about 1.3 times higher than those obtained with a glucose standard.

Carbohydrates with a high GI are rapidly converted into glucose resulting in a spike in blood glucose. Foods with a low GI break down more slowly, and release glucose gradually into the blood stream. This results in a lower insulin demand. Carbohydrates fall into three types based on their GI values (Brennan *et al.* 2005).

- Low GI: less than 55.
- Intermediate GI: between 56 and 69.
- High GI: higher than 70.

Globally there are significant differences in how the GI of foods is used. The concept is more widely understood among consumers in Australia, New Zealand and South Africa, where labeling regulations regarding these foods have been introduced and more low-GI products are available. Australian food companies, for example, can participate in the GI Symbol Program (run by Glycemic Index Limited, a nonprofit company formed by the University of Sydney, Diabetes Australia and the Juvenile Diabetes Research Foundation). Foods that meet specific nutritional criteria and for which an approved method of measuring GI has been used, can display a symbol and list the GI value on their labels. The US and European markets lag in communication and product introduction in the GI area. However, the concept is gaining acceptance in Europe, especially in Germany and the United Kingdom. In the United States the GI concept is still met with caution; recently, however, products have been introduced for this purpose. In addition, some companies have started promoting the concepts in the media. The Dietary Guidelines Advisory Committee of the Department of Health and Human Services in the United States decided to pass over GI in its most recent guidelines, saying: 'Although the use of food with a low-glycemic index may reduce postprandial glucose, there is not sufficient evidence of long-term benefit to recommend general use of diets that have a low glycemic index.' However, interest in Europe and the United States has increased significantly in the past year or so, as evidenced by the introduction of many whole grain products – including whole-grain breads that have a comparatively low GI value. Bread is the most widely consumed cereal-based product in these countries

The reluctance to embrace GI may be due to the complicated nature of the GI concept. GI is not a simple measurement like fat content or calories. A number of factors and characteristics influence a food's glycemic response. These include:

- differences in composition of foods due to growing conditions, processing effects, etc.
- the amount of carbohydrates and the type of carbohydrates such as monosaccharides, the type of starch, starch entrapment, interactions with other food components, etc.

- The type and level of resistant starch (RS), degree of gelatinization.
- Mechanical and thermal processing, particle size (faster absorption of smaller particles).
- Cellular structure, other food and meal components – especially protein, fat and dietary fiber (which can inhibit or decrease the rate of carbohydrate absorption).
- Individual differences in intestinal absorption, intestinal content, food preparation habits, timing of previous meal, etc.

For example, rice might be generally considered a high-GI ingredient. However, researchers across the globe have measured a large range of GI values, occurring due to botanical differences in rice from country to country. High-amylose grains and starches might have a GI in the high 30s to 50s, while other varieties might score over 100. This might allow the use of weighted average calculations that provide a good correlation between meal GI and the observed glycemic responses of meals of equal nutrient composition. (Values for each food equal the proportion of total glycemic carbohydrate multiplied by the food GI. The sum of these values is the mean GI.) But again, a value of the GI for every food in the diet or meal needs to have been assigned, and the accuracy of the calculation depends on the accuracy of the GI values ascribed to foods.

21.2.2 Glycemic load

The GI does not necessarily tell a food's whole story. More recently, attention has focused on the concept of glycemic load (GL), partly derived from the paper of Salmeron *et al.* (1997). The concept of GL differs from that of the GI, in that the GL refers to the impact of the total food on glucose production.

GL takes into account both the amount of carbohydrate in the food and the impact of that carbohydrate on blood sugar levels. A food's GL is determined by multiplying its GI by the amount of available carbohydrates in the food. Both a low-GI/high-carbohydrate food and a high-GI/low-carbohydrate food can have the same GL, depending on the amount consumed.

$$\text{GL} = \text{GI of food product} \times \text{carbohydrate content}$$

This value may be of particular use when trying to evaluate the glycemic response of foods within a meal setting, as it takes into account the relative amount of carbohydrate. The lower the GL, the less likely it is that blood sugar will spike. GL is categorized as follows.

- High GL: 20 and higher.
- Moderate GL: 11–19.
- Low GL: 10 or less. (These valves are less than the GI. Carbohydrate content must be less than 1. This can be clarified by stating the units for carbohydrate content.)

Another concept, the glycemic potency and glycemic glucose equivalent (GGE) of food was introduced by Monro (2005) to guide food exchanges within equicarbohydrate food categories, and it expresses the glycemic potency of the available carbohydrate component in a food relative to that of glucose. As GI is a relative value based on 'available carbohydrate', it cannot guide food choice for glycemic control unless the foods are equal in available carbohydrate. Glycemic glucose equivalent (GGE) overcomes these limitations of GI. The GGE content of an amount of food is the weight of glucose (g) that would induce a glycemic response equal to that induced by the food. Few studies have compared GI and GGE as guides to food choice for glycemic control, but in a direct test of the predictive validity of GGE in a group of foods of differing carbohydrate and GI, GGE predicted glycemic potency well, whereas GI was unrelated to glycemic effect. As a general guide to food choices for the control of glycemia, GI does not satisfy the criteria of predictive validity, accuracy, safety, ease of use, flexibility, sufficiency and compatibility as does GGE, per Monro. In summary, GGE may be more difficult to use, but product designers may find that using the concept as a roadmap could lead to better way of predicting the actual glycemic response of a product, leading to a healthier bottom line as well as a healthier consumer.

Confounding factors aside, in the current environment it is best for the food industry to formulate with ingredients that have or contribute to a low GI or GL. In general, that means avoiding high levels of 'glycemic carbohydrates' – ingredients high in sugar, paying particular attention to glucose, and high in starch, especially amylopectin, or any other compound that easily converts to glucose. It bears repeating: formulating for glycemic response is not a simple matter. A food scientist cannot just plug a formula into a database and come up with a valid number – at least not as the science now stands. However, the industry can find justification for taking the concept into account.

'Carbohydrates in human nutrition' (FAO/WHO 1998), a report of a Joint FAO/WHO Expert Consultation, recommends the following role of GI in food choice:

> That for healthy food choices, both the chemical composition and physiologic effects of food carbohydrates be considered, because the chemical nature of the carbohydrates in foods does not completely describe their physiological effects. In making food choices, the glycemic index may be used as a useful indicator of the impact of foods on the integrated response of blood glucose. Clinical application includes diabetes and impaired glucose tolerance. It is recommended that the glycemic index be used to compare foods of similar composition within food groups. Also published glycemic response data may be supplemented where possible with tests of local foods as normally prepared, because of the important effects that food variety and cooking can have on glycemic responses.

21.3 Carbohydrate digestibility

21.3.1 Introduction

Carbohydrates supply most of the energy derived from foods in the diets of humans. In developed countries, carbohydrates supply about 50% of total caloric intake compared with about 70% in the developing economies, and this figure is declining as the developing economies grow and prosper. According to the FAO/ WHO report (FAO/WHO 1998) on 'Carbohydrates in human nutrition', carbohydrates have a wide range of physiological effects that may be important to health, such as: provision of energy, effects on satiety/gastric emptying, control of blood sugar and insulin metabolism, protein glycosylation, cholesterol and triglyceride metabolism, bile acid dehydroxylation, fermentation in the colon that produces short-chain fatty acids, bowel habit/laxation and effects on microflora of the large bowel. These are diversified and essential physiological steps in the functioning of healthy individuals.

Carbohydrates from different sources and manufactured into consumer products by many diverse processes, fall somewhere on the continuum of readily digestible/available carbohydrates to slowly digestible/available carbohydrates to resistant carbohydrates/starch (RS). Readily digestible carbohydrates (examples of which are freshly cooked starch-rich carbohydrate food materials) typically have a high GI, whereas slowly digestible and resistant carbohydrates (for example, the starch present in raw potato or unripe banana, and starches from peas, beans and yams) have lower GI values. GI is dependent on the rate at which carbohydrate is hydrolyzed, and its subsequent absorption and metabolism. Carbohydrates differ physiologically in that some (e.g. glucose, fructose, sucrose, maltose, lactose, cooked starch) are hydrolyzed and absorbed from the small intestine and are then metabolized in the body tissues; some are incompletely hydrolyzed and/or absorbed and metabolized (e.g. polyols such as isomalt, sorbitol, xylitol); some are absorbed, not metabolized and excreted via the urine (e.g. erythritol, mannitol); some pass through the small intestine unchanged and are fermented completely or partially by bacteria in the colon (e.g. polydextrose, pectin, fructo-oligosaccharides, inulin, resistant maltodextrin, RS); and some pass through the digestive tract unchanged and are barely fermented (e.g. cellulose).

Resistant starches are further divided as follows:

- RS1 describes physically inaccessible starch that is locked within cell walls such as those found in partially milled grains, seeds and legumes.
- RS2 refers to native RS granules like those typically found in bananas, raw potatoes and high-amylose maize starch.
- RS3 is retrograded or crystalline non-granular starch, like the starch that is cooked and cooled, and retrograded high-amylose maize starch.
- RS4 is a relatively new classification that refers to specific chemical modification of starch or re-polymerized dextrinized starch such as chain linkage altered dextrins.

Figure 21.1 is a representation of the RS1, RS2, RS3 and RS4 starches.

Diets that are composed of slowly digestible and resistant carbohydrates have been successfully applied as therapy in the treatment of diabetes mellitus and other conditions exhibiting derangements in carbohydrate and lipid metabolism (Brand-Miller 1994). In these studies, the major dietary alterations were to the starch-containing foods, with the substitution of slowly digested low-GI products – such as pasta, whole grain cereal and legumes – for rapidly digested high-GI products, such as bread, breakfast cereals and potatoes. The properties of some of these cereal products were reported recently in a publication by Englyst and his colleagues (Englyst *et al.* 2003). This publication also introduces the terms 'readily available glucose' and 'slowly available glucose' to describe the physiological significance of glucose with regards to glycemic response.

Starch that is resistant to digestion by human alpha-amylases (starch hydrolyzing enzymes), is described as resistant starch (RS). RS has been defined as 'the sum of starch and products of starch degradation not absorbed in the small intestine of healthy individuals' (Englyst *et al.* 1992). Several people have defined RS as a physiological concept rather than a chemical entity. This is mainly due to the fact that it is the sum of starch and starch digestion products not absorbed in the small intestine, and these products may or may not be fermented in large intestine. RS is one of the components of dietary fibers and is classified as a part of total dietary fiber when measured by the Association of Official Analytical Chemists (AOAC) Method 991.43 and the American Association of Cereal Chemists (AACC) Method 32-07.

	RS1: physically inaccessible starch • e.g. partly milled grains and seeds and legumes
	RS2: granular starch/native starch granule • e.g. native uncooked potato starch and green banana
	RS3: retrograded starch, mainly retrograded amylose • e.g. cooled cooked potato, bread, corn flakes, high-amylose starch
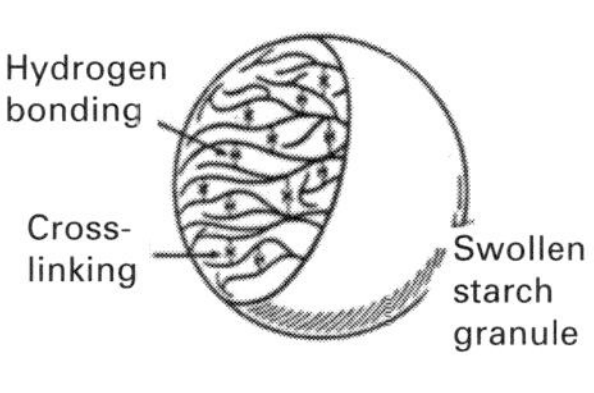	RS4: chemically modified to inhibit or limit swelling and reduce amylase digestion and pyrodextrins that are produced with a combination of heat and acid treatment that alters the glycosidic linkage and purified by methods such as ion exchange followed by filtration. Cross-links between amylase and amylopectin chains 'spot-weld' the chains to make swelling more difficult and make the chains indigestible by stomach enzymes

Fig. 21.1 Representation of the RS1, RS2, RS3 and RS4 starches.

Even though RS escapes digestion in the small intestine, it may be fermented in the large intestine by colonic microflora. In the last decade, there has been an increased interest in the nutritional implications of RS, not only because of its decreased caloric content, but also due to the fact that RS may have a similar physiological effect to dietary fiber. The possibility of altering the RS content in ingredients by different processing treatments has also gained the attention of food technologists.

21.3.2 Starch digestibility and glycemic index

Several publications report the relationship between the rate of starch digestion and GI. GI has been established by investigations of *in vitro* amylolytic hydrolysis (O'Dea *et al.* 1981; Jenkins *et al.* 1982; Englyst *et al.* 1992). Digestibility and GI have been reported for a range of products – such as bakery products, breakfast cereals and biscuits – using the terms readily available and, slowly available glucose, as explained by Englyst *et al.* (2003). The rate and extent of starch digestion is influenced by botanical origin, as this determines the amylose:amylopectin ratio and the structural type of the starch granule (Gallant *et al.* 1992). The other important factor is food processing, which determines the extent of starch gelatinization, particle size and the integrity of the plant cell wall. These physico-chemical variables of starch-containing foods are difficult to characterize in a quantitative manner that relates to their likely physiological fate. Effects of such physico-chemical characteristics on the rate and extent of carbohydrate digestion can be measured, and this can then be used to provide a description of nutritionally important aspects of the food.

The glycemic carbohydrate fraction that is available for absorption in the small intestine is measured as the sum of sugars, maltodextrins, syrups and starch, excluding RS. The glycemic carbohydrates can be divided into rapidly available carbohydrates and slowly available carbohydrates, to reflect the likely rate of release and absorption of glucose (Englyst *et al.*, 2003). The type of foods and consumption habits influence gastrointestinal events and post-absorptive metabolism and their influence on GI. Starch digestibility is affected by several factors such as the ratio of amylose:amylopectin, particle size, starch–protein interaction, physical form, method and time of cooking, encapsulation of starch in protein or lipid matrices, etc. (Thorne *et al.* 1983). Legumes contain twice as much protein and 5–10% more amylose than cereal grains, which affects starch digestibility. Gastric emptying is affected by food particle size and fat content as well as by viscous fiber, which also limits enzymatic hydrolysis in the small intestine by restricting access to the food bolus (Jenkins *et al.* 1981). Post-absorptive factors that can influence GI include the identity of the sugar moieties, which are metabolized differently, and the insulinotropic effect of protein, which can increase the clearance rate of circulating glucose. This emphasizes the fact that GI values do not represent a direct measure of carbohydrate absorption from the small intestine. Rather, the GI values are determined by the combined effects of all the properties of a food that influence the rate of influx and removal of glucose from the circulation. A better understanding of the mechanisms involved should provide insight into the concept of GI, and help to establish whether different types of low-GI diets are equally beneficial to health.

Several university, government and industry research groups continue to work in the area of starch digestibility with different approaches. These include product and process developments for RS and slowly digestible starch. Slowly digestible starch, which undergoes slow but complete digestion in the small intestine, is a very interesting area due to its benefits in controlled and sustained release of glucose into the blood stream. This is important for people with Type II diabetes who suffer from poor insulin response after consumption of readily digestible starch.

21.4 Dietary fiber and glycemic index

The work of Burkitt *et al.* (1972, 1975), during the 1960s and 1970s, pioneered our understanding of the role that dietary fiber could play in the prevention of diabetes, obesity, heart diseases, large bowel disease and colon cancer. Dietary fibers and their effect on animal and human nutrition have since been studied extensively by Fahey and co-workers (Fahey *et al.* 2004), among others. Fahey *et al.* have published an excellent review of the benefits of moderate to high consumption of dietary fiber. These include physiological responses such as fecal bulking, production of short-chain fatty acids, enhanced colonic function and a positive influence on the distribution of the colonic microflora, leading to decreased food intake and improved appetite regulation. Foods rich in dietary fiber also have a low GI and dietary fiber ingredients, such as the RS described above, are very useful tools for product developers formulating low-GI foods. Definitions of dietary fiber vary. AACC International defines dietary fiber as:

> the edible parts of plants or analogous carbohydrates that are resistant to digestion and absorption in the human small intestine with complete or partial fermentation in the large intestine. Dietary fiber includes polysaccharides, oligosaccharides, lignin and associated plant substances. Dietary fibers promote beneficial physiological effects including laxation, and/or cholesterol attenuation, and/or blood glucose attenuation.

As the definition suggests, dietary fibers are mainly derived from plant material, and are mainly found in cell wall material. Dietary fibers include non-starch polysaccharides, celluloses, lignins, hemicelluloses that come from plant cell wall materials such as cereal brans, and non-digestible oligosaccharides (RS, inulin, polydextrose and others). Their common characteristic is that they escape digestion in the small intestine and reach the large intestine where they undergo fermentation; hence their effects on metabolism and disease regulation are intrinsically linked to their physico-chemical properties as they pass through the gastrointestinal tract. The physico-chemical properties of dietary fibers are dominated by the conformation of the individual polysaccharide chains (ordered, disordered, random coil, chain geometry), and the way in which the polysaccharide chains of different dietary fibers interact with one another and other food components. It is these conformations that affect their hydration characteristics, and hence their solubility (Brennan *et al.*, 2005).

Based on their solubility, dietary fibers have been divided into soluble and insoluble fibers. In many cases, the physiological activity of dietary fibers is largely determined on the basis of solubility. Soluble dietary fibers are used in the food industry to modify or control the viscous properties of liquid and semi-liquid food products, and to alter their textural characteristics. The majority of soluble dietary fibers have the ability to form gels and alter the viscosity of products. This effect has led to them being known as 'gums' or 'hydrocolloids' within the food ingredient sector. The ability to form gels and viscous networks is not only important for the processing properties of fibers when used in food products, but may also affect their nutritional characteristics. For instance, numerous studies have illustrated the potential of these components to increase the viscosity of digesta when ingested. This in turn may explain their observed effects on carbohydrate metabolism (Mann 2001).

Research has shown that insoluble dietary fiber may reduce the gastrointestinal transit time (Harris *et al.* 2000) and that the dietary fiber containing high-viscosity fiber reduces starch digestibility allowing portions of the starch to escape to the large intestine. When in the small intestine, dietary fibers are thought to increase digesta viscosity to strengthen the so-called unstirred water layer in the gut, which potentially leads to a higher diffusion barrier, and also to binding of enzymes nonspecifically, hence reducing their activity. This, in itself has a direct influence on the rate of digestion and effectiveness of nutrient absorption (Endress *et al.* 2001). Such effects include moderation of postprandial glucose and insulin response, reduction in total and low-density lipoprotein (LDL) cholesterol and regulation of appetite (Davidson *et al.* 1988).

According to Schaafsma (2004), the generally accepted beneficial effects of dietary fiber on health are:

- improvement of colonic function (generic claim; transit time reduction, decrease of constipation, fecal bulking);
- lowering of blood LDL cholesterol (restricted to soluble fibers with water-holding capacity);
- reduction of the blood glucose level (GI restricted to soluble fibers with water-holding capacity).

Recent publicity and the trends in documenting the beneficial health effects of high-fiber, low-GI, whole grain diets are very positive and many new ingredients and products are being introduced. Recently, there has been a tremendous interest, on the part of industry, on the role of dietary fiber in regulating weight control and the glycemic response of individuals. There has been a surge in the number of introductions of high-fiber, whole grain and low-GI food ingredients and foods. Nevertheless, consumers consume too little dietary fiber and must increase their intake of this type of food further to meet the recommended daily intake for dietary fiber.

21.5 Ingredients for low glycemic index foods

According to the 2005 Health Focus® Trend Report (Health Focus International, 2005), about one in four American shoppers have decreased their consumption of high-GI carbohydrates within the past 2 years, and one in three have decreased their consumption of carbohydrates. At the same time, approximately one in ten are increasing their consumption of low-GI carbohydrates and 'better-for-you' carbohydrates. The demand for low- or reduced GI foods is expected to significantly expand over the next few years, presenting significant opportunities for new product development. There are major challenges in formulating reduced-GI foods. These include: (a) determining the analytical method for measuring the glycemic impact of the food; and (b) determining the intended marketing claim or health benefit to be communicated to consumers. These two strategic decisions will affect the selection of ingredients in food formulations and the number of products introduced.

There are many ingredients that are marketed to support the development of low-GI foods. In the United States, a recent search in one of the nutrition trade magazines lists 87 suppliers of such ingredients. The majority of ingredients available for the development of low-GI foods are from plant sources and are dietary fibers or derivatives of various carbohydrate polymers. The selection of a native material or derivative of the native material in insoluble or soluble form will depend on:

- the type of food to be developed, processing methods and the desired health effects based on the needs of consumers;
- the desired taste and aesthetic choices based on consumer preferences;
- ingredient availability and costs.

The next section outlines some of the challenges faced by the food technologists in designing appropriate food products and the selection of appropriate dietary fiber and low-GI ingredients. The range of ingredients, their composition, occurrence, food applications and physiological effects are briefly listed in Table 21.1 (Inglett and Carriere 2001, Nelson 2001, Tungland *et al.* 2002, Fahey *et al.*, 2004).

Table 21.1 Major commercial dietary fibers and non- or controlled digestibility low-glycemic index polysaccharides (Inglett and Carriere 2001, Nelson 2001, Tunland *et al.* 2002, Fahey 2004)

Name	Occurrence properties and composition	Food applications	Physiological effects
Cereal (wheat, corn, rice, etc.) fibers and brans	Cellulose, lignin and hemicelluloses. Mainly insoluble with minor soluble components	Bakery, breakfast cereals, snack/nutrition bars, fiber gels in fat replacement	Laxation, fecal bulk, lowering of cholesterol and low GI, contain phytosterols
Soy and pea fibers	Cellulosic and non-cellulosic material, mainly insoluble with minor soluble components	High-fiber and -protein bread, nutritional beverages, Japanese noodles	Laxation, fecal bulk, lowering of cholesterol and low GI, iron source, phytosterols
Oat and barley fibers	Insoluble cellulose, lignin and hemicellulose, and source of soluble fibers and beta-glucan	Bakery, breakfast cereals, soups, sauces, nutrition bars, pizza crust, etc., soluble fibers in many other food segments	Laxation, fecal bulk, lowering of cholesterol and low GI, beta-glucan has FDA nutrition claim for lowering of cholesterol and CHD
Cotton seed fibers	Cellulose, lignin and hemicellulose, mainly insoluble	Bakery	Laxation, fecal bulk, lowering of cholesterol and low GI
Cellulose, microcrystalline cellulose	Beta-1–4 glucan, insoluble, non-caloric, MCC increase water-holding capacity, insoluble	Fiber fortification in bakery products, beverages in combination with inulin and other soluble fibers, sauces and dressings	Laxation, fecal bulk, lowering of cholesterol and low GI
Resistant starch	Alpha-1–4 and 1 6 glucose polymers. RS1 to RS4, see text for detail, insoluble	Bakery products, batters, snack/nutrition bars, products where high water holding of insoluble and soluble fibers needs water balance	Fecal bulk, prebiotic, fermentation in colon to produce SCFAs, low GI

Pectins	Fruits and vegetables (apples, citrus, sugar beets, etc.), alpha-1–4–D-galacturonic acid backbone/ neutral sugar side chains, soluble	James, jellies, dairy, bakery products, beverages, snacks	Fermented on lower intestine, decrease intestine transit time, low GI
Carrageenan	From red algae; mixture of sulfated polysaccharides – alpha–D-galactose and 3,6 anhydro-D-galactose, soluble gum	Bakery, dairy, confectionery, desserts, processed meats	Increases viscosity, decreases gastric emptying and transit time, fermented in large intestine to SCFAs, low GI
Acacia	Dried exudates from acacia tree; highly branched beta-D-galactose and alpha-1–4 galacturonic acid backbone, with L-rhamnose and ester groups, soluble gum	Confectionery, dairy, spray-dried flavors, desserts, jellies, good emulsifier properties for many applications	Prebiotic, low GI
Xanthan	Produced by fermentation using *Xanthomonas campestris*, alpha-1–4-D-glucose, beta-D-glucuronic acid and β-D-mannose, soluble gum	Dairy products, beverages, sauces, gravies, fillings, salad dressings, etc.	Increases viscosity, produces SCFAs in large intestine, low GI
Gellan	Beta-1–4-D-glucose backbone, beta-D-glucuronic acid and D-rhamnose	Produced by fermentation using *Pseudomonas elodea*, soluble gums	Increases viscosity, produces SCFAs in large intestine, low GI
Arabinogalactan	Extracted from the pulp of western larch tree; highly branched beta-1–3 and beta-1–6-D-arabinose and galactose in 6:1 ratio, soluble	Dietary supplements, functional and health foods	Fermented in large intestine, increase *Lactobacillus* population, immunological properties, low GI

Table 21.1 Continued

Name	Occurrence properties and composition	Food applications	Physiological effects
Inulin	Beta-D-2, 1-fructofuranosyl, alpha-D-glucopyranoside, extracted from chicory and Jerusalem antichoke, soluble	Dressings, sauces, dairy, bakery, snack/nutrition bars beverages, fat replacement	Effects microbial population, SCFA production, enhanced calcium absorption, low GI
Fructosaccharides and oligosaccharides	Produced by transglucosylation of beta-fructoside of *Aspergillus niger*, also by partial enzymatic degradation of inulin; beta-fructofuranosyl, alpha-D-glucopyranoside with DP of 3–5, soluble	Fillings, icings, yoghurts, desserts, dietary supplements, prebiotic functional foods	Fermentation and production of SCFAs in large intestine, enhanced calcium absorption, low GI
Citrus and apple fibers	Cellulose, lignin and hemicellulose; mainly insoluble with minor soluble components	Bakery, breakfast cereals, fiber gels in fat replacement	Laxation, fecal bulk, lowering of cholesterol and low GI, contain phytosterols
Polydextrose	Produced by vacuum thermal polymerization of glucose, sorbitol and citric acid; mixed and random glycosidic 1–6-glucose linkage	Low-calorie bulking agent, sugar and fat replacement; bakery, snack bars, confectionery, dairy, sauces, fillings, etc.	Fermented; produce SCFAs in large intestine, stool bulking, low GI
Beta-glucan	From cereals, main sources oats and barley; beta-1–4-D-glucose and beta-1–3-D-glucose branched, soluble fiber	Several food segments, bakery, dairy, beverages breakfast cereals, snack/nutrition bars, functional and health foods	Fermented in large intestine, strong butyrate (SCFA) production, lowers blood lipids and low GI, FDA health claim for cholesterol reduction

Curdlan	Produced by fermentation using *Alcaligenes faecalis*; beta-1–3-D-glucose, insoluble	Sauces, dressings, processed meats, seafood products, desserts, jellies	High water holding, prebiotic, low GI
Chitin/chitosan	Produced by alkaline degradation of crustacean chitin, found in fungi cell walls of *Zygomycetes* sp., insoluble	High viscosity may limit applications	Interferes with intestinal absorption of cholesterol, low GI
Resistant maltodextrins	Produced by dextrinization/pyrolysis of starch with HCl and heat; mixed alpha-1–4, 1–6, 1–2 and 1–3 glucosidal bonds, soluble	Many food segments, bakery, dairy, beverages, snack/nutrition bars	Low-viscosity fiber source, partially digestible, laxation, colonic fermentation, effects on lipid levels
Isomalto-oligosaccharides	Occurs in soy sauce, sake, honey, produced by transglucylation of glucose residue, alpha-1–4 and alpha-1–6-linked branched oligosaccharides, soluble	Dairy, bakery, dessert gels, processed meats, low-sugar products	Increases viscosity, reduces gastric emptying and intestinal transit time, fermented to SCFAs, low GI
Guar gum oligosaccharides	Produced by partial hydrolysis of guar gum, galactomannan, soluble	Beverages, dairy foods, functional and health foods	Fermented by colonic bacteria, lowers lipids, low GI
Alginate oligosaccharides	Produced by enzymatic degradation of alginate, high in mannuronic acid and glucuronic acid, soluble	Diet supplements, health and functional foods	Effect on intestinal microflora, low GI

Table 21.1 Continued

Name	Occurrence properties and composition	Food applications	Physiological effects
Sucromalt	Produced from high-maltose syrup and sucrose using dextransucrase enzyme produced by *Leuconostoc*, alpha-1–4- and 1–6-glucose oligosaccharide	Slowly but fully digestible sweetener; sports beverages, yoghurt, snack bars	Slowly digestible, controlled release of glucose in the blood, low GI
Konjac flour	Obtained from tubers of *Amorphophallus*; beta-D-1–4 glucose and mannose with acetyl groups; soluble hydrocolloid	Bakery, snacks, also fat replacement	Increase fecal bulk and viscosity, low-GI

CHD, coronary heart disease; SCFAs, short-chain fatty acids; MCC, microcrystalline cellulose.

21.6 Effects of processing on the properties of dietary fiber ingredients and formulation challenges

Dietary fiber components provide many functional properties depending on their source, the processing technologies utilized in their production and their chemical/physical attributes. To complicate the matter further, the polysaccharides that make up dietary fibers in nature are heterogeneous and their structure–function relationships are difficult to establish. The oligosaccharides derived from polysaccharides or manufactured by fermentation are fairly well defined making them easier to incorporate into food systems. The type of processing utilized to produce high-fiber ingredients greatly influences the functionality of these ingredients. The majority of processes involve: milling, acid, alkali and enzyme treatments, extrusion and dehydration. Milling processes separate the bran, germ and the endosperm from the grain. The degree of separation affects particle size and the release of starch from the protein/lipid matrices which increases starch digestibility and GI. The recent trend of formulating foods with whole grain aims to maintain the physical state of the starch in these matrices, thereby controlling the digestibility of starch and the GI. The whole grains concept offers a range of health-promoting components including dietary fiber, protein, lipids, slowly digestible starch and important vitamins, along with phytonutrients.

Cereal and fruit fibers can be separated and treated to enhance their functional properties to improve their usage in different food systems. One technology produces a commercial functional fiber gel (cellulosic fiber gel) from plant fibers that can replace flour and also fat in bakery products (Inglett and Carriere 2001 (Chapter 21)). A two-stage process with thermal alkaline degradation is used with high shear in the first stage followed by alkaline peroxide treatment with shear in the second stage. The purified gel can be dried to produce a versatile, white, powdered ingredient for use in many food segments. Similar new technologies in combination with or without enzymes are utilized to produce fiber-rich products, with low GI values, that offer options to food formulators. Bleaching by peroxide or hypochlorite is practiced to eliminate dark colors and to oxidize tannins and lignins. Grinding to suitable particle size is essential to improve mouth feel or eliminate grittiness. Grinding can further increase water-holding capacity which can have positive or negative effects on food texture, and water balance has to be managed by formulations that include RS or low-viscosity oligosaccharides to build solids. Enzyme treatment, extrusion, dehydration and roasting are other treatments used to make dietary fiber easier to use in food formulations, depending on the desired food sensory properties.

The primary properties of isolated/treated fiber ingredients and their derivatives, polysaccharides or oligosaccharides, relate to their solubility, viscosity and gelation-forming ability, water-binding capacity, oil-binding capacity, and mineral and organic molecule-binding capacity. Aesthetic and functional properties desired by consumers affect the selection and processing of dietary fiber ingredients for the final food. These functional properties also influence the food product's properties during its processing and its final product quality and characteristics. The food technologist is faced with a myriad of choices among dietary fibers and derived carbohydrates and many (sometimes conflicting) claims are made regarding the ingredients' technical functions – such as viscosity, water holding, texture, etc. and nutritional claims about the digestibility and GI among others. The challenge is to sift through this information and develop a good knowledge that can support the development of good tasting, stable and economic production of high-fiber, low-GI foods that can satisfy internal marketing requirements and ultimately the consumer needs and preferences. The following guide is offered as an aid to the systematic gathering of information for appropriate ingredient selection and development of a product concept.

1 Marketing/formulation requirements:
   (a) product properties – structural, organoleptic;
   (b) shelf stability – high/low temperature storage, pH, flavor stability, moisture/oil migration, microbial stability, etc.;
   (c) aesthetic properties – clarity, opacity, texture (pulpiness, smoothness, graininess, etc.), color, flavor, consistency, mouth feel, flowability;
   (d) product form – dry, semi-moist/solid, liquid, gelled, expanded, rigid, rubbery;
   (e) utilization – ready-to-eat, require additional preparation, microwave, bake, fry, reconstitute, etc.;
   (f) economics – cost of ingredients, food distribution, packaging, storage, distribution;
   (g) regulatory considerations, health claims.
2 Production requirements:
   (a) processing temperature;
   (b) holding time;
   (c) process equipment;
   (d) process parameters;
   (e) process economics;
   (f) packaging, storage, transportation and distribution (retail, food service, etc.).

Figure 21.2 provides a graphic representation of some of the challenges and essentials involved in the successful development and marketing of the high-fiber, low-GI food ingredients and foods discussed above.

21.6.1 Properties of high-fiber, low glycemic index ingredients

Solubility

The solubility of fiber is influenced by primary structural features of the polymeric backbone. More branching generally results in greater solubility (for example, gum acacia), as does the presence of ionizing groups (for

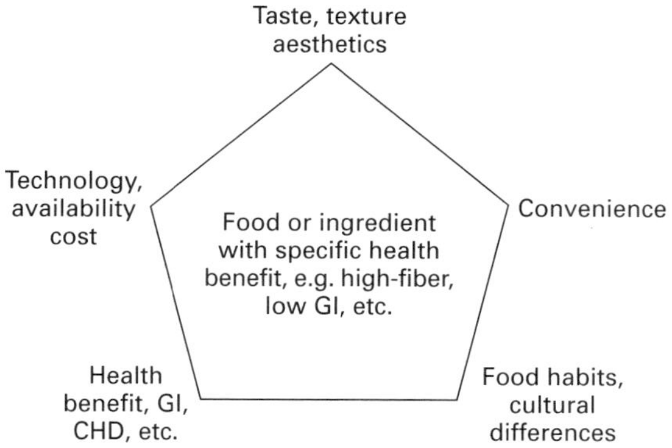

Fig. 21.2 Essentials for the successful marketing of high-fiber or functional foods. CHD, coronary heart disease.

example, pectin methoxylation), and the potential for inter-unit positional bonding (for example, alpha-glucans with mixed alpha, 1–3 and alpha, 1–4 linkages). Alterations of the monosaccharide units or their molecular form and structure further increase solubility (for example, gum acacia, arabinogalactan and xanthan gum). The term 'soluble fiber' does not necessarily mean that the fiber dissolves in water in a similar fashion to glucose. It means that it forms a colloidal dispersion in water. For example, beta-glucan is not completely soluble in water because it contains starch and other insoluble materials. Gum acacia and pectins are, on the other hand, soluble because of extreme branching of the polysaccharide. Solubility is important when formulating beverages or for clarity in food products.

Viscosity

High-fiber ingredients contribute to the viscosity of foods and the rheology of food systems. The molecular weight or chain length of the fiber increases the viscosity of the fiber in solution or dispersion. When formulating food with high fiber content, the required fiber concentration, the process temperature, pH, shear conditions of processing and ionic strength substantially determine the type and sources of fiber used. Long-chain polymers, such as the gums (for example, guar gum, locust bean gum, tragacanth gum and so on), bind significant water and exhibit high solution viscosity. These are used as thickening agents in foods at low concentrations. Cereal-based ingredients – such as oat fiber, wheat bran and fruit-based fibers – can also affect viscosity, Recent developments in fiber gel technologies have produced cereal and fruit fibers with controlled viscosity and water-holding properties. These new cereal and fruit fibers are becoming good economical alternatives to gums and oligosaccharides, which are usually used when low-viscosity fibers are required. The new processing technologies used in the manufacture of the new cereal and fruit fibers have enhanced their ability to be used at high levels that may provide significant benefits as a fermentable food source for colon microbiota. In general, however, it is the highly soluble fibers (those that are highly branched or are relatively short-chain polymers, such as gum arabic, isolated arabinogalactans, inulin and oligosaccharides) that have low viscosities. These low-viscosity fibers have low GI and are generally used to modify texture or rheology, manage water migration, influence the solution properties of the food system, and improve the marketability of the food product as a health-promoting or functional food product. These fiber sources can be used in food products at relatively high levels, as they typically enhance the food product's taste, mouth feel and shelf-life without significantly altering the specific application characteristics. For example, they can be used in sugar-free and fat-free products, increasing the potential for a high-fiber claim.

Gelation

A gel is formed due to the association of polymer units that form a network of junction zones. Gelation properties are an important attribute of some fiber ingredients as a means of adding form or structure to various food products and meeting the required aesthetic properties of foods. A gel formed by this process encapsulates water and other components in solution to form a firm three-dimensional structure. The type of gel-forming ingredient, its concentration, temperature, the presence of ions (for example, calcium), pH and the presence of other rheology modifiers in the food system influence the strength and stability of gels. Plant gums are used as gelling agents due to their high viscosity and gelling ability. Gums – including guar gum, gum arabic, karaya, lambda carrageenan and tragacanth gum are – typically used as rheology modifiers or stabilizers in food systems. Particle gel-forming polymers, such as starch and inulin, are typically used at much higher concentrations than the molecular gel-forming plant gums listed above. Starches and inulin are used with the gums in systems to influence the rheology and overall texture of food systems.

Water-holding and -binding capacity

Fiber ingredients interact with water in different ways depending on their source and the way in which the fibers have been produced; this variation dictates how the fiber is used and how it functions in a food system. Two terms are used to describe a fiber ingredient's properties in relation to water: 'water-holding capacity' (WHC) and 'water-binding capacity' (WBC). Other terms include: water uptake, hydration, adsorption, absorption, binding, and water holding. Solubility is important when formulating beverages or for clarity in food products affected by properties expressed as WBC or WHC. The two are used interchangeably, but the terms are differentiated based on the ability of a fiber to retain water under stress. WHC refers to the amount of water the gel system retains within its structure without pressure or stress, while WBC refers to the amount of water the gel system retains after it is stressed, as following centrifugation. The WBC probably has greater practicality, because food manufacturing processes typically use some form of physical stress (for example, extrusion, mixing or kneading, homogenization and so on). The source of the fiber determines the physico-chemical properties of the fiber ingredient – such as fiber length, particle size and porosity – which influence the WBC and its use and the conditions of its use in food product development. Other factors in the food system can also influence WBC, such as pH, ionic strength, concentration of the fiber component, and interaction with other water-binding ingredients (i.e., sugar, starches, etc.).

Oil-binding capacity

Many dietary fibers are fat and/or oil dispersible, and some also bind oil. Oil-binding capacity is in part related to the physical material properties of the fiber, largely a function of the porosity of the fiber structure rather than the affinity of the fiber molecule for oil. When fibers are hydrated, the water occupies the fiber pores, significantly reducing oil pick up in batters and breadings of fried foods while maintaining crispiness of foods (Tungland and Meyer 2002) Some dietary fibers, for example, fruit and vegetable fibers, have uronic acids and carboxyl groups attached to the monosaccharide components of the fibers. This property allows binding with cations such as calcium, cadmium, zinc and copper. Some dietary fibers have also been shown to absorb organic molecules (for example, lignin binds bile acids and wheat bran binds certain carcinogens such as benzopyrazine). The binding properties to minerals and organic acids vary depending on the type and source of fiber and factors such as pH.

The functional properties of dietary fibers discussed above are successfully utilized by food technologists to develop novel, good-tasting foods for today's convenience- and health-driven consumer. Newly developed ingredients such as Sucromalt, RS, and the slowly or non-digestible starches and oligosaccharides offer new options to food technologists to enable them to meet growing consumer demands for taste, convenience and healthy foods. New processes for producing functional fibers will enhance the use of cereal brans and other polysaccharides as we move forward.

21.7 Summary

The future of dietary fiber and low-GI ingredients will continue to evolve as we learn more about human physiology and the benefits of such ingredients. New and effective analytical methods must continuously be developed to support these advances and health claims related to new ingredients and formulations. Cereals play a major role in our diets and will continue to dominate new developments in dietary fiber and low-GI ingredients. Several universities, government and industry research groups continue to work in the area of starch digestibility with different approaches. This includes process developments for oligosaccharides and other novel polysaccharides with controlled digestibility, RS and SDS. Plant and animal science – along with biotechnology, enzyme development, nutrigenomics and other novel approaches – will offer a growing number of more healthful options to the consumers. Food companies are turning to functional foods as a promising growth area and this emphasis has enhanced interest in two of the most actively researched and debated issues in the food industry: the role the dietary fiber and the concept of GI. There is increasing evidence that the dietary fibers, non-digestible and slowly digestible carbohydrates and oligosaccharides are linked to a reduction of the effects of serious diseases such as diabetes and obesity. The concerted actions by food scientists, fiber suppliers, food companies and regulatory agencies will reward innovative fiber manufacturers and food companies for some years to come. There has been a wave of new ingredient introductions that are not substantiated by sound science. In the future, food ingredient suppliers and developers must be truthful in their claims of benefits and functional attributes. Regulatory agencies must be proactive in the development of guidelines, definitions, and appropriate checks and balances to protect consumers, who are confused by the bombardment with new and sometimes conflicting information. A sustained and economically cost-effective supply of these ingredients will be essential to fulfill the needs of large segments of the population worldwide.

21.8 References

Brand-Miller, J. C. Importance of glycemic index in diabetes. *American Journal of Clinical Nutrition* **59**, Suppl. 747S–752S, 1994.

Brennan, C. S. Dietary fiber, glycemic index and diabetes. *Molecular Nutrition and Food Research*, **49**, 560–570, 2005.

Burkitt, D. P., Walker, A. R. P. and Painter, N. S. Effect of dietary fiber on stools and transit times and the role in the causation of disease. *Lancet* **2** (7792), 1408–1412, 1972.

Burkitt, D. P. and Howel, H. C. (eds). *Refined Carbohydrate Foods and disease: Some Implications of Dietary Fiber*. Academic Press, London, 1975.

Davidson, M. H. and McDonald, A. Fiber: forms and functions. *Nutrition Research* **18**, 617–624, 1988.

Diplock, A. T., Aggett, P. J., Ashwell, M., Bornet, F., Fern, E. B. and Roberfroid, M. B. Scientific concepts of functional foods in Europe: consensus document. *British Journal of Nutrition* **81**, Suppl. 1, S1–S27.

Endress, H. U. and Fisher, J. Fibers and fiber blends for individual needs: a physiological and technological approach. In: McCleary, B. V. and Prosky, L. (eds), *Advanced Dietary Fiber Technology*. Blackwell Science Ltd, Oxford, pp. 283–298, 2001.

Englyst, H. N., Kingmann, S. M. and Cummings, J. H. Classification and measurement of nutritionally important starch fractions. *Journal of Clinical Nutrition*, **46**, Suppl. 2, S30–S50, 1992.

Englyst, K. N., Vinoy, S., Englyst, H. N. and Lang, V. Glycemic index of cereal products explained by their content of rapidly and slowly available glucose. *British Journal of Nutrition* **89**, **3**, 329–39, 2003.

Fahey, G. C. Jr., Flickinger, E. A., Grieshop, C. M. and Swanson, K. S. The role of dietary fibre in companion animal nutrition. In: van der Kamp, J. W., Asp, N. G., Miller Jones,

J. and Schaafsma, G. (eds), *Dietary Fibre: Bio-active Carbohydrates for Food and Feed.* Wageningen Academic Publishers, The Netherlands, pp. 295–328, 2004.

FAO/WHO. FAO/WHO Report 66. Carbohydrates in human nutrition. FAO, Rome, 1998.

Gallant, D. J., Bouchet, A. and Parez, S. Physical characterization of starch granules and susceptibility to enzymatic degradation. *European Journal of Clinical Nutrition* **46**, S3–S16, 1992.

Harris, P. J., Tasman-Jones, C., Ferguson, L. R. Effects of two contrasting dietary fibers on starch digestion, short-chain fatty acid production and transit time in rats. *Journal of the Science of Food and Agriculture* **80**, 2089–2095, 2000.

Health Focus International. HealthFocus® Trend Report: The national study of public attitudes and actions toward shopping and eating. Health Focus International®, St Petersburg, FL. 2005.

Inglett, G. E. and Carriere, C. J. Cellulosic fiber gels prepared from cell wall of maize hulls. *Cereal Chemistry* **78**, 471–475, 2001.

Jenkins, D. J. A., Wolever, T. M. S. and Taylor, R. H. Glycemic index of foods: a physiological basis for carbohydrate exchange. *American Journal of Clinical Nutrition* **134**, 362–366, 1981.

Mann, J. Dietary fiber and diabetes revisited. *European Journal of Clinical Nutrition* **55**, 919–921, 2001.

Mehta, R. S. Dietary fiber benefits. *Cereal Foods World* **50**, 66–71, 2005.

Monro J. Expressing glycemic potency of foods. *Proceedings of the Nutrition Society* **64**, 115–122, 2005.

Nelson, A. L. Properties of high fiber ingredients. *Cereal Foods World* **46**, 93–97, 2001.

O'Dea, K., Snow, P. and Nestle, P. Rate of starch hydrolysis in vitro as predictor of metabolic responses to complex carbohydrates in vivo. *American Journal of Clinical Nutrition* **34**, 1991–1993, 1981.

Prochaska, L. J., Nguyen, X. T., Donat, N. and Piekutowski, W. V. Effect of food processing on the thermodynamic and nutritive value of foods: literature and database survey. *Medical Hypotheses* **54**, 254–262, 2000.

Salmeron, J., Ascherio, A., Rimm, E. B., Colditz, G. A. *et al.* Dietary fiber, glycemic load, and risk of NIDDM in men. *Diabetes Care* **20**, 545–550, 1997.

Schaafsma, G. Health claims, options for dietary fibre. In: van der Kamp, J. W., Asp, N. G., Miller. Jones, J. and Schaafsma, G. (eds), *Dietary Fibre: Bio-Active Carbohydrates for Food and Feed.* Wageningen Academic Publishers, The Netherlands, 2004.

Thorne M. J., Nicholas, P. L. and Thornburn, A. W. Factors affecting starch digestibility and the glycemic response with special reference to legumes. *American Journal of Clinical Nutrition* **38**, 481–488, 1983.

Tungland, B. C. and Meyer, D. Non-digestible oligo and polysaccharides (dietary fibers): their physiology and role in human health and food. *Comprehensive Reviews in Food Science and Food Safety* **1**, 73–92, 2002.

22

Methods to slow starch digestion rate in functional cereal products

G. Zhang, M. Venkatachalam and B. R. Hamaker, Purdue University, USA

22.1 Introduction

Glycemic index (GI), measured as the area under the curve (AUC) of postprandial glycemic response for a given amount of available carbohydrate in a test food compared with a reference food (usually glucose or white bread) (Wolever *et al.*, 1991), is a quantitative measurement of dietary carbohydrate quality (Jenkins *et al.*, 1981). Foods with a high GI (GI > 70), or high glycemic load (a value that takes into account the amount of such carbohydrate consumed in a diet), produce a high glycemic response which may relate to a number of disease and pre-disease conditions such as diabetes and pre-diabetes, cardiovascular disease, cancer, and obesity (Ludwig, 2002; Brand-Miller, 2003; Aston, 2006).

The range of GI values observed from a GI table published by Foster-Powell *et al.* (2002) indicates different degrees of carbohydrate digestion and absorption of various foods in the human digestive system, although most industrialized food products belong to the category of high-GI foods. Starch, being the principal component in cereals and tubers, is the major carbohydrate related to postprandial glycemia and GI of foods. However, depending on the food source, other carbohydrates also contribute to GI, such as sucrose, lactose, high-fructose corn syrup, and maltodextrins. Carbohydrates modified for lower GI would be desirable as ingredients in functional cereal and other products for certain groups (e.g. diabetics) and possibly for other improvements in general health. The current understanding of GI, particularly in relation to slowly digestible cereal starches – in terms of their structure and manufacturing approaches, is the focus of this chapter.

From a nutritional standpoint, Englyst *et al.* (1992) classified starch into three categories based on digestion time in their *in vitro* digestion assay: rapidly digestible starch (RDS), 20 min; slowly digestible starch (SDS), between 20 and 120 min; and resistant starch (RS), undigested at 120 min. Foods with a high proportion of RDS have a high GI value based on a correlative relationship between RDS and GI (Englyst *et al.*, 1996). The rapid increase in blood glucose level from RDS triggers the secretion of insulin from pancreas β-cells in order to promote glucose uptake by muscle and adipose tissues to maintain blood glucose homeostasis; if the increase in postprandial glycemia is pronounced, a hypoglycemic episode is often generated between 1 and 2 hours after consumption of RDS. In contrast, SDS is that portion of starch that is digested more slowly than RDS, implying that it is digested throughout the small intestine to provide a slow and prolonged release of glucose. Such a moderated and controlled release of glucose may place less stress on the blood glucose regulatory system. RS is that portion of starch not digested in the small intestine. However, it is digested by colonic microflora enzymes and then fermented to produce short-chain fatty acids (acetic, propionic, and butyric) that have been shown to be beneficial to colonic health (Bird *et al.*, 2000). Thus, the nutritional quality of starch-based foods is attributed to their relative amounts of RDS, SDS, and RS. A low-GI starch ingredient should contain lower amounts of RDS and a higher proportion of SDS and perhaps RS.

The opportunity to alter the GI of cereal foods, and specifically the SDS property, can be viewed from a number of perspectives. The most obvious is the form in which the food is eaten. A simple example would be that white bread made from refined wheat flour is rapidly digested and absorbed, perhaps even more so than glucose itself, while the same flour made into pasta is digested at a much slower rate. This difference can be attributed to the 'food form', where the former example has starch that is readily accessible to digestive enzymes, and the latter is less accessible due to a combination of a protein matrix effect and high density of the food. The GI, and slow digestion property, of starchy foods also changes depending on food processing conditions (e.g. degree of starch gelatinization related to temperature and water content, shear, cooling rate, and time) and storage (factors related to starch retrogradation) (Björck *et al.*, 2000; Fernandes *et al.*, 2005). Starch, being composed of the macromolecules of amylose and amylopectin, can also be altered structurally to moderate its digestion rate. In this regard, changes to its molecular architecture at various stages of food production can affect its susceptibility to digestive enzymes. Additionally, other non-starch components in the food can lower starch digestion rate by influencing gastric emptying and luminal viscosity, or can inhibit amylolytic enzymes, all of which affect postprandial glycemia.

Based on the current *Dietary Guidelines for Americans* (USDA, 2005), dietary carbohydrates should provide 45–65% of the total calories. Dietary carbohydrates, structurally, can be divided into human enzyme-digestible monosaccharides, disaccharides, oligosaccharides, and polysaccharides (i.e. starch, maltodextrins), and the indigestible – though potentially fermentable – dietary fiber constituents including some oligosaccharides (e.g. inulin) and non-starch polysaccharides (such as pectins, arabinoxylans, cellulose). Although SDS have a moderated glycemic response and may provide extended energy release, the majority of starches in cooked and processed foods are rapidly digested leading to a high postprandial glycemia. Thus, a challenge facing researchers in the public and private sectors is to create such SDS with low GI either as ingredients for typically high-GI processed cereal foods or new functional cereal products, or, through genetic breeding, to be naturally present in cereal grains. Some examples and basic theories will be discussed in this section.

Lowering the amount of RDS and increasing the proportion of SDS and RS, particularly SDS, is the fundamental basis for a starch-based, low-GI cereal food. RDS has no special structural component associated with its fast digestion and generally can be viewed as readily accessible to starch-degrading enzymes. RS, on the other hand, is resistant to digestion due to physical inaccesibility, chemical modifications or its highly ordered crystalline structure. The starch structure or other factors leading to SDS, however, are not well understood. The complexity of factors leading to the creation of the 'SDS state', that is the food form wherein starch has a slow digestion property, is shown in Fig. 22.1. Further investigation of the requirements of the SDS state is needed for its better understanding and preparation. This is particularly so because SDS, compared with RS, is more related to the glycemic response as it alone both reduces the initial postprandial glycemic peak and provides extended glucose release over time, for perhaps the optimal glycemic response.

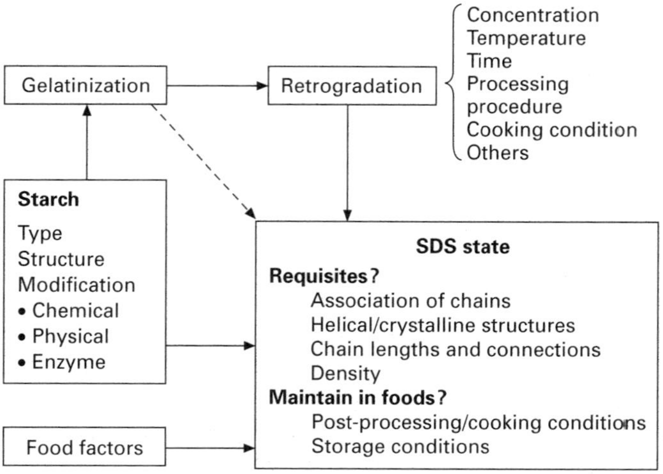

Fig. 22.1 Factors leading to the slowly digestible starch (SDS) state.

22.2 Slowly digestible and resistant characteristics of raw starches

22.2.1 Native starches

Native starch naturally exists in the form of starch granules with different shapes, sizes, and surface properties. X-ray diffraction can be used to classify starches into A-, B-, and C-(combination of A and B) types based on the packing of the semicrystalline structures of starch granules. Most cereal starches (maize, wheat, rice) are of the A-type, tuber starches (potato, yam) are B-type, and legume starches (kidney bean, soybean) normally belong to the C-type. Some A-type starches (maize, sorghum, millets, and large granules of wheat, rye, and barley at the equatorial groove) have surface pores connected to interior cavities through channels (Fannon *et al.*, 1992). There are no such surface pores in B-type starch granules. This macrostructural difference between native starch granules is important during digestion as starch-digesting enzymes enter the channels and digest them in an inside-out manner. As far as the nutritional quality of starch is concerned, native A-type starches inherently have a high amount of SDS while native B-type starches are essentially resistant (high RS) (Ferguson *et al.*, 2000), based on the *in vitro* Englyst assay (Englyst *et al.*, 1992). In an *in vivo* study, Seal *et al.* (2003) clearly showed the changes of blood glucose levels after consumption of native maize starch (Fig. 22.2), producing a typical profile of SDS. There have been two patents (Axelsen and Smith, 2001; Qi and Tester, 2005) on SDS in which native A-type starches were used as the sole ingredient in the preparation of medical foods for treatment of glycogen storage disease (GSD1) (Chen *et al.*, 1984) and other diabetic conditions.

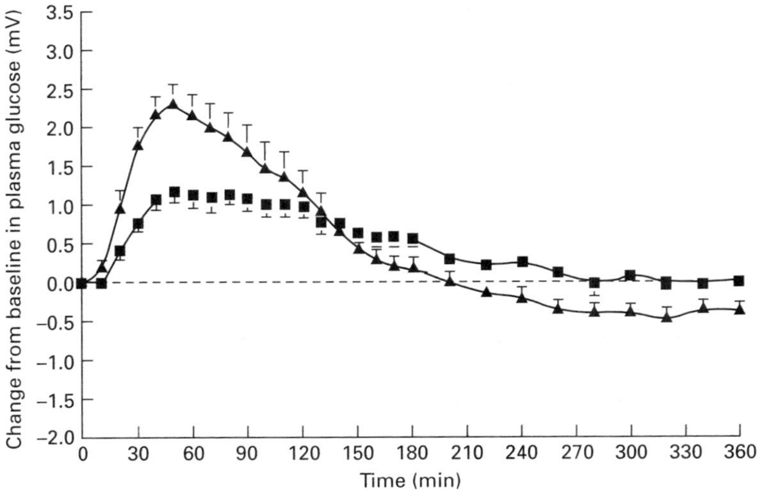

Fig. 22.2 Change from baseline of plasma glucose concentrations with time after consumption of 50 g rapidly hydrolysed (▲) and 50 g slowly hydrolysed (■) native maize starch in healthy subjects. From Seal *et al.* (2003).

Our own investigation of native starch digestion properties shows that A-type native maize starch is a near ideal SDS source (Zhang *et al.*, 2006a,b). This is because native granules, at any given time point of digestion in the Englyst assay, provide similar RDS, SDS, and RS amounts, and thus continue to provide slow release of glucose over an extended period of time. It is well known that the semicrystalline structure of starch granules is composed of alternating concentric layers of ordered and dense crystalline regions and less ordered amorphous regions (also called amorphous background) from the hilum to the surface of the granules. Based on our investigation (Zhang *et al.*, 2006a), the surface pores and interior channels are the starting points for enzymic digestion, and gradually the amorphous background and dense crystalline regions are evenly digested by enlarging the interior channels through a side-by-side digestion mechanism from the inside of the granule to the outside. This is the mechanism for the slow digestion property of A-type maize starch and could conceivably be replicated in other food materials to create a similar digestion effect. B-type starches have a somewhat similar granular structure with different crystalline and amorphous arrangements, but there are no surface pores or interior cavities. Thus, digestion occurs through pitting from the outside of the starch granules. Additionally, B-type crystallites are somewhat resistant to digestion compared with A-type crystallites (Gérard *et al.*, 2000). Therefore B-type native starch is largely RS (~70%) due to the non-porous surface and the nature of the B-type crystallites.

22.2.2 Hydrothermal modification

Hydrothermal modification is one of the commonly used physical methods to modify the functional properties of starch granules while maintaining granular structure. Three parameters are varied in hydrothermal treatment of starch granules: temperature, moisture, and time. Hydrothermal treatments can be divided into the two general areas of annealing and heat-moisture treatment. Annealing is usually performed at excess (>66%) or intermediate water contents (40–55% w/w), while heat–moisture treatment is defined for low moisture conditions (<35% w/w). The temperature range used is generally between the glass transition temperature (T_g), the transition point between the glassy and rubbery state of starch, and the gelatinization temperature at which irreversible loss in crystallinity occurs. The time can be varied from hours to a week. Detailed conditions for a variety of starches can be found in the review by Jacobs and Delcour (1998), or can be determined based on the T_g and gelatinization temperature of a specific starch sample.

Hydrothermal treatment results in changes in the crystalline and amorphous regions of starch granules as well as some molecular interaction between these two regions. It results in an increase in the perfection of crystallites, alteration of crystalline packing from B- or C-type to A-type, and increased interaction of molecules within the amorphous and crystalline regions.

Hydrothermal treatment can be used as a way to increase the slow digestion property of native starch granules resulting in lower GI. Shin *et al.* (2005) reported that hydrothermal treatment of sweet potato converted its C-type structure to A-type, and the SDS content of the treated granule increased by 200% compared with native starch granules. Jolly-Zarrouk *et al.* (2006) developed an extended energy beverage by incorporating slowly digestible potato starch (32% SDS) made by hydrothermal treatment (32% moisture, 105 °C). Due to the broad range of physicochemical changes caused by hydrothermal treatment, more studies are needed to investigate systematically the effect of hydrothermal treatment on the slow digestion property and related low glycemic response of starches.

22.3 Starch structural modification

The structure, and perhaps size, of starch molecules can fundamentally affect digestion properties in cooked and processed foods. This can be viewed in terms of both inherent starch structural properties and retrogradation-related structural differences (double helical structures and crystallites) that affect enzyme binding and rate of digestion. Therefore, an understanding of starch structure is critical for the moderation of glycemic response and in order to manufacture starch ingredients with enhanced health value for the food industry. Starch is composed of macromolecules of essentially linear amylose comprising α-1,4-linked D-glucopyranosyl units and the highly branched amylopectin (often > 1 million glucose units) in which linear α-1,4-linked D-glucopyranosyl chains are joined through α-1,6 linkages. Starch structural modification therefore can be viewed as a key strategy to achieve SDS. Structural modification of starch molecules can be achieved by genetic, enzymatic, physical, and chemical means.

22.3.1 Amylose/amylopectin ratio

The amylose to amylopectin ratio is one of the main parameters measured in starch quality evaluation. In diluted starch solution (< 1% w/v) containing both amylose and amylopectin, amylose can be easily digested by α-amylase, while amylopectin is digested more slowly due to its branching structure (Park and Rollings, 1994); however, in a concentrated starch system, especially with a higher ratio of amylose/amylopectin, a firm gel will form and digestibility is decreased due to the large amount of retrograded starch formed by amylose; much, if not most, of the highly retrograded amylose is RS. Studies have shown that the amylose/amylopectin ratio is an important determinant of starch quality and amylose content is highly correlated with RS content. Thus, foods containing a substantial amount of high-amylose starches generally produce a moderate glycemic response due to their significant amounts of RS (Miller *et al.*, 1992).

The amount of amylose and amylopectin in starches is genetically controlled through the enzymes involved in starch biosynthesis. Traditional breeding has been successfully used to produce starches (especially in maize) with a wide range of amylose to amylopectin ratios, such as waxy starch (essentially pure amylopectin) and high-amylose starches (50–70% amylose). Today, the ratio of amylose to amylopectin can be manipulated through modern genetic techniques, such as genetic engineering, to up- or down-regulate genes related to amylose and amylopectin synthesis. Modern techniques such as transposon insertion, site-directed mutagenesis, antisense inactivation, and TILLING (targeting induced local lesions in genomes) of mutagenized populations – as well as producing transgenic lines with the addition of foreign genes – are being used to produce starch mutants with the desired properties. High-amylose starches are an ideal starting material to make heat-stable RS (Brown, 2004) to decrease the postprandial glycemic response of processed foods. Conceivably, novel mutants could cause changes in the amount of amylose, and branching structure of amylopectin to produce starches with higher amounts of SDS.

22.3.2 Amylopectin fine structure

Amylopectin is the major component, by weight, of normal starch granules (~75%), and the cluster model (Fig. 22.3) is the widely accepted model used

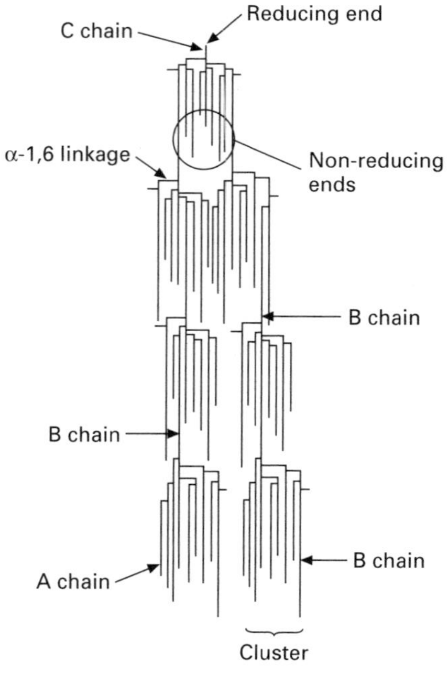

Fig. 22.3 The cluster model of amylopectin structure including A, B, and C chains. From Manners (1989).

to describe its fine structure. The branched chains are classified as A chains (no further branching point), B chains (with one or more branching points), and the C chain – the only chain with a reducing end. The fine structure of amylopectin mainly refers to the chain length distribution, the branching density, internal chain lengths, and cluster repeating distance. The relationship between amylopectin fine structure and starch functionality has been well studied, but its effect on the nutritional quality of starch has not been as well investigated. Recent research in our laboratory (unpublished data, 2006) shows that both high amounts of short chains (principally the short A chains) and high amounts of long chains (intermediate to long B chains) tend to result in more SDS, as revealed in the parabolic relationship between the weight ratio of the short-chain fraction (degree of polymerization, DP < 13) to the long-chain fraction (DP ≥ 13) and SDS amount (Fig. 22.4). Mechanistically, the molecular structure of amylopectin with more short chains (high branch density) determines the comparably high SDS amount, while starch retrogradation is related to the slow digestion property of amylopectin with a greater proportion of longer chains. Another of our studies (Benmoussa *et al.*, 2007) supports the view that a comparably low concentration of very short linear chains increases retrogradation as well as SDS. Consistently, Shi and Seib (1995) showed that B1 chains (with DP 16–30) are more associated with starch retrogradation and RS content. The structure of amylopectin is more critical than amylose for SDS, as the latter is closely related to type III RS in processed foods.

The fine structure of amylopectin is regulated by soluble starch synthase, branching enzyme isoforms, and debranching enzymes. Variations in activities, or mutations, in these enzymes can produce a range of amylopectin with different fine structures. If the relationship between nutritional quality of

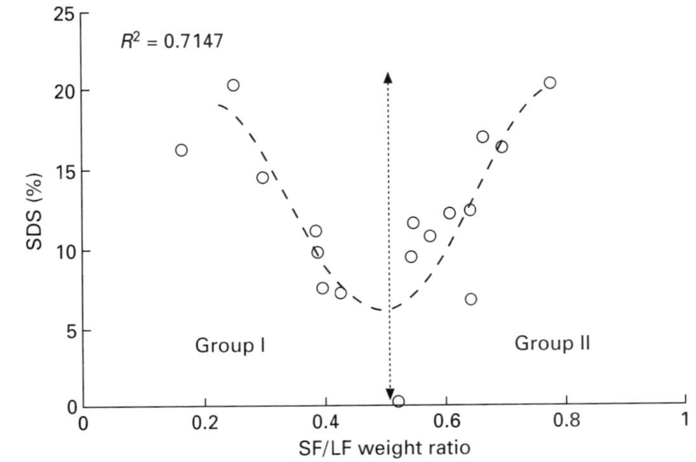

Fig. 22.4 A parabolic relationship between slowly digestible starch percentage and the weight ratio of the amylopectin short chain (SF) to long chain fraction (LF).

starch and amylopectin fine structure (distribution patterns of branches and chain length) is thoroughly understood, genetic engineering or other manipulation methods could be used to produce improved starches with high-SDS and low-GI properties.

22.3.3 Enzyme modification

The molecular structure of amylose and amylopectin is one of the essential intrinsic factors associated with the nutritional quality of starch, as described above. Enzyme modification is an alternative way to change the structure of starch molecules to achieve appropriate digestion or glycemic properties. Theoretically speaking, all the enzymes involved in starch biosynthesis and degradation can be used to modify starch structure, but in practice only commercially available enzymes are used – such as α-amylases, β-amylases, debranching enzymes (isoamylase and pullulanase), transglucosidases, and glycosyltransferases (CGTase). These enzymes can decrease the molecular weight of starch molecules, remove outer chains of amylopectin, or add new structures to existing chains. The purpose of using enzyme modification to affect GI is to obtain an optimum structure that will have low-GI properties through an increase in SDS. The approach may be counter to the commonly known enzyme modifications used to increase starch functionality – such as texture, gel clarity, and pasting properties. Therefore, better definition of the starch molecular structure needed to produce SDS is still of importance in order to enzymically modify starch for improved glycemic properties.

It has been shown that amylose with DP ~ 100 rather easily forms RS (Eerlingen *et al.*, 1993), and amylopectin hinders the association of amylose to form RS (Berry, 1986). RS can be used to decrease the glycemic response of a food; however, it does not, by definition, provide slow and prolonged release of glucose. Starch can be modified using pullulanase or isoamylase, and α-amylase, to debranch amylopectin and shorten amylose, respectively, so that the RS formation can be increased significantly. Related to this, several forms of enzyme-modified RS have been patented and used commercially (National Starch (Chiu *et al.*, 1994; Shi and Trzasko, 1997) and Roquette, France (Nutriose – a new dextrin)).

For SDS, partially debranched waxy starch using pullulanase has been used to make SDS (Shin *et al.*, 2004). Liu *et al.* (2006) filed a patent on a 'slowly digestible carbohydrate', or oligosaccharide, obtained during starch saccharification. Han *et al.* (2006) recently reported a method for producing a starch with some SDS properties, along with RS, using α-amylase digestion of partially retrograded normal maize starch. In a recent investigation in our laboratory (Ao *et al.*, 2007), a combination of maltogenic α-amylase, β-amylase, and transglucosidase treatment of normal corn starch was used to form starch with an increased proportion of SDS and RS, which can decrease postprandial glycemia. These studies suggest that the structure of SDS could be imperfect crystallites, and amylopectin with high branch density or extra long chains might be the structural basis for forming an imperfect and entangled structure with substantial density or juncture zones to slow down enzyme digestion.

22.3.4 Other physical modification (retrogradation)

When starch granules are cooked during food processing, the crystalline structure is lost and amylose and amylopectin are dispersed to a certain extent (gelatinization) depending on water content, temperature, time, and shear of the process. In order to produce low-GI foods in this situation, physical modification (i.e. starch concentration, processing condition, and storage method) might be used to moderate starch digestion properties. Modifications of starch gelatinization and retrogradation are the main categories of physical modification used to achieve low-GI benefits in cooked and processed starchy foods.

As described above, native starch (especially A-type) is an ideal SDS, and if incomplete gelatinization can be achieved through lowering temperature, decreasing water content, or shortening processing time, the nutritional benefit of SDS could be partially retained. However, the organoleptic properties may not be suitable for consumption and this needs to be studied further. The effects of hydrothermal modification on the nutritional properties of starch may work in this situation to achieve a low-GI food product.

Extensive research has shown that water content, temperature of storage, and the existence of other ingredients affect starch retrogradation. Most of the RS in food is produced through retrogradation, particularly by amylose retrogradation which occurs very rapidly. Alternatively, controlled retrogradation could probably be used to alter the amounts of SDS. Shi *et al.* (2003) and Shin *et al.* (2004) showed that less retrogradation with partially debranched starch ingredients makes it possible to produce SDS. Perhaps SDS, in these cases, is due to the formation of imperfect crystallites.

22.3.5 Chemical modification

In order to improve functionality and create value-added starch-based products, native starches are additionally subjected to various chemical modifications to overcome some of the structural and rheological problems that they inherently have during food processing and storage. These include: loss in viscosity; acid, shear, and/or heat stability; pasting; thickening; syneresis; and retrogradation. Although most chemical treatments have intended to improve the functionality of the starches in foods, some recent work has focused on treatments for creating SDS and RS starches. Many chemical modifications essentially result in the creation of RS (Wolf *et al.*, 1999). Chemically modified starch derivatives such as citrate starches (Wepner *et al.*, 1999) and cross-linked starches (Woo and Seib, 2002) have been shown to decrease starch digestion rate depending on the type and degree of modification, and the extent of gelatinization. Han and BeMiller (2007) demonstrated high SDS contents in 2-octen-1-ylsuccinic anhydride (OSA)-esterified waxy starch, and relatively high SDS and RS contents in cross-linked hydroxypropylated and acetylated waxy starches. Although human trials of OSA-modified starch showed a noticeable decrease of glycemic response (Wolf *et al.*, 2001), a toxicological study is needed to ensure the safety and health benefits following long-term and elevated consumption. On the other hand, the elucidation of the mechanism for the slow digestion properties of these chemically modified starches may shed some insight on novel ways to make safe and low-GI starch materials. Recently, Shin *et al.*, (2007) showed that citric acid treatment of rice starch at a high temperature (128.4°C) forms cross-links that substantially increased the content of SDS and RS, and reduced the glycemic response as effectively as a commercial RS. The acid treatment also showed decreased blue value (iodine binding ability) and viscosity as well as a reduced gel-forming ability.

22.4 Influence of other food components on starch digestion rate

The presence of other major and minor food components besides starch, has also been shown to influence the glycemic response of carbohydrates in foods. Various factors at both the macroscopic and microscopic levels of foods can affect digestion variables such as decrease in gastric emptying, creation of matrix barriers to access starch during digestion, and interactions between digestive enzymes and other food components.

22.4.1 Protein–starch interaction

Extruded pasta products represent an excellent example of the effect of proteins in slowing starch digestion rate. Several studies have demonstrated that digestion of pasta in both healthy and diabetic subjects is characterized by low glycemic and insulinemic responses (Parillo *et al.*, 1985; Wolever *et al.*, 1986; Granfeldt and Björck, 1991). *In vitro* digestion studies on pasta (Colonna *et al.*, 1990; Fardet *et al.*, 1998, 1999) have shown that restricted swelling and the entrapment or encapsulation of starch in structured protein network-associated dense food results in decreased enzyme accessibility, and hence lowers its digestionrate and extent. Furthermore, pre-incubation of pasta with pepsin (proteolytic enzyme) enhanced starch digestibility, thereby providing support for the importance of protein components in slowing starch digestion. Even in gelatinized flour pastes in cereals such as sorghum and maize, entrapment of starch in protein web-like structures has been observed to impede starch digestibility (Bugusu, 2003).

22.4.2 Lipid complexation with starch

The complexation of amylose with the free fatty acids and monoglycerides in foods has been recognized to occur at various stages of food preparation and storage, and even during the digestion process. Such a complex not only results in significant changes in the physicochemical and functional properties of starch – such as a change in the starch X-ray diffraction pattern to 'V' type, reduced solubility, increased gelatinization temperature, and retarded retrogradation during storage (Eliasson *et al.*, 1981; Biliaderis and Galloway, 1989) – but has also been shown to be less digestible in various *in vitro* and *in vivo* models (Holm *et al.*, 1983; Seneviratne and Biliaderis, 1991; Murray *et al.*, 1998). A greater resistance to digestion of amylose–lipid complexes was observed when amylose was complexed with long-chain, saturated monoglycerides compared with complexes with shorter chain, unsaturated monoglycerides (Eliasson and Krog, 1985). Murray *et al.* (1998) incorporated a mix of monostearate and monopalmitate complexed with debranched amylopectin to manufacture the V-complex in an experimental diet in dogs and compared their digestion to control and RS. The authors found that the ileal and total tract digestibilities of carbohydrate for the V-complex treatment were intermediate (digestibilities were ranked as control > V-complex > RS). The same group of authors also reported that dogs that consumed a V-complex-containing diet had lower carbohydrate digestibility and subsequently lower serum glucose and insulin responses than dogs fed a maltodextrin-containing control diet (Patil *et al.*, 1998). Further human studies seem to be warranted to show this effect and its physiological consequences.

Apart from the starch–lipid complexation that could modify the glycemic response, several studies have shown that co-intake of fat along with carbohydrates in a mixed meal can affect postprandial glucose response. It is believed that fat may reduce postprandial glucose level by decreasing the rate of gastric emptying, at least in part related to increased stimulation of the gastrointestinal hormones (such as glucose-dependent insulin-releasing polypeptide (GIP) and glucagon-like polypeptide-1 (GLP-1)) (Morgan, 1998). Several issues, including dosage levels of fat affecting the glucose response, have been described by Owen and Wolever (2003). The authors showed that fat intake along with carbohydrates in normal healthy subjects, in a dose–dependent relationship, could decrease the glycemic response; however fat consumption in a normal range (17–44% energy) does not significantly affect glycemic response. As pointed out by Owen and Wolever (2003), it is important to note that individuals with diabetes or insulin resistance should not add fat to carbohydrate meals to prevent high blood glucose surges, because studies have shown that fat addition to carbohydrate does not affect the glycemic responses in subjects with Type-2 diabetes (Gannon *et al.* 1993). Various factors including the type and the amount of fat consumed, the type of the carbohydrate eaten with the fat, and the health status of the subjects consuming the food need to be taken into account when evaluating postprandial glucose and insulinemic response to a mixed carbohydrate–lipid meal.

22.4.3 Dietary fiber

The extent and mechanism by which non-digestible carbohydrates (dietary fiber) in foods influence glycemic response are the subject of much debate. It may relate to the 'second meal' effect on blood glucose and insulin response or a direct effect on starch digestion rate, or may, depending on the type of fiber consumed and other factors, not have any effect on glycemia. Björck and Elmståhl (2003) recommend that low-GI cereal foods should generally be formulated with dietary fiber since they complement each other, and probably have a synergistic effect physiologically. Insoluble fiber may have a role in reducing GI (Wolever, 1990), although a clearer relationship is found with incorporation of soluble fibers (guar gum, psyllium, β-glucans, pectin) (Jenkins *et al.* 1978; Wood *et al.*, 1990; Brennan, 2005). Several studies have shown that, when RS is used in a food product, it not only serves as a source of fermentable fiber but also lowers glycemic response (Björck *et al.*, 2000).

As discussed in earlier chapters, the concept of whole grain foods (breads, pasta, cereals) is gaining momentum and consumption of a number of grains and grain extracts has been reported to control or improve glucose tolerance and reduce insulin resistance (Hallfrisch and Behall, 2000). The structure and composition of the grain – including particle size, amount and type of fiber, viscosity effects, and amylose and amylopectin content – affect the metabolism of carbohydrates from grains. The use of cereal viscous fibers in lowering the glycemic response has been related to reduced gastric emptying (Jenkins *et al.*, 1978; Braaten *et al.*, 1991). For example, a highly viscous β-glucan-containing barley genotype (Prowashonupana) has been demonstrated to lower the glycemic response of breakfast foods (breads, porridges) significantly in healthy and diabetic volunteers (Liljeberg *et al.*, 1996; Rendell *et al.*, 2005). In the future, it may be possible to select suitable cereal genotypes for the preparation of tailor-made foods with defined glycemic and other nutritional attributes based on both starch type and structure, matrix effect, and dietary fiber composition.

22.4.4 Other constituents

The rate of starch digestion is also influenced by several minor plant constituents (usually referred to as anti-nutrients) such as phytates, phenolic compounds (tannins), saponins, lectins, and several enzyme inhibitors. These components interfere with the catalytic activity of the α-glucosidase enzymes through different mechanisms, thereby limiting their action (Thompson, 1988). Fish and Thompson (1991) showed that lectin and tannic acid (from red kidney bean) individually can inhibit the starch digestive enzymes, salivary and pancreatic α-amylases; however, a combination of the two anti-nutrients abolished their inhibitory activity. The authors concluded that the effect of individual anti-nutrients may not necessarily relate to the effects observed upon consumption of mixture of a anti-nutrients as normally observed in foods. Although a majority of these minor components could potentially influence glycemic response, their practical significance has usually been limited by the fact that they are removed or destroyed at various stages of food preparation and consumption.

Recently, there has been a renewed interest in the use of inhibitors of the human α-glucosidases to moderate carbohydrate digestion and its associated insulin response for treatment of non-insulin-dependent (Type-2) diabetes. Acarbose, an oligosaccharide formed by strains of the genus *Actinoplanes*, functions as an inhibitor of both α-amylase and the mucosal α-glucosidases (sucrase-isomaltase and maltase-glucoamylase) (Hiele *et al.*, 1992). It has been effective in the treatment of diabetes as it slows down digestion of disaccharides and starch (Chiasson *et al.*, 1994; Conniff *et al.*, 1994) and is used in some countries to treat diabetes. However, it has frequently been shown to have gastrointestinal side effects due to malabsorption of disaccharides. Starch blockers, purified amylase inhibitors from Great Northern beans (phaseolamin), have also shown promise in glucose homeostatis (Boivin *et al.*, 1988) and are marketed as dietary supplements (Phase 2 Starch Neutralizer™; Udani *et al.*, 2004). Unlike these blockers, which need to be ingested in large quantities (4–6 g per meal) to show significant effects, trestatin (a mixture of complex oligosaccharides produced microbiologically) has been proven to be a potent inhibitor (3–6 mg per 75 g starch) of pancreatic α-amylase in several *in vitro* and *in vivo* studies (Golay *et al.*, 1991). Glycemic and insulinemic responses in healthy and diabetic volunteers were moderated after consumption of breads that had had trestatin added during processing without serious gastrointestinal side effects. Although the use of enzyme inhibitors has shown promise, further studies are needed to evaluate the efficacy of these inhibitors after addition to starch-based processed foods, the dose–response relationship, and more importantly the long-term tolerance and side effects of such inhibitors.

The presence of organic acids or acid salts, such as those produced during sourdough fermentation or added during baked food preparation, has been seen to influence glycemic and insulinemic responses (Liljeberg *et al.*, 1995; Liljeberg and Björck, 1996, 1998). For example, consumption of sourdough bread (with lactic acid produced during fermentation) or breads with added calcium lactate, or sodium propionate, significantly reduced glycemic and insulinemic indexes compared with wholemeal bread in the absence of these acids (Liljeberg *et al.*, 1995). Intake of bread with a high concentration of sodium propionate not only lowered postprandial blood glucose and insulin responses, but also significantly prolonged the duration of satiety compared with all other breads. *In vitro* digestion of these breads, however, only showed a significant decrease in the rate of amylase digestion in bread containing lactic acid, probably by decreasing the pH of the reaction system. The authors concluded that the effect of acid salts such as sodium propionate on metabolic responses and satiety was due to effects other than starch hydrolysis, such as delayed gastric emptying (Liljeberg and Björck, 1996). Similar effects were observed upon addition of vinegar to a starchy meal (Liljeberg and Björck, 1998). The potential of fermentative processes, or processes that incorporate organic acids, to improve the nutritional features of carbohydrates needs to be considered.

Recent studies in our laboratory (unpublished data, 2006) show that entrapment of starch in an alginate–calcium ion biopolymer matrix effectively creates a barrier (cooked in the entrapped form) to digestion by amylases and provides a slow glucose release profile. Scanning electron microscope images of the cooked starch microspheres showed that the gelatinized starch trapped in the biopolymer matrix represents a highly dense food form that is gradually digested by the amylases from the periphery towards the center of the sphere. Various factors – including biopolymer type (alginate or blend of alginate with other polymers, such as gellan gum, chitosan, or carrageenan), biopolymer concentration, microsphere size, and calcium ion concentration – could be used to obtain biopolymer-entrapped starches with the desired slow digestion profiles. Such microspheres not only lowered glycemic response (as indicated by initial clinical studies), but may also serve as novel starch ingredients providing extended release of glucose in food products.

22.5 Slowly digestible starch and low glycemic index cereal foods: future trends

Strategies for producing slowly digesting/low-GI functional cereal-based foods could include creating proper food forms and incorporating particular ingredients into foods such as: starches with the slow digesting property that extend glucose release, some non-digesting carbohydrates (dietary fiber, RS) including viscosity-increasing polysaccharides that delay gastric emptying or decrease digestive enzyme access, organic acids and their salts to slow down gastric emptying, or anti-nutritional agents that inhibit digestion of starch, and other low-glycemic carbohydrates. Most importantly, human trials must show what levels of such carbohydrate-based ingredients or foods are needed to elicit a positive physiological or metabolic response.

Food form will continue to be a major issue in the development of slow digesting cereal foods. Highly organized, dense food forms impede starch digestion and, thereby, lower the glycemic response of starchy foods. An organized food form could simply preserve the crystalline order of starch (prevent complete gelatinization) during food processing or provide a barrier to digestive enzymes. Besides pasta (described under protein–starch interactions, Section 22.4.1), whole kernel foods, wherein the cellular layers surrounding the starch granules are intact, also present an example of an organized food form with low GI and probable slow digestion property. Whole wheat flour bread, in which some of the grain structure persists, has been reported to provide a lower glycemic response than white bread (Liljeberg *et al.*, 1992). Similarly, legumes cooked under mild heat processing conditions (such as boiling), that had an intact cellular structure, had lower GI compared with legumes cooked under harsher conditions (pressure cooking) or milled prior to cooking (Golay *et al.*, 1986; Wolever *et al.*, 1987; Tovar *et al.*, 1992). Beverages, from the point of view of providing extended energy release, represent a special challenge that some companies are addressing, because soluble glucose-containing oligosaccharides, maltodextrins, and starches tend to be rapidly digested. There is a need for a more systematic research approach to understand how glycemic carbohydrates and food form (type of matrix, concentration, cooling profile, storage conditions) influence GI and glucose release profiles, and to establish their potential health and well-being outcomes.

22.6 References

Ao Z, Simsek S, Zhang G, Venkatachalam M, Reuhs B, Hamaker B R (2007) 'Starch with slow digestion property produced by altering branch density, chain length and crystalline structure', *J Agric Food Chem*, **55**, 4540–4547.

Aston L M (2006) 'Glycaemic index and metabolic disease risk', *Proc Nutr Soc*, **65**, 125–134.

Axelsen M, Smith U (2001) 'Treatment for diabetes', US Patent 6,316,427.

Benmoussa M, Moldenhauer K A K, Hamaker B R (2007) 'Rice amylopectin fine structure variability affects starch digestion properties', *J Agric Food Chem*, **55**, 1475–1479.

Berry C S (1986) 'Resistant starch formation and measurement of starch that survives exhaustive digestion with amylolytic enzymes during the determination of dietary fiber', *J Cereal Sci*, **4**, 301–314.

Biliaderis C G, Galloway G (1989) 'Crystallization behavior of amylose-V complexes: structure–property relationships', *Carbohydr Res*, **189**, 31–48.

Bird A R, Brown I L, Topping D L (2000) 'Starches, resistant starches, the gut microflora and human health', *Curr Issues Intest Microbiol*, **1**, 25–37.

Björck I, Elmståhl H L (2003) 'The glycaemic index: importance of dietary fibre and other food properties', *Proc Nutr Soc*, **62**, 201–206.

Björck I, Liljeberg H, Ostman E (2000) 'Low glycaemic-index foods', *Br J Nutr*, **83**, S149–155.

Boivin M, Flourie B, Rizza R A, Go V L, DiMagno E P (1988) 'Gastrointestinal and metabolic effects of amylase inhibition in diabetics', *Gastroenterology*, **94**, 387–394.

Braaten J T, Wood P, Scott F W, Riedel K D, Poste L, Collins W (1991) 'Oat gum lowers glucose and insulin after an oral glucose load', *Am J Clin Nutr*, **53**, 1425–1430.

Brand-Miller J C (2003) 'Glycemic load and chronic disease', *Nutr Rev*, **61**, S49–55.

Brennan C A (2005) 'Dietary fibre, glycaemic response, and diabetes', *Mol Nutr Food Res*, **49**, 560–570.

Brown I L (2004) 'Applications and uses of resistant starch', *J AOAC Int*, **87**, 727–732.

Bugusu B A (2003) 'Understanding the basis of the slow starch digestion characteristic of sorghum porridges and how to manipulate starch digestion rate', PhD Thesis, Purdue University, West Lafayette, Indiana, USA.

Chen Y T, Cornbalth M, Sidbury J B (1984), 'Corn starch therapy in type I glycogen-storage disease', *N Engl J Med*, **310**, 171–175.

Chiasson J L, Josse R G, Hunt J A, Palmason C, Rodger N W, Ross S A, Ryan E A, Tan MH, Wolever T M (1994) 'The efficacy of acarbose in the treatment of patients with non-insulin-dependent diabetes mellitus. A multicenter controlled clinical trial', *Ann Intern Med*, **121**, 928–935.

Chiu C W, Henley M, Altieri P (1994) 'Process for making amylase resistant starch from high amylose starch', US Patent 5,281,276, National Starch and Chemical Investment Holding Corp.

Colonna P, Barry J L, Cloarec D, Bornet F, Gouilloud S, Galmiche J P (1990) 'Enzymic susceptibility of starch from pasta', *J Cereal Sci*, **11**, 59–70.

Conniff R F, Shapiro J A, Seaton T B (1994) 'Long-term efficacy and safety of acarbose in the treatment of obese subjects with non-insulin-dependent diabetes mellitus', *Arch Intern Med*, **154**, 2442–2448.

Eerlingen R C, Deceuninck M, Delcour J A (1993) 'Enzyme-resistant starch. II. Influence of amylose chain length on resistant starch formation', *Cereal Chem*, **70**, 345–350.

Eliasson A-C, Krog N (1985) 'Physical properties of amylose–monoglyceride complexes', *J Cereal Sci*, **3**, 239–248.

Eliasson A-C, Carlson T L-G, Larsson K, Miezis Y (1981) 'Some effects of starch lipids on the thermal and rheological properties of wheat starch', *Starch/Stärke*, **33**, 130–134.

Englyst H N, Kingman S M, Cummings J H (1992) 'Classification and measurement of nutritionally important starch fractions', *Eur J Clin Nutr*, **46**, S33–50.

Englyst H N, Veenstra J, Hudson G J (1996) 'Measurement of rapidly available glucose (RAG) in plant foods: a potential in vitro predictor of the glycaemic response', *Br J Nutr*, **75**, 327–337.

Fannon J E, Hauber R J, BeMiller J N (1992) 'Surface pores of starch granules', *Cereal Chem*, **69**, 284–288.

Fardet A, Hoebler C B, Baldwin P M, Bouchet B, Gallant D J, Barry J L (1998) 'Involvement of the protein network in the in vitro degradation of starch from spaghetti and lasagna: a microscopic and enzymatic study', *J Cereal Sci*, **27**, 133–145.

Fardet A, Abecassis J, Hoebler C, Baldwin P M, Buleon A, Berot S, Barry J L, Ferguson L R (1999) 'Influence of technological modifications of the protein network from pasta on in vitro starch degradation', *J Cereal Sci*, **30**, 133–145.

Ferguson L R, Tasman-Jones C, Englyst H, Harris P J (2000) 'Comparative effects of three resistant starch preparations on transit time and short-chain fatty acid production in rats', *Nutr Cancer*, **36**, 230–237.

Fernandes G, Velangi A, Wolever T M S (2005) 'Glycemic index of potatoes commonly consumed in North America', *Am Diet Assoc*, **105**, 557–562.

Fish B C, Thompson L U (1991) 'Lectin-tannin interactions and their influence on pancreatic amylase activity and starch digestibility', *J Agric Food Chem*, **39**, 727–731.

Foster-Powell K, Holt S H, Brand-Miller J C (2002) 'International table of glycemic index and glycemic load values', *Am J Clin Nutr*, **76**, 55–56.

Gannon M C, Ercan N, Westphal S A, Nuttall F Q (1993) 'Effect of added fat on plasma glucose and insulin response to ingested potato in individuals with NIDDM', *Diabetes Care*, **16**, 874–880.

Gérard C, Planchot V, Colonna P, Bertoft E (2000) 'Relationship between branching density and crystalline structure of A- and B-type maize mutant starches', *Carbohydr Res*, **326**, 130–144.

Golay A, Coulston A, Hollenbeck C B, Kaiser L L, Wursch P, Reaven G M (1986) 'Comparison of metabolic effects of white beans processed into two different physical forms', *Diabetes Care*, **9**, 260–266.

Golay A, Schneider H, Temler E, Felber J P (1991) 'Effect of trestatin, an amylase inhibitor, incorporated into bread, on glycemic responses in normal and diabetic patients', *Am J Clin Nutr*, **53**, 61–65.

Granfeldt Y, Björck I (1991) 'Glycemic response to starch in pasta: a study of mechanism of limited enzyme availability', *J Cereal Sci*, **14**, 47–61.

Hallfrisch J, Behall K M (2000) 'Mechanisms of the effects of grains on insulin and glucose responses', *J Am Coll Nutr*, **19**, 320S–325S.

Han J-A, BeMiller J N (2007) 'Preparation and physical characteristics of slowly digesting modified food starches', *Carbohydr Polym*, **67**, 366–374.

Han X Z, Ao Z, Janaswamy S, Jane J L, Chandrasekaran R, Hamaker B R (2006) 'Development of a low glycemic maize starch: preparation and characterization', *Biomacromolecules,* **7**, 1162–1168.

Hiele M, Ghoos Y, Rutgeerts P, Vantrappen G (1992) 'Effects of acarbose on starch hydrolysis. Study in healthy subjects, ileostomy patients, and in vitro', *Dig Dis Sci*, **37**, 1057–1064.

Holm J, Björck I, Ostrowska S, Eliasson A C, Asp N G, Larsson K, Lundquist I (1983) 'Digestibility of amylose–lipid complexes in vitro and in vivo', *Starch/Stärke*, **35**, 294–297.

Jacobs H, Delcour J A (1998) 'Hydrothermal modifications of granular starch, with retention of the granular structure – a review', *J Agric Food Chem*, **46**, 2895–2905.

Jenkins D J, Wolever T M, Leeds A R, Gassull M A, Haisman P, Dilawari J, Goff D V, Metz G L, Alberti K G (1978) 'Dietary fibres, fibre analogues, and glucose tolerance: importance of viscosity', *Br Med J*, **1**, 1392–1394.

Jenkins D J, Wolever T M, Taylor R H, Barker H, Fielden H, Baldwin J M, Bowling A C, Newman H C, Jenkins A L, Goff D V (1981) 'Glycemic index of foods: a physiological basis for carbohydrate exchange', *Am J Clin Nutr*, **34**, 362–366.

Jolly-Zarrouk L M B, Fischer A M, Merinat S J, Robin F, Lehmann U (2006) 'Extended energy beverages', Patent WO 2006114191.

Liljeberg H G, Björck I M (1996) 'Delayed gastric emptying rate as a potential mechanism for lowered glycaemia after eating sourdough bread: studies in humans and rats using test products with added organic acids or an organic salt', *Am J Clin Nutr*, **64**, 886–893.

Liljeberg H, Björck I (1998) 'Delayed gastric emptying rate may explain improved glycaemia in healthy subjects to a starchy meal with added vinegar', *Eur J Clin Nutr*, **52**, 368–371.

Liljeberg H G, Granfeldt Y E, Björck I M (1992) 'Metabolic response to starch in bread containing intact kernels versus milled flour', *Eur J Clin Nutr*, **46**, 561–575.

Liljeberg H G, Lonner C H, Björck I M (1995) 'Sourdough fermentation or addition of organic acids or corresponding salts to bread improves nutritional properties of starch in healthy humans', *J Nutr*, **125**, 1503–1511.

Liljeberg H G, Granfeldt Y E, Björck I M (1996) 'Products based on a high-fiber barley genotype, but not on common barley or oats, lower post-prandial glucose and insulin responses in healthy humans', *J Nutr*, **126**, 458–466.

Liu C, Armentrout R W, Hoffman A J (2006) 'Slowly digestible carbohydrate', US Patent 20060210696.

Ludwig D S (2002) 'The glycemic index: physiological mechanisms relating to obesity, diabetes, and cardiovascular disease', *JAMA*, **287**, 2414–2423.

Manners D J (1989) 'Recent developments in our understanding of amylopectin structure', *Carbohydr Polym*, **11**, 87–112.

Miller J B, Pang E, Bramall L (1992) 'Rice: a high or low glycemic index food?', *Am J Clin Nutr*, **56**, 1034–1036.

Morgan L M (1998) 'The role of gastrointestinal hormones in carbohydrate and lipid metabolism and homeostasis: effects of gastric inhibitory polypeptide and glucagon-like peptide-1', *Biochem Soc Trans*, **26**, 216–222.

Murray S, Patil A, Fahey G, Merchen N, Wolf B, Lai S-C, Gareb K (1998) 'Apparent digestibility of a debranched amylopectin–lipid complex and resistant starch incorporated into enteral formulas fed to ileal-cannulated dogs', *J Nutr*, **128**, 2032–2035.

Owen B, Wolevar T M S (2003) 'Effect of fat on glycaemic responses in normal subjects: a dose–response study', *Nutr Res*, **23**, 1341–1347.

Parillo M, Giacco R, Riccardi G, Pacioni D, Rivellese A (1985) 'Different glycaemic responses to pasta, bread, and potatoes in diabetic patients', *Diabet Med*, **2**, 374–377.

Park J T, Rollings J E (1994) 'Effects of substrate branching characteristics on kinetics of enzymatic depolymerization of mixed linear and branched polysaccharides. 1. Amylose/amylopectin alpha-amylolysis', *Biotechnol Bioeng*, **44**, 792–800.

Patil A R, Murray S M, Fahey G C, Merchen N R, Wolf B W, Lai C, Garleb K A (1998) 'Apparent digestibility and glycemic responses to a debranched amylopectin–lipid complex and resistant starch incorporated into enteral formulas fed to dogs cannulated in the ileum', *FASEB J*, **12**, A210 (abstract).

Qi X and Tester R (2005) 'Compositions and uses thereof', Patent WO 2005/044284 Al.

Rendell M, Vanderhoof J, Venn M, Shehan M, Arndt E, Rao C, Gill G, Newman R, Newman C (2005) 'Effect of a barley breakfast cereal on blood glucose and insulin response in normal and diabetic patients', *Plant Foods Human Nutr*, **60**, 63–67.

Seal C J, Daly M E, Thomas L C, Bal W, Birkett A M, Jeffcoat R, Mathers J C (2003) 'Postprandial carbohydrate metabolism in healthy subjects and those with type 2 diabetes fed starches with slow and rapid hydrolysis rates determined in vitro', *Br J Nutr*, **90**, 853–64.

Seneviratne H D, Biliaderis C G (1991) 'Action of α-amylases on amylose–lipid complex superstructures', *J Cereal Sci*, **13**, 129–143.

Shi Y-C, Seib P A (1995) 'Fine structure of maize starches from four *wx*-containing genotypes of the W64A inbred line in relation to gelatinization and retrogradation', *Carbohydr Polym*, **26**, 141–147.

Shi Y C, Trzasko P T (1997) 'Process for producing amylose resistant granular starch', US Patent 5,593,503, National Starch and Chemical Investment Holding Corp.

Shi Y C, Cui X, Birkett A M, Thatcher M G (2003) 'Slowly digestible starch product', US Patent Application 20030219520, 20030215562.

Shin M, Song J, Seib P A (2000) 'In vitro digestibility of cross-linked starches–RS4', *Starch/Starke*, **56**, 478–483.

Shin S I, Choi H J, Chung K M, Hamaker B R, Park K H, Moon T W (2004) 'Slowly digestible starch from debranched waxy sorghum starch: preparation and properties', *Cereal Chem*, **81**, 404–408.

Shin S I, Kim H J, Ha H J, Lee S H, Moon T W (2005) 'Effect of hydrothermal treatment on formation and structural characteristics of slowly digestible non-pasted granular sweet potato starch', *Starch/Stärke*, **57**, 421–430.

Shin S I, Lee C J, Kim D-I, Lee H A, Cheong, J-J, Chung, K M, Baik M-Y, Park C S, Kim C H, Moon T W (2007) 'Formation, characterization, and glucose response in mice to rice starch with low digestibility produced by citric acid treatment', *J Cereal Sci*, **45**, 24–33.

Thompson L U (1988) 'Antinutrients and blood glucose', *Food Technol*, **42**, 123–132.

Tovar J, Granfeldt Y, Björck I (1992) 'Effects of processing on blood glucose and insulin responses to starch in legumes', *J Agric Food Chem*, **40**, 1846–1851.

Udani J, Hardy M, Madsen D (2004) 'Blocking carbohydrate absorption and weight loss: a clinical trial using Phase 2 brand proprietary fractionated white bean extract', *Alter Med Rev*, **9**, 63–69.

USDA (United States Department of Agriculture) (2005) 'Carbohydrates' Chapter 7 in *Dietary Guidelines for Americans*, http://www.health.gov/dietaryguidelines/dga2005/document/.

Wepner B, Berghofer E, Miesenberger E, Tiefenbacher K, Ng P N K (1999) 'Citrate starch-application as resistant starch in different food systems', *Starch/Starke*, **51**, 354–361.

Wolever T M (1990) 'Relationship between dietary fiber content and composition in foods and the glycemic index', *Am J Clin Nutr*, **51**, 72–75.

Wolever T M S, Jenkins D J A, Thompson L U, Wong G S, Josse R G (1987) 'Effect of canning on the blood glucose response to beans in patients with type 2 diabetes', *Hum Nutr Clin Nutr*, **41**, 135–140.

Wolever T M, Jenkins D J, Jenkins A L, Josse R G (1991) 'The glycemic index: methodology and clinical implications', *Am J Clin Nutr*, **54**, 846–854.

Wolever T, Jenkins D J, Kalmusky J, Giordano C, Giudici S, Jenkins A L, Thompson L U, Wong G S, Josse R G (1986) 'Glycemic response to pasta: effect of surface area, degree of cooking, and protein enrichment', *Diab Care*, **9**, 401–404.

Wolf B W, Laura L B, Fahey G C (1999) 'Effects of chemical modification on in vitro rate and extent of food starch digestion: an attempt to discover a slowly digested starch', *J Agric Food Chem*, **47**, 4178–4183.
Wolf B W, Wolever T M S, Bolognesi C, Zinker B A, Garleb K A, Firkins J L (2001) 'Glycemic response to a food starch esterified by 1-octenyl succinic anhydride in humans', *J Agric Food Chem*, **49**, 2674–2678.
Woo K S, Seib P A (2002) 'Cross-linked resistant starch. Preparation and properties', *Cereal Chem*, **79**, 819–825.
Wood P J, Braaten J T, Fraser W S, Riedel D, Poste L M (1990) 'Comparison of viscous properties of oat and guar gum and the effects of these and oat bran on glycemic index', *J Agric Food Chem*, **38**, 753–757.
Zhang G, Ao Z, Hamaker B R (2006a) 'Slow digestion property of native cereal starches', *Biomacromolecules*, **7**, 3252–3258.
Zhang G, Venkatachalam, M, Hamaker B R (2006b) 'Structural basis for the slow digestion property of native cereal starch', *Biomacromolecules*, **7**, 3259–3266.

Index